The Finite Element Method for Three-dimensional Thermomechanical Applications

The Finite Element Method for Three-dimensional Thermomechanical Applications

Guido Dhondt
Munich, Germany

John Wiley & Sons, Ltd

Telephone (+44) 1243 779777

Email (for orders and customer service enquiries): cs-books@wiley.co.uk
Visit our Home Page on www.wileyeurope.com or www.wiley.com

Other Wiley Editorial Offices

John Wiley & Sons Inc., 111 River Street, Hoboken, NJ 07030, USA

Jossey-Bass, 989 Market Street, San Francisco, CA 94103-1741, USA

Wiley-VCH Verlag GmbH, Boschstr. 12, D-69469 Weinheim, Germany

John Wiley & Sons Australia Ltd, 33 Park Road, Milton, Queensland 4064, Australia

John Wiley & Sons (Asia) Pte Ltd, 2 Clementi Loop #02-01, Jin Xing Distripark, Singapore 129809

John Wiley & Sons Canada Ltd, 22 Worcester Road, Etobicoke, Ontario, Canada M9W 1L1

Wiley also publishes its books in a variety of electronic formats. Some content that appears in print may not be available in electronic books.

British Library Cataloguing in Publication Data

A catalogue record for this book is available from the British Library

ISBN 0-470-85752-8

Produced from LaTeX files supplied by the author, typeset by Laserwords Private Limited, Chennai, India

This book is printed on acid-free paper responsibly manufactured from sustainable forestry in which at least two trees are planted for each one used for paper production.

To my wife Barbara and my children Jakob and Lea

Contents

Preface

In 1998, in times of ever increasing computer power, I had the unusual idea of writing my own finite element program, with just 20-node brick elements for elastic fracture-mechanics calculations. Especially with the program FEAP as a guide, it proved exceedingly simple to get a program with these minimal requirements to run. However, time has shown that this was only the beginning of a long and arduous journey. I was soon joined by my colleague Klaus Wittig, who had written a fast postprocessor for visualizing the results of several other finite element programs and who thought of expanding his program with preprocessing capabilities. He also brought along quite a few ideas for the solver. Coming from a modal-analysis department, he suggested including frequency and linear dynamic calculations. Furthermore, since he was interested in running real-size engine models, he required the code to be not only correct but also fast. This really meant that the code was to be competitive with the major commercial finite element codes. In terms of speed, the mathematical linear equation solver plays a dominant role. In this respect, we were very lucky to come across SPOOLES for static problems and ARPACK for eigenvalue problems, both excellent packages that are freely available on the Internet. I think it was at that time that we decided that our code should be free. The term "free" here primarily means freedom of thought as proclaimed by the GNU General Public License. We had profited enormously from the free equation solvers; why would not others profit from our code?

The demands on the code, but, primarily, also our eagerness to include new features, grew quickly. New element types were introduced. Geometric nonlinearity was implemented, hyperelastic constitutive relations and viscoplasticity followed. We selected the name CalculiX®, and in December 2000 we put the code on the web. Major contributions since then include nonlinear dynamics, cyclic symmetry conditions, anisotropic viscoplasticity and heat transfer. The comments and enthusiasm from users all over the world encourage us to proceed. But above all, the conviction that one cannot master a theory without having gone through the agony of implementing it ever anew drives me to go on.

This book contains the theory that was used to implement CalculiX®. This implies that the topics treated are ready to be coded, and, with a few exceptions, their practical implementation can be found in the CalculiX® code (www.calculix.de). One of the criteria for including a subject in CalculiX® or not is its industrial relevance. Therefore, topics such as cyclic symmetry or multiple point constraints, which are rarely treated in textbooks, are covered in detail. As a matter of fact, multiple point constraints constitute a very versatile workhorse in any industrial finite element application. Conditions such as rigid body motion, the application of a mean rotation, or the requirement that a node has to stay in a plane defined by three other moving nodes are readily formulated as nonlinear multiple point constraints. Clearly, new theories have to face several barriers before being accepted in an industrial environment. This especially applies to material models because of the enormous cost of the parameter identification through testing. Nevertheless, a couple of newer models in the area of anisotropic hyperelasticity and single-crystal viscoplasticity are covered, since they are the prototypes of new constitutive developments and because of the analytical insight they produce.

Although the applications are very practical, the theory cannot be developed without a profound knowledge of continuum mechanics. Therefore, a lot of emphasis is placed on the introduction of kinematic variables, the formulation of the balance laws and the derivation of the constitutive theory. The kinematic framework of a theory is its foundation. Among the kinematic tensors, the deformation gradient plays a special role, as amply demonstrated by the multiplicative decomposition used in viscoplastic theories. The balance equations in their weak form are the governing equations of the finite element method. Finally, the constitutive theory tells us what kind of conditions must be fulfilled by a material law to make sense physically. The knowledge of these rules is a prerequisite for the skillful description of new kinds of materials. This is clearly shown in the treatment of hyperelastic and viscoplastic materials, both in their isotropic and anisotropic form.

The only prerequisite for reading this book is a profound mathematical background in tensor analysis, matrix algebra and vector calculus. The book is largely self-contained, and all other knowledge is introduced within the text. It is oriented toward

1. graduate students working in the finite element field, enabling them to acquire a profound background,
2. researchers in the field, as a reference work,
3. practicing engineers who want to add special features to existing finite element programs and who have to familiarize themselves with the underlying theory.

This book would not have been possible without the help of several people. First, I would like to thank two teachers of mine: Lic. Antoine Van de Velde, for introducing me to the fascinating world of calculus, and Professor A. Cemal Eringen, for acquainting me with continuum mechanics. Readers of his numerous publications will doubtless recognize his stamp on my thinking. Further, I am very indebted to my colleague and friend Klaus Wittig; together we have developed the CalculiX® code in a rare symbiosis. His encouragement and the ever new demands on the code were instrumental in the growth of CalculiX®. I would also like to thank all the colleagues who read portions of the text and gave valuable comments: Dr Bernard Fedelich (Bundesanstalt für Materialforschung), Dr Hans-Peter Hackenberg (MTU Aero Engines), Dr Stefan Hartmann (University of Kassel), Dr Manfred Köhl (MTU Aero Engines), Dr Joop Nagtegaal (ABAQUS®), Dr Erhard Reile (MTU Aero Engines), Dr Harald Schönenborn (MTU Aero Engines) and others. Last but not least, I am very grateful to my wife Barbara and my children Jakob and Lea, who bravely endured my mental absence of the last few months.

Nomenclature

$\boldsymbol{A}$, A^{KL}	kinematic internal variable in material coordinates
$\boldsymbol{A}$, A_{MN}	thermal strain tensor per unit temperature
$\boldsymbol{A}$, $\boldsymbol{a}$, A^K, a^k	acceleration vector
A	deformed area of the body
A_0	undeformed area of the body
$A = \sigma\epsilon$	radiation coefficient
$\{A\}$	global acceleration vector
$\boldsymbol{b}$, b^{kl}	left Cauchy–Green tensor
$\boldsymbol{b}^{\mathrm{e}-1}$	inverse left elastic Cauchy–Green tensor or elastic Finger tensor
$\boldsymbol{C}^{\mathrm{p}}$, C^{p}_{KL}	right plastic Cauchy–Green tensor
$\mathrm{Cof}\boldsymbol{E}$	cofactor matrix of a second rank tensor E
cofactor E_{KL}	cofactor of tensor component E_{KL}
$[C]$	global capacity matrix
c	specific heat
c_0	speed of light in vacuum
c_{p}	specific heat at constant pressure
c_{v}	specific heat at constant volume
$\boldsymbol{d}$, d_{kl}	deformation rate tensor
$\mathrm{d}\boldsymbol{A}$, $\mathrm{d}A_K$	infinitesimal area one-form in material coordinates
$\mathrm{d}\boldsymbol{a}$, $\mathrm{d}a_k$	infinitesimal area one-form in spatial coordinates
$\det\boldsymbol{E}$	determinant of a second rank tensor E

dev $\boldsymbol{\sigma}$	deviatoric tensor of a second rank tensor $\boldsymbol{\sigma}$
$\mathrm{d}S$	infinitesimal length in material coordinates
$\mathrm{d}s$	infinitesimal length in spatial coordinates
$\mathrm{d}V$	infinitesimal volume in material coordinates
$\mathrm{d}v$	infinitesimal volume in spatial coordinates
$\mathrm{d}\boldsymbol{X}, \mathrm{d}X^K$	infinitesimal length vector in material coordinates
$\mathrm{d}\boldsymbol{x}, \mathrm{d}x^k$	infinitesimal length vector in spatial coordinates
$\mathrm{d}\Sigma$	infinitesimal length in the intermediate configuration
$\mathrm{d}\omega$	infinitesimal spatial angle
$\tilde{\boldsymbol{E}}, \tilde{E}_{KL}$	infinitesimal strain tensor in material coordinates
$\mathcal{E}$	total internal energy in the body
$\boldsymbol{E}, E_{KL}$	Lagrange strain tensor
E	Young's modulus
E	total emissive power
E_{b}	total emissive power of a blackbody
E_{λ}	spectral, hemispherical emissive power
$\tilde{\boldsymbol{e}}, \tilde{e}_{kl}$	infinitesimal strain tensor in spatial coordinates
$\boldsymbol{e}, e_{kl}$	Euler strain tensor
e_{LMP}, e^{LMP}	alternating symbols
$\boldsymbol{F}, F^k{}_K$	deformation gradient
F_{ij}	viewfactor: fraction of the radiation power leaving surface i that is intercepted by surface j
$\{F\}$	global force vector
$\{F\}_{\mathrm{e}}$	element force vector
$\boldsymbol{f}, f^k, f^K$	force per unit mass
$\boldsymbol{G}, \boldsymbol{G}^{\flat}, G_{KL}$	covariant metric tensor in the reference system
$\boldsymbol{G}, \boldsymbol{G}^{\sharp}, G^{KL}$	contravariant metric tensor in the reference system

$\boldsymbol{G}^K$	contravariant curvilinear basis vectors in the reference system
$\boldsymbol{G}_K$	covariant curvilinear basis vectors in the reference system
G	hemispherical irradiation power
$\boldsymbol{g}, \boldsymbol{g}^\flat, g_{kl}$	covariant metric tensor in the spatial system
$\boldsymbol{g}, \boldsymbol{g}^\sharp, g^{kl}$	contravariant metric tensor in the spatial system
$g^{Kk}, g^K{}_k, g^k{}_K$	shifters
$\boldsymbol{g}^k$	contravariant curvilinear basis vectors in the spatial system
$\boldsymbol{g}_k$	covariant curvilinear basis vectors in the spatial system
h	Planck constant
h	convection coefficient
h	heat generation per unit mass
$\mathbb{I}_A$	unit tensor of rank four where the unit tensor $\boldsymbol{I}$ is replaced by the tensor $\boldsymbol{A}$
$\mathbb{I}_I$	unit tensor of rank four
$\boldsymbol{I}, I_{KL}, I^{KL}, \delta^K{}_L$	metric tensor in rectangular coordinates in the reference system
$\boldsymbol{I}^K, \boldsymbol{I}_K$	rectangular basis vectors in the reference system
I_E	spectral, directional radiation intensity
$I_{E,\mathrm{b}}$	spectral intensity of blackbody radiation
I_I	spectral, directional irradiation intensity
I_{kd}	kth invariant of the deformation rate tensor
I_{kE}	kth invariant of the Lagrangian strain tensor
$\overline{I}_k, I_{k\overline{C}}$	kth invariant of the reduced Cauchy–Green tensor
I_k, I_{kC}	kth invariant of the Cauchy–Green tensor
$I_{k\sigma}$	kth invariant of the Cauchy tensor
$\boldsymbol{i}, i_{kl}, i^{kl}, \delta^k{}_l$	metric tensor in rectangular coordinates in the spatial system
$\boldsymbol{i}^k, \boldsymbol{i}_k$	rectangular basis vectors in the spatial system
$\boldsymbol{J}, J^K$	Jacobian vector

J	Jacobian determinant of the deformation
J	radiosity
J^*	Jacobian of the global–local transformation
J_k, J_{kC}	kth invariant of the Cauchy–Green tensor of the form tr$\boldsymbol{C}^k$
$\mathcal{K}$	total kinetic energy in the body
K	bulk modulus
$[K]$	global stiffness matrix
$[K]_{\mathrm{e}}$	element stiffness matrix
k	Boltzmann constant
$[L]_{\mathrm{e}}$	element localization matrix
$\boldsymbol{l}$, l_{kl}	velocity gradient
$\boldsymbol{M}_i = \boldsymbol{N}_i \otimes \boldsymbol{N}_i$, M_i^{KL}	contravariant structural tensors in material coordinates
$\boldsymbol{M}^i = \boldsymbol{N}^i \otimes \boldsymbol{N}^i$, M^i_{KL}	covariant structural tensors in material coordinates
$[M]$	global mass matrix
$[M]_{\mathrm{e}}$	element mass matrix
$\dot{m}_{ij}$	absolute value of the mass flow between node i and node j
$\boldsymbol{N}_i$, N_i^K	ith normalized eigenvector in material coordinates
$\boldsymbol{N}^i$, N^i_K	ith normalized eigen-one-form in material coordinates
$\boldsymbol{N}$, N_K	normalized area one-form in material coordinates
$\boldsymbol{n}_i$, n_i^k	ith normalized eigenvector in spatial coordinates
$\boldsymbol{n}^i$, n^i_k	ith normalized eigen-one-form in spatial coordinates
$\boldsymbol{n}$, n_k	normalized area one-form in spatial coordinates
$\boldsymbol{P}$, P^{Kk}	first Piola–Kirchhoff stress tensor
P	radiation power
p	pressure
$\boldsymbol{Q}$	internal dynamic variable in material coordinates

$\boldsymbol{Q}, Q^{K'}_{L}$	orthogonal transformation matrix
$\boldsymbol{Q}, Q^{K}, \boldsymbol{Q}^{\theta}$	heat vector in material coordinates
$\{Q\}$	global heat flux vector
$\{Q\}_e$	element heat flux vector
$\boldsymbol{q}, \boldsymbol{q}^{i}$	internal dynamic variable in spatial coordinates
$\boldsymbol{q}, q^{k}, \boldsymbol{q}^{\theta}$	heat vector in spatial coordinates
$\tilde{\boldsymbol{R}}, \tilde{R}_{KL}$	infinitesimal rotation tensor in material coordinates
$\boldsymbol{R}, R^{k}{}_{L}$	rotation tensor
R	specific gas constant
$\boldsymbol{S}, S^{K}$	entropy vector in material coordinates
$\boldsymbol{S}, S^{KL}$	second Piola–Kirchhoff stress tensor
$\boldsymbol{s}, s^{k}$	entropy vector in spatial coordinates
$\boldsymbol{T}^{K}$	traction vector on a surface with normal parallel to $\boldsymbol{G}^{K}$
$\boldsymbol{T}_{(\boldsymbol{N})}, T^{K}_{(\boldsymbol{N})}$	traction vector on a surface with normal $\boldsymbol{N}$ in material coordinates
T	relative temperature
$\{T\}$	global temperature vector
$\{T\}_{\mathrm{e}}$	element temperature vector
$\boldsymbol{t}^{k}$	traction vector on a surface with normal parallel to $\boldsymbol{g}^{k}$
$\boldsymbol{t}_{(\boldsymbol{n})}, t^{k}_{(\boldsymbol{n})}$	traction vector on a surface with normal $\boldsymbol{n}$ in spatial coordinates
$\mathrm{tr}\boldsymbol{E}$	trace of a second rank tensor E
$\boldsymbol{U}, U^{K}_{L}$	right stretch tensor
$\boldsymbol{U}, \boldsymbol{u}, U^{K}, u^{k}$	displacement vector
U	volumetric free energy potential
$\{U\}$	global displacement vector
$\{U\}_e$	element displacement vector
$\boldsymbol{V}, V^{k}_{l}$	left stretch tensor
$\boldsymbol{V}, \boldsymbol{v}, V^{K}, v^{k}$	velocity vector

V	deformed volume of the body
V_0	undeformed volume of the body
V_{0e}	undeformed volume of a finite element
$\{V\}$	global velocity vector
$\mathcal{W}$	total rate of work in the body
$\boldsymbol{w}, w_{kl}$	spin tensor
$\boldsymbol{X}, X^K$	position vector in material coordinates
$\boldsymbol{x}, x^k$	position vector in spatial coordinates
$\boldsymbol{\alpha}, \alpha^{kl}$	kinematic internal variable in spatial coordinates
α	total, hemispherical absorptivity
$\boldsymbol{\beta}, \beta^{KL}$	thermal stress tensor per unit temperature
$\boldsymbol{\gamma}, \gamma^{KL}$	residual stress tensor
$\boldsymbol{\gamma}(\xi, \eta, \zeta)$	vector of local coordinates
$\dot{\gamma}$	consistency parameter
$\delta^K{}_L$	mixed-variant metric tensor in the reference system
$\delta^k{}_l$	mixed-variant metric tensor in the spatial system
δT	temperature perturbation
$\delta \boldsymbol{U}, \delta U_K$	displacement perturbation
$\boldsymbol{\epsilon}, \epsilon_{kl}$	infinitesimal strain tensor in spatial coordinates
$\boldsymbol{\epsilon}^{\mathrm{e}}, \epsilon^e_{kl}$	infinitesimal elastic strain tensor in spatial coordinates
$\boldsymbol{\epsilon}^{\mathrm{p}}, \epsilon^p_{kl}$	infinitesimal plastic strain tensor in spatial coordinates
ϵ	emissivity
$\epsilon_{\lambda,\omega}$	spectral, directional emissivity
ε	energy density
ζ	local coordinate
η	entropy per unit mass
η	local coordinate

θ	absolute temperature
θ_e	absolute environmental temperature
θ_{ref}	reference temperature
$\kappa, \kappa^K, \kappa^{KL}, \kappa^{KLM}$	conduction coefficients
Λ_{iE}	ith eigenvalue of the Lagrangian strain tensor
Λ_{iS}	ith eigenvalue of the second Piola–Kirchhoff stress tensor
Λ_i, Λ_{iC}	ith eigenvalue of the Cauchy–Green tensor
λ	Lamé constant
λ_i	principal stretches, eigenvalues of $\boldsymbol{F}$
$\lambda_{i\sigma}$	ith eigenvalue of the Cauchy stress tensor
λ_v	fluid constant
μ	Lamé constant
μ_v	fluid constant
ν	Poisson coefficient
Ξ, Ξ^{KL}	relative stress tensor in material coordinates
ξ	local coordinate
ρ	mass density in the spatial configuration
ρ	total, hemispherical reflectivity
ρ_0	mass density in the material configuration
$\Sigma_0, \Sigma^{KL}, \Sigma^{KLMN}$	free energy coefficients
$\Sigma = \rho_0\psi$	free energy per unit volume in the reference configuration
$\boldsymbol{\sigma}, \sigma^{kl}$	Cauchy stress tensor
σ	Stefan–Boltzmann constant
τ	total, hemispherical transmissivity
$\varphi_i(\xi, \eta, \zeta)$	shape functions
ψ	free energy per unit mass
ω	circular frequency
∇	spatial gradient
∇_0	material gradient

1

Displacements, Strain, Stress and Energy

1.1 The Reference State

Continuum mechanics deals with the change of field variables due to external actions. Examples of field variables are displacements, stresses, temperatures and magnetic induction. Actions include mechanical forces, heating, and so on. In general, a reference state is chosen with respect to which the change of field variables is measured. Let the fields of interest be defined in the reference state in a set of points, the so-called material points, occupying a volume V_0 with a surface A_0 in Eucledian space $\mathbb{R}^3$ (Figure 1.1). Assume that the reference space is described by a set of curvilinear coordinates $\left\{X^K\right\}_{K=1,2,3}$ related to a rectangular system $\{Z^K\}_{K=1,2,3}$ by

$$Z^K = Z^K(X^1, X^2, X^3). \tag{1.1}$$

Coordinates in the reference state are also called *material coordinates*. Consider an infinitesimal vector $\mathrm{d}\boldsymbol{X}$. One can write

$$\mathrm{d}\boldsymbol{X} = \frac{\partial \boldsymbol{X}}{\partial Z^K} \mathrm{d}Z^K \tag{1.2}$$

(summation over repeated indices).

$$\boldsymbol{I}_K = \frac{\partial \boldsymbol{X}}{\partial Z^K} \tag{1.3}$$

is a set of basis vectors in the rectangular system. Accordingly, $\boldsymbol{I}_K, K = 1, 2, 3$ do not depend on Z^K. In an analogous way, one can write

$$\mathrm{d}\boldsymbol{X} = \frac{\partial \boldsymbol{X}}{\partial X^K} \mathrm{d}X^K. \tag{1.4}$$

The Finite Element Method for Three-dimensional Thermomechanical Applications Guido Dhondt
 ISBN: 0-470-85752-8

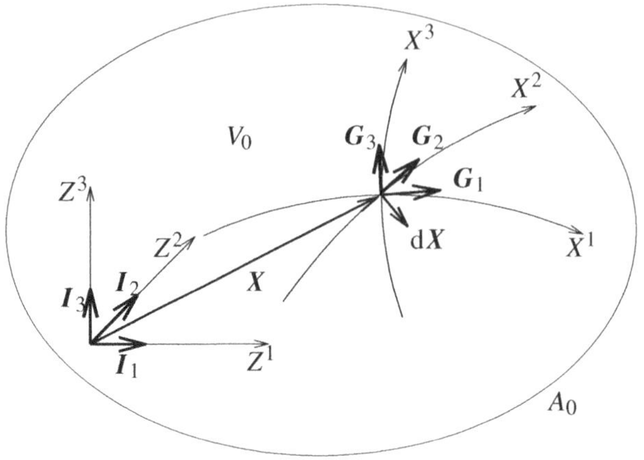

Figure 1.1 Material coordinate systems

The vectors

$$\boldsymbol{G}_K = \partial \boldsymbol{X} / \partial X^K \tag{1.5}$$

constitute a basis in the curvilinear coordinate system. One can write (compare Equation (1.2) with Equation (1.4))

$$\boldsymbol{G}_K \, \mathrm{d}X^K = \boldsymbol{I}_L \, \mathrm{d}Z^L \tag{1.6}$$

or

$$\boldsymbol{G}_K = \frac{\partial Z^L}{\partial X^K} \boldsymbol{I}_L. \tag{1.7}$$

The size $\mathrm{d}S$ of a vector $\mathrm{d}\boldsymbol{X}$ is defined as

$$\mathrm{d}S^2 := \mathrm{d}\boldsymbol{X} \cdot \mathrm{d}\boldsymbol{X} \tag{1.8}$$

where the "·" denotes the inner product of two vectors (also called the *dot product* or the *contraction of two vectors*). In rectangular coordinates, one finds (substitute Equation (1.2) into Equation (1.8))

$$\begin{aligned} \mathrm{d}S^2 &= \boldsymbol{I}_K \, \mathrm{d}Z^K \cdot \boldsymbol{I}_L \, \mathrm{d}Z^L \\ &= \mathrm{d}Z^K \, \mathrm{d}Z^L \boldsymbol{I}_K \cdot \boldsymbol{I}_L \\ &=: \mathrm{d}Z^K \, \mathrm{d}Z^L I_{KL}. \end{aligned} \tag{1.9}$$

The metric tensor I_{KL} takes the value 1 for $K = L$ and 0 for $K \neq L$. In curvilinear coordinates, one obtains (substitute Equation (1.4) into Equation (1.8)),

$$\begin{aligned} \mathrm{d}S^2 &= \boldsymbol{G}_K \, \mathrm{d}X^K \cdot \boldsymbol{G}_L \, \mathrm{d}X^L \\ &= \mathrm{d}X^K \, \mathrm{d}X^L \boldsymbol{G}_K \cdot \boldsymbol{G}_L \\ &=: \mathrm{d}X^K \, \mathrm{d}X^L G_{KL} \end{aligned} \tag{1.10}$$

G_{KL} is called the *metric tensor* for the coordinate system $\{X^K\}$. In general, $G_{KL} \neq 0$ for $K \neq L$, and $G_{KL} \neq 1$ for $K = L$. Thus, the basis vectors $\boldsymbol{G}_K$ are not necessarily orthonormal. Using the set $\{\boldsymbol{G}_K\}$, one can define another set $\{\boldsymbol{G}^L\}$ through the relations

$$\boldsymbol{G}_K \cdot \boldsymbol{G}^L = \delta_K^{\ L} \tag{1.11}$$

where $\delta_K^{\ L} = 0$ for $K \neq 0$ and $\delta_K^{\ L} = 1$ for $K = L$. In modern Riemannian geometry, $\{\boldsymbol{G}^L\}$ are called *one-forms* (or *covariant tensors of rank 1*). They map the vectors $\{\boldsymbol{G}_K\}$ (which are also called *contravariant tensors of rank 1*) into a scalar by Equation (1.11). $\{\boldsymbol{G}^L\}$ forms a basis for the vector space of one-forms and is also called the *dual basis of* $\{\boldsymbol{G}_k\}$. If $\boldsymbol{\alpha}$ is a one-form, one writes

$$\boldsymbol{\alpha} = \alpha_K \boldsymbol{G}^L. \tag{1.12}$$

The dot product of a vector $\boldsymbol{V}$ and a one-form $\boldsymbol{\alpha}$ is defined by

$$\boldsymbol{V} \cdot \boldsymbol{\alpha} = V^K \boldsymbol{G}_K \cdot \alpha_L \boldsymbol{G}^L = V^K \alpha_K \tag{1.13}$$

through Equation (1.11). In the same way, the dot product of two vectors and two one-forms yields

$$\boldsymbol{V} \cdot \boldsymbol{W} = V^K \boldsymbol{G}_K \cdot W^L \boldsymbol{G}_L = V^K W^L G_{KL} \tag{1.14}$$

$$\boldsymbol{\alpha} \cdot \boldsymbol{\beta} = \alpha_K \boldsymbol{G}^K \cdot \beta_L \boldsymbol{G}^L = \alpha_K \beta_L G^{KL} \tag{1.15}$$

where G^{KL} is defined by

$$G^{KL} := \boldsymbol{G}^K \cdot \boldsymbol{G}^L. \tag{1.16}$$

Notice that in Equations (1.13), (1.14) and (1.15) the same symbol is used for the dot product. The context shows whether a (covariant or contravariant) metric tensor is needed. Multiplying a vector $\boldsymbol{V}$ with the one-form $\boldsymbol{G}^L$ yields

$$\boldsymbol{V} \cdot \boldsymbol{G}^L = V^K \boldsymbol{G}_K \cdot \boldsymbol{G}^L = V^K \delta_K^{\ L} = V^L. \tag{1.17}$$

Thus, the components V^L of $\boldsymbol{V}$ can be obtained by taking the scalar product of $\boldsymbol{V}$ with the basis one-form $\boldsymbol{G}^L$. Hence,

$$\boldsymbol{V} = (\boldsymbol{V} \cdot \boldsymbol{G}^L)\boldsymbol{G}_L. \tag{1.18}$$

Similar statements to Equation (1.17) and Equation (1.18) can be made on the basis of one-forms:

$$\boldsymbol{\alpha} \cdot \boldsymbol{G}_L = \alpha_L \tag{1.19}$$

$$\boldsymbol{\alpha} = (\boldsymbol{\alpha} \cdot \boldsymbol{G}_L)\boldsymbol{G}^L. \tag{1.20}$$

Although the separation of tensors of rank one into vectors and one-forms is instructive from a theoretical point of view, there is no reason why a vector cannot be written in terms of a contravariant basis or a one-form in terms of a covariant basis. Substituting $\boldsymbol{G}^K$ in Equation (1.18) and $\boldsymbol{G}_K$ in Equation (1.19), one obtains

$$\boldsymbol{G}^K = G^{KL}\boldsymbol{G}_L \tag{1.21}$$

$$\boldsymbol{G}_K = G_{KL}\boldsymbol{G}^L. \tag{1.22}$$

The operation in Equation (1.21) and in Equation (1.22) is called *raising* and *lowering* of the index respectively. As we will see later on, some fields are naturally represented by covariant tensors (such as the Lagrangian strain and normals on a plane), whereas others are predestinate for a contravariant representation (such as stresses and normals in a direction). They can be viewed as dual fields.

1.2 The Spatial State

Because of the actions, the body B is mapped from its reference state into some other state, a spatial state. Let the spatial state be described by rectangular coordinates $\{z^k\}$ and curvilinear coordinates $\{x^k\}$. These coordinates are called *spatial coordinates*. The same definitions of the reference state apply to the spatial state, for instance,

$$\mathrm{d}s^2 = \mathrm{d}x^k\,\mathrm{d}x^l g_{kl} \tag{1.23}$$

where g_{kl} is the metric tensor of the spatial state. Within the theory of continuum mechanics, one tries to predict the spatial state from the reference state and the actions on it. Since

$$\boldsymbol{g}_k = \frac{\partial \boldsymbol{x}}{\partial x^k} \tag{1.24}$$

one can write

$$\begin{aligned} \mathrm{d}\boldsymbol{x} &= \mathrm{d}x^k \boldsymbol{g}_k = \frac{\partial x^k}{\partial X^K}\,\mathrm{d}X^K \boldsymbol{g}_k \\ &= x^k_{,K}\,\mathrm{d}X^K \boldsymbol{g}_k. \end{aligned} \tag{1.25}$$

This reveals that the spatial state can be predicted from the material state if $x^k_{,K}$ is known.

Defining the dyadic product of two vectors $\boldsymbol{a}$ and $\boldsymbol{b}$, written as $\boldsymbol{a} \otimes \boldsymbol{b}$ such that

$$(\boldsymbol{a} \otimes \boldsymbol{b}) \cdot \boldsymbol{c} = \boldsymbol{a}(\boldsymbol{b} \cdot \boldsymbol{c}) \tag{1.26}$$

and

$$\boldsymbol{c} \cdot (\boldsymbol{a} \otimes \boldsymbol{b}) = (\boldsymbol{c} \cdot \boldsymbol{a})\boldsymbol{b} \tag{1.27}$$

for an arbitrary vector $\boldsymbol{c}$, and similar for two one-forms or a vector and a one-form, one finds that

$$\begin{aligned} \mathrm{d}\boldsymbol{x} &= x^k_{,K}\boldsymbol{g}_k(\boldsymbol{G}^K \cdot \mathrm{d}\boldsymbol{X}) \\ &= x^k_{,K}(\boldsymbol{g}_k \otimes \boldsymbol{G}^K) \cdot \mathrm{d}\boldsymbol{X}. \end{aligned} \tag{1.28}$$

Defining the deformation gradient $\boldsymbol{F}$ as

$$\boldsymbol{F} = x^k_{,K}(\boldsymbol{g}_k \otimes \boldsymbol{G}^K) \tag{1.29}$$

Equation (1.28) is transformed into

$$\mathrm{d}\boldsymbol{x} = \boldsymbol{F} \cdot \mathrm{d}\boldsymbol{X}. \tag{1.30}$$

This shows that the deformation gradient is the Jacobian matrix of the motion from the material into the spatial state.

The dyadic product of two vectors, of two one-forms and of a vector and a one-form is called a *contravariant tensor of rank two*, a *covariant tensor of rank two* and a *mixed-variant tensor of rank two* respectively. If

$$\boldsymbol{a} = a_{KL}\boldsymbol{G}^K \otimes \boldsymbol{G}^L \tag{1.31}$$

then one can also write (Equation (1.21))

$$\boldsymbol{a} = a_{KL}G^{KM}G^{LN}\boldsymbol{G}_M \otimes \boldsymbol{G}_N = a^{MN}\boldsymbol{G}_M \otimes \boldsymbol{G}_N \tag{1.32}$$

where $a^{MN} = a_{KL}G^{KM}G^{LN}$ is obtained by raising the indices. Equation (1.31) is the covariant expansion of $\boldsymbol{a}$, Equation (1.32) is the contravariant expansion. To emphasize this, the notation $\boldsymbol{a}^\flat$ will be used for the covariant expansion and $\boldsymbol{a}^\sharp$ for the contravariant one (this agrees with recent literature, see (Marsden and Hughes 1983), (Holzapfel 2000)). Accordingly,

$$\boldsymbol{a}^\flat = a_{KL}\boldsymbol{G}^K \otimes \boldsymbol{G}^L \tag{1.33}$$

$$\boldsymbol{a}^\sharp = a^{KL}\boldsymbol{G}_K \otimes \boldsymbol{G}_L. \tag{1.34}$$

$\boldsymbol{F}$ is called a *mixed-variant two-point tensor* since it is the dyadic product of basis vectors belonging to different states (the material and the spatial state).

Notice that the dot on the right-hand side and on the left-hand side of Equation (1.26) have a different meaning: the dot on the right-hand side denotes the contraction of two vectors already encountered in Equation (1.8). The dot on the left-hand side symbolizes the contraction of a tensor of rank two and a vector. Whereas the contraction of two vectors is commutative, the contraction of a tensor of rank two and a vector is not

$$(\boldsymbol{a} \otimes \boldsymbol{b}) \cdot \boldsymbol{c} = \boldsymbol{a}(\boldsymbol{b} \cdot \boldsymbol{c}) \neq (\boldsymbol{c} \cdot \boldsymbol{a})\boldsymbol{b} = \boldsymbol{c} \cdot (\boldsymbol{a} \otimes \boldsymbol{b}). \tag{1.35}$$

However,

$$(\boldsymbol{a} \otimes \boldsymbol{b}) \cdot \boldsymbol{c} = \boldsymbol{a}(\boldsymbol{b} \cdot \boldsymbol{c}) = (\boldsymbol{c} \cdot \boldsymbol{b})\boldsymbol{a} = \boldsymbol{c} \cdot (\boldsymbol{b} \otimes \boldsymbol{a}) = \boldsymbol{c} \cdot (\boldsymbol{a} \otimes \boldsymbol{b})^{\mathrm{T}} \tag{1.36}$$

where

$$(\boldsymbol{a} \otimes \boldsymbol{b})^{\mathrm{T}} := \boldsymbol{b} \otimes \boldsymbol{a} \tag{1.37}$$

is the transpose of $\boldsymbol{a} \otimes \boldsymbol{b}$.

The length $\mathrm{d}s$ of a vector $\mathrm{d}\boldsymbol{x}$ in the spatial state satisfies

$$\begin{aligned}
\mathrm{d}s^2 &= \mathrm{d}\boldsymbol{x} \cdot \mathrm{d}\boldsymbol{x} \\
&= x^k_{,K}\, \mathrm{d}X^K \boldsymbol{g}_k \cdot x^l_{,L}\, \mathrm{d}X^L \boldsymbol{g}_l \\
&= x^k_{,K} x^l_{,L}\, \mathrm{d}X^K\, \mathrm{d}X^L \boldsymbol{g}_k \cdot \boldsymbol{g}_l \\
&= x^k_{,K} x^l_{,L} g_{kl}\, \mathrm{d}X^K\, \mathrm{d}X^L.
\end{aligned} \tag{1.38}$$

Defining the right Cauchy–Green tensor by

$$\boldsymbol{C} := C_{KL} \boldsymbol{G}^K \otimes \boldsymbol{G}^L \tag{1.39}$$

where

$$C_{KL} = x^k_{,K} x^l_{,L} g_{kl} \tag{1.40}$$

one obtains

$$\mathrm{d}s^2 = C_{KL}\, \mathrm{d}X^K\, \mathrm{d}X^L. \tag{1.41}$$

Comparing Equation (1.23) and Equation (1.41), one notices that for the calculations of $\mathrm{d}s^2$, the tensor $\boldsymbol{C}$ is the equivalent of $\boldsymbol{g}$ in the reference frame. One also says that $\boldsymbol{C}$ is the pullback of $\boldsymbol{g}$ and, equivalently, $\boldsymbol{g}$ is the push-forward of $\boldsymbol{C}$. Equation (1.41) also shows that the Cauchy–Green tensor is positive definite. Furthermore, it satisfies

$$\boldsymbol{C} = \boldsymbol{F}^{\mathrm{T}} \cdot \boldsymbol{F} \tag{1.42}$$

where $\boldsymbol{F}^{\mathrm{T}}$ is the transpose of $\boldsymbol{F}$ defined by

$$\boldsymbol{F}^{\mathrm{T}} := x^k_{,K} (\boldsymbol{G}^K \otimes \boldsymbol{g}_k). \tag{1.43}$$

Indeed, since $(\boldsymbol{a} \otimes \boldsymbol{b}) \cdot (\boldsymbol{c} \otimes \boldsymbol{d}) = (\boldsymbol{a} \otimes \boldsymbol{d}) \boldsymbol{b} \cdot \boldsymbol{c}$, one finds

$$\begin{aligned}
\boldsymbol{F}^{\mathrm{T}} \cdot \boldsymbol{F} &= x^k_{,K} x^l_{,L} (\boldsymbol{G}^K \otimes \boldsymbol{g}_k) \cdot (\boldsymbol{g}_l \otimes \boldsymbol{G}^L) \\
&= x^k_{,K} x^l_{,L} g_{kl} \boldsymbol{G}^K \otimes \boldsymbol{G}^L.
\end{aligned} \tag{1.44}$$

The stretch in a direction $\boldsymbol{N} = (\mathrm{d}X^K/\mathrm{d}S) \boldsymbol{G}_K$ is defined by

$$\begin{aligned}
\lambda_{(\boldsymbol{N})} = \frac{\mathrm{d}s}{\mathrm{d}S} &= \sqrt{C_{KL} \frac{\mathrm{d}X^K}{\mathrm{d}S} \frac{\mathrm{d}X^L}{\mathrm{d}S}} \\
&= \sqrt{C_{KL} N^K N^L}
\end{aligned} \tag{1.45}$$

where $N^K = \mathrm{d}X^K/\mathrm{d}S$. Thus, $\lambda_{(\boldsymbol{N})}$ is the change of length of an infinitesimal vector in direction $\boldsymbol{N}$ in the reference state.

If the mapping $\boldsymbol{x}(\boldsymbol{X})$ is one to one, it can be inverted to yield $\boldsymbol{X}(\boldsymbol{x})$. Since matter cannot disappear, the Jacobian determinant

$$J := \det(x^k_{,K}) = \det \boldsymbol{F} \tag{1.46}$$

cannot be zero and the mapping is one to one. Assuming the transformation to be continuous, this means that J must be either everywhere positive or everywhere negative. Since $J = 1$ for the identical transformation, it is everywhere positive.

$\mathrm{d}S^2$ can also be written as

$$\begin{aligned} \mathrm{d}S^2 &= \mathrm{d}X^K\,\mathrm{d}X^L G_{KL} = X^K_{,k} X^L_{,l}\,\mathrm{d}x^k\,\mathrm{d}x^l G_{KL} \\ &= (b^{-1})_{kl}\,\mathrm{d}x^k\,\mathrm{d}x^l \end{aligned} \tag{1.47}$$

where

$$(b^{-1})_{kl} := X^K_{,k} X^L_{,l} G_{KL}. \tag{1.48}$$

The tensor $\boldsymbol{b}$ (the inverse of $\boldsymbol{b}^{-1}$) is called the *left Cauchy–Green tensor* and satisfies

$$b^{kl} = x^k_{,K} x^l_{,L} G^{KL} \tag{1.49}$$

or, equivalently,

$$\boldsymbol{b} = \boldsymbol{F} \cdot \boldsymbol{F}^{\mathrm{T}}. \tag{1.50}$$

Consequently,

$$\boldsymbol{b}^{-1} = \boldsymbol{F}^{-\mathrm{T}} \cdot \boldsymbol{F}^{-1} \tag{1.51}$$

where

$$\boldsymbol{F}^{-1} = X^K_{,k} \boldsymbol{G}_K \otimes \boldsymbol{g}^k \tag{1.52}$$

is the inverse of the deformation gradient and

$$\boldsymbol{F}^{-\mathrm{T}} = X^K_{,k} \boldsymbol{g}^k \otimes \boldsymbol{G}_K. \tag{1.53}$$

Equation (1.47) shows that, with respect to $\mathrm{d}S^2$, $\boldsymbol{b}^{-1}$ plays in the spatial state the role that is assumed by $\boldsymbol{G}$ in the reference state, that is,

$$G_{KL}\,\mathrm{d}X^K\,\mathrm{d}X^L = (b^{-1})_{kl}\,\mathrm{d}x^k\,\mathrm{d}x^l. \tag{1.54}$$

Therefore, $\boldsymbol{b}^{-1}$ is called the *push-forward* of $\boldsymbol{G}$ and equivalently $\boldsymbol{G}$ is called the *pullback* of $\boldsymbol{b}^{-1}$. Equation (1.41) and Equation (1.47) can also be written as

$$\mathrm{d}s^2 = \mathrm{d}\boldsymbol{X} \cdot \boldsymbol{C} \cdot \mathrm{d}\boldsymbol{X} \tag{1.55}$$

and

$$\mathrm{d}S^2 = \mathrm{d}\boldsymbol{x} \cdot \boldsymbol{b}^{-1} \cdot \mathrm{d}\boldsymbol{x}. \tag{1.56}$$

Since J is the determinant of $x^k_{,K}$, one also has

$$\frac{\partial J}{\partial x^k_{,K}} = \mathrm{cofactor}(x^k_{,K}). \tag{1.57}$$

The cofactor of $x^k_{,K}$ is defined as the *determinant of the matrix one obtains after deleting row k and column K in* $x^k_{,K}$ *(this is the so-called minor determinant of* $x^k_{,K}$*), multiplied by* $(-1)^{k+K}$. Equation (1.57) is easily derived by recalling that the determinant of a matrix can be obtained by taking the dot product of any row with the row of the corresponding cofactors, for example, if the first row is used,

$$J = x^1_{,1}\text{cofactor }(x^1_{,1}) + x^1_{,2}\text{cofactor }(x^1_{,2}) + x^1_{,3}\text{cofactor }(x^3_{,3}). \tag{1.58}$$

$X^K_{,k}$ is the inverse of $x^k_{,K}$. Accordingly,

$$X^K_{,k} = \frac{1}{J}\text{cofactor }(x^k_{,K}). \tag{1.59}$$

Indeed, the inverse of a matrix $\boldsymbol{M}$ satisfies (Greenberg 1978)

$$(M^{-1})^{KL} = \frac{1}{\det \boldsymbol{M}}\text{cofactor }(M^{LK}). \tag{1.60}$$

Comparing Equation (1.57) with Equation (1.59), one finds

$$\frac{\partial J}{\partial x^k_{,K}} = J X^K_{,k}. \tag{1.61}$$

This relationship will be needed for the time derivative of J.

So far, only length changes were considered. Since the determinant of a map describes its volume change, one can write

$$\mathrm{d}v = J\,\mathrm{d}V \tag{1.62}$$

where $\mathrm{d}v$ and $\mathrm{d}v$ are infinitesimal volume elements in the reference and spatial configuration respectively. Denoting an infinitesimal surface element in the reference configuration by the one-form $\mathrm{d}\boldsymbol{A}$ orthogonal to the surface element and with size equal to the area of the surface, and similarly for the spatial configuration, one obtains for Equation (1.62),

$$\mathrm{d}\boldsymbol{a}\cdot\mathrm{d}\boldsymbol{x} = J\,\mathrm{d}\boldsymbol{A}\cdot\mathrm{d}\boldsymbol{X} \tag{1.63}$$

or

$$\mathrm{d}\boldsymbol{a}\cdot\boldsymbol{F}\cdot\mathrm{d}\boldsymbol{X} = J\,\mathrm{d}\boldsymbol{A}\cdot\mathrm{d}\boldsymbol{X}. \tag{1.64}$$

Since this applies to an arbitrary vector $\mathrm{d}\boldsymbol{X}$, one finds

$$\mathrm{d}\boldsymbol{a} = J\,\mathrm{d}\boldsymbol{A}\cdot\boldsymbol{F}^{-1} \tag{1.65}$$

or

$$\mathrm{d}\boldsymbol{a} = J\boldsymbol{F}^{-\mathrm{T}}\cdot\mathrm{d}\boldsymbol{A}. \tag{1.66}$$

This is feasible since it expresses that for isochoric (volume-preserving, $J = 1$) motion, the surface change is inversely proportional to the length change.

1.3 Strain Measures

Physically, we are interested in the change from $\mathrm{d}\boldsymbol{X}$ to $\mathrm{d}\boldsymbol{x}$ and not as much in the actual size of $\mathrm{d}\boldsymbol{x}$. After all, assuming the body to be stress-free at the outset of the calculation, it is the change of $\mathrm{d}\boldsymbol{X}$ that generates the stress field in a mechanical problem. The vector connecting the initial position of a material particle at $\boldsymbol{X}$ to its new position $\boldsymbol{x}$ at time t is called the *displacement* $\boldsymbol{U}(\boldsymbol{X}, t)$ of that particle at time t. One can write (see Figure 1.2)

$$\boldsymbol{u} = \boldsymbol{U} = \boldsymbol{o} + \boldsymbol{x} - \boldsymbol{X} \tag{1.67}$$

$\boldsymbol{o}$ is the vector connecting the spatial frame of reference $\{\boldsymbol{g}_k\}$ with the material frame $\{\boldsymbol{G}_K\}$. Since the displacement connects a material vector with a spatial vector, it does not uniquely belong to the material nor to the spatial frame, and both upper case notation $\boldsymbol{U}$ and lower case notation $\boldsymbol{u}$ will be used. The component notation yields

$$\boldsymbol{U} = U^K \boldsymbol{G}_K \tag{1.68}$$

and

$$\boldsymbol{u} = u^k \boldsymbol{g}_k. \tag{1.69}$$

The difference between $\mathrm{d}S^2$ and $\mathrm{d}s^2$ can be written as (Equation (1.41))

$$\mathrm{d}s^2 - \mathrm{d}S^2 = (C_{KL} - G_{KL})\,\mathrm{d}X^K\,\mathrm{d}X^L \tag{1.70}$$

as well as (Equation (1.47))

$$\mathrm{d}s^2 - \mathrm{d}S^2 = (g_{kl} - b_{kl}^{-1})\,\mathrm{d}x^k\,\mathrm{d}x^l. \tag{1.71}$$

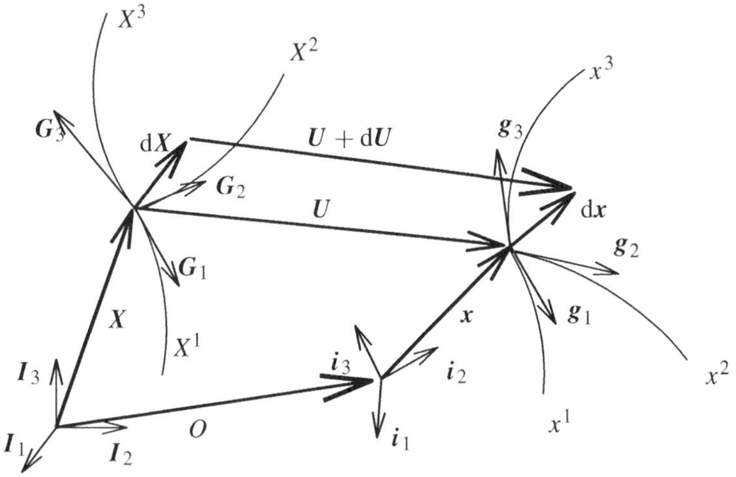

Figure 1.2 Displacement vectors

Now, the Lagrangian strain tensor $\boldsymbol{E}$ (also sometimes called the *Green–Lagrange strain tensor*) is defined by

$$E_{KL} := \tfrac{1}{2}(C_{KL} - G_{KL}) \tag{1.72}$$

and the Eulerian strain tensor $\boldsymbol{e}$ (also sometimes called the *Euler–Almansi strain tensor*) by

$$e_{kl} := \tfrac{1}{2}(g_{kl} - b_{kl}^{-1}). \tag{1.73}$$

Accordingly,

$$\mathrm{d}s^2 - \mathrm{d}S^2 = 2E_{KL}\,\mathrm{d}X^K\,\mathrm{d}X^L \tag{1.74}$$

$$= 2e_{kl}\,\mathrm{d}x^k\,\mathrm{d}x^l. \tag{1.75}$$

$\boldsymbol{E}$ and $\boldsymbol{e}$ are second-order tensors and can be interpreted as measures for the change of length in a body. Using Equation (1.67) one can write

$$\begin{aligned} \mathrm{d}s^2 - \mathrm{d}S^2 &= \mathrm{d}\boldsymbol{x}\cdot\mathrm{d}\boldsymbol{x} - \mathrm{d}\boldsymbol{X}\cdot\mathrm{d}\boldsymbol{X} \\ &= (\mathrm{d}\boldsymbol{U} + \mathrm{d}\boldsymbol{X})\cdot(\mathrm{d}\boldsymbol{U} + \mathrm{d}\boldsymbol{X}) - \mathrm{d}\boldsymbol{X}\cdot\mathrm{d}\boldsymbol{X} \\ &= \mathrm{d}\boldsymbol{U}\cdot\mathrm{d}\boldsymbol{X} + \mathrm{d}\boldsymbol{X}\cdot\mathrm{d}\boldsymbol{U} + \mathrm{d}\boldsymbol{U}\cdot\mathrm{d}\boldsymbol{U} \\ &= (\boldsymbol{U}_{,K}\cdot\boldsymbol{X}_{,L} + \boldsymbol{X}_{,K}\cdot\boldsymbol{U}_{,L} + \boldsymbol{U}_{,K}\cdot\boldsymbol{U}_{,L})\,\mathrm{d}X^K\,\mathrm{d}X^L. \end{aligned} \tag{1.76}$$

Since $\boldsymbol{U} = U^M\boldsymbol{G}_M$ and $\mathrm{d}\boldsymbol{X} = \mathrm{d}X^N\boldsymbol{G}_N$, one finds

$$\begin{aligned} \frac{\partial\boldsymbol{U}}{\partial X^K} &= \frac{\partial}{\partial X^K}(U^M\boldsymbol{G}_M) = \frac{\partial U^M}{\partial X^K}\boldsymbol{G}_M + U^M\frac{\partial\boldsymbol{G}_M}{\partial X^K} \\ &= \frac{\partial U^M}{\partial X^K}\boldsymbol{G}_M + U^M\frac{\partial^2 Z^L}{\partial X^K\partial X^M}\boldsymbol{I}_L \\ &= \frac{\partial U^M}{\partial X^K}\boldsymbol{G}_M + U^M\frac{\partial^2 Z^L}{\partial X^K\partial X^M}\frac{\partial X^N}{\partial Z^L}\boldsymbol{G}_N \\ &= \left(\frac{\partial U^M}{\partial X^K} + U^N\frac{\partial^2 Z^L}{\partial X^K\partial X^N}\frac{\partial X^M}{\partial Z^L}\right)\boldsymbol{G}_M \\ &=: U^M_{;K}\boldsymbol{G}_M \end{aligned} \tag{1.77}$$

and

$$\frac{\partial\boldsymbol{X}}{\partial X^L} = \boldsymbol{G}_L. \tag{1.78}$$

$U^M_{;K}$ is the covariant derivative of U^M and can also be written as

$$U^M_{;K} = U^M_{,K} + U^N\begin{Bmatrix} M \\ KN \end{Bmatrix} \tag{1.79}$$

where

$$\left\{ \begin{matrix} M \\ KN \end{matrix} \right\} := \frac{\partial^2 Z^L}{\partial X^K \partial X^N} \frac{\partial X^M}{\partial Z^L} \tag{1.80}$$

are called the *Christoffel symbols* of the second kind. Hence,

$$\mathrm{d}s^2 - \mathrm{d}S^2 = (U^M_{;K} G_{LM} + U^M_{;L} G_{KM} + U^M_{;K} U^N_{;L} G_{MN})\, \mathrm{d}X^K\, \mathrm{d}X^L. \tag{1.81}$$

Comparison of Equation (1.74) with Equation (1.81) finally yields

$$2E_{KL} = U^M_{;K} G_{LM} + U^M_{;L} G_{KM} + U^M_{;K} U^N_{;L} G_{MN}. \tag{1.82}$$

Similarly, one finds

$$2e_{kl} = u^m_{;k} g_{lm} - u^m_{;l} g_{km} + u^m_{;k} u^n_{;l} g_{mn}. \tag{1.83}$$

It is important to note that the extra term in Equation (1.77) derives from the fact that $\boldsymbol{G}_M$ is not necessarily constant in space. The expression $U^M_{;K}$ is also called the *covariant derivative* covariant derivative of $\boldsymbol{U}$ (Eringen 1980). For rectangular coordinates, the unit vectors do not vary in space and Equations (1.82) and (1.83) reduce to

$$2E_{KL} = U^M_{,K} G_{LM} + U^M_{,L} G_{KM} + U^M_{,K} U^N_{,L} G_{MN} \tag{1.84}$$

and

$$2e_{kl} = u^m_{,k} g_{lm} - u^m_{,l} g_{km} + u^m_{,k} u^n_{,l} g_{mn}. \tag{1.85}$$

Furthermore, the distinction between $\{\boldsymbol{G}^K\}$ and $\{\boldsymbol{G}_K\}$ fades since both bases are identical, and G_{KL} is the unit tensor. Consequently, Equations (1.84) and (1.85) can be further simplified to

$$2E_{KL} = U_{K,L} + U_{L,K} + U_{M,K} U_{M,L} \tag{1.86}$$

and

$$2e_{kl} = u_{k,l} + u_{l,k} - u_{m,k} u_{m,l}. \tag{1.87}$$

The above equations establish a relationship between displacements and strains. This relationship is nonlinear owing to the last terms in Equations (1.82) to (1.87). In problems with small deformations, the nonlinear terms are frequently neglected, leading to the linear strain $\tilde{E}_{KL}$, in rectangular coordinates:

$$\tilde{E}_{KL} := \tfrac{1}{2}(U_{K,L} + U_{L,K}). \tag{1.88}$$

Defining the infinitesimal rotation as

$$\tilde{R}_{KL} := \tfrac{1}{2}(U_{K,L} - U_{L,K}) \tag{1.89}$$

Equation (1.86) can be rewritten as

$$E_{KL} = \tilde{E}_{KL} + \tfrac{1}{2}(\tilde{E}_{MK} + \tilde{R}_{MK})(\tilde{E}_{ML} + \tilde{R}_{ML}) \tag{1.90}$$

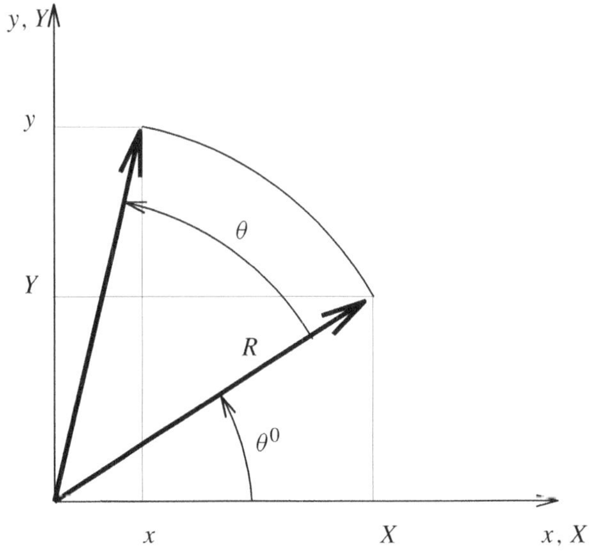

Figure 1.3 Finite rotation of a rod

showing that for the linear strain to be a good approximation for the actual strain, both the linear strain and the linear rotation must be small. Accordingly, for a rod freely rotating about one of its ends, linear strains are a poor approximation of the real strains. This is easily shown. Consider a rod of length R rotating about the origin (Figure 1.3). The original position is

$$X = R\cos\theta^0$$
$$Y = R\sin\theta^0. \tag{1.91}$$

The final position is characterized by

$$x = R\cos(\theta^0 + \theta)$$
$$y = R\sin(\theta^0 + \theta). \tag{1.92}$$

Consequently, the displacements amount to

$$U_X = x - X = X(\cos\theta - 1) - Y\sin\theta$$
$$U_Y = y - Y = X\sin\theta + Y(\cos\theta - 1). \tag{1.93}$$

The infinitesimal strains yield

$$\tilde{E}_{XX} = U_{X,X} = \cos\theta - 1$$
$$\tilde{E}_{YY} = V_{Y,Y} = \cos\theta - 1$$
$$\tilde{E}_{XY} = (U_{X,Y} + U_{Y,X})/2 = 0 \tag{1.94}$$

which shows that $\tilde{E}_{XX}$ and $\tilde{E}_{YY}$ are generally not zero. Since a rigid body motion must not generate strains, this clearly shows that the infinitesimal strains are not suited for finite rotations. The Lagrangian strain tensor, on the other hand, vanishes. For instance,

$$\begin{aligned} E_{XX} &= U_{X,X} + (U_{X,X}^2 + U_{Y,X}^2)/2 \\ &= (\cos\theta - 1) + \left[(\cos\theta - 1)^2 + \sin^2\theta\right]/2 = 0. \end{aligned} \tag{1.95}$$

This is especially important for slender structures such as shells and beams in which strains are usually small but rotations can be large.

1.4 Principal Strains

An infinitesimal vector $\mathrm{d}\boldsymbol{X}$ with size $\mathrm{d}S$ is transformed by the motion of the body into $\mathrm{d}\boldsymbol{x}$ with size $\mathrm{d}s$ satisfying

$$\mathrm{d}s^2 - \mathrm{d}S^2 = E_{KL}\,\mathrm{d}X^K\,\mathrm{d}X^L \tag{1.96}$$

or

$$\frac{\mathrm{d}s^2 - \mathrm{d}S^2}{\mathrm{d}S^2} = E_{KL}N^K N^L \tag{1.97}$$

where

$$N^K := \frac{\mathrm{d}X^K}{\mathrm{d}S} \tag{1.98}$$

is a unit vector satisfying

$$N^K N^L G_{KL} = \frac{\mathrm{d}X^K \mathrm{d}X^L G_{KL}}{\mathrm{d}S^2} = 1. \tag{1.99}$$

The expression in Equation (1.97) is a measure for the relative change of length of a fiber originally parallel to $\boldsymbol{N}$. The question we want to look into now is the following: in which directions is this change of length maximal? This boils down to maximizing Equation (1.97) subject to the constraint Equation (1.99). The variables are the components of $\boldsymbol{N}$. Following the usual procedure of calculus, finding an extremum reduces to setting the derivative of the target function with respect to the variables to zero. The target function is

$$F(\boldsymbol{N}) = E_{KL}N^K N^L - \Lambda_E(N^K N^L G_{KL} - 1) \tag{1.100}$$

where Λ_E is a Lagrange multiplier. Hence,

$$\frac{\partial}{\partial N^M}[E_{KL}N^K N^L - \Lambda_E(N^K N^L G_{KL} - 1)] = 0$$

$$\Updownarrow$$

$$E_{ML}N^L + E_{KM}N^K - \Lambda_E N^L G_{ML} - \Lambda_E N^K G_{KM} = 0$$

$$\Updownarrow$$

$$(E_{KM} - \Lambda_E G_{KM})N^K = 0. \tag{1.101}$$

This is a classical eigenvalue problem (generalized for curvilinear coordinates). Since $\boldsymbol{E}$ is a symmetric tensor, the eigenvalues Λ_E are real and the corresponding eigenvectors are mutually orthogonal (or can be made orthogonal). Indeed, suppose that Λ_E is a complex eigenvalue, then

$$E_{KM}N^K = \Lambda_E G_{KM}N^K. \tag{1.102}$$

Premultiplying with the complex conjugate of $\boldsymbol{N}$ yields

$$\overline{N}^M E_{KM}N^K = \Lambda_E \overline{N}^M G_{KL}N^K. \tag{1.103}$$

Since $\boldsymbol{E}$ is symmetric and real, one obtains

$$\begin{aligned}\overline{\overline{N}^M E_{KM}N^K} &= N^M E_{KM}\overline{N}^K = N^K E_{MK}\overline{N}^M \\ &= N^K E_{KM}\overline{N}^M \end{aligned} \tag{1.104}$$

and similar for $\boldsymbol{G}$. Consequently, $\overline{N}^M E_{KM}N^K$ and $\overline{N}^M G_{KM}N^K$ are real and Λ_E must be real because of Equation (1.103) and the positive-definiteness of $\boldsymbol{G}$. Because of Equation (1.102), the eigenvectors $\boldsymbol{N}$ are real too.

To prove that the eigenvectors are mutually orthogonal, consider two distinct eigenvalues Λ_1 and Λ_2 with two corresponding eigenvectors $\boldsymbol{N}_1$ and $\boldsymbol{N}_2$. Then

$$E_{KM}N_1^K = \Lambda_1 G_{KM}N_1^K \tag{1.105}$$

and

$$E_{KM}N_2^K = \Lambda_2 G_{KM}N_2^K. \tag{1.106}$$

Multiplying Equation (1.105) with N_2^M and Equation (1.106) with N_1^M and subtracting both yields

$$N_2^M E_{KM}N_1^K - N_1^M E_{KM}N_2^K = \Lambda_1 N_2^M G_{KM}N_1^K - \Lambda_2 N_2^M G_{MK}N_1^K \tag{1.107}$$

or

$$N_2^M (E_{KM} - E_{MK})N_1^K = \Lambda_1 N_2^M G_{KM}N_1^K - \Lambda_2 N_1^M G_{KL}N_2^K. \tag{1.108}$$

Since both $\boldsymbol{E}$ and $\boldsymbol{G}$ are symmetric, this yields

$$0 = (\Lambda_1 - \Lambda_2)N_2^M G_{KM}N_1^K. \tag{1.109}$$

Λ_1 and Λ_2 are assumed to be distinct, which means

$$N_1^K G_{KM}N_2^M = 0 \tag{1.110}$$

or

$$\boldsymbol{N}_1 \cdot \boldsymbol{N}_2 = 0. \tag{1.111}$$

This completes the proof.

The eigenvalues are the solution of a third-order nonlinear equation expressing that the determinant of the matrix in Equation (1.101) has to satisfy

$$\det(E_{KM} - \Lambda_E G_{KM}) = 0 \Leftrightarrow \det(E^K{}_M - \Lambda_E \delta^K{}_M) = 0 \tag{1.112}$$

for the equation to have nontrivial solutions. Since the extremal strains have a physical relevance and are independent of the coordinate system, the coefficients of Equation (1.112) are invariants. Indeed, Equation (1.112) can be written as

$$-\Lambda_E^3 + I_{1E}\Lambda_E^2 - I_{2E}\Lambda_E + I_{3E} = 0 \tag{1.113}$$

where

$$I_{1E} = \delta^K{}_L E^L{}_K = \mathrm{tr}\boldsymbol{E} \tag{1.114}$$

$$I_{2E} = \frac{1}{2}\left[I_{1E}^2 - \mathrm{tr}(\boldsymbol{E}^2)\right] \tag{1.115}$$

$$I_{3E} = \det \boldsymbol{E} = \frac{1}{3!} e_{LMP} e^{KNQ} E^L{}_K E^M{}_N E^P{}_Q \tag{1.116}$$

are the first, second and third invariant of $\boldsymbol{E}$. The expression tr$\boldsymbol{E}$ stands for the trace of $\boldsymbol{E}$, e_{LMP} and e^{KNQ} are the alternating symbols: $e_{KLM} = 1$ for $KLM = 123$ or any cyclic rotation thereof, $e_{KLM} = -1$ for $KLM = 321$ or any cyclic rotation, else $e_{KLM} = 0$. The eigenvalues are called *principal strains* and the corresponding direction $\boldsymbol{N}_i$ are called *principal directions*. They are obtained by solving Equation (1.101) in which the solutions of Equation (1.113) are substituted. For the solution of Equation (1.113), which is a cubic equation, see (Simo and Hughes 1997) or (Abramowitz and Stegun 1972).

Since $\boldsymbol{E}$ and $\boldsymbol{C}$ differ by the metric tensor, Equation (1.101) can also be written as

$$(C_{KM} - \Lambda_C G_{KM})N^K = 0 \tag{1.117}$$

where

$$\Lambda_C = 2\Lambda_E + 1. \tag{1.118}$$

Consequently, the eigenvectors of $\boldsymbol{C}$ and $\boldsymbol{E}$ are the same and the eigenvalues are directly related by Equation (1.118). In what follows, Λ_i denotes the eigenvalues of $\boldsymbol{C}$, that is, $\Lambda_i = \Lambda_{iC}$. The calculation of the eigenvectors $\boldsymbol{N}_i$ is somewhat tedious. Sometimes, it is more advantageous to calculate the tensors $\boldsymbol{N}^i \otimes \boldsymbol{N}^i$, which play the role of a tensorial basis. Here, $\boldsymbol{N}^i$ (index up) are the one-forms obtained by raising the index of $\boldsymbol{N}$:

$$\boldsymbol{N}^i = \boldsymbol{G}^\flat \cdot \boldsymbol{N}_i \tag{1.119}$$

and satisfy (Equation (1.110)

$$\boldsymbol{N}^i \cdot \boldsymbol{N}_j = \delta^i{}_j. \tag{1.120}$$

The one-forms $\{\boldsymbol{N}^i\}$ are the dual basis of the vectors $\{\boldsymbol{N}_i\}$.

Theorem 1.4.1 *Let $\boldsymbol{C}$ be a symmetric covariant second-order tensor in R^3, Λ_i its eigenvalues and $\boldsymbol{N}_i$ the corresponding eigenvectors, then*

$$\boldsymbol{C} = \sum_{i=1}^{3} \Lambda_i \boldsymbol{M}^i \tag{1.121}$$

where

$$\boldsymbol{M}^i = \boldsymbol{N}^{\underline{i}} \otimes \boldsymbol{N}^{\underline{i}} \tag{1.122}$$

and $\boldsymbol{N}^i$ are the one-forms dual to $\boldsymbol{N}_i$.

Proof.

$$\begin{aligned}
\sum_i \Lambda_i \boldsymbol{M}^i \cdot \boldsymbol{N}_l &= \sum_i \Lambda_i (\boldsymbol{N}^i \otimes \boldsymbol{N}^i) \cdot \boldsymbol{N}_l \\
&= \sum_i \Lambda_i \boldsymbol{N}^i (\boldsymbol{N}^i \cdot \boldsymbol{N}_l) \\
&= \sum_i \Lambda_i \boldsymbol{N}^i \delta^i{}_l \\
&= \Lambda_{\underline{l}} \boldsymbol{N}^{\underline{l}} = \Lambda_{\underline{l}} \boldsymbol{G}^\flat \cdot \boldsymbol{N}_{\underline{l}}, \forall l
\end{aligned} \tag{1.123}$$

where $\boldsymbol{G}^\flat$ is the covariant metric tensor and an underscore or a summation sign remove implicit summation. Consequently, $\boldsymbol{C}$ and $\sum_i \Lambda_i \boldsymbol{M}^i$ have the same eigenvalues and eigenvectors and are identical. Since $\boldsymbol{C}$ is a covariant tensor, it is logical that it is made up of one-forms and not of vectors.

An interesting property is

$$\begin{aligned}
\boldsymbol{C} \cdot \boldsymbol{C} &= \left(\sum_i \Lambda_i \boldsymbol{M}^i\right) \cdot \left(\sum_j \Lambda_j \boldsymbol{M}^j\right) \\
&= \sum_i \sum_j \Lambda_i \Lambda_j (\boldsymbol{N}^i \otimes \boldsymbol{N}^i) \cdot (\boldsymbol{N}^j \otimes \boldsymbol{N}^j) \\
&= \sum_i \sum_j \Lambda_i \Lambda_j (\boldsymbol{N}^i \otimes \boldsymbol{N}^j)(\boldsymbol{N}^i \cdot \boldsymbol{N}^j) \\
&= \sum_i \Lambda_i^2 (\boldsymbol{N}^i \otimes \boldsymbol{N}^i) = \sum_i \Lambda_i^2 \boldsymbol{M}^i.
\end{aligned} \tag{1.124}$$

This property allows for the following simple calculation of $\boldsymbol{M}_i$. Since

$$\left\{\begin{array}{rcl}
\boldsymbol{M}^1 + \boldsymbol{M}^2 + \boldsymbol{M}^3 &=& \boldsymbol{G} \\
\Lambda_1 \boldsymbol{M}^1 + \Lambda_2 \boldsymbol{M}^2 + \Lambda_3 \boldsymbol{M}^3 &=& \boldsymbol{C} \\
\Lambda_1^2 \boldsymbol{M}^1 + \Lambda_2^2 \boldsymbol{M}^2 + \Lambda_3^2 \boldsymbol{M}^3 &=& \boldsymbol{C}^2
\end{array}\right. \tag{1.125}$$

one obtains

$$\boldsymbol{M}^i = \frac{1}{D}\left[\boldsymbol{C}^2 - (I_{1C} - \Lambda_i)\boldsymbol{C} + I_{3C}\Lambda_i^{-1}\boldsymbol{G}\right] \tag{1.126}$$

where

$$D_i = (\Lambda_{\underline{i}} - \Lambda_j)(\Lambda_{\underline{i}} - \Lambda_k) \tag{1.127}$$

for $j, k \neq i$.

If two eigenvalues are identical, for example, $\Lambda = \Lambda_1 = \Lambda_2 \neq \Lambda_3$ one obtains instead of Equation (1.125),

$$\left\{ \begin{array}{rcl} (\boldsymbol{M}^1 + \boldsymbol{M}^2) + \boldsymbol{M}^3 & = & \boldsymbol{G} \\ \Lambda(\boldsymbol{M}^1 + \boldsymbol{M}^2) + \Lambda_3 \boldsymbol{M}^3 & = & \boldsymbol{C} \\ \Lambda^2(\boldsymbol{M}^1 + \boldsymbol{M}^2) + \Lambda_3^2 \boldsymbol{M}^3 & = & \boldsymbol{C}^2 \end{array} \right. \tag{1.128}$$

Discarding the third equation, one finds

$$\boldsymbol{M}^1 + \boldsymbol{M}^2 = \frac{\Lambda_3 \boldsymbol{G} - \boldsymbol{C}}{\Lambda_3 - \Lambda} \tag{1.129}$$

$$\boldsymbol{M}^3 = \frac{\boldsymbol{C} - \Lambda \boldsymbol{G}}{\Lambda_3 - \Lambda}. \tag{1.130}$$

This means that $\boldsymbol{M}^1$ and $\boldsymbol{M}^2$ are not known individually, only their sum can be derived. For three equal eigenvalues, the set in Equation (1.125) reduces to the first equation (Itskov 2001). The tensors $\boldsymbol{M}^1$, $\boldsymbol{M}^2$ and $\boldsymbol{M}^3$ are sometimes called *structural tensors*. They are genuine tensors of rank two subject to Equation (1.122) and the normality condition of $\boldsymbol{N}^i$.

Notice that

$$\boldsymbol{C} \cdot \boldsymbol{M}_i = \boldsymbol{C} \cdot (\boldsymbol{N}_{\underline{i}} \otimes \boldsymbol{N}_{\underline{i}}) = (\boldsymbol{C} \cdot \boldsymbol{N}_{\underline{i}})\boldsymbol{N}_{\underline{i}} = \Lambda_{\underline{i}}(\boldsymbol{G} \cdot \boldsymbol{N}_{\underline{i}})\boldsymbol{N}_{\underline{i}} = \Lambda_{\underline{i}} \boldsymbol{G} \cdot \boldsymbol{M}_{\underline{i}} \tag{1.131}$$

and

$$\boldsymbol{C} : \boldsymbol{M}_i = \boldsymbol{C} : (\boldsymbol{N}_{\underline{i}} \otimes \boldsymbol{N}_{\underline{i}}) = \boldsymbol{N}_{\underline{i}} \cdot \boldsymbol{C} \cdot \boldsymbol{N}_{\underline{i}} = \boldsymbol{N}_{\underline{i}} \Lambda_{\underline{i}} \cdot (\boldsymbol{G} \cdot \boldsymbol{N}_{\underline{i}}) = \Lambda_i \tag{1.132}$$

since Equation (1.117) is equivalent to

$$\boldsymbol{C} \cdot \boldsymbol{N}_i = \Lambda_{\underline{i}} \boldsymbol{G} \cdot \boldsymbol{N}_{\underline{i}} \tag{1.133}$$

and the double contraction or inner product of two second-order tensors $\boldsymbol{a} \otimes \boldsymbol{b}$ and $\boldsymbol{c} \otimes \boldsymbol{d}$ is defined by

$$(\boldsymbol{a} \otimes \boldsymbol{b}) : (\boldsymbol{c} \otimes \boldsymbol{d}) = (\boldsymbol{a} \cdot \boldsymbol{c})(\boldsymbol{b} \cdot \boldsymbol{d}) = \operatorname{tr}\left[(\boldsymbol{a} \otimes \boldsymbol{b})^{\mathrm{T}} \cdot (\boldsymbol{c} \otimes \boldsymbol{d})\right]. \tag{1.134}$$

One finds that the eigenvalues λ_i of $\boldsymbol{F}$ satisfy

$$\lambda_i = \sqrt{\Lambda_i} \tag{1.135}$$

because of Equation (1.42). Defining

$$\boldsymbol{n}_i := \boldsymbol{F} \cdot \boldsymbol{N}_{\underline{i}} / \lambda_{\underline{i}} \tag{1.136}$$

one can write

$$\boldsymbol{F} = \sum_i \lambda_i (\boldsymbol{n}_i \otimes \boldsymbol{N}^i). \tag{1.137}$$

The normals $\boldsymbol{N}_i$ along the principal directions in the material frame are mapped into the normals $\boldsymbol{n}_i$ in the spatial frame, strained by an amount λ_i. Notice that Equation (1.136) actually defines the right-hand side of the eigenvalue problem for $\boldsymbol{F}$. Furthermore, since $\boldsymbol{F}$ is a two-point tensor, it cannot map a vector into a multiple of itself.

Not only are $\{\boldsymbol{N}_i\}$ mutually orthogonal but $\{\boldsymbol{n}_i\}$ are also a mutually orthogonal set of vectors. Indeed,

$$\begin{aligned}\boldsymbol{n}_i \cdot \boldsymbol{n}_j &= \frac{1}{\lambda_{\underline{i}}\lambda_{\underline{j}}}\boldsymbol{N}_{\underline{i}} \cdot \boldsymbol{F}^{\mathrm{T}} \cdot \boldsymbol{F} \cdot \boldsymbol{N}_{\underline{j}} \\ &= \frac{1}{\lambda_{\underline{i}}\lambda_{\underline{j}}}\boldsymbol{N}_{\underline{i}} \cdot \boldsymbol{C} \cdot \boldsymbol{N}_{\underline{j}} \\ &= \frac{\lambda_{\underline{j}}}{\lambda_{\underline{i}}}\boldsymbol{N}_{\underline{i}} \cdot \boldsymbol{N}_{\underline{j}} = \frac{\lambda_{\underline{j}}}{\lambda_{\underline{i}}}\delta_{\underline{i}\underline{j}}.\end{aligned} \tag{1.138}$$

Hence, in each material point, there exist three mutually orthogonal vectors, the deformation of which is extremal and yields again three mutually orthogonal vectors. The vectors $\{\boldsymbol{n}_i\}$ are the eigenvectors of the inverse of the left Cauchy–Green tensor $\boldsymbol{b}^{-1}$. Indeed, substituting Equation (1.136) into Equation (1.117) yields

$$\boldsymbol{C} \cdot \boldsymbol{F}^{-1} \cdot \boldsymbol{n}_i = \Lambda_C \boldsymbol{G} \cdot \boldsymbol{F}^{-1} \cdot \boldsymbol{n}_i. \tag{1.139}$$

Substituting $\boldsymbol{C}$ (Equation (1.42)) leads to

$$\boldsymbol{F}^{\mathrm{T}} \cdot \boldsymbol{g} \cdot \boldsymbol{n}_i = \Lambda_C \boldsymbol{G} \cdot \boldsymbol{F}^{-1} \cdot \boldsymbol{n}_i \tag{1.140}$$

or

$$\frac{1}{\Lambda_C}\boldsymbol{g} \cdot \boldsymbol{n}_i = \boldsymbol{F}^{-\mathrm{T}} \cdot \boldsymbol{G} \cdot \boldsymbol{F}^{-1} \cdot \boldsymbol{n}_i \tag{1.141}$$

which is equivalent to

$$\frac{1}{\Lambda_C}\boldsymbol{g} \cdot \boldsymbol{n}_i = \boldsymbol{b}^{-1} \cdot \boldsymbol{n}_i. \tag{1.142}$$

At this point, the polar decomposition theorem should be mentioned because of its physical relevance. It states that the deformation gradient $\boldsymbol{F}$ can be written as the product of an orthogonal matrix $\boldsymbol{R}$ and a symmetric tensor $\boldsymbol{U}$, called the *right-stretch tensor*. Accordingly,

$$\boldsymbol{F} = \boldsymbol{R} \cdot \boldsymbol{U} \tag{1.143}$$

where

$$\boldsymbol{R}^{\mathrm{T}} = \boldsymbol{R}^{-1} \tag{1.144}$$

and

$$\boldsymbol{U} = \boldsymbol{U}^{\mathrm{T}}. \tag{1.145}$$

Since

$$\begin{aligned} \boldsymbol{C} &= \boldsymbol{F}^{\mathrm{T}} \cdot \boldsymbol{F} = \boldsymbol{U}^{\mathrm{T}} \cdot \boldsymbol{R}^{\mathrm{T}} \cdot \boldsymbol{R} \cdot \boldsymbol{U} \\ &= \boldsymbol{U}^{\mathrm{T}} \cdot \boldsymbol{U} = \boldsymbol{U} \cdot \boldsymbol{U} = \boldsymbol{U}^2 \end{aligned} \tag{1.146}$$

$\boldsymbol{U}$ and $\boldsymbol{F}$ have the same eigenvalues equal to the square root of the eigenvalues of $\boldsymbol{C}$. Since $\boldsymbol{C}$ is positive-definite (Equation (1.41)), $\boldsymbol{U}$ is also positive-definite. Furthermore, the eigenvectors of $\boldsymbol{C}$ and $\boldsymbol{U}$ are identical. We have

$$\boldsymbol{U} = \sum_i \lambda_i \boldsymbol{N}^i \otimes \boldsymbol{N}^i = \sum_i \sqrt{\Lambda_i} \boldsymbol{N}^i \otimes \boldsymbol{N}^i \tag{1.147}$$

and

$$\boldsymbol{R} = \sum_i \boldsymbol{n}_i \otimes \boldsymbol{N}_i. \tag{1.148}$$

Indeed,

$$\begin{aligned} \boldsymbol{R} \cdot \boldsymbol{U} &= \sum_i (\boldsymbol{n}_i \otimes \boldsymbol{N}_i) \sum_j \lambda_j (\boldsymbol{N}^j \otimes \boldsymbol{N}^j) \\ &= \sum_i \sum_j \lambda_j \boldsymbol{n}_i \otimes \boldsymbol{N}^j (\boldsymbol{N}_i \cdot \boldsymbol{N}^j) \\ &= \sum_i \lambda_i \boldsymbol{n}_i \otimes \boldsymbol{N}^i = \boldsymbol{F}. \end{aligned} \tag{1.149}$$

In a similar way, one can decompose $\boldsymbol{F}$ into

$$\boldsymbol{F} = \boldsymbol{V} \cdot \boldsymbol{R}. \tag{1.150}$$

$\boldsymbol{V}$ is the left-stretch tensor.

Equation (1.143) shows that the motion can be locally decomposed into a pure stretch along the principal directions followed by a rotation. It should be emphasized that a pure stretch is guaranteed for the principal directions only. For all other directions $\boldsymbol{N}$, the product $\boldsymbol{U} \cdot \boldsymbol{N}$ will involve some rotation, unless some of the principal values coincide. Furthermore, $\boldsymbol{R}$ is not constant in space. Consequently, $\boldsymbol{R}$ denotes a microscopic rotation in the material point of interest and not a macroscopic rotation.

1.5 Velocity

In most problems time is involved. The total time rate of change of a field is denoted by the total derivative D/Dt. It physically means that a material particle is followed while monitoring the change of some field at the momentaneous location of the moving particle. The partial derivative $\partial/\partial t$ is used when looking at the change in time of a field at a fixed spatial position. Both are related by (chain rule)

$$\frac{D}{Dt}\phi(\boldsymbol{x}, t) = \frac{\partial \phi}{\partial t} + \frac{\partial \phi}{\partial \boldsymbol{x}} \cdot \frac{\partial \boldsymbol{x}}{\partial t} \tag{1.151}$$

where $\phi(\boldsymbol{x}, t)$ is some field variable. The second term in Equation (1.151) is also called the *convective time rate of change* and is solely due to the nonzero velocity of the particle. The vector field

$$\boldsymbol{v} := \frac{\partial \boldsymbol{x}(\boldsymbol{X}, t)}{\partial t} \tag{1.152}$$

is the classical velocity of a particle originally at location $\boldsymbol{X}$. Applying Equation (1.151) to the particle acceleration, $\boldsymbol{a}$ defined by

$$\boldsymbol{a} := \frac{D\boldsymbol{v}}{Dt}(\boldsymbol{X}, t) \tag{1.153}$$

one finds

$$\boldsymbol{a} = \frac{\partial \boldsymbol{v}}{\partial t} + \frac{\partial \boldsymbol{v}}{\partial \boldsymbol{x}} \cdot \frac{\partial \boldsymbol{x}}{\partial t} \tag{1.154}$$

$$= \frac{\partial \boldsymbol{v}}{\partial t} + (\boldsymbol{v} \otimes \nabla) \cdot \boldsymbol{v} \tag{1.155}$$

where

$$\nabla := \frac{\partial}{\partial \boldsymbol{x}} \tag{1.156}$$

is a one-form. Writing

$$\boldsymbol{v} = v^k \boldsymbol{g}_k \tag{1.157}$$

and

$$\nabla = \boldsymbol{g}^l \frac{\partial}{\partial x^l} \tag{1.158}$$

leads to

$$\begin{aligned} \nabla \otimes \boldsymbol{v} = (\boldsymbol{v} \otimes \nabla)^{\mathrm{T}} &= \boldsymbol{g}^l \otimes \frac{\partial}{\partial x^l}(v^k \boldsymbol{g}_k) \\ &= v^k_{;l} \boldsymbol{g}^l \otimes \boldsymbol{g}_k \end{aligned} \tag{1.159}$$

and, consequently,

$$\begin{aligned} a^k \boldsymbol{g}_k &= \frac{\partial v^k}{\partial t} \boldsymbol{g}_k + v^k_{;l} v^m (\boldsymbol{g}_k \otimes \boldsymbol{g}^l) \cdot \boldsymbol{g}_m \\ &= \left(\frac{\partial v^k}{\partial t} + v^k_{;l} v^l \right) \boldsymbol{g}_k \end{aligned} \tag{1.160}$$

or, in rectangular coordinates

$$a^k = \frac{\partial v^k}{\partial t} + v^k_{,l} v^l. \tag{1.161}$$

The change of length in time is given by

$$\frac{D}{Dt}\mathrm{d}s^2 = \frac{D}{Dt}\mathrm{d}\boldsymbol{x}\cdot\mathrm{d}\boldsymbol{x} = \left(\frac{D}{Dt}\mathrm{d}\boldsymbol{x}\right)\cdot\mathrm{d}\boldsymbol{x} + \mathrm{d}\boldsymbol{x}\cdot\left(\frac{D}{Dt}\mathrm{d}\boldsymbol{x}\right). \tag{1.162}$$

Since

$$\frac{D}{Dt}\mathrm{d}\boldsymbol{x} = \mathrm{d}\boldsymbol{v} = \frac{\partial \boldsymbol{v}}{\partial \boldsymbol{x}}\cdot\mathrm{d}\boldsymbol{x} = (\boldsymbol{v}\otimes\nabla)\cdot\mathrm{d}\boldsymbol{x} \tag{1.163}$$

one finds

$$\begin{aligned}\frac{D}{Dt}\mathrm{d}s^2 &= \mathrm{d}\boldsymbol{x}\cdot(\nabla\otimes\boldsymbol{v})\cdot\mathrm{d}\boldsymbol{x} + \mathrm{d}\boldsymbol{x}\cdot(\boldsymbol{v}\otimes\nabla)\cdot\mathrm{d}\boldsymbol{x}\\ &= \mathrm{d}\boldsymbol{x}\cdot(\nabla\otimes\boldsymbol{v} + \boldsymbol{v}\otimes\nabla)\cdot\mathrm{d}\boldsymbol{x}\\ &= 2\mathrm{d}\boldsymbol{x}\cdot\boldsymbol{d}\cdot\mathrm{d}\boldsymbol{x}\end{aligned} \tag{1.164}$$

where

$$\boldsymbol{d} := \tfrac{1}{2}(\nabla\otimes\boldsymbol{v} + \boldsymbol{v}\otimes\nabla)^{\flat} \tag{1.165}$$

is called the *deformation rate tensor*. One also defines the velocity gradient $\boldsymbol{l}$ and the spin tensor $\boldsymbol{w}$:

$$\boldsymbol{l} := (\boldsymbol{v}\otimes\nabla)^{\flat} \tag{1.166}$$

$$\boldsymbol{w} := \tfrac{1}{2}(\boldsymbol{l} - \boldsymbol{l}^{\mathrm{T}}). \tag{1.167}$$

Consequently, one obtains

$$\boldsymbol{d} = \tfrac{1}{2}(\boldsymbol{l} + \boldsymbol{l}^{\mathrm{T}}). \tag{1.168}$$

Equation (1.164) shows that $2\boldsymbol{d}$ plays a similar role for $D(\mathrm{d}s^2)/Dt$ as $\boldsymbol{g}$ for $\mathrm{d}s^2$.

Since

$$\mathrm{d}\boldsymbol{x} = \boldsymbol{F}\cdot\mathrm{d}\boldsymbol{X} \tag{1.169}$$

one finds by taking the total derivative of both sides

$$\begin{aligned}(\boldsymbol{v}\otimes\nabla)\cdot\mathrm{d}\boldsymbol{x} &= \dot{\boldsymbol{F}}\cdot\mathrm{d}\boldsymbol{X}\\ &= \dot{\boldsymbol{F}}\cdot\boldsymbol{F}^{-1}\cdot\mathrm{d}\boldsymbol{x}\end{aligned} \tag{1.170}$$

or

$$\boldsymbol{l} = (\dot{\boldsymbol{F}}\cdot\boldsymbol{F}^{-1})^{\flat} \tag{1.171}$$

where

$$\dot{(\ )} := \overline{(\ )}^{\cdot} := \frac{D}{Dt}(\ ). \tag{1.172}$$

In component notation, Equation (1.164) reads

$$\frac{D}{Dt}\,\mathrm{d}s^2 = 2\mathrm{d}x^k\,\mathrm{d}x^l\,d_{kl}. \tag{1.173}$$

Since

$$\mathrm{d}s^2 = (2E_{KL} + G_{KL})\,\mathrm{d}X^K\,\mathrm{d}X^L \tag{1.174}$$

one also finds

$$\frac{D}{Dt}\,\mathrm{d}s^2 = 2\dot{E}_{KL}\,\mathrm{d}X^K\,\mathrm{d}X^L. \tag{1.175}$$

Comparison of Equation (1.173) and Equation (1.175) leads to

$$\dot{E}_{KL} = d_{kl}x^k_{\ ,K}x^l_{\ ,L} \tag{1.176}$$

or

$$\dot{\boldsymbol{E}} = \boldsymbol{F}^{\mathrm{T}} \cdot \boldsymbol{d} \cdot \boldsymbol{F}. \tag{1.177}$$

Accordingly, the tensor $\dot{\boldsymbol{E}}$ is the pullback of $\boldsymbol{d}$ and equivalently $\boldsymbol{d}$ is the push-forward of $\dot{\boldsymbol{E}}$.

The time derivative of the Jacobian J can be derived as follows:

$$\begin{aligned}
\frac{DJ}{Dt} &= \frac{DJ}{Dx^k_{\ ,K}}\frac{Dx^k_{\ ,K}}{Dt} \\
&= JX^K_{\ ,k}\left(\frac{Dx^k}{Dt}\right)_{,K} \\
&= JX^K_{\ ,k}v^k_{\ ,K} \\
&= JX^K_{\ ,k}v^k_{\ ;l}x^l_{\ ,K} \\
&= Jv^k_{\ ;k}.
\end{aligned} \tag{1.178}$$

In this derivation, Equation (1.61) was used. The expression $v^k_{\ ;k}$ corresponds to the divergence of the velocity, also written as $\nabla \cdot \boldsymbol{v}$.

1.6 Objective Tensors

Observers are not always on the same place and they do not necessarily use the same time. Consequently, observations are made by people in totally different places characterized by local coordinate systems for time and space. In space, these coordinate systems are related by a translation described by a vector $\boldsymbol{c}(t)$ and a rotation defined by an orthogonal matrix

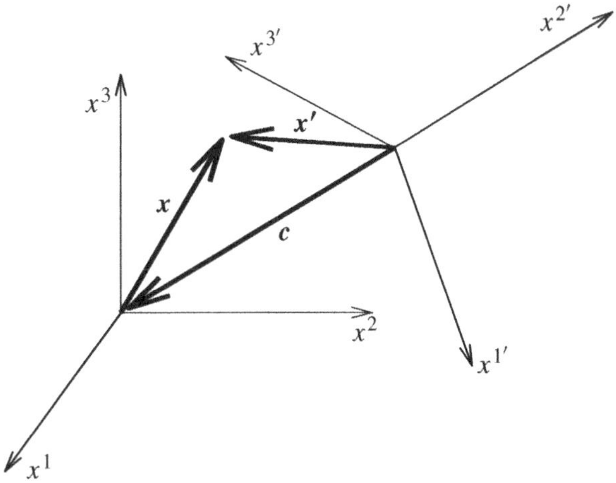

Figure 1.4 Frames of different observers

$\boldsymbol{Q}(t)$ (Figure 1.4). Notice that, since the observers generally move with a different speed, $\boldsymbol{c}$ and $\boldsymbol{Q}$ are a function of the time t. The different wall-clock time can be expressed by a shift of time. Hence,

$$\boldsymbol{x}'(\boldsymbol{X}, t') = \boldsymbol{c}(t) + \boldsymbol{Q}(t) \cdot \boldsymbol{x}(\boldsymbol{X}, t) \tag{1.179}$$

$$t' = t - a. \tag{1.180}$$

Since $\boldsymbol{Q}$ is an orthogonal matrix $\boldsymbol{Q}^{-1} = \boldsymbol{Q}^{\mathrm{T}}$ and $\det \boldsymbol{Q} = 1$. Here, only rigid body motions excluding reflections are considered and hence $\det \boldsymbol{Q} = 1$. The transformation in Equation (1.179) conserves the distance and angles. Indeed,

$$\mathrm{d}\boldsymbol{x}' = \boldsymbol{Q} \cdot \mathrm{d}\boldsymbol{x} \tag{1.181}$$

and consequently

$$(\mathrm{d}s')^2 = \mathrm{d}\boldsymbol{x}' \cdot \mathrm{d}\boldsymbol{x}' = \mathrm{d}\boldsymbol{x} \cdot \boldsymbol{Q}^{\mathrm{T}} \cdot \boldsymbol{Q} \cdot \mathrm{d}\boldsymbol{x} = \mathrm{d}\boldsymbol{x} \cdot \mathrm{d}\boldsymbol{x} = \mathrm{d}s^2 \tag{1.182}$$

and

$$\mathrm{d}\boldsymbol{x}' \cdot \mathrm{d}\boldsymbol{y}' = \mathrm{d}\boldsymbol{x} \cdot \boldsymbol{Q}^{\mathrm{T}} \cdot \boldsymbol{Q} \cdot \mathrm{d}\boldsymbol{y} = \mathrm{d}\boldsymbol{x} \cdot \mathrm{d}\boldsymbol{y}. \tag{1.183}$$

It is generally accepted that material properties should be independent of the coordinate frame of the observer. Hence, in describing these material properties, we would like to use quantities that ensure that the frame independence is guaranteed. For a time-independent rigid body motion, it is known that vectors $\boldsymbol{a}$ and second-order tensors $\boldsymbol{b}$ in the spatial description transform according to

$$\boldsymbol{a}' = \boldsymbol{Q} \cdot \boldsymbol{a} \tag{1.184}$$

and

$$\boldsymbol{b}' = \boldsymbol{Q} \cdot \boldsymbol{b} \cdot \boldsymbol{Q}^{\mathrm{T}}. \tag{1.185}$$

Requiring this to be true for time-dependent rigid motions guarantees the spatial frame indifference of any material law using such quantities. Vectors and tensors obeying Equation (1.184) and Equation (1.185) for time-dependent rigid body motions are called *objective*. From Equation (1.181), it is clear that $\mathrm{d}\boldsymbol{x}$ is objective while time-differentiation of Equation (1.179) reveals that the velocity $\boldsymbol{v}$ and the acceleration are not:

$$\boldsymbol{v}' = \dot{\boldsymbol{Q}} \cdot \boldsymbol{x} + \boldsymbol{Q} \cdot \boldsymbol{v} \tag{1.186}$$

$$\boldsymbol{a}' = \ddot{\boldsymbol{Q}} \cdot \boldsymbol{x} + 2\dot{\boldsymbol{Q}} \cdot \boldsymbol{v} + \boldsymbol{Q} \cdot \boldsymbol{a}. \tag{1.187}$$

Accordingly, $\boldsymbol{v}$ and $\boldsymbol{a}$ should not be used to describe material laws. That the acceleration is not objective is well known and is the reason for the Coriolis force in mechanics. Since the transformation in Equation (1.179) conserves the distance, one obtains (Equation (1.164)):

$$\begin{aligned}\frac{D}{Dt}(\mathrm{d}s')^2 &= 2\mathrm{d}\boldsymbol{x}' \cdot \boldsymbol{d}' \cdot \mathrm{d}\boldsymbol{x}' \\ &= 2\mathrm{d}\boldsymbol{x} \cdot \boldsymbol{Q}^{\mathrm{T}} \cdot \boldsymbol{d}' \cdot \boldsymbol{Q} \cdot \mathrm{d}\boldsymbol{x} \\ &= \frac{D}{Dt}\mathrm{d}s^2 = 2\mathrm{d}\boldsymbol{x} \cdot \boldsymbol{d} \cdot \mathrm{d}\boldsymbol{x}\end{aligned} \tag{1.188}$$

and consequently,

$$\boldsymbol{d} = \boldsymbol{Q}^{\mathrm{T}} \cdot \boldsymbol{d}' \cdot \boldsymbol{Q}. \tag{1.189}$$

This shows that the deformation rate tensor is objective. Notice that a second-order tensor $\boldsymbol{a}$, which maps an objective vector $\boldsymbol{b}$ into another objective vector $\boldsymbol{c}$, is objective. Indeed,

$$\boldsymbol{c}' = \boldsymbol{a}' \cdot \boldsymbol{b}' \tag{1.190}$$

implies

$$\boldsymbol{c} = (\boldsymbol{Q}^{\mathrm{T}} \cdot \boldsymbol{a}' \cdot \boldsymbol{Q}) \cdot \boldsymbol{b} \tag{1.191}$$

yielding

$$\boldsymbol{a} = \boldsymbol{Q}^{\mathrm{T}} \cdot \boldsymbol{a}' \cdot \boldsymbol{Q}. \tag{1.192}$$

The time derivative of an objective vector or tensor is generally not objective. Indeed, time differentiation of Equation (1.184) and Equation (1.185) yields

$$\dot{\boldsymbol{a}}' = \underline{\dot{\boldsymbol{Q}} \cdot \boldsymbol{a}} + \boldsymbol{Q} \cdot \dot{\boldsymbol{a}} \tag{1.193}$$

$$\dot{\boldsymbol{b}}' = \underline{\dot{\boldsymbol{Q}} \cdot \boldsymbol{b} \cdot \boldsymbol{Q}^{\mathrm{T}}} + \boldsymbol{Q} \cdot \dot{\boldsymbol{b}} \cdot \boldsymbol{Q}^{\mathrm{T}} + \underline{\boldsymbol{Q} \cdot \boldsymbol{b} \cdot \dot{\boldsymbol{Q}}^{\mathrm{T}}}. \tag{1.194}$$

The terms that are underlined are the reason for the lack of objectivity.

Finally, all vectors and tensors in the material description (such as $\boldsymbol{C}$) are objective since they are not influenced by a change of the spatial frame of reference.

1.7 Balance Laws

Balance laws are important statements describing the conservation of some physical quantities. These quantities and the conservation thereof will be defined in the present section.

1.7.1 Conservation of mass

Each object in space is assigned a strictly positive scalar quantity called the *mass*. The mass is assumed to be continuously distributed, which allows for the definition of density $\rho_0(\boldsymbol{X}, t)$ by letting the volume containing particle $\boldsymbol{X}$ go to zero:

$$\rho_0(\boldsymbol{X}) := \lim_{\Delta V_0 \to 0} \frac{\Delta M}{\Delta V_0}, \quad \boldsymbol{X} \in \Delta V_0. \tag{1.195}$$

ΔV_0 is the volume the particle occupies in the reference configuration at time $t = t_0$. The density can change during the motion of a body. The density of a particle at time t originally at $\boldsymbol{X}$ is

$$\rho(\boldsymbol{X}, t) := \lim_{\Delta V \to 0} \frac{\Delta M}{\Delta V}, \quad \boldsymbol{x}(\boldsymbol{X}, t) \in \Delta V. \tag{1.196}$$

ΔV is the volume the particle occupies at time t in the spatial configuration. The axiom of the conservation of mass now states that "the time rate of change of the total mass of a body is zero". Accordingly,

$$\frac{D}{Dt}\left(\int_V \rho \, \mathrm{d}v\right) = 0. \tag{1.197}$$

1.7.2 Conservation of momentum

The momentum (also called *linear momentum*) of an infinitesimal mass $\mathrm{d}m$ moving with a velocity $\boldsymbol{v}$ is defined as

$$\boldsymbol{v} \, \mathrm{d}m = \rho \boldsymbol{v} \, \mathrm{d}v. \tag{1.198}$$

The principle of conservation of momentum states that "the time rate of change of linear momentum is equal to the total force $\boldsymbol{F}$ acting on a body". Forces acting on a body are either body forces $\boldsymbol{F}_{\mathrm{b}}$ resulting from distant actions such as gravity, surface tractions $\boldsymbol{F}_{\mathrm{s}}$ resulting from immediate contact such as classical friction forces, or concentrated forces $\boldsymbol{F}_{\mathrm{c}}$. Enough continuity is assumed such that the body force per unit volume $\boldsymbol{f}$ and the force per unit area $\boldsymbol{t}_{(\boldsymbol{n})}$ can be defined as follows:

$$\mathrm{d}\boldsymbol{F}_{\mathrm{b}} =: \rho \boldsymbol{f} \, \mathrm{d}v \tag{1.199}$$

$$\mathrm{d}\boldsymbol{F}_{\mathrm{s}} =: \boldsymbol{t}_{(\boldsymbol{n})} \, \mathrm{d}a. \tag{1.200}$$

Accordingly,

$$\frac{D}{Dt}\int_V \rho \boldsymbol{v} \, \mathrm{d}v = \oint_A \boldsymbol{t}_{(\boldsymbol{n})} \, \mathrm{d}a + \int_V \rho \boldsymbol{f} \, \mathrm{d}v + \sum \boldsymbol{F}_{\mathrm{c}} \tag{1.201}$$

where A denotes the surface of the body at stake. This principle is also known as *Newton's second law*.

1.7.3 Conservation of angular momentum

The angular momentum of a particle with mass dm, velocity $\boldsymbol{v}$ and location $\boldsymbol{x}$ is defined as

$$\boldsymbol{x} \times \boldsymbol{v}\,\mathrm{d}m \tag{1.202}$$

where $\times$ symbolizes the vector product (also called the *cross product*) of two vectors. The vector product of two vectors $\boldsymbol{a}$ and $\boldsymbol{b}$ is a one-form $\boldsymbol{c}$ satisfying

$$\boldsymbol{c} \cdot \boldsymbol{a} = \boldsymbol{c} \cdot \boldsymbol{b} = 0 \tag{1.203}$$

and

$$\boldsymbol{c} \cdot \boldsymbol{c} = (\boldsymbol{a} \cdot \boldsymbol{a})(\boldsymbol{b} \cdot \boldsymbol{b}) - (\boldsymbol{a} \cdot \boldsymbol{b})^2. \tag{1.204}$$

Accordingly, $\boldsymbol{g}_i \times \boldsymbol{g}_j$ is proportional to $\boldsymbol{g}^k$. The proportionality constant λ can be determined from Equation (1.204):

$$\lambda^2 g^{kk} = g_{ii} g_{jj} - (g_{ij})^2 = \mathrm{cofactor}(g_{kk}). \tag{1.205}$$

Since $\boldsymbol{g}^\sharp$ is the inverse of $\boldsymbol{g}^\flat$, one finds

$$g^{kk} = \frac{\mathrm{cofactor}(g_{kk})}{\det \boldsymbol{g}^\flat} \tag{1.206}$$

leading to

$$\boldsymbol{g}_i \times \boldsymbol{g}_j = e_{ijk} \boldsymbol{g}^k \sqrt{\det \boldsymbol{g}^\flat}. \tag{1.207}$$

Similarly, the moment of a force $\boldsymbol{F}$ acting at a location $\boldsymbol{x}$ is defined as $\boldsymbol{x} \times \boldsymbol{F}$. The principle of conservation of angular momentum states that "the time rate of change of angular momentum is equal to the total moment due to forces and couples acting on the body". Hence,

$$\frac{D}{Dt} \int_V \rho \boldsymbol{x} \times \boldsymbol{v}\,\mathrm{d}v = \oint_A \boldsymbol{x} \times \boldsymbol{t}_{(\boldsymbol{n})}\,\mathrm{d}a + \int_V \rho \boldsymbol{x} \times \boldsymbol{f}\,\mathrm{d}v + \sum \boldsymbol{x} \times \boldsymbol{F}_{\mathrm{c}} + \sum \boldsymbol{M}_{\mathrm{c}}. \tag{1.208}$$

Here $\boldsymbol{M}_{\mathrm{c}}$ represents concentrated moments. It is assumed that there are no distributed moments, which essentially means that this treatise is limited to nonpolar theories. Readers interested in polar theories (used, for example, for the description of liquid crystals) are referred to (Eringen 1980).

1.7.4 Conservation of energy

This principle states that "the time rate of change of the sum of the kinetic energy $\mathcal{K}$ and internal energy $\mathcal{E}$ is equal to the sum of the work rate of all forces and couples $\mathcal{W}$ acting on the body and all other energies $\mathcal{U}$ that enter or leave the body per unit time". The total kinetic energy of a body is defined by

$$\mathcal{K} = \frac{1}{2} \int_V \rho \boldsymbol{v} \cdot \boldsymbol{v}\,\mathrm{d}v \tag{1.209}$$

and the rate of work of all forces and couples by

$$\mathcal{W} = \oint_A \boldsymbol{t}_{(\boldsymbol{n})} \cdot \boldsymbol{v}\, \mathrm{d}a + \int_V \rho \boldsymbol{f} \cdot \boldsymbol{v}\, \mathrm{d}v + \sum \boldsymbol{F}_\mathrm{c} \cdot \boldsymbol{v}_\mathrm{c} + \sum \boldsymbol{M}_\mathrm{c} \cdot \boldsymbol{\omega}_\mathrm{c} \tag{1.210}$$

where $\boldsymbol{\omega}_\mathrm{c}$ is the angular velocity of the particle $\boldsymbol{M}_c$ is acting.

The internal energy is a new quantity. It is assumed that it is continuously distributed such that the energy density or energy per unit mass ε can be defined as

$$\mathcal{E} = \int_V \rho \varepsilon\, \mathrm{d}v. \tag{1.211}$$

Other energies can, for example, be of thermal, chemical or electromagnetic origin. Here we limit the discussion to thermal energy. In that case, $\mathcal{U}$ amounts to

$$\mathcal{U} = -\oint_A \boldsymbol{q} \cdot \mathrm{d}\boldsymbol{a} + \int_V \rho h\, \mathrm{d}v + \sum H_\mathrm{c} \tag{1.212}$$

where $\boldsymbol{q}$ is the heat flux through area $\mathrm{d}\boldsymbol{a}$ (the minus sign implies that the body is losing energy if $\boldsymbol{q}$ points outwards), h is the body heat density and H_c is the heat due to concentrated heat sources. Consequently, the principle of conservation of energy reads

$$\begin{aligned} \frac{D}{Dt} \int_V \left(\rho \varepsilon + \frac{1}{2} \rho \boldsymbol{v} \cdot \boldsymbol{v} \right) \mathrm{d}v &= \oint_A (\boldsymbol{t}_{(\boldsymbol{n})} \cdot \boldsymbol{v} - \boldsymbol{q} \cdot \boldsymbol{n})\, \mathrm{d}a \\ &\quad + \int_V (\rho \boldsymbol{f} \cdot \boldsymbol{v} + \rho h)\, \mathrm{d}v + \sum H_\mathrm{c} + \sum \boldsymbol{F}_\mathrm{c} \cdot \boldsymbol{v}_\mathrm{c} + \sum \boldsymbol{M}_\mathrm{c} \cdot \boldsymbol{\omega}_\mathrm{c}. \end{aligned} \tag{1.213}$$

This is also called the *first law of thermodynamics*.

1.7.5 Entropy inequality

This principle, also called the *second law of thermodynamics* or Clausius–Duhem inequality, states that "the time rate of change of the entropy H of a body is never less than the sum of the entropy s entering the body through its surface and the entropy B generated by body sources". Hence,

$$\frac{DH}{Dt} \geq B + \oint_A \boldsymbol{s} \cdot \mathrm{d}\boldsymbol{a}. \tag{1.214}$$

Defining the entropy density η and the entropy source density b by

$$H = \int_V \rho \eta\, \mathrm{d}v \tag{1.215}$$

and

$$B = \int_V \rho b\, \mathrm{d}v \tag{1.216}$$

one finds

$$\frac{D}{Dt} \int_V \rho \eta\, \mathrm{d}v \geq \int_V \rho b\, \mathrm{d}v + \oint_A \boldsymbol{s} \cdot \mathrm{d}\boldsymbol{a}. \tag{1.217}$$

Notice that this is an inequality. If other phenomena are considered such as electromagnetic actions, additional laws apply. Here we concentrate on thermomechanical processes.

1.7.6 Closure

At first sight, the formulation of the balance laws does not look very promising for our primary goal, that is, the determination of $\boldsymbol{x}(\boldsymbol{X}, t)$. Indeed, a lot of extra unknowns have been defined: ρ, $\boldsymbol{t}_{(\boldsymbol{n})}$, ε, η, ... On the other hand, some of the new variables are formulated in terms of previously defined unknowns such as $\mathcal{K}(\boldsymbol{v})$. The relevance of the balance laws is based on the relationship they establish with the physical world through quantities such as $\boldsymbol{f}$ and h. They are fundamental axioms based on physical observations and as such indispensable. The extra unknowns will be taken care of later on by the material description (constitutive equations).

1.8 Localization of the Balance Laws

An important notion in the classical theory of continuum mechanics is the localization of the balance laws. In the previous section, the balance laws were formulated for finite bodies. The localization principle postulates that the balance laws are valid for any body, no matter how small. This strong assumption leads to a differential form of the balance laws. Nonlocal theories exist (Eringen 1976), which do not make this assumption but rather assume a sphere of influence for every point.

1.8.1 Conservation of mass

Since $\mathrm{d}v = J\,\mathrm{d}V$, one can write Equation (1.197) as

$$\frac{D}{Dt}\left(\int_{V_0} \rho J\,\mathrm{d}V\right) = 0. \tag{1.218}$$

V_0 is the volume of the mass at time $t = t_0$ and as such not dependent on time. Hence, Equation (1.218) is equivalent to

$$\int_{V_0} \frac{D}{Dt}(\rho J)\,\mathrm{d}V = 0. \tag{1.219}$$

Since this equation must be satisfied for any volume, the balance of mass yields

$$\frac{D}{Dt}(\rho J) = 0 \tag{1.220}$$

which can also be written as (Equation (1.178))

$$\rho\nabla\cdot\boldsymbol{v} + \frac{D\rho}{Dt} = 0 \tag{1.221}$$

or

$$\frac{\partial\rho}{\partial t} + \nabla\rho\cdot\boldsymbol{v} + \rho\nabla\cdot\boldsymbol{v} = 0 \tag{1.222}$$

which is equivalent to

$$\frac{\partial\rho}{\partial t} + \nabla\cdot(\rho\boldsymbol{v}) = 0. \tag{1.223}$$

1.8.2 Conservation of momentum

Equation (1.201) can be written as

$$\int_{V_0} \frac{D}{Dt}(\rho J \boldsymbol{v})\,\mathrm{d}V = \oint_A \boldsymbol{t}_{(\boldsymbol{n})}\,\mathrm{d}a + \int_{V_0} \rho J \boldsymbol{f}\,\mathrm{d}V + \sum \boldsymbol{F}_{\mathbf{c}}. \tag{1.224}$$

Before localization can be applied to Equation (1.224) the surface integral in the right-hand side has to be converted to a volume integral. To this end, the original conservation of momentum Equation (1.201) is applied to the volume in Figure 1.5. In addition, the mean value theorem is used, stating that for a continuous function ϕ in a domain Ω a point $\boldsymbol{x}^* \in \Omega$ exists such that

$$\int_{\Omega} \phi(\boldsymbol{x})\,\mathrm{d}\Omega = \phi(\boldsymbol{x}^*) \int_{\Omega} \mathrm{d}\Omega. \tag{1.225}$$

Hence

$$\frac{D}{Dt}(\rho^* \boldsymbol{v}^* \Delta v) = \boldsymbol{t}_{(\boldsymbol{n})} \Delta a + \boldsymbol{t}_{(-\boldsymbol{n}^k)} \Delta a_k + \rho^* \boldsymbol{f}^* \Delta v \tag{1.226}$$

assuming there are no point loads in the volume Δv. Newton's third law (action = reaction) dictates that

$$\boldsymbol{t}_{(-\boldsymbol{n}^k)} = -\boldsymbol{t}_{(\boldsymbol{n}^k)} \tag{1.227}$$

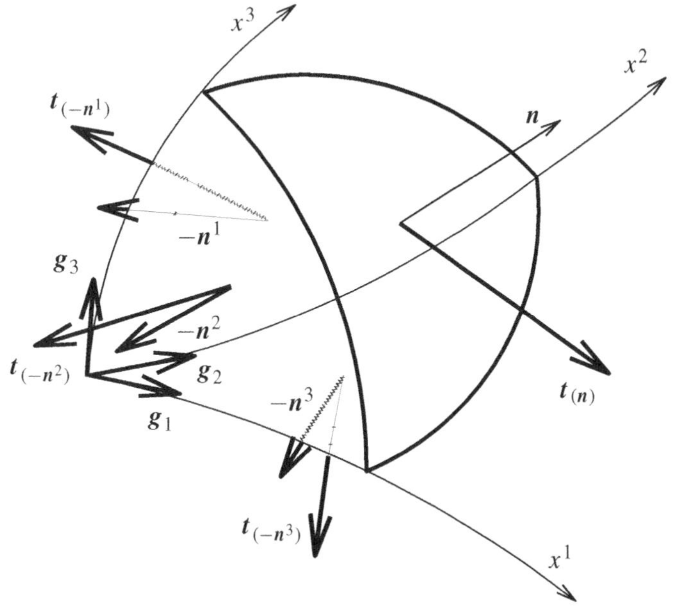

Figure 1.5 Equilibrium of an infinitesimal mass element

and Equation (1.226) reduces to

$$\frac{D}{Dt}(\rho^* \boldsymbol{v}^* \Delta v) = \boldsymbol{t}_{(\boldsymbol{n})}\Delta a - \boldsymbol{t}_{(\boldsymbol{n}^k)}\Delta a_k + \rho^* \boldsymbol{f}^* \Delta v. \tag{1.228}$$

Notice that $\boldsymbol{n}^1$, $\boldsymbol{n}^2$ and $\boldsymbol{n}^3$ are positive in the direction of $\boldsymbol{g}^1$, $\boldsymbol{g}^2$ and $\boldsymbol{g}^3$ respectively. Since in the limit $\Delta v \to 0$

$$\lim_{\Delta v \to 0} \frac{\Delta v}{\Delta a} = 0 \tag{1.229}$$

for an infinitesimal volume, Equation (1.226) reduces to

$$\boldsymbol{t}_{(\boldsymbol{n})}\Delta a = \boldsymbol{t}_{(\boldsymbol{n}^k)}\Delta a_k. \tag{1.230}$$

Since

$$\Delta a_k = n_k \Delta a \tag{1.231}$$

where n_k are the components of the one-form $\boldsymbol{n}$, that is, $\boldsymbol{n} = n_k \boldsymbol{g}^k$, one finds

$$\boldsymbol{t}_{(\boldsymbol{n})} = \boldsymbol{t}_{(\boldsymbol{n}^k)} n_k. \tag{1.232}$$

Notice that the normal to a surface is a one-form since the inner product with a length vector produces a scalar volume. Denoting the traction vector on a surface with unit normal $\boldsymbol{n}^k$ by $\boldsymbol{t}^k$, Equation (1.232) reads

$$\boldsymbol{t}_{(\boldsymbol{n})} = \boldsymbol{t}^k n_k. \tag{1.233}$$

Accordingly, the stress on a surface with normal $\boldsymbol{n}$ is a linear combination of the stresses on surfaces perpendicular to the coordinate axes. Substituting Equation (1.233) into Equation (1.224) and applying Cauchy's theorem , which reads

$$\oint_A \boldsymbol{t}^k n_k \, \mathrm{d}a = \int_V \boldsymbol{t}^k{}_{;k} \, \mathrm{d}v \tag{1.234}$$

one finds after localization

$$\frac{D}{Dt}(\rho J \boldsymbol{v}) = \boldsymbol{t}^k{}_{;k} J + \rho \boldsymbol{f} J \tag{1.235}$$

at locations without concentrated forces. Applying the balance of mass yields

$$\rho \frac{D\boldsymbol{v}}{Dt} = \boldsymbol{t}^k{}_{;k} + \rho \boldsymbol{f} \tag{1.236}$$

or

$$\rho \left[\frac{\partial \boldsymbol{v}}{\partial t} + (\boldsymbol{v} \otimes \nabla) \cdot \boldsymbol{v} \right] = \boldsymbol{t}^k{}_{;k} + \rho \boldsymbol{f}. \tag{1.237}$$

1.8.3 Conservation of angular momentum

Localization of Equation (1.208) at points without concentrated forces nor moments, taken Equation (1.232) into account, yields

$$\frac{D}{Dt}(\rho J \boldsymbol{x} \times \boldsymbol{v}) = J(\boldsymbol{x} \times \boldsymbol{t}^k)_{,k} + \rho J \boldsymbol{x} \times \boldsymbol{f} \tag{1.238}$$

or

$$\frac{D}{Dt}(\rho J)\boldsymbol{x} \times \boldsymbol{v} + J\boldsymbol{x} \times \rho \frac{D\boldsymbol{v}}{Dt} = J\boldsymbol{x}_{,k} \times \boldsymbol{t}^k + J\boldsymbol{x} \times \boldsymbol{t}^k_{,k} + J\boldsymbol{x} \times \rho \boldsymbol{f}. \tag{1.239}$$

Using the balance of mass, Equation (1.220), and the balance of momentum, Equation (1.236), yields

$$\boldsymbol{g}_k \times \boldsymbol{t}^k = 0 \tag{1.240}$$

since $\boldsymbol{x}_{,k} = \boldsymbol{g}_k$. The meaning of Equation (1.240) will become clear in Section 1.9.

1.8.4 Conservation of energy

Similar operations as in the previous section convert Equation (1.213) into

$$\frac{D}{Dt}\left(\rho J \varepsilon + \frac{1}{2}\rho J \boldsymbol{v} \cdot \boldsymbol{v}\right) = J(\boldsymbol{v} \cdot \boldsymbol{t}^k)_{;k} - \nabla \cdot \boldsymbol{q} J + \rho J \boldsymbol{f} \cdot \boldsymbol{v} + \rho J h \tag{1.241}$$

or

$$\rho J \frac{D\varepsilon}{Dt} + J\boldsymbol{v} \cdot \rho \frac{D\boldsymbol{v}}{Dt} = J\boldsymbol{v} \cdot \boldsymbol{t}^k_{;k} + J\boldsymbol{v}_{;k} \cdot \boldsymbol{t}^k - J\nabla \cdot \boldsymbol{q} + J\boldsymbol{v} \cdot \rho \boldsymbol{f} + J\rho h. \tag{1.242}$$

Application of the balance of momentum finally leads to

$$\rho \frac{D\varepsilon}{Dt} = \boldsymbol{v}_{;k} \cdot \boldsymbol{t}^k - \nabla \cdot \boldsymbol{q} + \rho h. \tag{1.243}$$

Equation (1.243) shows that the change of the internal energy per unit of time is balanced by the stress power ($\boldsymbol{v}_{;k} \cdot \boldsymbol{t}^k$), the heat influx ($-\nabla \cdot \boldsymbol{q}$) and the heat source power (ρh).

1.8.5 Entropy inequality

Along the same lines Equation (1.217) is reduced to

$$\rho \frac{D\eta}{Dt} \geq \rho b + \nabla \cdot \boldsymbol{s}. \tag{1.244}$$

1.9 The Stress Tensor

In the previous section, it was explained that $\boldsymbol{t}^k(\boldsymbol{x})$ is the stress on an infinitesimal surface at $\boldsymbol{x}$ perpendicular to $\boldsymbol{g}_k$. The components σ^{kl} of $\boldsymbol{t}^k$ are defined by

$$\boldsymbol{t}^k = \sigma^{kl}\boldsymbol{g}_l. \tag{1.245}$$

Now, $\boldsymbol{t}_{(n)} = \boldsymbol{t}^k n_k$ can be rewritten as

$$\boldsymbol{t}_{(n)} = \sigma^{kl} n_k \boldsymbol{g}_l \tag{1.246}$$

or, since $n_k = \boldsymbol{g}_k \cdot \boldsymbol{n}$,

$$\boldsymbol{t}_{(n)} = \sigma^{kl} \boldsymbol{g}_l \cdot (\boldsymbol{g}_k \cdot \boldsymbol{n}) = (\sigma^{kl} \boldsymbol{g}_l \otimes \boldsymbol{g}_k) \cdot \boldsymbol{n} \tag{1.247}$$

which shows that σ^{kl} is a second-order contravariant tensor (the so-called Cauchy stress tensor) and that the stress vector on an infinitesimal surface perpendicular to $\boldsymbol{n}$ can be obtained by the scalar product of the transpose of the stress tensor at that point with $\boldsymbol{n}$, in component notation:

$$t^l_{(n)} = \sigma^{kl} n_k. \tag{1.248}$$

The Cauchy stress is also called the *true stress* since it is defined in the spatial state of reference. It is the stress the deformed state truly experiences.

An important property of σ^{kl} follows from Equation (1.240) in component notation:

$$e_{ijl} g_k{}^j \sigma^{kl} = 0 \tag{1.249}$$

where e_{ijl} is the alternating symbol. Since $g_k{}^j = \delta_k{}^j$ one finds

$$\sigma^{kl} = \sigma^{lk} \tag{1.250}$$

that is, the stress tensor is symmetric. Letting

$$\boldsymbol{\sigma} := \sigma^{kl} \boldsymbol{g}_k \otimes \boldsymbol{g}_l \tag{1.251}$$

Equation (1.250) is equivalent to

$$\boldsymbol{\sigma} = \boldsymbol{\sigma}^{\mathrm{T}} \tag{1.252}$$

and $\boldsymbol{t}_{(n)} = \boldsymbol{\sigma}^{\mathrm{T}} \cdot \boldsymbol{n}$, Equation (1.247), is transformed into $\boldsymbol{t}_{(n)} = \boldsymbol{\sigma} \cdot \boldsymbol{n}$.

For the special case of $\boldsymbol{t}^k$, Equation (1.247) reduces to

$$\boldsymbol{t}^k = \boldsymbol{\sigma}^{\mathrm{T}} \cdot \boldsymbol{g}^k. \tag{1.253}$$

The term $\boldsymbol{v}_{,k} \cdot \boldsymbol{t}^k$ in the energy balance, Equation (1.243), becomes (see also Equation (1.163))

$$\begin{aligned} \boldsymbol{v}_{,k} \cdot \boldsymbol{t}^k &= \boldsymbol{g}_k \cdot (\nabla \otimes \boldsymbol{v}) \cdot \boldsymbol{\sigma}^{\mathrm{T}} \cdot \boldsymbol{g}^k \\ &= (\boldsymbol{v} \otimes \nabla) : \boldsymbol{\sigma}^{\mathrm{T}} = (\boldsymbol{v} \otimes \nabla) : \boldsymbol{\sigma} \end{aligned} \tag{1.254}$$

yielding for the complete energy equation

$$\rho \frac{D\varepsilon}{Dt} = (\boldsymbol{v} \otimes \nabla) : \boldsymbol{\sigma} - \nabla \cdot \boldsymbol{q} + \rho h. \tag{1.255}$$

Using the definition in Equation (1.166)

$$\rho \frac{D\varepsilon}{Dt} = \boldsymbol{l} : \boldsymbol{\sigma} - \nabla \cdot \boldsymbol{q} + \rho h \tag{1.256}$$

or

$$\rho \frac{D\varepsilon}{Dt} = \boldsymbol{d} : \boldsymbol{\sigma} - \nabla \cdot \boldsymbol{q} + \rho h. \tag{1.257}$$

Since $\boldsymbol{\sigma}$ is symmetric, all its eigenvalues are real. The meaning of the eigenvalues can be clarified by looking for the maximum normal stress in a point. Since $\boldsymbol{t}_{(\boldsymbol{n})} = \boldsymbol{n} \cdot \boldsymbol{\sigma}$, the normal stress σ on an infinitesimal surface with normal $\boldsymbol{n}$ is given by

$$\sigma = \boldsymbol{n} \cdot \boldsymbol{\sigma} \cdot \boldsymbol{n}. \tag{1.258}$$

Maximizing σ with the constraint $\|\boldsymbol{n}\| = \boldsymbol{n} \cdot \boldsymbol{g}^{\sharp} \cdot \boldsymbol{n} = 1$ yields

$$\frac{\partial}{\partial \boldsymbol{n}} \left[\boldsymbol{n} \cdot \boldsymbol{\sigma} \cdot \boldsymbol{n} - \lambda (\boldsymbol{n} \cdot \boldsymbol{g}^{\sharp} \cdot \boldsymbol{n}) \right] = 0. \tag{1.259}$$

$\boldsymbol{g}^{\sharp}$ is the contravariant metric tensor whose components g^{kl} satisfy

$$g^{kl} = \boldsymbol{g}^{k} \cdot \boldsymbol{g}^{l}. \tag{1.260}$$

Equation (1.259) leads to the eigenvalue problem

$$(\boldsymbol{\sigma} - \lambda \boldsymbol{g}^{\sharp}) \cdot \boldsymbol{n} = 0. \tag{1.261}$$

Similar to Equation (1.121), one can write

$$\boldsymbol{\sigma} = \sum_{i=1}^{3} \lambda_{i\sigma} (\boldsymbol{n}_i \otimes \boldsymbol{n}_i) \tag{1.262}$$

where $\boldsymbol{n}_i$ are the complementary basis vectors to the eigen one-forms of $\boldsymbol{\sigma}$. However, contrary to $\boldsymbol{C}$ the tensor $\boldsymbol{\sigma}$ is not positive definite, since σ in Equation (1.258) can be negative (pressure). In general, the stress eigenvectors do not coincide with the strain eigenvectors. Consequently, $\boldsymbol{n}_i$ in Equation (1.262) is usually distinct from $\boldsymbol{n}_i$ in Equation (1.136).

The force on an infinitesimal area $\mathrm{d}a$ can be written as

$$\begin{aligned} \mathrm{d}\boldsymbol{F} = \boldsymbol{t}_{(\boldsymbol{n})}\, \mathrm{d}a &= \boldsymbol{\sigma} \cdot \boldsymbol{n}\, \mathrm{d}a \\ &= \boldsymbol{\sigma} \cdot \mathrm{d}\boldsymbol{a} \\ &= \boldsymbol{\sigma} \cdot J \boldsymbol{F}^{-\mathrm{T}} \cdot \mathrm{d}\boldsymbol{A} \\ &= J\boldsymbol{\sigma} \cdot (\boldsymbol{F}^{-\mathrm{T}} \cdot \boldsymbol{N})\, \mathrm{d}A \\ &=: \boldsymbol{T}_{(\boldsymbol{N})}\, \mathrm{d}A \end{aligned} \tag{1.263}$$

where Equation (1.65) was used. The vector $\boldsymbol{T}_{(\boldsymbol{N})}$ represents an equivalent stress vector on the surface in the reference configuration and satisfies

$$\boldsymbol{T}_{(\boldsymbol{N})} = J\boldsymbol{\sigma} \cdot \boldsymbol{F}^{-\mathrm{T}} \cdot \boldsymbol{N}. \tag{1.264}$$

Defining the Piola–Kirchhoff tensor of the first kind by an expression similar to Equation (1.247):

$$\boldsymbol{T}_{(\boldsymbol{N})} := \boldsymbol{P}^{\mathrm{T}} \cdot \boldsymbol{N} \tag{1.265}$$

one finds

$$\boldsymbol{P} = J\boldsymbol{F}^{-1} \cdot \boldsymbol{\sigma}. \tag{1.266}$$

Notice that $\boldsymbol{P}$ is a two-point tensor, in component notation:

$$P^{Kk} = JX^{K}{}_{,l}\sigma^{lk}. \tag{1.267}$$

The tensor $\boldsymbol{P}$ is not symmetric. Indeed, $\boldsymbol{\sigma} = \boldsymbol{\sigma}^{\mathrm{T}}$ is equivalent to

$$\boldsymbol{F} \cdot \boldsymbol{P} = \boldsymbol{P}^{\mathrm{T}} \cdot \boldsymbol{F}^{\mathrm{T}}. \tag{1.268}$$

To remediate this, a Piola–Kirchhoff stress tensor of the second kind, $\boldsymbol{S}$, is defined by

$$\boldsymbol{S} := \boldsymbol{P} \cdot \boldsymbol{F}^{-\mathrm{T}} = J\boldsymbol{F}^{-1} \cdot \boldsymbol{\sigma} \cdot \boldsymbol{F}^{-\mathrm{T}}. \tag{1.269}$$

This tensor is symmetric and satisfies

$$\boldsymbol{S} = S^{KL}\boldsymbol{G}_K \otimes \boldsymbol{G}_L. \tag{1.270}$$

One also defines the Kirchhoff stress τ by

$$\boldsymbol{\tau} := J\boldsymbol{\sigma}. \tag{1.271}$$

Equation (1.257) can now also be written as

$$\rho_0 \frac{D\varepsilon}{Dt} = \boldsymbol{d} : \boldsymbol{\tau} - J\nabla \cdot \boldsymbol{q} + \rho_0 h. \tag{1.272}$$

In the balance equations in the previous section, a couple of quantities were defined on surfaces in the spatial configuration such as the heat vector $\boldsymbol{q}$. Similar to the derivation in Equation (1.263), an equivalent quantity in the reference configuration is defined by

$$\boldsymbol{q} \cdot \mathrm{d}\boldsymbol{a} = \boldsymbol{Q} \cdot \mathrm{d}\boldsymbol{A} \tag{1.273}$$

yielding

$$\boldsymbol{Q} = J\boldsymbol{q} \cdot \boldsymbol{F}^{-\mathrm{T}}. \tag{1.274}$$

Analogously, one defines

$$\boldsymbol{S} = J\boldsymbol{s} \cdot \boldsymbol{F}^{-\mathrm{T}} \tag{1.275}$$

for the entropy flux. Do not confuse the infinitesimal length $\mathrm{d}S$, Equation (1.8), with the Piola–Kirchhoff stress tensor of the second kind $\boldsymbol{S}$, Equation (1.269), and the entropy vector $\boldsymbol{S}$, Equation (1.275). The context should clarify what is meant.

1.10 The Balance Laws in Material Coordinates

In Sections 1.7 and 1.8, the balance laws were derived in spatial coordinates. In some cases (think of objective quantities), it is advantageous to work in material coordinates. The material form can be derived by starting over with the global form, or by simply converting spatial quantities into material quantities.

1.10.1 Conservation of mass

The spatial form $D(\rho J)/Dt$ can be trivially converted by integration into

$$\rho J = \rho_0. \tag{1.276}$$

Substitution into the spatial form yields

$$\frac{D}{Dt}(\rho_0) = 0. \tag{1.277}$$

1.10.2 Conservation of momentum

Substitution of Equation (1.253) into Equation (1.236) yields the spatial form

$$(\boldsymbol{\sigma}^{\mathrm{T}} \cdot \boldsymbol{g}^k)_{;k} + \rho \boldsymbol{f} = \rho \frac{D\boldsymbol{v}}{Dt} \tag{1.278}$$

or

$$\left[\sigma^{ml}(\boldsymbol{g}_l \otimes \boldsymbol{g}_m) \cdot \boldsymbol{g}^k)\right]_{;k} + \rho \boldsymbol{f} = \rho \frac{D\boldsymbol{v}}{Dt} \tag{1.279}$$

yielding

$$(\sigma^{kl} \boldsymbol{g}_l)_{;k} + \rho \boldsymbol{f} = \rho \frac{D\boldsymbol{v}}{Dt} \tag{1.280}$$

in component form

$$\sigma^{kl}{}_{;k} + \rho f^l = \rho \frac{Dv^l}{Dt}. \tag{1.281}$$

Notice that the semicolon in Equation (1.281) stands for the covariant differentiation of a second-order tensor ($\boldsymbol{g}_m$ in Equation (1.280) is not constant in space). For second-order contravariant tensors, one finds

$$\begin{aligned}
\boldsymbol{\sigma}_{,m} &= (\sigma^{kl} \boldsymbol{g}_k \otimes \boldsymbol{g}_l)_{,m} \\
&= \sigma^{kl}{}_{,m} \boldsymbol{g}_k \otimes \boldsymbol{g}_l + \sigma^{kl} \frac{\partial \boldsymbol{g}_k}{\partial x^m} \otimes \boldsymbol{g}_l + \sigma^{kl} \boldsymbol{g}_k \otimes \frac{\partial \boldsymbol{g}_l}{\partial x^m} \\
&= \sigma^{kl}{}_{,m} \boldsymbol{g}_k \otimes \boldsymbol{g}_l + \sigma^{kl} \frac{\partial^2 z^p}{\partial x^m \partial x^k} \frac{\partial x^q}{\partial z^p} \boldsymbol{g}_q \otimes \boldsymbol{g}_l + \sigma^{kl} \boldsymbol{g}_k \otimes \frac{\partial^2 z^p}{\partial x^m \partial x^l} \frac{\partial x^q}{\partial z^p} \boldsymbol{g}_q \\
&= \left[\sigma^{kl}{}_{,m} + \sigma^{ql} \frac{\partial^2 z^p}{\partial x^m \partial x^q} \frac{\partial x^k}{\partial z^p} + \sigma^{kq} \frac{\partial^2 z^p}{\partial x^m \partial x^q} \frac{\partial x^l}{\partial z^p}\right] \boldsymbol{g}_k \otimes \boldsymbol{g}_l.
\end{aligned} \tag{1.282}$$

Hence,

$$\sigma^{kl}{}_{;m} = \sigma^{kl}{}_{,m} + \sigma^{ql} \begin{Bmatrix} k \\ mq \end{Bmatrix} + \sigma^{kq} \begin{Bmatrix} l \\ mq \end{Bmatrix} \tag{1.283}$$

where the braces denote the Christoffel symbols of the second kind (cf Equation (1.80)).

Substituting Equation (1.267) and Equation (1.276) into Equation (1.281) yields

$$J(J^{-1}x^k_{,K}P^{Kl})_{;k} + \rho_0 f^l = \rho_0 \frac{Dv^l}{Dt} \tag{1.284}$$

or

$$x^k_{,K}(P^{Kl})_{;k} + \rho_0 f^l = \rho_0 \frac{Dv^l}{Dt} \tag{1.285}$$

since $(J^{-1}x^k_{,K})_{,k} = 0$. This identity can be obtained by realizing that J^{-1} is the Jacobian determinant of the inverse deformation $\boldsymbol{X}(\boldsymbol{x})$. Consequently, Equation (1.59) becomes

$$\begin{aligned} x^k_{,K} &= \frac{1}{J^{-1}}\text{cofactor }(X^K_{,k}) \\ &= \tfrac{1}{2}Je^{klm}e_{KLM}X^L_{,l}X^M_{,m} \end{aligned} \tag{1.286}$$

Accordingly, focusing on rectangular coordinates for simplicity,

$$(J^{-1}x^k_{,K})_{,k} = \tfrac{1}{2}e^{klm}e_{KLM}(X^L_{,lk}X^M_{,m} + X^L_{,l}X^M_{,mk}) = 0 \tag{1.287}$$

since by switching l and k or m and k the permutation symbols change sign. Another derivation is given by (Ogden 1984). Equation (1.285) finally yields

$$P^{Kl}_{;K} + \rho_0 f^l = \rho_0 \frac{Dv^l}{Dt}. \tag{1.288}$$

Notice that the first term involves covariant differentiation of a contravariant two-point tensor. One finds (Eringen 1975)

$$\begin{aligned} (P^{Kk}\boldsymbol{G}_K \otimes \boldsymbol{g}_k)_{,L} &= P^{Kk}_{,L}\boldsymbol{G}_K \otimes \boldsymbol{g}_k + P^{Kk}\frac{\partial \boldsymbol{G}_K}{\partial X^L} \otimes \boldsymbol{g}_k + P^{Kk}\boldsymbol{G}_K \otimes \frac{\partial \boldsymbol{g}_k}{\partial x^l}x^l_{,L} \\ &= \left(P^{Kk}_{,L} + P^{Kk}\frac{\partial^2 Z^M}{\partial X^L \partial X^N}\frac{\partial X^K}{\partial Z^M} + P^{Kn}\frac{\partial^2 z^m}{\partial x^l \partial x^n}\frac{\partial x^k}{\partial z^m}x^l_{,L}\right)\boldsymbol{G}_K \otimes \boldsymbol{g}_k. \end{aligned} \tag{1.289}$$

Accordingly,

$$P^{Kk}_{;L} = P^{Kk}_{,L} + P^{Nk}\begin{Bmatrix} K \\ LN \end{Bmatrix} + P^{Kn}\begin{Bmatrix} k \\ ln \end{Bmatrix}x^l_{,L}. \tag{1.290}$$

In vector form, Equation (1.288) yields

$$\nabla_0 \cdot \boldsymbol{P} + \rho_0 \boldsymbol{f} = \rho_0 \frac{D\boldsymbol{v}}{Dt} \tag{1.291}$$

or

$$\nabla_0 \cdot (\boldsymbol{S} \cdot \boldsymbol{F}^{\mathrm{T}}) + \rho_0 \boldsymbol{f} = \rho_0 \frac{D\boldsymbol{v}}{Dt} \tag{1.292}$$

where ∇_0 represents the gradient in material coordinates.

1.10.3 Conservation of angular momentum

The material form of $\boldsymbol{\sigma} = \boldsymbol{\sigma}^{\mathrm{T}}$ was derived in Section 1.8 and can be written as

$$\boldsymbol{F} \cdot \boldsymbol{P} = \boldsymbol{P}^{\mathrm{T}} \cdot \boldsymbol{F}^{\mathrm{T}} \tag{1.293}$$

or

$$\boldsymbol{S} = \boldsymbol{S}^{\mathrm{T}} \tag{1.294}$$

depending on whether the first or second Piola–Kirchhoff stress tensor is used.

1.10.4 Conservation of energy

Since

$$\boldsymbol{q} = J^{-1}\boldsymbol{Q} \cdot \boldsymbol{F}^{\mathrm{T}} \tag{1.295}$$

$$\boldsymbol{\sigma} = J^{-1}\boldsymbol{F} \cdot \boldsymbol{S} \cdot \boldsymbol{F}^{\mathrm{T}} \tag{1.296}$$

$$\boldsymbol{d} = \boldsymbol{F}^{-\mathrm{T}} \cdot \dot{\boldsymbol{E}} \cdot \boldsymbol{F}^{-1} \tag{1.297}$$

Equation (1.257) now yields

$$\rho_0 \frac{D\varepsilon}{Dt} = \dot{\boldsymbol{E}} : \boldsymbol{S} - \nabla_0 \cdot \boldsymbol{Q} + \rho_0 h. \tag{1.298}$$

Because of the expressions for work power in Equation (1.272) and Equation (1.298), it is said that $(\boldsymbol{d}, \boldsymbol{\tau})$ and $(\dot{\boldsymbol{E}}, \boldsymbol{S})$ are conjugate pairs in the spatial and material description respectively. Indeed,

$$\boldsymbol{d} : \boldsymbol{\tau} = \dot{\boldsymbol{E}} : \boldsymbol{S}. \tag{1.299}$$

Recall that $\boldsymbol{d}$ is the push-forward tensor of $\dot{\boldsymbol{E}}$. Equivalently, $\boldsymbol{\tau}$ is called the *push-forward* of $\boldsymbol{S}$ and equivalently $\boldsymbol{S}$ is the *pullback* of $\boldsymbol{\tau}$. One obtains

$$\boldsymbol{S} = \boldsymbol{F}^{-1} \cdot \boldsymbol{\tau} \cdot \boldsymbol{F}^{-\mathrm{T}}. \tag{1.300}$$

1.10.5 Entropy inequality

Along the same lines, one obtains for the material equivalent of Equation (1.244)

$$\rho_0 \frac{D\eta}{Dt} \geq \rho_0 b + \nabla_0 \cdot \boldsymbol{S} \tag{1.301}$$

where $\boldsymbol{S}$ is the entropy flux vector in material coordinates.

The entropy inequality plays an important role in the derivation of admissible constitutive equations. For thermal processes, the entropy influx and source can be written as the corresponding heat influx and source, divided by the absolute temperature θ . Consequently,

$$\boldsymbol{S} = -\frac{\boldsymbol{Q}}{\theta} + \boldsymbol{S}_1 \tag{1.302}$$

$$b = \frac{h}{\theta} + b_1 \tag{1.303}$$

where $\boldsymbol{S}_1$ and b_1 are the entropy influx and source due to other processes respectively. For simple thermomechanical processes that are considered here, $\boldsymbol{S}_1$ and b_1 are zero. Accordingly, the entropy inequality reads

$$\rho_0 \frac{D\eta}{Dt} \geq \rho_0 \frac{h}{\theta} - \nabla_0 \cdot \frac{\boldsymbol{Q}}{\theta} \tag{1.304}$$

or

$$\rho_0 \frac{D\eta}{Dt} \geq \rho_0 \frac{h}{\theta} - \frac{1}{\theta} \nabla_0 \cdot \boldsymbol{Q} + \frac{1}{\theta^2} \boldsymbol{Q} \cdot \nabla_0 \theta. \tag{1.305}$$

Solving the energy balance Equation (1.298) for $\nabla_0 \cdot \boldsymbol{Q}$ and substituting into Equation (1.305) yields

$$\rho_0 \left(\dot{\eta} - \frac{\dot{\varepsilon}}{\theta} \right) + \frac{1}{\theta} \dot{\boldsymbol{E}} : \boldsymbol{S} - \frac{1}{\theta^2} \boldsymbol{Q} \cdot \nabla_0 \theta \geq 0. \tag{1.306}$$

This is the preferred form that will be used for the derivation of the constitutive equations.

1.11 The Weak Form of the Balance of Momentum

In this section, an alternative form of the balance of momentum will be derived, which will form the basis for much of the finite element formulation to follow in subsequent chapters. In the material formulation, the weak form is generally known as the *principle of virtual work* and in the spatial description it is known as the *virtual power principle*. It will be shown that the strong form deduced so far, Equation (1.288), is completely equivalent to the weak form by first deriving the weak form from the strong form and subsequently the strong form from the weak form. General curvilinear coordinates are assumed throughout. To obtain the equations in rectangular coordinates, replace the covariant differentiation by partial differentiation.

1.11.1 Formulation of the boundary conditions (material coordinates)

The balance of momentum in material form

$$P^{Kk}{}_{;K} + \rho_0 f^k = \rho_0 \frac{D^2 u^k}{Dt^2} \tag{1.307}$$

will be supplemented here with boundary conditions. Suppose that the material volume V_0 is surrounded by a surface A_0 consisting of internal surfaces A_{0i}, surfaces on which the displacements are described A_{0u} and surfaces on which the traction is defined, A_{0t}. Accordingly,

$$\boldsymbol{T}^{+}_{(\boldsymbol{N}^{+})} + \boldsymbol{T}^{-}_{(\boldsymbol{N}^{-})} = 0 \quad \text{on } A_{0i} \tag{1.308}$$

$$\boldsymbol{u} = \overline{\boldsymbol{u}} \quad \text{on } A_{0u} \tag{1.309}$$

$$\boldsymbol{T}_{(\boldsymbol{N})} = \overline{\boldsymbol{T}}_{(\boldsymbol{N})} \quad \text{on } A_{0t}. \tag{1.310}$$

At internal surfaces the material is connected, but it might change its properties, for example, due to a change of material. Equation (1.308) is equivalent to Newton's third law: action equals reaction. The plus and minus sign denote the two sides of the internal surface. Since

$$\boldsymbol{T}_{(N)} = \boldsymbol{T}^K N_K \tag{1.311}$$

and $\boldsymbol{N}^- = -\boldsymbol{N}^+$, Equation (1.308) also reads

$$(\boldsymbol{T}^{K+} - \boldsymbol{T}^{K-})N_K^+ =: \left[\boldsymbol{T}^K\right] N_K^+ = 0. \tag{1.312}$$

This is also called the *traction continuity condition*. Note that Equation (1.307) is equivalent to

$$P^{Kk}{}_{;K}\boldsymbol{g}_k + \rho_0 f^k \boldsymbol{g}_k = \rho_0 \frac{D^2u^k}{Dt^2}\boldsymbol{g}_k \tag{1.313}$$

and hence,

$$P^{Kk}{}_{;K} g_k{}^L + \rho_0 f^k g_k{}^L = \rho_0 \frac{D^2u^k}{Dt^2} g_k{}^L \tag{1.314}$$

where $g_k{}^L := \boldsymbol{g}_k \cdot \boldsymbol{G}^L$ are called *shifters* since they move quantities from one coordinate system into another. Indeed, for a vector $\boldsymbol{v}$ one has

$$\boldsymbol{v} = v^k \boldsymbol{g}_k \Rightarrow \boldsymbol{v} \cdot \boldsymbol{G}^L = v^k g_k{}^L \Rightarrow V^L = v^k g_k{}^L. \tag{1.315}$$

Accordingly, Equation (1.307) is equivalent to

$$P^{Kk}{}_{;K} g_k{}^L + \rho_0 f^L = \rho_0 \frac{D^2u^L}{Dt^2}. \tag{1.316}$$

1.11.2 Deriving the weak form from the strong form (material coordinates)

Let us consider an infinitesimal perturbation of the displacement field $\delta\boldsymbol{u}$ with components δu_k satisfying the geometric boundary conditions in Equation (1.309). Accordingly,

$$\delta\boldsymbol{u} = 0 \quad \text{on } A_{0u}. \tag{1.317}$$

Taking the scalar product of the vector Equation (1.307) with the one-form $\delta\boldsymbol{u}$ and integrating over the material volume leads to

$$\int_{V_0} \left[P^{Kk}{}_{;K} + \rho_0 \left(f^k - \frac{D^2u^k}{Dt^2} \right) \right] \delta u_k \, \mathrm{d}V = 0. \tag{1.318}$$

Since (the usual differentiation rules also apply to covariant differentiation)

$$P^{Kk}{}_{;K}\delta u_k = (P^{Kk}\delta u_k)_{;K} - P^{Kk}\delta u_{k;K} \tag{1.319}$$

and applying Cauchy's theorem, Equation (1.234), one obtains

$$\int_{A_0} P^{Kk} N_K \delta u_k \, \mathrm{d}A - \int_{V_0} \left\{ P^{Kk} \delta u_{k;K} - \left[\rho_0 \left(f^k - \frac{D^2 u^k}{Dt^2} \right) \right] \delta u_k \right\} \mathrm{d}V = 0 \quad (1.320)$$

or

$$\int_{V_0} P^{Kk} \delta u_{k;K} \, \mathrm{d}V = \int_{A_{0t}} \overline{T}^K_{(N)} \delta U_K \, \mathrm{d}A + \int_{V_0} \rho_0 f^K \delta U_K \, \mathrm{d}V - \int_{V_0} \rho_0 \frac{D^2 u^K}{Dt^2} \delta U_K \, \mathrm{d}V \quad (1.321)$$

since $T^{K+}_{\ \ (N^+)} + T^{K-}_{\ \ (N^-)} = 0$ on A_{0i}, $\delta U_K = 0$ on A_{0u} and $\boldsymbol{T}_{(N)} = \overline{\boldsymbol{T}}_{(N)}$ on A_{0t}. Through the relationship

$$P^{Kk} = S^{KL} x^k_{\ ,L} \quad (1.322)$$

one obtains

$$P^{Kk} \delta u_{k;K} = S^{KL} x^k_{\ ,L} \delta x_{k,K} = S^{KL} x^k_{\ ,L} \delta x^m_{\ ,K} g_{km} - S^{KL} \delta E_{KL}. \quad (1.323)$$

Indeed,

$$\begin{aligned} S^{KL} \delta E_{KL} &= S^{KL} \delta \ \tfrac{1}{2} x^k_{\ ,L} x^m_{\ ,K} g_{km} - \tfrac{1}{2} G_{KL} \big) \\ &= \tfrac{1}{2} S^{KL} (\delta x^k_{\ ,L} x^m_{\ ,K} + x^k_{\ ,L} \delta x^m_{\ ,K}) g_{km} \\ &= S^{KL} x^k_{\ ,L} \delta x^m_{\ ,K} g_{km} \end{aligned} \quad (1.324)$$

since both S^{KL} and g_{km} are symmetric. Notice that covariant differentiation does not apply to $\boldsymbol{x}$. Indeed, from

$$\boldsymbol{u} = \boldsymbol{o} + \boldsymbol{x} - \boldsymbol{X} \quad (1.325)$$

one obtains

$$\boldsymbol{u}_{,K} = \boldsymbol{x}_{,K} - \boldsymbol{G}_K \quad (1.326)$$

leading to (see Equation (1.25))

$$u^k_{\ ;K} = x^k_{\ ,K} - g^k_K. \quad (1.327)$$

Concluding, Equation (1.321) can also be written as

$$\int_{V_0} S^{KL} \delta E_{KL} \, \mathrm{d}V = \int_{A_{0t}} \overline{T}^K_{(N)} \delta U_K \, \mathrm{d}A + \int_{V_0} \rho_0 f^K \delta U_K \, \mathrm{d}V - \int_{V_0} \rho_0 \frac{D^2 U^K}{Dt^2} \delta U_K \, \mathrm{d}V. \quad (1.328)$$

The left-hand side is called the *internal virtual work*, the first term on the right-hand side is the virtual work due to external tractions, the second term is due to distributed forces and the last term is due to inertia. Notice that, although all quantities in Equation (1.328) are expressed in terms of material coordinates, some are defined as a function of their spatial counterparts such as $\overline{\boldsymbol{T}}_{(N)}$ and $\boldsymbol{f}$ through $\overline{\boldsymbol{T}}_{(N)}(\boldsymbol{X}, t) \, \mathrm{d}A = \overline{\boldsymbol{t}}_{(\boldsymbol{n})}(\boldsymbol{X}, t) \, \mathrm{d}a$ and $f^K(\boldsymbol{X}, t) \boldsymbol{G}_K = f^k(\boldsymbol{X}, t) \boldsymbol{g}_k$. Hence, both $\overline{\boldsymbol{T}}_{(N)}$ and f^K are a function of the deformation. For example, if a rotating body expands because of centrifugal loads, f^K changes.

1.11.3 Deriving the strong form from the weak form (material coordinates)

Starting from the weak form in Equation (1.321) and applying Equation (1.319) and Cauchy's theorem one obtains

$$\int_{A_0} P^{Kk} N_K \delta u_k \,\mathrm{d}A - \int_{V_0} P^{Kk}_{\ ;K} \delta u_k \,\mathrm{d}V - \int_{A_{0t}} \overline{T}^k_{(N)} \delta u_k \,\mathrm{d}A - \int_{V_0} \rho_0 \left(f^k - \frac{D^2 u^k}{Dt^2} \right) \delta u_k \,\mathrm{d}V = 0. \quad (1.329)$$

Since $P^{Kk} N_K = T^k_{(N)}$, $A_0 = A_{0u} \cup A_{0t} \cup A_{0i}$ and $\delta \boldsymbol{u} = 0$ for A_{0u}, one obtains

$$\int_{A_{0t}} T^k_{(N)} - \overline{T}^k_{(N)} \Big) \delta u_k \,\mathrm{d}A + \int_{A_{0i}} T^{k+}_{(N^+)} + T^{k-}_{(N^-)} \Big) \delta u_k \,\mathrm{d}A - \int_{V_0} \left(P^{Kk}_{\ ;K} + \rho_0 f^k - \rho_0 \frac{D^2 u^k}{Dt^2} \right) \delta u_k \,\mathrm{d}V = 0. \quad (1.330)$$

So far we only specified $\delta \boldsymbol{u}$ to be a virtual displacement field satisfying the geometric boundary conditions. Now, we require Equation (1.330) to be valid not only for one special $\delta \boldsymbol{u}$ but also for any $\delta \boldsymbol{u}$ satisfying $\delta \boldsymbol{u} = 0$ on A_{0u}. Because of the arbitrariness of $\delta \boldsymbol{u}$, the functional analysis density theorem applies (for a proof, the reader is referred to (Belytschko *et al.* 2000)) requiring the coefficients of δu_k in each term in Equation (1.330) to be zero. Accordingly,

$$T^k_{(N)} - \overline{T}^k_{(N)} = 0 \quad \text{on } A_{0t} \quad (1.331)$$

$$T^{k+}_{(N^+)} + T^{k-}_{(N^-)} \quad \text{on } A_{0i} \quad (1.332)$$

$$P^{Kk}_{\ ;K} + \rho_0 f^k = \rho_0 \frac{D^2 u^k}{Dt^2} \quad \text{on } V_0 \quad (1.333)$$

which is the strong form.

1.11.4 The weak form in spatial coordinates

Here again, the starting point is the strong form, Equation (1.281)

$$(\sigma^{kl} \boldsymbol{g}_l)_{,k} + \rho f^l \boldsymbol{g}_l = \rho \frac{Dv^l}{Dt} \boldsymbol{g}_l \quad (1.334)$$

subject to

$$t^{k+}_{(\boldsymbol{n}^+)} + t^{k-}_{(\boldsymbol{n}^-)} \quad \text{on } A_i \quad (1.335)$$

$$\boldsymbol{u} = \overline{\boldsymbol{u}} \quad \text{on } A_u \quad (1.336)$$

$$t^k_{(\boldsymbol{n})} - \overline{t}^k_{(\boldsymbol{n})} = 0 \quad \text{on } A_t. \quad (1.337)$$

In spatial coordinates, a virtual velocity field $\delta\boldsymbol{v}$ is selected satisfying $\delta\boldsymbol{v} = 0$ on A_u. The reason for this will become clear in the derivation. Scalar multiplication yields

$$\int_V \left[(\sigma^{kl}\boldsymbol{g}_l)_{,k} \cdot \delta\boldsymbol{v} + \rho f^l \boldsymbol{g}_l \cdot \delta\boldsymbol{v}\right] \mathrm{d}v = \int_V \rho \frac{Dv^l}{Dt} \boldsymbol{g}_l \cdot \delta\boldsymbol{v}\, \mathrm{d}v. \tag{1.338}$$

Since

$$(\sigma^{kl}\boldsymbol{g}_l)_{,k} \cdot \delta\boldsymbol{v} = (\sigma^{kl}\boldsymbol{g}_l \cdot \delta\boldsymbol{v})_{,k} - \sigma^{kl}\boldsymbol{g}_l \cdot \delta\boldsymbol{v}_{,k} \tag{1.339}$$

and

$$\delta\boldsymbol{v}_{,k} = (\delta v_m \boldsymbol{g}^m)_{,k} = \delta v_{m;k}\boldsymbol{g}^m \tag{1.340}$$

one obtains

$$\int_V \left[(\sigma^{kl}\delta v_l)_{,k} - \sigma^{kl}\delta v_{l;k} + \rho\left(f^l - \frac{Dv^l}{Dt}\right)\delta v_l\right] \mathrm{d}v = 0 \tag{1.341}$$

or

$$\int_A \sigma^{kl} n_k \delta v_l\, \mathrm{d}a - \int_V \left[\sigma^{kl}\delta d_{kl} - \rho\left(f^l - \frac{Dv^l}{Dt}\right)\delta v_l\right] \mathrm{d}v = 0. \tag{1.342}$$

Indeed, $d_{kl} = (v_{k;l} + v_{l;k})/2$ and $\sigma^{kl} = \sigma^{lk}$. Taking into account the boundary conditions finally yields

$$\int_V \sigma^{kl}\delta d_{kl}\, \mathrm{d}v = \int_{A_t} \bar{t}^l_{(\boldsymbol{n})} \delta v_l\, \mathrm{d}a + \int_V \rho f^l \delta v_l\, \mathrm{d}v - \int_V \rho \frac{Dv^l}{Dt} \delta v_l\, \mathrm{d}v. \tag{1.343}$$

Equation (1.53) expresses the principle of virtual power. Notice that the principle of virtual work is of no avail here since the expression $\sigma^{kl}\delta u_{l;k}$ cannot be simplified because of the presence of nonlinear terms in the definition of the Eulerian strain measure. Accordingly, the spatial description implies a rate formulation and necessitates a thorough discussion of objective rate tensors. This can be largely avoided by using the material description.

Naturally, the strong form can also be obtained starting from the weak form. Interested readers are referred to (Belytschko *et al.* 2000).

1.12 The Weak Form of the Energy Balance

We start from the strong form expressed by Equation (1.298)

$$\rho_0 \frac{D\varepsilon}{Dt} = \dot{\boldsymbol{E}} : \boldsymbol{S} - \nabla_0 \cdot \boldsymbol{Q} + \rho_0 h \quad \text{on } V \tag{1.344}$$

completed by appropriate boundary conditions: we assume that the temperature T and the flux $\boldsymbol{Q}$ are known on A_{0T} and on A_{0Q} respectively. Furthermore, the flux normal to an interface A_i is continuous.

$$T = \overline{T} \quad \text{on } A_{0T} \tag{1.345}$$

$$\boldsymbol{Q} = \overline{\boldsymbol{Q}} \quad \text{on } A_{0Q} \tag{1.346}$$

$$\boldsymbol{Q}^+ \cdot \boldsymbol{N}^+ + \boldsymbol{Q}^- \cdot \boldsymbol{N}^- = 0 \quad \text{on } A_{0i}. \tag{1.347}$$

To obtain the weak form, we consider again an infinitesimal perturbation of the independent variable field T, satisfying the "geometric" boundary conditions in Equation (1.345). Hence,

$$\delta T = 0 \quad \text{on } A_{0T}. \tag{1.348}$$

Multiplying Equation (1.343) with δT and integrating over V, one obtains

$$\int_{V_0} \rho_0 \frac{D\varepsilon}{Dt} \delta T \, \mathrm{d}V = \int_{V_0} (\dot{\boldsymbol{E}} : \boldsymbol{S} - \nabla_0 \cdot \boldsymbol{Q} + \rho_0 h) \delta T \, \mathrm{d}V. \tag{1.349}$$

The second term on the right can be written as

$$\begin{aligned} -\int_{V_0} \nabla_0 \cdot \boldsymbol{Q} \delta T \, \mathrm{d}V &= -\int_{V_0} Q^K{}_{,K} \delta T \, \mathrm{d}V \\ &= -\int_{V_0} \left[(Q^K \delta T)_{,K} - Q^K \delta T_{,K} \right] \mathrm{d}V \\ &= -\int_{A_{0Q}} \delta T Q^K N_K \, \mathrm{d}A + \int_{V_0} Q^K \delta T_{,K} \, \mathrm{d}V. \end{aligned} \tag{1.350}$$

This step is essential to reduce the degree of differentiation of T in the resulting equation and is similar to Equations (1.318) to (1.320) for the balance of momentum. Indeed, the constitutive equations in Section 1.13 will show that $\boldsymbol{Q} \sim -\nabla_0 T$. Consequently, $\nabla_0 \cdot \boldsymbol{Q} \sim -\nabla_0 \cdot \nabla_0 T = -\nabla_0^2 T$ and $\nabla_0 \cdot \boldsymbol{Q}\ \delta T$ is the product of two terms of which the first one is twice differentiated, the second one not at all. On the other hand, both terms in $Q^K \delta T_{,K}$ are differentiated only once. This implies that the shape functions in the finite element formulation can have a lesser degree of smoothness and still comply with Equation (1.350).

Substitution of Equation (1.350) into Equation (1.349) yields

$$-\int_{V_0} \boldsymbol{Q} \cdot \delta \nabla_0 T \, \mathrm{d}V = \int_{V_0} \dot{\boldsymbol{E}} : \boldsymbol{S} \delta T \, \mathrm{d}V - \int_{A_{0Q}} \boldsymbol{Q} \cdot \boldsymbol{N} \delta T \, \mathrm{d}A + \int_{V_0} \rho_0 \left(h - \frac{D\varepsilon}{Dt} \right) \delta T \, \mathrm{d}V. \tag{1.351}$$

This equation is the analogue of Equation (1.328). Similar to what was said in Section 1.11.3, the strong form can be derived from the weak form if one allows the temperature perturbation to be absolutely general provided the "geometric" boundary conditions are satisfied.

1.13 Constitutive Equations

1.13.1 Summary of the balance equations

In Section 1.10, the balance equations were derived in material coordinates. They amount to

$$\rho J = \rho_0 \quad \text{(1 equation)} \tag{1.352}$$

$$\nabla_0 (\boldsymbol{S} \cdot \boldsymbol{F}^{\mathrm{T}}) + \rho_0 \boldsymbol{f} = \rho_0 \dot{\boldsymbol{v}} \quad \text{(3 equations)} \tag{1.353}$$

$$\boldsymbol{S} = \boldsymbol{S}^{\mathrm{T}} \quad \text{(3 equations)} \tag{1.354}$$

$$\rho_0\dot{\varepsilon} = \dot{\boldsymbol{E}} : \boldsymbol{S} - \nabla_0 \cdot \boldsymbol{Q} + \rho_0 h \quad \text{(1 equation)} \tag{1.355}$$

$$\rho_0\left(\dot{\eta} - \frac{\dot{\varepsilon}}{\theta}\right) + \frac{1}{\theta}\dot{\boldsymbol{E}} : \boldsymbol{S} - \frac{1}{\theta^2}\boldsymbol{Q} \cdot (\nabla_0\theta) \geq 0. \tag{1.356}$$

In sum, there are eight equations and one inequality. The unknowns are ρ (1), J (1), $\boldsymbol{S}$ (9), $\boldsymbol{F}$ (9), $\boldsymbol{v}$ (3), ε (1), $\boldsymbol{E}(6)$, $\boldsymbol{Q}$ (3), η (1) and θ (1) which yields 35 unknowns. The variables J, $\boldsymbol{F}$, $\boldsymbol{v}$ and $\boldsymbol{E}$ can be reduced to $\boldsymbol{x}$ (3 unknowns) since

$$J = \det(x^k_{,K}) \tag{1.357}$$

$$\boldsymbol{F} = x^k_{,K}\boldsymbol{g}_k \otimes \boldsymbol{G}^K \tag{1.358}$$

$$\boldsymbol{v} = \dot{\boldsymbol{x}} \tag{1.359}$$

$$\boldsymbol{E} = \tfrac{1}{2}(x^k_{,K}x^l_{,L}g_{kl} - G_{KL})\boldsymbol{G}^K \otimes \boldsymbol{G}^L. \tag{1.360}$$

In that way, 19 unknowns remain. Accordingly, we need another 11 equations to solve the problem for $\boldsymbol{x}(\boldsymbol{X}, t)$ and $\theta(\boldsymbol{X}, t)$. This is not surprising, since the material properties were not considered so far. All balance equations apply to steel as well as to wood, water or air. It is well known that these materials behave quite differently and it is the task of the constitutive equations to describe these different kinds of behavior. It looks like a huge task to tackle but luckily there are some physical principles that may guide us. Here I wish to adhere to a simplified form of the axiomatic formulation found in (Eringen 1980) since it leads us in a systematic way to the constitutive equations of widely different materials.

1.13.2 Development of the constitutive theory

The constitutive equations bridge the gap between physically observable quantities (independent variables in the constitutive equations) and the quantities arising in the balance laws (dependent variables in the constitutive equations). For thermomechanical processes, the observable quantities are the location $\boldsymbol{x}(\boldsymbol{X}, t)$ and the temperature $\theta(\boldsymbol{X}, t)$. All other variables such as the stress $\boldsymbol{S}$, the flux $\boldsymbol{Q}$, the internal energy ε and the entropy η are measured indirectly by the effect they produce on the displacements and the temperature. For instance, the stress is usually measured through strain gauges. Accordingly, the value of the dependent variables ($\boldsymbol{S}$, $\boldsymbol{Q}$, ε, and η – the density ρ is not considered as a dependent variable but rather immediately eliminated through Equation (1.352)) at $\boldsymbol{X}$ at time t is assumed to be a function of the value of the independent variables ($\boldsymbol{x}$, θ) at all former times and in the complete body. This can be written in the form of the following functionals:

$$\boldsymbol{S}(\boldsymbol{X}, t) = \boldsymbol{S}[\boldsymbol{x}(\boldsymbol{X}', t'), \theta(\boldsymbol{X}', t'), \boldsymbol{X}, t] \tag{1.361}$$

$$\boldsymbol{Q}(\boldsymbol{X}, t) = \boldsymbol{Q}[\boldsymbol{x}(\boldsymbol{X}', t'), \theta(\boldsymbol{X}', t'), \boldsymbol{X}, t] \tag{1.362}$$

$$\varepsilon(\boldsymbol{X}, t) = \varepsilon[\boldsymbol{x}(\boldsymbol{X}', t'), \theta(\boldsymbol{X}', t'), \boldsymbol{X}, t] \tag{1.363}$$

$$\eta(\boldsymbol{X}, t) = \eta[\boldsymbol{x}(\boldsymbol{X}', t'), \theta(\boldsymbol{X}', t'), \boldsymbol{X}, t] \tag{1.364}$$

$$t' \leq t, \boldsymbol{X}' \in V_0.$$

Hence, *a priori* it is assumed that the deformation and temperature in the complete body at all former times can have an impact on the value of any dependent variable at

X and t. This formulation includes memory effects (e.g. viscosity) and nonlocal effects (atomic forces).

There are two major postulates that must be obeyed by the constitutive equations. First, there is the principle of objectivity, which states that the constitutive equations must not depend on the spatial motion of the observer. This principle has already been briefly discussed in Section 1.6. There, it was emphasized that only objective tensors should be used in constitutive equations. What does this translate to in Equations (1.361) to (1.364)? Since the left-hand side of these equations is formulated in terms of material quantities, objectivity is no problem. What about the right-hand side? In general, a time-dependent rigid body motion combined with a time-shift maps $\boldsymbol{x}(\boldsymbol{X}, t)$ into

$$\overline{\boldsymbol{x}}(\boldsymbol{X}', \overline{t'}) = \boldsymbol{Q}(t') \cdot \boldsymbol{x}(\boldsymbol{X}', t') + \boldsymbol{b}(t') \tag{1.365}$$

$$\boldsymbol{Q} \cdot \boldsymbol{Q}^{\mathrm{T}} = \boldsymbol{Q}^{\mathrm{T}} \cdot \boldsymbol{Q} = \boldsymbol{I}, \quad \overline{t'} = t' - a. \tag{1.366}$$

This mapping can be split into a time-dependent translation, a time-shift and a time-dependent rotation.

A time-dependent translation must not change the constitutive equation. Taking the translation to be $\boldsymbol{x}(\boldsymbol{X}, t')$ one obtains, Equation (1.361),

$$\boldsymbol{S}(\boldsymbol{X}, t) = \boldsymbol{S}[\boldsymbol{x}(\boldsymbol{X}', t') - \boldsymbol{x}(\boldsymbol{X}, t'), \theta(\boldsymbol{X}', t'), \boldsymbol{X}, t], \tag{1.367}$$

which means that only the relative position with respect to $\boldsymbol{x}(\boldsymbol{X}, t')$ is kept.
A time shift must not influence Equation (1.367) either. Taking the shift to be t one obtains

$$\boldsymbol{S}(\boldsymbol{X}, t) = \boldsymbol{S}[\boldsymbol{x}(\boldsymbol{X}', t' - t) - \boldsymbol{x}(\boldsymbol{X}, t' - t), \theta(\boldsymbol{X}', t' - t), \boldsymbol{X}, 0]. \tag{1.368}$$

Consequently, the explicit dependence on t drops out:

$$\boldsymbol{S}(\boldsymbol{X}, t) = \boldsymbol{S}[\boldsymbol{x}(\boldsymbol{X}', t') - \boldsymbol{x}(\boldsymbol{X}, t'), \theta(\boldsymbol{X}', t'), \boldsymbol{X}]. \tag{1.369}$$

Finally, a time-dependent rotation of the spatial frame of reference should also leave the constitutive equation unaltered. One obtains

$$\boldsymbol{S}(\boldsymbol{X}, t) = \boldsymbol{S}\left[\boldsymbol{Q}(t) \cdot (\boldsymbol{x}(\boldsymbol{X}', t') - \boldsymbol{x}(\boldsymbol{X}, t')), \theta(\boldsymbol{X}', t'), \boldsymbol{X}\right] \tag{1.370}$$

for an arbitrary rotation $\boldsymbol{Q}(t)$.

The second postulate states that the constitutive equations must be form-invariant with respect to a certain class of rotations $\boldsymbol{Q}$ and translations $\boldsymbol{B}$ of the material frame, which are the result of material symmetries and material homogeneities. A lot of materials exhibit symmetries with respect to a specific class of rotations due to the intrinsic crystallographic structure. For example, single crystals are frequently orthotropic, which means that mutually orthogonal planes exist in the material frame with respect to which the material properties are symmetric. A usual assumption in polycrystals is the form-invariance with respect to the full group of rotations: this means that the material properties are independent of the direction and this is called *isotropy*. For a homogeneous material, the properties are not changed by an arbitrary translation. Once the classes $\{\boldsymbol{Q}\}$ and $\{\boldsymbol{B}\}$ are determined on the basis of observations of the material behavior, the constitutive equations should be form-invariant with respect to transformations of the type

$$\overline{\boldsymbol{X}} = \boldsymbol{Q} \cdot \boldsymbol{X} + \boldsymbol{B}, \quad \boldsymbol{Q}^{\mathrm{T}} \cdot \boldsymbol{Q} = \boldsymbol{Q} \cdot \boldsymbol{Q}^{\mathrm{T}} = \boldsymbol{I}, \quad \det \boldsymbol{Q} = \pm 1 \tag{1.371}$$

mapping the material frame $\boldsymbol{X}$ into $\overline{\boldsymbol{X}}$. Notice that the axiom of objectivity involves transformations of the spatial frame, whereas the axiom of material invariance concerns transformations of the material frame.

In order to obtain further simplifications, the expressions for the independent quantities are expanded in a Taylor series. Taylor expansion of $\boldsymbol{x}(\boldsymbol{X}', t') - \boldsymbol{x}(\boldsymbol{X}, t')$ in space yields

$$\begin{aligned} \boldsymbol{x}(\boldsymbol{X}', t') - \boldsymbol{x}(\boldsymbol{X}, t') = {} & \boldsymbol{x}_{,K_1}(\boldsymbol{X}, t') \left. X'^{K_1} - X^{K_1}\right) \\ & + \frac{1}{2!} \boldsymbol{x}_{,K_1K_2}(\boldsymbol{X}, t') \left. X'^{K_1} - X^{K_1}\right) \left(X'^{K_2} - X^{K_2}\right) + \cdots \end{aligned} \tag{1.372}$$

Similarly,

$$\begin{aligned} \theta(\boldsymbol{X}', t') = {} & \theta(\boldsymbol{X}, t') + \theta_{,K_1}(\boldsymbol{X}, t') \left. X'^{K_1} - X^{K_1}\right) \\ & + \frac{1}{2!} \theta_{,K_1K_2}(\boldsymbol{X}, t') \left. X'^{K_1} - X^{K_1}\right) \left(X'^{K_2} - X^{K_2}\right) + \cdots \end{aligned} \tag{1.373}$$

and Equation (1.98) can be replaced by

$$\begin{aligned} \boldsymbol{S}(\boldsymbol{X}, t) = \boldsymbol{S}[& \boldsymbol{Q}(t) \cdot \boldsymbol{x}_{,K_1}(\boldsymbol{X}, t'), \boldsymbol{Q}(t) \cdot \boldsymbol{x}_{,K_1K_2}(\boldsymbol{X}, t'), \ldots, \\ & \theta(\boldsymbol{X}, t'), \theta_{,K_1}(\boldsymbol{X}, t'), \theta_{,K_1K_2}(\boldsymbol{X}, t'), \ldots, \boldsymbol{X}]. \end{aligned} \tag{1.374}$$

Notice that the dependent variables are explicitly dependent on $\theta(\boldsymbol{X}, t')$ but not on $\boldsymbol{x}(\boldsymbol{X}, t')$. Materials satisfying Equation (1.374) are said to be of mechanical grade N and thermal grade M if the spatial derivatives are at most of Nth order and the thermal derivatives of at most Mth order.

Taylor expanding the remaining independent variables in time yields

$$\boldsymbol{x}_{,K_1}(\boldsymbol{X}, t') = \boldsymbol{x}_{,K_1}(\boldsymbol{X}, t) + \dot{\boldsymbol{x}}_{,K_1}(\boldsymbol{X}, t)(t' - t) + \frac{1}{2!} \ddot{\boldsymbol{x}}_{,K_1}(\boldsymbol{X}, t)(t' - t)^2 + \cdots \tag{1.375}$$

and similar for the other variables. Hence, one can replace Equation (1.374) by

$$\begin{aligned} \boldsymbol{S}(\boldsymbol{X}, t) = \boldsymbol{S}[& \boldsymbol{Q}(t) \cdot \boldsymbol{x}_{,K_1}(\boldsymbol{X}, t), \boldsymbol{Q}(t) \cdot \dot{\boldsymbol{x}}_{,K_1}(\boldsymbol{X}, t), \ldots, \\ & \boldsymbol{Q}(t) \cdot \boldsymbol{x}_{,K_1K_2}(\boldsymbol{X}, t), \boldsymbol{Q}(t) \cdot \dot{\boldsymbol{x}}_{,K_1K_2}(\boldsymbol{X}, t), \ldots, \\ & \ldots \\ & \boldsymbol{Q}(t) \cdot \boldsymbol{x}_{,K_1K_2\ldots K_N}(\boldsymbol{X}, t), \boldsymbol{Q}(t) \cdot \dot{\boldsymbol{x}}_{,K_1K_2\ldots K_N}(\boldsymbol{X}, t), \ldots, \\ & \theta(\boldsymbol{X}, t), \dot{\theta}(\boldsymbol{X}, t), \ddot{\theta}(\boldsymbol{X}, t), \ldots, \\ & \theta_{,K_1}(\boldsymbol{X}, t), \dot{\theta}_{,K_1}(\boldsymbol{X}, t), \ddot{\theta}_{,K_1}(\boldsymbol{X}, t), \ldots, \\ & \ldots \\ & \theta_{,K_1K_2\ldots K_M}(\boldsymbol{X}, t), \dot{\theta}_{,K_1K_2\ldots K_M}(\boldsymbol{X}, t), \ddot{\theta}_{,K_1K_2\ldots K_M}(\boldsymbol{X}, t), \ldots, \boldsymbol{X}]. \end{aligned} \tag{1.376}$$

In what follows, we will concentrate on materials of mechanical grade 1 and thermal grade 1. Hence,

$$\begin{aligned} S(X,t) = S[&Q(t)\cdot x_{,K}(X,t), Q(t)\cdot \dot{x}_{,K}(X,t), \dots, \\ &\theta(X,t), \dot{\theta}(X,t), \ddot{\theta}(X,t), \dots, \\ &\theta_{,K}(X,t), \dot{\theta}_{,K}(X,t), \ddot{\theta}_{,K}(X,t), \dots, X]. \end{aligned} \quad (1.377)$$

The principle of objectivity implies that the right-hand side of Equation (1.377) must be invariant with respect to spatial rotations. This means that the list of independent variables can be replaced by the invariants of $\{x_{,K}, \dot{x}_{,K}, \ddot{x}_{,K}, \dots\}$ with respect to an arbitrary rotation. The theory of invariants (Spencer 1971) shows that an integrity basis for the invariants of the above set subject to proper transformations (i.e. $\det Q = +1$) consists of the scalar product of any two vectors in the set, for example,

$$x_{,K}\cdot x_{,L} = x^k_{,K} x^l_{,L} g_{kl} = C_{KL} \quad (1.378)$$

and triple products of the form

$$e_{klm} x^k_{,K} x^l_{,L} x^m_{,M}. \quad (1.379)$$

For $K \neq L$, $K \neq M$ and $L \neq M$ the expression in Equation (1.379) is the Jacobian determinant J. For $K = L$, $K = M$ or $K = M$ Equation (1.379) is zero since this amounts to the determinant of a matrix with two equal rows or columns. Consequently, the dependence on $\{x_{,K}, \dot{x}_{,K}, \ddot{x}_{,K}, \dots\}$ in Equation (1.377) can be replaced by a dependence on

$$\{C_{KL}, \dot{C}_{KL}, \ddot{C}_{KL}, \dots, J, \dot{J}, \ddot{J}, \dots\} \quad (1.380)$$

or

$$\{C_{KL}, \dot{C}_{KL}, \ddot{C}_{KL}, \dots, \rho^{-1}, \dot{\rho}, \ddot{\rho}, \dots\} \quad (1.381)$$

since $J = \rho_0\rho^{-1}$. Equation (1.377) now reads

$$\begin{aligned} S(X,t) = S[&C_{KL}(X,t), \dot{C}_{KL}(X,t), \dots, \\ &\rho^{-1}(X,t), \dot{\rho}(X,t), \dots, \\ &\theta(X,t), \dot{\theta}(X,t), \ddot{\theta}(X,t), \cdots, \\ &\theta_{,K}(X,t), \dot{\theta}_{,K}(X,t), \ddot{\theta}_{,K}(X,t), \dots, X]. \end{aligned} \quad (1.382)$$

This also applies to Q, ε and η yielding the missing 11 equations.

1.14 Elastic Materials

1.14.1 General form

Elastic materials are defined as *materials without memory*. Consequently, the time derivatives are dropped in Equation (1.382) and one obtains

$$S(X,t) = S[C(X,t), \rho^{-1}(X,t), \theta(X,t), \nabla_0\theta(X,t), X] \quad (1.383)$$

and similarly (dropping the dependence),

$$\boldsymbol{Q} = \boldsymbol{Q}[\boldsymbol{C}, \rho^{-1}, \theta, \nabla_0\theta, \boldsymbol{X}] \tag{1.384}$$

$$\varepsilon = \varepsilon[\boldsymbol{C}, \rho^{-1}, \theta, \nabla_0\theta, \boldsymbol{X}] \tag{1.385}$$

$$\eta = \eta[\boldsymbol{C}, \rho^{-1}, \theta, \nabla_0\theta, \boldsymbol{X}]. \tag{1.386}$$

Since $\det \boldsymbol{C} = J^2$ and $\rho J = \rho_0$ (balance of mass) the explicit dependence on ρ is dropped. The balance of momentum, the balance of energy and the entropy inequality remain to be satisfied. It is amazing that the entropy inequality, being an inequality, plays an extremely important role in the derivation of the material laws. To see this, we first define the free energy $\psi(\boldsymbol{X}, t)$ to simplify the calculations:

$$\psi := \varepsilon - \theta\eta. \tag{1.387}$$

Since

$$\dot{\psi} = \dot{\varepsilon} - \dot{\theta}\eta - \theta\dot{\eta} \tag{1.388}$$

the entropy inequality now reads

$$-\frac{\rho_0}{\theta}(\dot{\psi} + \dot{\theta}\eta) + \frac{1}{\theta}\boldsymbol{S} : \dot{\boldsymbol{E}} - \frac{1}{\theta^2}\boldsymbol{Q} \cdot \nabla_0\theta \geq 0. \tag{1.389}$$

Notice that because of Equation (1.385) and Equation (1.386), dropping the dependence on ρ^{-1}

$$\psi = \psi[\boldsymbol{C}, \theta, \nabla_0\theta, \boldsymbol{X}]. \tag{1.390}$$

Accordingly,

$$\dot{\psi} = \frac{\partial \psi}{\partial \boldsymbol{C}} : \dot{\boldsymbol{C}} + \frac{\partial \psi}{\partial \theta}\dot{\theta} + \frac{\partial \psi}{\partial \nabla_0\theta} \cdot \dot{\overline{\nabla_0\theta}}. \tag{1.391}$$

Substituting Equation (1.391) into Equation (1.389) and noting that $\dot{\boldsymbol{E}} = \dot{\boldsymbol{C}}/2$ yields

$$\frac{1}{\theta}\left(-\rho_0\frac{\partial \psi}{\partial \boldsymbol{C}} + \tfrac{1}{2}\boldsymbol{S}\right) : \dot{\boldsymbol{C}} - \frac{\rho_0}{\theta}\left(\frac{\partial \psi}{\partial \theta} + \eta\right)\dot{\theta} - \frac{\rho_0}{\theta}\frac{\partial \psi}{\partial \nabla_0\theta} \cdot \dot{\overline{\nabla_0\theta}} - \frac{1}{\theta^2}\boldsymbol{Q} \cdot \nabla_0\theta \geq 0. \tag{1.392}$$

Since $\boldsymbol{S}$, $\boldsymbol{Q}$, ψ and η are not a function of $\dot{\boldsymbol{C}}$ nor $\dot{\theta}$ nor $\dot{\overline{\nabla_0\theta}}$, Equation (1.392) is linear in $\dot{\boldsymbol{C}}, \dot{\theta}$ and $\nabla_0\dot{\theta}$. Hence, for Equation (1.392) to be valid for any $\dot{\boldsymbol{C}}$, $\dot{\theta}$ or $\nabla_0\dot{\theta}$, the coefficients of these terms must be zero. Defining $\Sigma := \rho_0\psi$ one obtains

$$\boldsymbol{S} = 2\rho_0\frac{\partial \psi}{\partial \boldsymbol{C}} = 2\frac{\partial \Sigma}{\partial \boldsymbol{C}} = \frac{\partial \Sigma}{\partial \boldsymbol{E}} \tag{1.393}$$

$$\eta = -\frac{\partial \psi}{\partial \theta} = -\frac{1}{\rho_0}\frac{\partial \Sigma}{\partial \theta} \tag{1.394}$$

$$\frac{\partial \psi}{\partial \nabla_0\theta} = \frac{\partial \Sigma}{\partial \nabla_0\theta} = 0 \tag{1.395}$$

and the entropy inequality reduces to

$$-\boldsymbol{Q} \cdot \nabla_0 \theta \geq 0. \tag{1.396}$$

Consequently, for elastic materials there exists a function $\Sigma(\boldsymbol{C}, \theta, \boldsymbol{X})$ such that $\boldsymbol{S}$ and η can be obtained by partial differentiation. ε satisfies

$$\varepsilon(\boldsymbol{C}, \theta, \boldsymbol{X}) = \frac{\Sigma}{\rho_0} + \theta\eta. \tag{1.397}$$

The only dependent variable that depends on $\nabla_0\theta$ is $\boldsymbol{Q}$. Equation (1.396) requires that $\boldsymbol{Q}$ is at least linear in $\nabla_0\theta$, that is,

$$\boldsymbol{Q} = -\kappa(\boldsymbol{C}, \theta, \nabla_0\theta, \boldsymbol{X}) \cdot \nabla_0\theta. \tag{1.398}$$

The entropy inequality has dictated the shape of nearly all variables! The only equations left to satisfy are the balance of momentum and the balance of energy. Summarizing,

$$\boldsymbol{S} = \frac{\partial \Sigma}{\partial \boldsymbol{E}}(\boldsymbol{C}, \theta, \boldsymbol{X}) \tag{1.399}$$

$$\eta = -\frac{1}{\rho_0}\frac{\partial \Sigma}{\partial \theta}(\boldsymbol{C}, \theta, \boldsymbol{X}) \tag{1.400}$$

$$\varepsilon = \frac{\Sigma}{\rho_0} + \theta\eta \tag{1.401}$$

$$\boldsymbol{Q} = \boldsymbol{Q}(\boldsymbol{C}, \theta, \nabla_0\theta, \boldsymbol{X}). \tag{1.402}$$

Elastic materials in this general form are also called *hyperelastic materials*. Σ is sometimes called the *stored energy function* (Ciarlet 1993).

1.14.2 Linear elastic materials

Special forms arise if we linearize $\boldsymbol{S}$ with respect to $\boldsymbol{E}$ and $\boldsymbol{Q}$ with respect to $\boldsymbol{E}$ and $\nabla_0\theta$ ($\boldsymbol{C}$ and $\boldsymbol{E}$ are equivalent independent variables). To obtain a linear relation between $\boldsymbol{S}$ and $\boldsymbol{E}$, we expand Σ about $\boldsymbol{E} = 0$ and truncate the series after the quadratic terms:

$$\Sigma \sim \Sigma_0(\theta, \boldsymbol{X}) + \Sigma^{KL}(\theta, \boldsymbol{X})E_{KL} + \tfrac{1}{2}\Sigma^{KLMN}(\theta, \boldsymbol{X})E_{KL}E_{MN}, \quad \|\boldsymbol{E}\| \to 0 \tag{1.403}$$

while $\boldsymbol{Q}$ is expanded at $\nabla_0\theta = 0$, $\boldsymbol{E} = 0$ and the linear terms are kept

$$\boldsymbol{Q} \sim -\kappa^K(\theta, \boldsymbol{X}) - \kappa^{KL}(\theta, \boldsymbol{X})\theta_{,L} - \kappa^{KLM}(\theta, \boldsymbol{X})E_{LM}, \quad \|\nabla_0\theta\| \to 0, \|\boldsymbol{E}\| \to 0. \tag{1.404}$$

Because of the symmetry of $\boldsymbol{E}$ one finds

$$\Sigma^{KL} = \Sigma^{LK} \tag{1.405}$$

$$\Sigma^{KLMN} = \Sigma^{LKMN} = \Sigma^{KLNM} = \Sigma^{MNKL} \tag{1.406}$$

$$\kappa^{KLM} = \kappa^{KML}. \tag{1.407}$$

Applying Equation (1.399) yields

$$S^{KL}(\theta, \boldsymbol{X}) = \Sigma^{KL}(\theta, \boldsymbol{X}) + \Sigma^{KLMN}(\theta, \boldsymbol{X})E_{MN}. \tag{1.408}$$

Physical observations and the second law of thermodynamics, cf Equations (1.396) and (1.398), dictate that there is no heat flux if the temperature gradient is zero. This leads to (see Equation (1.404))

$$\kappa^{K} = 0, \quad \kappa^{KLM} = 0 \tag{1.409}$$

and

$$Q^{K} = -\kappa^{KL}(\theta, \boldsymbol{X})\theta_{,L}. \tag{1.410}$$

The entropy inequality now amounts to

$$\kappa^{KL}\theta_{,K}\theta_{,L} \geq 0 \tag{1.411}$$

which means that the symmetric part of κ^{KL} must be positive definite. The physical meaning of κ^{KL} is the conduction coefficient matrix.

The term $\Sigma^{KL}(\theta, \boldsymbol{X})$ in Equation (1.408) contains the thermal stress. Let the temperature θ_{ref} represent a homogeneous temperature distribution without any thermal stresses. Then one can write

$$\Sigma^{KL}(\theta, \boldsymbol{X}) = \gamma^{KL}(\boldsymbol{X}) - \beta^{KL}(\theta, \boldsymbol{X})(\theta - \theta_{\text{ref}}). \tag{1.412}$$

γ^{KL} are residual stresses from other sources and β^{kl} is the compressive stress rise per unit temperature increase if no expansion is allowed. Furthermore, we define α^{KL} assuming Σ^{KLMN} to be invertible:

$$\beta^{KL}(\theta, \boldsymbol{X}) =: \Sigma^{KLMN}(\theta, \boldsymbol{X})\alpha_{MN}(\theta, \boldsymbol{X}). \tag{1.413}$$

Hence,

$$\begin{aligned} S^{KL}(\theta, \boldsymbol{X}) &= \gamma^{KL}(\boldsymbol{X}) - \beta^{KL}(\theta, \boldsymbol{X})(\theta - \theta_{\text{ref}}) + \Sigma^{KLMN}(\theta, \boldsymbol{X})E_{MN} \\ &= \gamma^{KL}(\boldsymbol{X}) + \Sigma^{KLMN}(\theta, \boldsymbol{X})\left[E_{MN} - \alpha_{MN}(\theta, \boldsymbol{X})(\theta - \theta_{\text{ref}})\right]. \end{aligned} \tag{1.414}$$

The tensor $\boldsymbol{\alpha}$ contains the expansion coefficients. Now, let us expand $\Sigma_0(\theta, \boldsymbol{X})$ in Equation (1.403) about θ_{ref}:

$$\Sigma_0(\theta, \boldsymbol{X}) = \rho_0(\boldsymbol{X})\psi_0(\boldsymbol{X}) - \rho_0(\boldsymbol{X})\eta_0(\boldsymbol{X})(\theta - \theta_{\text{ref}}) - \frac{\rho_0(\boldsymbol{X})c(\theta, \boldsymbol{X})}{2\theta_{\text{ref}}}(\theta - \theta_{\text{ref}})^2. \tag{1.415}$$

Notice that the equality sign applies, since $c(\theta, \boldsymbol{X})$ in the last term may depend on θ. Dropping the dependence on $\boldsymbol{X}$ in the notation and defining $T := \theta - \theta_{\text{ref}}$ yields

$$\Sigma = \rho_0\psi_0 - \rho_0\eta_0 T - \frac{\rho_0 c(\theta)}{2\theta_{\text{ref}}}T^2 + [\gamma^{KL} - \beta^{KL}(\theta)T]E_{KL} + \frac{1}{2}\Sigma^{KLMN}(\theta)E_{KL}E_{MN} \tag{1.416}$$

and

$$\eta = \eta_0 + \frac{c(\theta)T}{\theta_{\mathrm{ref}}} + \frac{\beta^{KL}(\theta)}{\rho_0}E_{KL} + \frac{c'(\theta)T^2}{2\theta_{\mathrm{ref}}} + \frac{\beta^{KL'}(\theta)}{\rho_0}TE_{KL} + \frac{1}{2}\Sigma^{KLMN'}(\theta)E_{KL}E_{MN}. \tag{1.417}$$

The last three terms are due to the temperature dependence of the coefficients. Since $\rho_0\varepsilon = \Sigma + \rho_0\theta\eta$, one obtains

$$\begin{aligned}\rho_0\varepsilon = {} & \rho_0\psi_0 + \rho_0\eta_0\theta_{\mathrm{ref}} + \rho_0 c(\theta)\left(T + \frac{T^2}{2\theta_{\mathrm{ref}}}\right) \\ & + [\gamma^{KL} + \beta^{KL}(\theta)\theta_{\mathrm{ref}}]E_{KL} + \frac{1}{2}\Sigma^{KLMN}(\theta)E_{KL}E_{MN} + \\ & \rho_0\frac{\theta c'(\theta)T^2}{2\theta_{\mathrm{ref}}} + \theta\beta^{KL'}(\theta)TE_{KL} + \frac{1}{2}\rho_0\theta\Sigma^{KLMN'}(\theta)E_{KL}E_{MN}.\end{aligned} \tag{1.418}$$

From Equation (1.418), it follows that c is the specific heat. Substituting the above equations into the energy balance, Equation (1.355), is quite a tedious task. Generally, the derivative of the coefficients with respect to the temperature can be neglected (the coefficients, however, are still a function of temperature). Furthermore, discarding the quadratic T term leads to

$$\rho_0\varepsilon = \rho_0\psi_0 + \rho_0\eta_0\theta_{\mathrm{ref}} + \rho_0 c(\theta)T + [\gamma^{KL} + \beta^{KL}(\theta)\theta_{\mathrm{ref}}]E_{KL} + \tfrac{1}{2}\Sigma^{KLMN}(\theta)E_{KL}E_{MN} \tag{1.419}$$

and for the stress

$$S^{KL} = [\gamma^{KL} - \beta^{KL}(\theta)T] + \Sigma^{KLMN}(\theta)E_{MN}. \tag{1.420}$$

Substitution into Equation (1.355) finally yields (after further linearization: $\theta\dot{E}_{KL} \approx \theta_{\mathrm{ref}}\dot{E}_{KL}$)

$$\rho_0 c(\theta)\dot{T} + \beta^{KL}(\theta)\theta_{\mathrm{ref}}\dot{E}_{KL} - (\kappa^{KL}(\theta)\theta_{,L})_{;K} - \rho_0 h = 0. \tag{1.421}$$

This is the classical heat equation for linear elastic materials.

If in Equation (1.420)

$$\begin{aligned}\gamma^{KL} &= 0 \quad \text{for } K \neq L \\ \beta^{KL} &= 0 \quad \text{for } K \neq L \\ \Sigma^{KLMN} &= 0 \quad \text{for } K \neq L \text{ and } M = N\end{aligned} \tag{1.422}$$

one obtains $S^{KL} = 0$, $K \neq L$ if $E_{KL} = 0$, $K \neq L$ and vice versa. If this is true for arbitrary orientations of the axes as in the case of isotropic materials, then the principal axes of $\boldsymbol{E}$ are also principal axes of $\boldsymbol{S}$. Indeed, take the principal axes of $\boldsymbol{E}$ as a local rectangular coordinate system. This means that $E_{KL} = 0$, $K \neq L$ and consequently $S^{KL} = 0$, $K \neq L$: the shear stress is zero. Accordingly,

$$\boldsymbol{E} = \sum_i \Lambda_{iE}\boldsymbol{N}^i \otimes \boldsymbol{N}^i \tag{1.423}$$

$$\boldsymbol{S} = \sum_i \Lambda_{iS}\boldsymbol{N}_i \otimes \boldsymbol{N}_i. \tag{1.424}$$

Furthermore, since

$$\boldsymbol{F} = \sum_i \sqrt{\Lambda_{iC}} \boldsymbol{n}_i \otimes \boldsymbol{N}^i \tag{1.425}$$

and

$$\boldsymbol{\sigma} = J^{-1} \boldsymbol{F} \cdot \boldsymbol{S} \cdot \boldsymbol{F}^{\mathrm{T}} \tag{1.426}$$

one finds

$$\begin{aligned} \boldsymbol{\sigma} &= J^{-1} \left(\sum_i \sqrt{\Lambda_{iC}} \boldsymbol{n}_i \otimes \boldsymbol{N}^i \right) \cdot \left(\sum_j \Lambda_{jS} \boldsymbol{N}_j \otimes \boldsymbol{N}_j \right) \cdot \left(\sum_k \sqrt{\Lambda_{kC}} \boldsymbol{N}^k \otimes \boldsymbol{n}_k \right) \\ &= J^{-1} \sum_i \Lambda_{iC} \Lambda_{iS} \boldsymbol{n}_i \otimes \boldsymbol{n}_i \end{aligned} \tag{1.427}$$

which yields for the true principal stresses

$$\lambda_{i\sigma} = J^{-1} \Lambda_{\underline{i}C} \Lambda_{\underline{i}S} = J^{-1} (2\Lambda_{\underline{i}E} + 1) \Lambda_{\underline{i}S}. \tag{1.428}$$

Since $J, \Lambda_{Ci} > 0$, the true stress and the second Piola–Kirchhoff stress have the same sign.

1.14.3 Isotropic linear elastic materials

For a linear elastic material, we found

$$\Sigma = \rho_0 \psi_0 - \rho_0 \eta_0 T - \frac{\rho_0 c(\theta)}{2\theta_{\mathrm{ref}}} T^2 + [\gamma^{KL} - \beta^{KL}(\theta) T] E_{KL} + \frac{1}{2} \Sigma^{KLMN}(\theta) E_{KL} E_{MN} \tag{1.429}$$

$$S^{KL} = [\gamma^{KL} - \beta^{KL}(\theta) T] + \Sigma^{KLMN}(\theta) E_{MN} \tag{1.430}$$

$$Q^K = -\kappa^{KL}(\theta) \theta_{,L}. \tag{1.431}$$

Isotropy means that the material data are independent of the direction in the material frame of reference. Hence, a transformation $\boldsymbol{Q}$ such that

$$\boldsymbol{X}' = \boldsymbol{Q} \cdot \boldsymbol{X}, \quad \boldsymbol{Q}^{\mathrm{T}} \cdot \boldsymbol{Q} = \boldsymbol{Q} \cdot \boldsymbol{Q}^{\mathrm{T}} = 1, \quad \det \boldsymbol{Q} = \pm 1 \tag{1.432}$$

must leave the constitutive equations invariant. Under such a transformation, second-order and fourth-order tensors transform according to

$$\gamma'^{KL} = \gamma^{MN} Q^K{}_M Q^L{}_N \tag{1.433}$$

$$\Sigma'^{KLMN} = \Sigma^{PQRS} Q^K{}_P Q^L{}_Q Q^M{}_R Q^N{}_S. \tag{1.434}$$

One can show that for this to be true for an arbitrary rotation, the tensors must satisfy

$$\gamma^{KL} = \gamma G^{KL} \tag{1.435}$$

$$\Sigma^{KLMN} = \lambda G^{KL} G^{MN} + \mu (G^{KM} G^{LN} + G^{KN} G^{LM}) \tag{1.436}$$

and similarly for the other tensors

$$\beta^{KL} = \beta G^{KL} \tag{1.437}$$

$$\kappa^{KL} = \kappa G^{KL}. \tag{1.438}$$

Since

$$\text{tr}\boldsymbol{E} = G^{KL}E_{KL} \tag{1.439}$$

is the trace of the tensor $\boldsymbol{E}$, one finds

$$\Sigma(\theta) = \rho_0\psi_0 - \rho_0\eta_0 T - \frac{\rho_0 c(\theta)}{2\theta_{\text{ref}}}T^2 + [\gamma - \beta(\theta)T]\text{tr}\boldsymbol{E} + \tfrac{1}{2}\lambda(\theta)(\text{tr}\boldsymbol{E})^2 + \mu(\theta)\text{tr}(\boldsymbol{E}^2) \tag{1.440}$$

$$S^{KL} = [\gamma - \beta(\theta)T]G^{KL} + \lambda(\theta)(\text{tr}\boldsymbol{E})G^{KL} + 2\mu(\theta)E_{MN}G^{KM}G^{LN} \tag{1.441}$$

$$Q^K = -\kappa(\theta)\theta_{,L}G^{KL}. \tag{1.442}$$

The energy equation reduces to

$$\rho c(\theta)\dot{T} + \beta(\theta)\theta_{\text{ref}}\dot{E}^K{}_K - (G^{KL}\kappa(\theta)\theta_{,L})_{;K} - \rho_0 h = 0. \tag{1.443}$$

The kind of material described by Equations (1.440) to (1.443) is also called a *St Venant–Kirchhoff material*.

The first and second invariant of a tensor $\boldsymbol{E}$ are defined by

$$I_{1E} = \text{tr}\boldsymbol{E} \tag{1.444}$$

$$I_{2E} = \frac{1}{2}[I_{1E}^2 - \text{tr}(\boldsymbol{E}^2)]. \tag{1.445}$$

Consequently, the free energy Σ can also be written as

$$\Sigma = \rho_0\psi_0 - \rho_0\eta_0 T - \frac{\rho_0 c(\theta)}{2\theta_{\text{ref}}}T^2 + [\gamma - \beta(\theta)T]I_{1E} + \frac{1}{2}[\lambda(\theta) + 2\mu(\theta)]I_{1E}^2 - 2\mu(\theta)I_{2E}. \tag{1.446}$$

$\lambda(\theta)$ and $\mu(\theta)$ are called *Lamé's constants*, $\kappa(\theta)$ is the conduction coefficient, $c(\theta)$ is the specific heat and $\beta(\theta)$ satisfies (substitute Equation (1.436) into Equation (1.413))

$$\beta(\theta) = [3\lambda(\theta) + 2\mu(\theta)]\alpha(\theta) \tag{1.447}$$

where $\alpha(\theta)$ is the isotropic expansion coefficient. The thermal stress now yields

$$S^{KL} = -[3\lambda(\theta) + 2\mu(\theta)]\alpha(\theta)TG^{KL}. \tag{1.448}$$

This stress is needed to suppress

$$E_{KL} = \alpha(\theta)TG_{KL} \tag{1.449}$$

which is the strain resulting from the temperature change.

Finally, it should be noted that frequently other elastic constants are used instead of the Lamé's constants λ and μ, the latter of which is also called the *shear modulus*. The Poisson coefficient ν and Young's modulus E satisfy

$$\mu = \frac{E}{2(1+\nu)} \tag{1.450}$$

$$\lambda = \frac{\nu E}{(1+\nu)(1-2\nu)} \tag{1.451}$$

which can be inverted to yield

$$\nu = \frac{\lambda}{2(\lambda+\mu)} \tag{1.452}$$

$$E = \frac{\mu(3\lambda+2\mu)}{\lambda+\mu}. \tag{1.453}$$

Another frequently used constant is the bulk modulus K. For linearized strains, it will be proven in the next section that K is the ratio of the hydrostatic pressure p to the volume reduction it produces. The following relations apply

$$K = \lambda + \frac{2}{3}\mu \tag{1.454}$$

$$\nu = \frac{3K-2\mu}{6K+2\mu}. \tag{1.455}$$

1.14.4 Linearizing the strains

So far, we consistently used the Lagrange strain tensor $\boldsymbol{E}$. Equation (1.82) shows that $\boldsymbol{E}$ is not linear in the displacement $\boldsymbol{U}$. To obtain a truly linear theory, the quadratic terms in $\boldsymbol{E}$ are dropped and one obtains the infinitesimal strains $\tilde{\boldsymbol{E}}$, Equation (1.88):

$$\tilde{E}_{KL} = \tfrac{1}{2}(U_{K;L} + U_{L;K}). \tag{1.456}$$

Recall that the infinitesimal rotation is defined by

$$\tilde{R}_{KL} = \tfrac{1}{2}(U_{K;L} - U_{L;K}). \tag{1.457}$$

Equation (1.90) has shown that E_{KL} can only be replaced by $\tilde{E}_{KL}$ if both the strain and the rotations are small. The same applies to e_{kl} and $\tilde{e}_{kl}$. Under the above assumptions, one can write

$$E_{KL} \sim \tilde{E}_{KL} \quad \|\tilde{\boldsymbol{E}}\|, \|\tilde{\boldsymbol{R}}\| \to 0 \tag{1.458}$$

$$e_{kl} \sim \tilde{e}_{kl} \quad \|\tilde{\boldsymbol{E}}\|, \|\tilde{\boldsymbol{R}}\| \to 0. \tag{1.459}$$

To derive further simplifications we start from Equation (1.67):

$$x^k \boldsymbol{g}_k = (X^L + U^L)\boldsymbol{G}_L - \boldsymbol{o}. \tag{1.460}$$

Taking the derivative with respect to K yields

$$x^k_{,K}\boldsymbol{g}_k = (\delta^L{}_K + U^L{}_{;K})\boldsymbol{G}_L \tag{1.461}$$

leading to

$$x^k_{,K} = (\delta^L{}_K + \tilde{E}^L{}_K + \tilde{R}^L{}_K)\boldsymbol{G}_L \cdot \boldsymbol{g}^k \tag{1.462}$$

or

$$x^k_{,K} = (\delta^L{}_K + \tilde{E}^L{}_K + \tilde{R}^L{}_K)g_L{}^k \tag{1.463}$$

where

$$g_L{}^k = \boldsymbol{G}_L \cdot \boldsymbol{g}^k = \boldsymbol{g}^k \cdot \boldsymbol{G}_L = g^k{}_L. \tag{1.464}$$

In a similar way, one arrives at

$$X^K{}_{,k} = (\delta^l{}_k - \tilde{e}^l{}_k - \tilde{r}^l{}_k)g^K{}_l. \tag{1.465}$$

For small strains and rotations, Equation (1.463) and Equation (1.465) reduce to

$$x^k_{,K} \sim g^k{}_K, \quad \|\tilde{\boldsymbol{E}}\|, \|\tilde{\boldsymbol{R}}\| \to 0 \tag{1.466}$$

$$X^K{}_{,k} \sim g^K{}_k, \quad \|\tilde{\boldsymbol{E}}\|, \|\tilde{\boldsymbol{R}}\| \to 0. \tag{1.467}$$

From Equation (1.352) and Equation (1.353), one finds

$$E_{KL} = e_{kl}x^k_{,K}x^l_{,L} \tag{1.468}$$

which reduces by the use of Equations (1.458), (1.459), (1.466) and (1.467) to

$$\tilde{E}_{KL} \sim \tilde{e}_{kl}g^k{}_K g^l{}_L, \quad \|\tilde{\boldsymbol{E}}\|, \|\tilde{\boldsymbol{R}}\| \to 0. \tag{1.469}$$

On the basis of Equation (1.457), a similar relationship applies to the infinitesimal rotation

$$\tilde{R}_{KL} \sim \tilde{r}_{kl}g^k{}_K g^l{}_L, \quad \|\tilde{\boldsymbol{E}}\|, \|\tilde{\boldsymbol{R}}\| \to 0. \tag{1.470}$$

Furthermore, $J = \det x^k_{,K}$. Substituting Equation (1.463) and linearizing yields

$$\begin{aligned} J \sim \tfrac{1}{3!}e^{KLM}e_{klm}[&g^k{}_K g^l{}_L g^m{}_M + g^k{}_K g^l{}_L(\tilde{E}^R{}_M + \tilde{R}^R{}_M)g^m{}_R \\ &+ g^k{}_K g^m{}_M(\tilde{E}^Q{}_L + \tilde{R}^Q{}_L)g^l{}_Q + g^l{}_L g^m{}_M(\tilde{E}^P{}_K + \tilde{R}^P{}_K)g^m{}_P], \quad \|\tilde{\boldsymbol{E}}\|, \|\tilde{\boldsymbol{R}}\| \to 0 \end{aligned} \tag{1.471}$$

where e^{KLM} and e_{klm} are alternating symbols. This is equivalent to

$$J \sim 1 + \tilde{E}^K{}_K \sim 1 + \tilde{e}^k{}_k, \quad \|\tilde{\boldsymbol{E}}\|, \|\tilde{\boldsymbol{R}}\| \to 0. \tag{1.472}$$

Substituting Equation (1.463) and Equation (1.472) into the relationship between the Cauchy stress and the second Piola–Kirchhoff stress leads to

$$\sigma^{kl} = J^{-1}S^{KL}x^k_{,K}x^l_{,L} \tag{1.473}$$

and linearizing yields

$$\sigma^{kl} \sim S^{KL}[g^k{}_K g^l{}_L(1-\tilde{e}^m{}_m) + (\tilde{e}^k{}_m + \tilde{r}^k{}_m)g^m{}_K g^l{}_L + (\tilde{e}^l{}_m + \tilde{r}^l{}_m)g^k{}_K g^m{}_L], \quad \|\tilde{\boldsymbol{E}}\|, \|\tilde{\boldsymbol{R}}\| \to 0. \tag{1.474}$$

The inverse of Equation (1.474) amounts to

$$S^{KL} \sim \sigma^{kl}[g^K{}_k g^L{}_l(1+\tilde{e}^m{}_m) - g^L{}_l g^K{}_m(\tilde{e}^m{}_k + \tilde{r}^m{}_k) - g^K{}_k g^L{}_m(\tilde{e}^m{}_l + \tilde{r}^m{}_l)], \quad \|\tilde{\boldsymbol{E}}\|, \|\tilde{\boldsymbol{R}}\| \to 0. \tag{1.475}$$

Substituting the above relations into Equation (1.420) yields a linearized expression for the stress:

$$\sigma^{kl} \sim \gamma^{kl}(1-\tilde{e}^m{}_m) + \gamma^{ml}(\tilde{e}^k{}_m + \tilde{r}^k{}_m) + \gamma^{km}(\tilde{e}^l{}_m + \tilde{r}^l{}_m) - \beta^{kl}T + \sigma^{klmn}\tilde{e}_{mn}, \quad \|\tilde{\boldsymbol{E}}\|, \|\tilde{\boldsymbol{R}}\| \to 0 \tag{1.476}$$

where

$$\gamma^{kl} = \gamma^{KL} g^k{}_K g^l{}_L \tag{1.477}$$

$$\beta^{kl} = \beta^{KL} g^k{}_K g^l{}_L \tag{1.478}$$

$$\sigma^{klmn} = \Sigma^{KLMN} g^k{}_K g^l{}_L g^m{}_M g^n{}_N. \tag{1.479}$$

In a similar way, by combining Equation (1.274) and Equation (1.410) one arrives at

$$q^k = -J^{-1}\kappa^{KL}\theta_{,l}x^k_{,K}x^l_{,L}. \tag{1.480}$$

Linearizing yields

$$q^k \sim -\kappa^{kl}\theta_{,l}, \quad \|\tilde{\boldsymbol{E}}\|, \|\tilde{\boldsymbol{R}}\| \to 0 \tag{1.481}$$

where

$$\kappa^{kl} = \kappa^{KL} g^k{}_K g^l{}_L. \tag{1.482}$$

Analogous considerations lead to

$$\rho_0\varepsilon \sim \rho_0\psi_0 + \rho_0\eta_0\theta_{\text{ref}} + \rho_0 c(\theta)T + [\gamma^{kl} + \beta^{kl}\theta_{\text{ref}}]\tilde{e}_{kl} + \frac{1}{2}\sigma^{klmn}\tilde{e}_{kl}\tilde{e}_{mn}, \quad \|\tilde{\boldsymbol{E}}\|, \|\tilde{\boldsymbol{R}}\| \to 0 \tag{1.483}$$

$$\eta = \eta_0 + \frac{cT}{\theta_{\text{ref}}} + \frac{\beta^{kl}}{\rho_0}\tilde{e}_{kl}, \quad \|\tilde{\boldsymbol{E}}\|, \|\tilde{\boldsymbol{R}}\| \to 0 \tag{1.484}$$

$$\Sigma = \rho_0\psi_0 - \rho_0\eta_0 T - \frac{\rho_0 c}{2\theta_{\text{ref}}}T^2 + [\gamma^{kl} - \beta^{kl}T]\tilde{e}_{kl} + \frac{1}{2}\sigma^{klmn}\tilde{e}_{kl}\tilde{e}_{mn}, \quad \|\tilde{\boldsymbol{E}}\|, \|\tilde{\boldsymbol{R}}\| \to 0. \tag{1.485}$$

The balance equations now read

$$\sigma^{kl}{}_{;k} + \rho(f^l - \dot{v}^l) = 0 \tag{1.486}$$

$$\rho_0 c\dot{T} + \beta^{kl}\theta_{\text{ref}}\dot{\tilde{e}}_{kl} - (\kappa^{kl}T_{,l})_{;k} - \rho_0 h = 0 \tag{1.487}$$

$$\kappa^{kl}T_{,k}T_{,l} \geq 0. \tag{1.488}$$

The derivation for isotropic materials runs along the same lines and yields for $\|\tilde{\boldsymbol{E}}\|, \|\tilde{\boldsymbol{R}}\| \to 0$

$$\sigma^{kl} \sim \gamma(1 - \tilde{e}^m{}_m)g^{kl} - \beta T g^{kl} + \lambda\tilde{e}^m{}_m g^{kl} + 2(\mu + \gamma)\tilde{e}^{kl} \tag{1.489}$$

$$q^k \sim -\kappa T_{,l} g^{kl} \tag{1.490}$$

$$\rho_0\varepsilon \sim \rho_0\psi_0 + \rho_0\eta_0\theta_{\text{ref}} + \rho_0 c T + [\gamma + \beta\theta_{\text{ref}}]\tilde{e}^m{}_m + \frac{1}{2}(\lambda + 2\mu)I_{1\tilde{e}}^2 - 2\mu I_{2\tilde{e}} \tag{1.491}$$

$$\eta = \eta_0 + \frac{cT}{\theta_{\text{ref}}} + \frac{\beta}{\rho_0}I_{1\tilde{e}} \tag{1.492}$$

$$\Sigma = \rho_0\psi_0 - \rho_0\eta_0 T - \frac{\rho_0 c}{2\theta_{\text{ref}}}T^2 + [\gamma - \beta T]I_{1\tilde{e}} + \frac{1}{2}(\lambda + 2\mu)I_{1\tilde{e}}^2 - 2\mu I_{2\tilde{e}} \tag{1.493}$$

$$\sigma^{kl}{}_{;k} + \rho(f^l - \dot{v}^l) = 0 \tag{1.494}$$

$$\rho_0 c\dot{T} + \beta\theta_{\text{ref}}\dot{I}_{1\tilde{e}} - (\kappa T_{,l}g^{kl})_{;k} - \rho_0 h = 0 \tag{1.495}$$

$$\kappa T_{,k}T_{,l}g^{kl} \geq 0. \tag{1.496}$$

For materials without residual stress and $T = 0$ Equation (1.489) reduces to

$$\sigma^{kl} \sim \lambda\tilde{e}^m{}_m g^{kl} + 2\mu\tilde{e}^{kl}. \tag{1.497}$$

Hence,

$$\sigma^k{}_k \sim (3\lambda + 2\mu)\tilde{e}^m{}_m. \tag{1.498}$$

For a uniform pressure p we have

$$\sigma^k{}_k = 3p \tag{1.499}$$

and (see Equation (1.472)),

$$\tilde{e}^m{}_m \sim J - 1 \sim \frac{\mathrm{d}v - \mathrm{d}V}{\mathrm{d}V}, \tag{1.500}$$

which is the volume change. Hence,

$$p = (\lambda + \frac{2}{3}\mu)\frac{\mathrm{d}v - \mathrm{d}V}{\mathrm{d}V} \tag{1.501}$$

from which Equation (1.454) follows.

Summarizing, in the small deformation theory, the strain tensors $\boldsymbol{E}$ and $\boldsymbol{e}$ are replaced by their infinitesimal counterparts $\tilde{\boldsymbol{E}}$ and $\tilde{\boldsymbol{e}}$. This is only justified for small strains together with small rotations. Therefore, it is better to use the expression *small deformation theory* rather than infinitesimal strain theory. Using the infinitesimal strains and rotations, the constitutive equations and balance laws can be simplified. Notice that the term "infinitesimal" does not apply to other quantities such as stresses. Equations (1.474) and (1.475) show that also in the linear strain theory the second Piola–Kirchhoff and Cauchy stress both exist and are generally different. The derived equations are valid in the spatial frame of reference.

1.14.5 Isotropic elastic materials

In this section, we start again from Equation (1.399) to Equation (1.402) and assume that Σ is isotropic in $\boldsymbol{C}$ but that the resulting stress $\boldsymbol{S}$ is not necessarily linear in $\boldsymbol{E}$. This covers the large family of so-called isotropic hyperelastic models such as neo–Hooke, Mooney–Rivlin, Ogden and many others, used for materials such as rubber and hyperfoam. Because of the isotropy, Σ can only be a function of the invariants of $\boldsymbol{C}$. These will be denoted in the present context by I_1, I_2 and I_3 (dropping the index C for convenience). Accordingly,

$$\Sigma = \Sigma(I_1, I_2, I_3, \theta, \boldsymbol{X}) \tag{1.502}$$

where

$$I_1 = \text{tr}(\boldsymbol{C}) \tag{1.503}$$

$$I_2 = \tfrac{1}{2}[I_1^2 - \text{tr}(\boldsymbol{C}^2)] \tag{1.504}$$

$$I_3 = \det \boldsymbol{C}. \tag{1.505}$$

Consequently, Equation (1.399),

$$\boldsymbol{S} = 2\left[\frac{\partial \Sigma}{\partial I_1}(I_1, I_2, I_3, \theta, \boldsymbol{X})\frac{\partial I_1}{\partial \boldsymbol{C}} + \frac{\partial \Sigma}{\partial I_2}(I_1, I_2, I_3, \theta, \boldsymbol{X})\frac{\partial I_2}{\partial \boldsymbol{C}} + \frac{\partial \Sigma}{\partial I_3}(I_1, I_2, I_3, \theta, \boldsymbol{X})\frac{\partial I_3}{\partial \boldsymbol{C}}\right]. \tag{1.506}$$

Since

$$\frac{\partial I_1}{\partial C_{KL}} = \frac{\partial C_{MN} G^{MN}}{\partial C_{KL}} = G^{KL} \tag{1.507}$$

$$\begin{aligned}\frac{\partial I_2}{\partial C_{KL}} &= \frac{1}{2}\frac{\partial}{\partial C_{KL}}\left[I_1^2 - C_{PQ}C_{MN}G^{PN}G^{QM}\right] \\ &= \frac{1}{2}\left[2I_1 G^{KL} - C_{MN}G^{KN}G^{LM} - C_{PQ}C^{PL}C^{QK}\right] \\ &= I_1 G^{KL} - C_{MN}G^{KN}G^{LM}\end{aligned} \tag{1.508}$$

$$\frac{\partial I_3}{\partial C_{KL}} = \text{cofactor}(C_{KL}) = \text{cofactor}(C_{LK}) = I_3(C^{-1})^{KL} \tag{1.509}$$

we obtain,

$$S = 2\left[\frac{\partial \Sigma}{\partial I_1}(I_1, I_2, I_3, \theta, X)G^\sharp + \frac{\partial \Sigma}{\partial I_2}(I_1, I_2, I_3, \theta, X)(I_1 G^\sharp - C^\sharp) + \frac{\partial \Sigma}{\partial I_3}(I_1, I_2, I_3, \theta, X)I_3 C^{-1}\right]. \quad (1.510)$$

Here, the θ-dependence is not specified yet. Whether the function $\Sigma(I_1, I_2, I_3, \theta, X)$ has to satisfy specific requirements to make sense physically will be discussed in Chapter 4 on hyperelastic materials. Since $C^\sharp$ and C^{-1} have the same eigenvectors and the eigenvectors of $C^\sharp$ are not modified by adding or subtracting a multiple of $G^\sharp$, Equation (1.510) shows that S has the same eigenvectors as $C^\sharp$. Consequently, for an isotropic elastic material the principal second Piola–Kirchhoff stress directions coincide with the principal stretch directions.

1.15 Fluids

Solids and fluids are two major classes of materials. Fluids include both liquids and gases. There are several ways in which a fluid can be described. Assume that there is no gravity. Then, the stress in a liquid at rest is zero. If you stir the liquid and wait until there is no motion the stress will again be zero. If you take a container filled with gas at a given pressure, stir the gas without increasing the external pressure and wait till there is no motion the stress reduces to the hydrostatic pressure before stirring. Consequently, the deformation of liquid materials does not induce stress as long as the liquid is at rest and the density is unchanged. Accordingly, the deformation gradient for quasistatic deformations leaving the density unchanged reduces to the shift operator (Eringen 1980):

$$x^k_{,K} = g^k{}_K. \quad (1.511)$$

In a similar way, one arrives at the following simplifications:

$$C_{KL} = x^k_{,K} x^l_{,L} g_{kl} = G_{KL} \quad (1.512)$$

$$\dot{C}_{KL} = 2d_{kl} x^k_{,K} x^l_{,L} = 2d_{kl} g^k{}_K g^l{}_L \quad (1.513)$$

$$\theta_{,K} = \theta_{,k} g^k{}_K. \quad (1.514)$$

Just as for elastic materials, we start from the material formulation of mechanical grade 1 and thermal grade 1, but now we keep the first time derivatives of the mechanical quantities as well:

$$S(X, t) = S(C, \dot{C}, \rho^{-1}, \dot{\rho}, \theta, \nabla_0 \theta, X). \quad (1.515)$$

Now, Equation (1.269) yields

$$\sigma^{kl} = J^{-1} x^k_{,K} S^{KL} x^l_{,L} = \frac{\rho}{\rho_0} S^{KL} g^k{}_K g^l{}_L. \quad (1.516)$$

Since (see Equation (1.178))

$$\dot{\rho} = \overline{\left(\frac{\rho_0}{J}\right)}^{\cdot} = -\frac{\rho_0}{J^2}\dot{J} = -\rho d^k_{\ k} \tag{1.517}$$

and

$$\frac{\partial}{\partial X^K} = \frac{\partial}{\partial x^k} g^k_{\ K} \tag{1.518}$$

the Cauchy stress takes the form

$$\boldsymbol{\sigma}(\boldsymbol{X}, t) = \boldsymbol{\sigma}(\boldsymbol{d}, \rho^{-1}, \theta, \nabla\theta, \boldsymbol{X}). \tag{1.519}$$

Since any configuration leaving the density unchanged is undeformed, $\boldsymbol{X}$ can be replaced by $\boldsymbol{x}$:

$$\boldsymbol{\sigma}(\boldsymbol{x}, t) = \boldsymbol{\sigma}(\boldsymbol{d}, \rho^{-1}, \theta, \nabla\theta, \boldsymbol{x}). \tag{1.520}$$

The principle of objectivity requires that Equation (1.520) does not change its form after applying an arbitrary time-dependent translation, for example, $\boldsymbol{x}(t)$:

$$\boldsymbol{\sigma}(\boldsymbol{d}, \rho^{-1}, \theta, \nabla\theta, \boldsymbol{x}) = \boldsymbol{\sigma}(\boldsymbol{d}, \rho^{-1}, \theta, \nabla\theta, \boldsymbol{0}) \tag{1.521}$$

and the explicit dependence on $\boldsymbol{x}$ drops out:

$$\boldsymbol{\sigma} = \boldsymbol{\sigma}(\boldsymbol{d}, \rho^{-1}, \theta, \nabla\theta) \tag{1.522}$$

and similar expressions for $\boldsymbol{q}$, ε and η. Just as in the derivation of the constitutive laws for elastic materials the entropy inequality plays a major role. The spatial formulation of Equation (1.389) reads

$$-\frac{\rho}{\theta}(\dot{\psi} + \dot{\theta}\eta) + \frac{1}{\theta}\boldsymbol{d} : \boldsymbol{\sigma} - \frac{1}{\theta^2}\boldsymbol{q} \cdot \nabla\theta \geq 0 \tag{1.523}$$

where $\psi = \varepsilon - \theta\eta$, Equation (1.387), and

$$\psi = \psi(\boldsymbol{d}, \rho^{-1}, \theta, \nabla\theta) \tag{1.524}$$

because of similar dependencies of ε and η. The time derivative of ψ reads

$$\dot{\psi} = \frac{\partial\psi}{\partial\boldsymbol{d}} : \dot{\boldsymbol{d}} + \frac{\partial\psi}{\partial\rho^{-1}} \cdot \overline{\rho^{-1}}^{\cdot} + \frac{\partial\psi}{\partial\theta}\dot{\theta} + \frac{\partial\psi}{\partial\nabla\theta} \cdot \overline{\nabla\theta}^{\cdot}. \tag{1.525}$$

Substituting Equation (1.525) into Equation (1.523) yields

$$-\frac{\rho}{\theta}\frac{\partial\psi}{\partial\boldsymbol{d}} : \dot{\boldsymbol{d}} - \frac{\rho}{\theta}\frac{\partial\psi}{\partial\rho^{-1}}\overline{\rho^{-1}}^{\cdot} - \frac{\rho}{\theta}\left(\frac{\partial\psi}{\partial\theta} + \eta\right)\dot{\theta}$$
$$- \frac{\rho}{\theta}\frac{\partial\psi}{\partial\nabla\theta} \cdot \overline{\nabla\theta}^{\cdot} + \frac{1}{\theta}\boldsymbol{d} : \boldsymbol{\sigma} - \frac{1}{\theta^2}\boldsymbol{q} \cdot \nabla\theta \geq 0. \tag{1.526}$$

Since (see Equation (1.517))

$$\dot{\overline{\rho^{-1}}} = \frac{1}{\rho} \boldsymbol{d} : \boldsymbol{g} \tag{1.527}$$

this is equivalent to

$$-\frac{\rho}{\theta}\frac{\partial \psi}{\partial \boldsymbol{d}} : \dot{\boldsymbol{d}} + \frac{1}{\theta}\boldsymbol{d} : \left(\boldsymbol{\sigma} - \frac{\partial \psi}{\partial \rho^{-1}}\boldsymbol{g}\right) - \frac{\rho}{\theta}\left(\frac{\partial \psi}{\partial \theta} + \eta\right)\dot{\theta} - \frac{\rho}{\theta}\frac{\partial \psi}{\partial \nabla\theta} \cdot \dot{\overline{\nabla\theta}} - \frac{1}{\theta^2}\boldsymbol{q} \cdot \nabla\theta \geq 0. \tag{1.528}$$

Since this equation is linear in the time derivatives, it can only be satisfied if the corresponding coefficients reduce to zero:

$$\frac{\partial \psi}{\partial \boldsymbol{d}} = 0 \tag{1.529}$$

$$\eta = -\frac{\partial \psi}{\partial \theta} \tag{1.530}$$

$$\frac{\partial \psi}{\partial \nabla\theta} = 0. \tag{1.531}$$

Hence,

$$\psi = \psi(\rho^{-1}, \theta) \tag{1.532}$$

and Equation (1.528) reduces to

$$\frac{1}{\theta}\boldsymbol{d} : \left(\boldsymbol{\sigma} - \frac{\partial \psi}{\partial \rho^{-1}}\boldsymbol{g}\right) - \frac{1}{\theta^2}\boldsymbol{q} \cdot \nabla\theta \geq 0. \tag{1.533}$$

Defining the pressure p by

$$p = -\frac{\partial \psi}{\partial \rho^{-1}} \tag{1.534}$$

and the dissipative stress by

$$\boldsymbol{t} := \boldsymbol{\sigma} + p\boldsymbol{g} \tag{1.535}$$

we finally arrive at the following equations:

$$p = -\frac{\partial \psi}{\partial \rho^{-1}}(\rho^{-1}, \theta) \tag{1.536}$$

$$\eta = -\frac{\partial \psi}{\partial \theta}(\rho^{-1}, \theta) \tag{1.537}$$

$$\boldsymbol{t} = \boldsymbol{t}(\boldsymbol{d}, \rho^{-1}, \theta, \nabla\theta) \tag{1.538}$$

$$\boldsymbol{q} = \boldsymbol{q}(\boldsymbol{d}, \rho^{-1}, \theta, \nabla\theta) \tag{1.539}$$

$$\varepsilon = \psi(\rho^{-1}, \theta) - \theta\frac{\partial \psi}{\partial \theta}(\rho^{-1}, \theta) \tag{1.540}$$

$$\boldsymbol{\sigma} = -p\boldsymbol{g} + \boldsymbol{t} \tag{1.541}$$

subject to

$$\frac{1}{\theta}\boldsymbol{d} : \boldsymbol{t} - \frac{1}{\theta^2}\boldsymbol{q} \cdot \nabla\theta \geq 0. \tag{1.542}$$

Equation (1.542) implies that $\boldsymbol{t}$ and $\boldsymbol{q}$ must be at least linear in $\boldsymbol{d}$ and $\nabla\theta$ respectively. Equation (1.538) and Equation (1.539) can be replaced by

$$\boldsymbol{t} = \boldsymbol{t}_L(\boldsymbol{d}, \rho^{-1}, \theta, \nabla\theta) : \boldsymbol{d} \tag{1.543}$$

$$\boldsymbol{q} = -\boldsymbol{\kappa}_L(\boldsymbol{d}, \rho^{-1}, \theta, \nabla\theta) \cdot \nabla\theta \tag{1.544}$$

where $\boldsymbol{t}_L$ is a fourth-order tensor, $\boldsymbol{\kappa}_L$ is a second-order tensor. Notice that the dissipative stress cannot be derived from a potential function, only the hydrostatic part p can. This is a major difference compared to elastic materials. Equation (1.542) is the fluid equivalent of Equation (1.396) for elastic materials.

Because of the principle of objectivity, Equation (1.543) can be further reduced to

$$\boldsymbol{t} = \alpha_0 \boldsymbol{g}^\sharp + \alpha_1 \boldsymbol{d}^\sharp + \alpha_2 (\boldsymbol{d}^2)^\sharp \tag{1.545}$$

where

$$\alpha_K(\rho^{-1}, \theta, \nabla\theta, I_{1d}, I_{2d}, I_{3d}) \tag{1.546}$$

Linearization yields

$$\alpha_0 = \lambda_v(\rho^{-1}, \theta, \nabla\theta) I_{1d} \tag{1.547}$$

$$\alpha_1 = 2\mu_v(\rho^{-1}, \theta, \nabla\theta) \tag{1.548}$$

$$\alpha_2 = 0 \tag{1.549}$$

and one arrives at the well-known stress expressions for linear Stokesian fluids:

$$\boldsymbol{\sigma} = (-p + \lambda_v \boldsymbol{g} : \boldsymbol{d})\boldsymbol{g} + 2\mu_v \boldsymbol{d}. \tag{1.550}$$

For details, the reader is referred to (Eringen 1980).

The energy equation, Equation (1.355), reads in spatial coordinates:

$$\rho\dot{\varepsilon} = \boldsymbol{d} : \boldsymbol{\sigma} - \nabla \cdot \boldsymbol{q} + \rho h. \tag{1.551}$$

Substitution of Equation (1.540) yields

$$\rho\left[-\left(p + \theta\frac{\partial^2\psi}{\partial\rho^{-1}\partial\theta}\right)\dot{\overline{\rho^{-1}}} - \theta\frac{\partial^2\psi}{\partial\theta^2}\dot{\theta}\right] = \boldsymbol{d} : \boldsymbol{\sigma} - \nabla \cdot \boldsymbol{q} + \rho h, \tag{1.552}$$

which reads by the use of Equation (1.527) and Equation (1.534):

$$\rho\theta\frac{\partial^2\psi}{\partial\theta^2}\dot{\theta} + \theta\frac{\partial^2\psi}{\partial\rho^{-1}\partial\theta}\boldsymbol{d} : \boldsymbol{g} + \boldsymbol{d} : \boldsymbol{t} - \nabla \cdot \boldsymbol{q} + \rho h = 0. \tag{1.553}$$

For most gases, $\boldsymbol{t} = 0$ is assumed (no stress dissipation) and Equation (1.553) reduces to

$$\rho\theta\frac{\partial^2\psi}{\partial\theta^2}\dot{\theta} + \theta\frac{\partial^2\psi}{\partial\rho^{-1}\partial\theta}\boldsymbol{d} : \boldsymbol{g} - \nabla \cdot \boldsymbol{q} + \rho h = 0. \tag{1.554}$$

2

Linear Mechanical Applications

2.1 General Equations

The basic equations for the finite element method are the weak formulation of the balance of momentum, Equation (1.328) and the weak formulation of the balance of energy, Equation (1.351). For mechanical applications in which the temperature is assumed to be known, only the balance of momentum is needed in order to determine the displacement fields:

$$\int_{V_0} S^{KL}\delta E_{KL}\,\mathrm{d}V = \int_{A_{0t}} \overline{T}^{K}_{(N)}\delta U_K\,\mathrm{d}A + \int_{V_0} \rho_0 f^K \delta U_K\,\mathrm{d}V - \rho_0 \int_{V_0} \frac{D^2 U^K}{Dt^2}\delta U_K\,\mathrm{d}V. \tag{2.1}$$

In the present chapter, primarily linear applications are envisaged. The term "linear" relates to the material, which is assumed to be linear elastic, and to the strain formulation. Consequently (see Equation (1.420)),

$$S^{KL} = [\gamma^{KL} - \beta^{KL}(\theta)T] + \Sigma^{KLMN}(\theta)E_{MN} \tag{2.2}$$

and

$$E_{KL} \approx \tilde{E}_{KL} = \tfrac{1}{2}(U_{K;L} + U_{L;K}) \tag{2.3}$$

or

$$E_{KL} \approx \tfrac{1}{2}(U_{K,L} + U_{L,K}) \tag{2.4}$$

for rectangular coordinates. In the rest of the chapter, rectangular coordinates will be assumed and the covariant differentiation will be replaced by simple differentiation. Now, Equation (2.1) can be written as

$$\begin{aligned}\int_{V_0} \delta\tilde{E}_{KL}\Sigma^{KLMN}(\theta)\tilde{E}_{MN}\,\mathrm{d}V &= \int_{A_{0t}} \overline{T}^{K}_{(N)}\delta U_K\,\mathrm{d}A + \int_{V_0} \rho_0 f^K\delta U_K\,\mathrm{d}V \\ &+ \int_{V_0} [\beta^{KL}(\theta)T - \gamma^{KL}]\delta\tilde{E}_{KL}\,\mathrm{d}V - \rho_0\int_{V_0}\frac{D^2U^K}{Dt^2}\delta U_K\,\mathrm{d}V.\end{aligned} \tag{2.5}$$

The Finite Element Method for Three-dimensional Thermomechanical Applications Guido Dhondt
 ISBN: 0-470-85752-8

Equation (2.5) shows that the residual and the thermal stresses can be considered as loads. Because of the symmetry relations satisfied by Σ^{KLMN}, β^{KL} and γ^{KL}, substitution of Equation (2.4) into Equation (2.5) yields

$$\int_{V_0} U_{M,N}\Sigma^{KLMN}(\theta)\delta U_{K,L}\,\mathrm{d}V = \int_{A_{0t}} \overline{T}^{K}_{(N)}\delta U_K\,\mathrm{d}A + \int_{V_0} \rho_0 f^K \delta U_K\,\mathrm{d}V$$
$$+ \int_{V_0} [\beta^{KL}(\theta)T - \gamma^{KL}]\delta U_{K,L}\,\mathrm{d}V - \rho_0 \int_{V_0} \frac{D^2U^K}{Dt^2}\delta U_K\,\mathrm{d}V. \quad (2.6)$$

Now, an assumption is made that can be considered as the quintessence of the finite element method. The volume V_0 is split in smaller volumes called "*finite*" elements:

$$V_0 = \sum_e V_{0e} \quad (2.7)$$

and the displacement field within each of these volumes is assumed to be a continuous function of the displacement in discrete points i, called "*nodes*":

$$\boldsymbol{U}(\boldsymbol{X}) = \sum_{i=1}^{N} \varphi_i(\boldsymbol{X})\boldsymbol{U}_i. \quad (2.8)$$

The functions φ_i are called *shape functions*.

In Equation (2.8), the position $\boldsymbol{X}$ is characterized by global coordinates. In practice, it is advantageous to define local coordinates (ξ, η, ζ) within each element satisfying $-1 \leq \xi, \eta, \zeta \leq 1$ (this applies to brick elements; the range for other types of elements will be discussed shortly) and to express both the global coordinates and the displacements as a function of discrete values at selected positions:

$$\boldsymbol{U}(\boldsymbol{X}) = \sum_{i=1}^{N} \varphi_i(\xi, \eta, \zeta)\boldsymbol{U}(\boldsymbol{X}_{\alpha_i}) \quad (2.9)$$

$$\boldsymbol{X} = \sum_{i=1}^{M} \psi_i(\xi, \eta, \zeta)\boldsymbol{X}_{\beta_i}. \quad (2.10)$$

If the discrete positions and the shape functions for $\boldsymbol{X}$ and $\boldsymbol{U}$ are the same, the formulation is called *isoparametric* (Zienkiewicz and Taylor 1989). Here, only isoparametric formulations will be considered. Accordingly,

$$\boldsymbol{U}(\boldsymbol{X}) = \sum_{i=1}^{N} \varphi_i(\xi, \eta, \zeta)\boldsymbol{U}(\boldsymbol{X}_i) \quad (2.11)$$

$$\boldsymbol{X} = \sum_{i=1}^{N} \varphi_i(\xi, \eta, \zeta)\boldsymbol{X}_i. \quad (2.12)$$

Equation (2.11) reads in component formulation

$$U_K(\boldsymbol{X}) = \sum_{i=1}^{N} \varphi_i(\xi, \eta, \zeta) U_{iK} \tag{2.13}$$

where U_{iK} is the component K of the displacement in node i. Hence,

$$U_{K,L}(\boldsymbol{X}) = \sum_{i=1}^{N} \varphi_{i,L}(\xi, \eta, \zeta) U_{iK} \tag{2.14}$$

where

$$\begin{aligned}\varphi_{i,L}(\xi, \eta, \zeta) &:= \frac{\partial \varphi_i}{\partial X^L}(\xi, \eta, \zeta) \\ &= \frac{\partial \varphi_i}{\partial \xi}\frac{\partial \xi}{\partial X^L} + \frac{\partial \varphi_i}{\partial \eta}\frac{\partial \eta}{\partial X^L} + \frac{\partial \varphi_i}{\partial \zeta}\frac{\partial \zeta}{\partial X^L}.\end{aligned} \tag{2.15}$$

The terms $\partial \varphi_i / \partial \xi$ are obtained through direct differentiation, while $\partial \xi / \partial X^L$ can be determined by inverting $\partial X^L / \partial \xi$:

$$\frac{\partial \xi}{\partial X^L} = \frac{1}{J^*}\text{cofactor}\left(\frac{\partial X^L}{\partial \xi}\right) \tag{2.16}$$

where

$$J^* := \det\left(\frac{\partial \boldsymbol{X}}{\partial \boldsymbol{\gamma}}\right), \quad \boldsymbol{\gamma}(\xi, \eta, \zeta) \tag{2.17}$$

is the Jacobian determinant of the transformation $\boldsymbol{X}(\boldsymbol{\gamma})$. The quantities $\frac{\partial X^L}{\partial \xi}$ are obtained through direct differentiation of Equation (2.12).

Splitting the integrals in Equation (1.8) across the elements e and using Equation (2.14) yields

$$\begin{aligned}\sum_e \int_{V_{0e}} \sum_{i=1}^{N} \sum_{j=1}^{N} \varphi_{j,N} \Sigma^{KLMN}(\theta) \varphi_{i,L} U_{jM} \delta U_{iK} \, \mathrm{d}V_e &= \sum_e \int_{A_{0e}} \sum_{i=1}^{N} \overline{T}^K_{(N)} \varphi_i \delta U_{iK} \, \mathrm{d}A_e \\ + \sum_e \int_{V_{0e}} \sum_{i=1}^{N} \rho_0 f^K \varphi_i \delta U_{iK} \, \mathrm{d}V_e &+ \sum_e \int_{V_{0e}} \sum_{i=1}^{N} [\beta^{KL}(\theta) T - \gamma^{KL}] \varphi_{i,L} \delta U_{iK} \, \mathrm{d}V_e \\ &- \sum_e \int_{V_{0e}} \sum_{i=1}^{N} \sum_{j=1}^{N} \rho_0 \varphi_i \varphi_j \frac{D^2 U_j^K}{Dt^2} \delta U_{iK} \, \mathrm{d}V_e\end{aligned} \tag{2.18}$$

or, removing everything that is not a function of space from the integrals

$$\sum_e \sum_{i=1}^{N} \sum_{j=1}^{N} \left[\int_{V_{0e}} \varphi_{j,N} \Sigma^{KLMN}(\theta) \varphi_{i,L} \, \mathrm{d}V_e \right] U_{jM} \delta U_{iK}$$
$$= \sum_e \sum_{i=1}^{N} \left[\int_{A_{0e}} \overline{T}^K_{(N)} \varphi_i \, \mathrm{d}A_e \right] \delta U_{iK} + \sum_e \sum_{i=1}^{N} \left[\int_{V_{0e}} \rho_0 f^K \varphi_i \, \mathrm{d}V_e \right] \delta U_{iK}$$
$$+ \sum_e \sum_{i=1}^{N} \left[\int_{V_{0e}} [\beta^{KL}(\theta) T - \gamma^{KL}] \varphi_{i,L} \, \mathrm{d}V_e \right] \delta U_{iK}$$
$$- \sum_e \sum_{i=1}^{N} \sum_{j=1}^{N} \left[\int_{V_{0e}} \rho_0 \varphi_i \varphi_j \, \mathrm{d}V_e \right] \frac{D^2 U_j^K}{Dt^2} \delta U_{iK}. \quad (2.19)$$

If we define for each element e a vector containing all displacements belonging to the element

$$\{U\}_e = \begin{bmatrix} U_{11} \\ U_{12} \\ U_{13} \\ U_{21} \\ \vdots \\ U_{N1} \\ U_{N2} \\ U_{N3} \end{bmatrix}_e \quad (2.20)$$

one can write for Equation (2.19)

$$\sum_e \delta \{U\}_e^{\mathrm{T}} [K]_e \{U\}_e = \sum_e \delta \{U\}_e^{\mathrm{T}} \{F\}_e - \sum_e \delta \{U\}_e^{\mathrm{T}} [M]_e \frac{D^2}{Dt^2} \{U\}_e \quad (2.21)$$

where the components of $[K]$, $\{F\}$ and $[M]$ satisfy

$$[K]_{e(iK)(jM)} = \int_{V_{0e}} \varphi_{i,L} \Sigma^{KLMN}(\theta) \varphi_{j,N} \, \mathrm{d}V_e \quad (2.22)$$

$$\{F\}_{e(iK)} = \int_{A_{0e}} \overline{T}^K_{(N)} \varphi_i \, \mathrm{d}A_e + \int_{V_{0e}} \rho_0 f^K \varphi_i \, \mathrm{d}V_e + \int_{V_{0e}} [\beta^{KL}(\theta) T - \gamma^{KL}] \varphi_{i,L} \, \mathrm{d}V_e \quad (2.23)$$

$$[M]_{e(iK)(jM)} = \int_{V_{0e}} \rho_0 \varphi_i \varphi_j \, \mathrm{d}V_e. \quad (2.24)$$

where $\left[K\right]_e$ is the element stiffness matrix, $\left\{F\right\}_e$ is the element force vector and $\left[M\right]_e$ is the element mass matrix. Notice that $\left[K\right]_e$ is symmetric because of the symmetry of Σ^{KLMN}, cf Equation (1.156). By defining a displacement vector $\left\{U\right\}$ containing all displacements of the complete model and a localization matrix $\left[L\right]_e$ localizing the degrees of freedom of element e in $\left\{U\right\}$:

$$\{U\}_e = [L]_e \{U\} \tag{2.25}$$

one obtains for Equation (2.21)

$$\delta \{U\}^{\mathrm{T}} \left(\sum_e [L]_e^{\mathrm{T}} [K]_e [L]_e\right) \{U\}$$
$$= \delta \{U\}^{\mathrm{T}} \left(\sum_e [L]_e^{\mathrm{T}} \{F\}_e\right) - \delta \{U\}^{\mathrm{T}} \left(\sum_e [L]_e^{\mathrm{T}} [M]_e [L]_e\right) \frac{D^2}{Dt^2} \{U\}. \tag{2.26}$$

Since Equation (2.26) must be satisfied for an arbitrary virtual displacement $\delta\left\{U\right\}$, one finally obtains

$$[K]\{U\} + [M]\frac{D^2}{Dt^2}\{U\} = \{F\} \tag{2.27}$$

where

$$[K] := \sum_e [L]_e^{\mathrm{T}} [K]_e [L]_e \tag{2.28}$$

$$[M] := \sum_e [L]_e^{\mathrm{T}} [M]_e [L]_e \tag{2.29}$$

$$\{F\} := \sum_e [L]_e^{\mathrm{T}} \{F\}_e \tag{2.30}$$

are the global stiffness matrix, global mass matrix and global force vector respectively. Notice how the introduction of the shape functions transformed the integral equation (2.1) into a set of linear algebraic equations over space, Equation (2.27). Instead of having to look for a solution everywhere in space, the unknowns are reduced to the displacements in a finite number of nodes. Equation (2.27) is the basic finite element equation to be solved for mechanical problems. It generally results in a system of hundreds of thousands of equations requiring special solution techniques (Sloan 1989), (Ashcraft *et al.* 1999).

2.2 The Shape Functions

Equation (2.8) shows that within each element the displacement field is assumed to be a continuous function of the displacements in the nodes. The question that arises is where

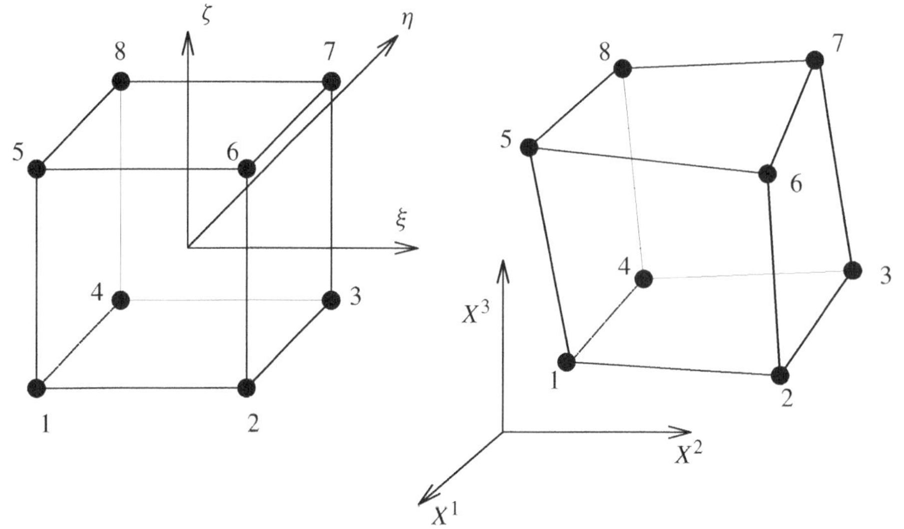

Figure 2.1 8-node brick element

those nodes are located and what the shape functions look like. There is no unique answer to this question. Rather, there are several schemes of which some have grown very popular through time. As mentioned in the previous section, we will concentrate on formulations in which the same shape functions are used for the displacements and the geometry (isoparametric formulation).

2.2.1 The 8-node brick element

The most popular element form is the brick shape, with local coordinates satisfying $-1 \leq \xi, \eta, \zeta \leq 1$. Figure 2.1 shows the 8-node brick element in local and global coordinates. At each vertex of the brick there is a node. If the shape functions are such that

$$\varphi_i(\xi_j, \eta_j, \zeta_j) = \delta_{ij} \tag{2.31}$$

then the Equations (2.11) and (2.12) are identically satisfied at the nodes. For the shape functions, one often takes polynomials because of their mathematically simple form. The lowest-order polynomial with eight unknown coefficients has the form

$$\varphi_i(\xi, \eta, \zeta) = a_i + b_i\xi + c_i\eta + d_i\zeta + e_i\xi\eta + f_i\xi\zeta + g_i\eta\zeta + h_i\xi\eta\zeta. \tag{2.32}$$

Equation (2.31) together with Equation (2.32) leads to eight equations in eight unknowns for each shape function. These sets of equations uniquely determine the coefficients. One obtains

$$\varphi_1 = (1 - \xi)(1 - \eta)(1 - \zeta)/8 \tag{2.33}$$

$$\varphi_2 = (1 + \xi)(1 - \eta)(1 - \zeta)/8 \tag{2.34}$$

$$\varphi_3 = (1+\xi)(1+\eta)(1-\zeta)/8 \tag{2.35}$$

$$\varphi_4 = (1-\xi)(1+\eta)(1-\zeta)/8 \tag{2.36}$$

$$\varphi_5 = (1-\xi)(1-\eta)(1+\zeta)/8 \tag{2.37}$$

$$\varphi_6 = (1+\xi)(1-\eta)(1+\zeta)/8 \tag{2.38}$$

$$\varphi_7 = (1+\xi)(1+\eta)(1+\zeta)/8 \tag{2.39}$$

$$\varphi_8 = (1-\xi)(1+\eta)(1+\zeta)/8. \tag{2.40}$$

The 8-node brick element is also called a *linear brick element* since the interpolation functions along any edge are linear (keep two of the three local coordinates constant with value ± 1). They look very attractive due to the simple shape functions and their intuitively attractive form for meshing purposes; however, the element exhibits marked problematic behavior such as shear locking, volumetric locking and hourglassing. This will be discussed in Section 2.5.

2.2.2 The 20-node brick element

This element has the same shape in local coordinates as the 8-node brick, but contains 20 nodes instead of 8. They are located at the vertices and in the middle of the edges (see Figure 2.2).

Using the following basis polynomials,

$$\begin{array}{cccccc} & & 1 & & & \\ \xi & & \eta & & \zeta & \\ \eta\zeta & & \xi\zeta & & \xi\eta & \\ \xi^2 & & \eta^2 & & \zeta^2 & \\ \eta\zeta^2 & \eta^2\zeta & \xi\zeta^2 & \xi^2\zeta & \xi\eta^2 & \xi^2\eta \\ & & \xi\eta\zeta & & & \\ \xi^2\eta\zeta & & \xi\eta^2\zeta & & \xi\eta\zeta^2 & \end{array} \tag{2.41}$$

one obtains for the shape functions, using Equation (2.31)

$$\varphi_1 = -(1-\xi)(1-\eta)(1-\zeta)(2+\xi+\eta+\zeta)/8 \tag{2.42}$$

$$\varphi_2 = -(1+\xi)(1-\eta)(1-\zeta)(2-\xi+\eta+\zeta)/8 \tag{2.43}$$

$$\varphi_3 = -(1+\xi)(1+\eta)(1-\zeta)(2-\xi-\eta+\zeta)/8 \tag{2.44}$$

$$\varphi_4 = -(1-\xi)(1+\eta)(1-\zeta)(2+\xi-\eta+\zeta)/8 \tag{2.45}$$

$$\varphi_5 = -(1-\xi)(1-\eta)(1+\zeta)(2+\xi+\eta-\zeta)/8 \tag{2.46}$$

$$\varphi_6 = -(1+\xi)(1-\eta)(1+\zeta)(2-\xi+\eta-\zeta)/8 \tag{2.47}$$

$$\varphi_7 = -(1+\xi)(1+\eta)(1+\zeta)(2-\xi-\eta-\zeta)/8 \tag{2.48}$$

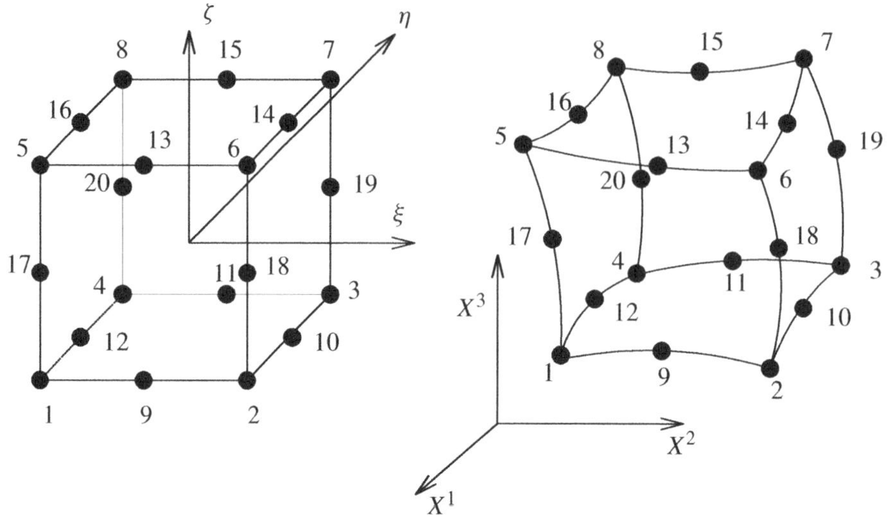

Figure 2.2 20-node brick element

$$\varphi_8 = -(1-\xi)(1+\eta)(1+\zeta)(2+\xi-\eta-\zeta)/8 \tag{2.49}$$

$$\varphi_9 = (1-\xi)(1+\xi)(1-\eta)(1-\zeta)/4 \tag{2.50}$$

$$\varphi_{10} = (1+\xi)(1-\eta)(1+\eta)(1-\zeta)/4 \tag{2.51}$$

$$\varphi_{11} = (1-\xi)(1+\xi)(1+\eta)(1-\zeta)/4 \tag{2.52}$$

$$\varphi_{12} = (1-\xi)(1-\eta)(1+\eta)(1-\zeta)/4 \tag{2.53}$$

$$\varphi_{13} = (1-\xi)(1+\xi)(1-\eta)(1+\zeta)/4 \tag{2.54}$$

$$\varphi_{14} = (1+\xi)(1-\eta)(1+\eta)(1+\zeta)/4 \tag{2.55}$$

$$\varphi_{15} = (1-\xi)(1+\xi)(1+\eta)(1+\zeta)/4 \tag{2.56}$$

$$\varphi_{16} = (1-\xi)(1-\eta)(1+\eta)(1+\zeta)/4 \tag{2.57}$$

$$\varphi_{17} = (1-\xi)(1-\eta)(1-\zeta)(1+\zeta)/4 \tag{2.58}$$

$$\varphi_{18} = (1+\xi)(1-\eta)(1-\zeta)(1+\zeta)/4 \tag{2.59}$$

$$\varphi_{19} = (1+\xi)(1+\eta)(1-\zeta)(1+\zeta)/4 \tag{2.60}$$

$$\varphi_{20} = (1-\xi)(1+\eta)(1-\zeta)(1+\zeta)/4. \tag{2.61}$$

The 20-node brick elements are also called *quadratic elements* because the interpolation along each edge is a quadratic function. Because of this, they can simulate curved boundaries by a piecewise-quadratic approximation. Quadratic brick elements are usually well behaved and in the author's opinion they should be preferred to linear brick elements.

A major disadvantage is that despite intensive research no satisfactory automatic meshers are yet available (Tautges 2001). Therefore, the efficient generation of a good-quality hexahedral mesh heavily relies on the expertise of the user.

If one of the faces of a 20-node brick element is collapsed, the element can simulate singular strain and stress fields (Dhondt 1993). This is used in special applications such as linear elastic fracture mechanics (Dhondt 2002).

2.2.3 The 4-node tetrahedral element

The 4-node tetrahedral element is characterized by linear interpolation functions within a tetrahedron (see Figure 2.3). The local coordinates are such that

$$0 \leq \xi, \eta, \zeta \leq 1 \tag{2.62}$$

$$\xi + \eta + \zeta \leq 1. \tag{2.63}$$

Nodes 2, 3 and 4 are characterized by $\xi = 1$, $\eta = 1$ and $\zeta = 1$ respectively. In the local coordinate system, the coordinates of a point P can be obtained by constructing the tetrahedra T_1, T_2 and T_3 extending from P to the faces $1-3-4$ (opposite node 2), $1-2-4$ (opposite node 3) and $1-2-3$ (opposite node 4) respectively. Denoting the volumes of T_1, T_2 and T_3 by V_1, V_2 and V_3 respectively, and the total volume by V one can write

$$\xi = V_1/V \tag{2.64}$$

$$\eta = V_2/V \tag{2.65}$$

$$\zeta = V_3/V. \tag{2.66}$$

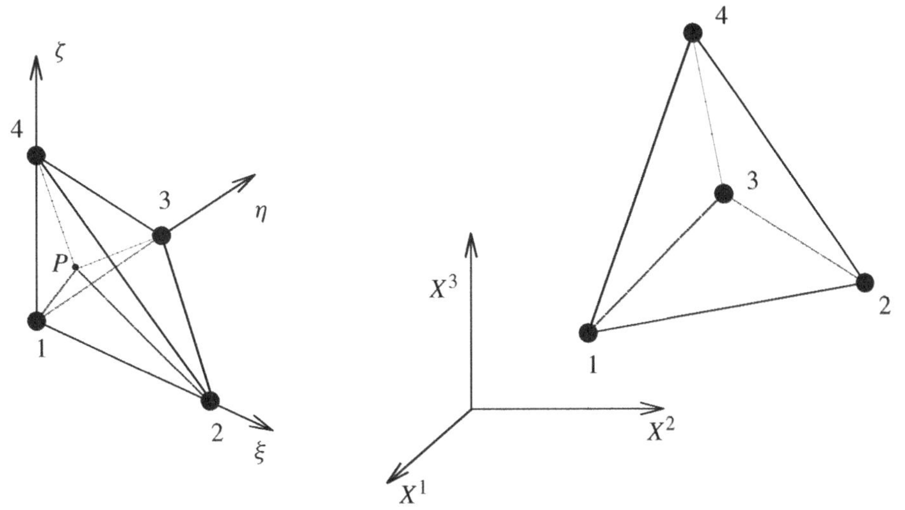

Figure 2.3 4-node tetrahedral element

The shape functions take the form

$$\varphi_1 = 1 - \xi - \eta - \zeta \tag{2.67}$$

$$\varphi_2 = \xi \tag{2.68}$$

$$\varphi_3 = \eta \tag{2.69}$$

$$\varphi_4 = \zeta. \tag{2.70}$$

The shape functions are exceedingly simple. However, for stress calculations the element is extremely stiff and lots of elements are needed to obtain acceptable results. It should generally be avoided.

2.2.4 The 10-node tetrahedral element

The 10-node tetrahedral element is characterized by quadratic interpolation functions within the element. The extra degrees of freedom are taken care of by introducing nodes in the middle of the element edges (Figure 2.4). The local coordinate system is the same as for the 4-node tetrahedral element. The shape functions take the form

$$\varphi_1 = [2(1 - \xi - \eta - \zeta) - 1][1 - \xi - \eta - \zeta] \tag{2.71}$$

$$\varphi_2 = (2\xi - 1)\xi \tag{2.72}$$

$$\varphi_3 = (2\eta - 1)\eta \tag{2.73}$$

$$\varphi_4 = (2\zeta - 1)\zeta \tag{2.74}$$

$$\varphi_5 = 4(1 - \xi - \eta - \zeta)\xi \tag{2.75}$$

$$\varphi_6 = 4\xi\eta \tag{2.76}$$

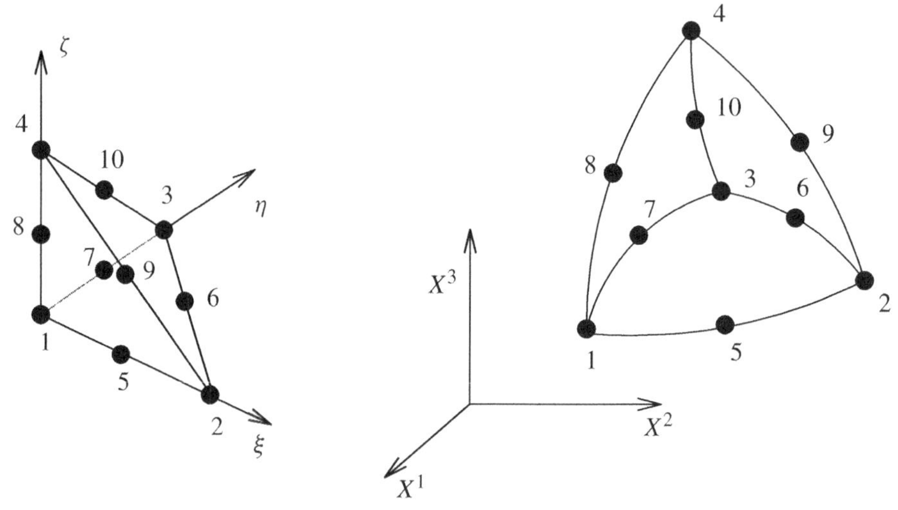

Figure 2.4 10-node tetrahedral element

$$\varphi_7 = 4(1 - \xi - \eta - \zeta)\eta \tag{2.77}$$

$$\varphi_8 = 4(1 - \xi - \eta - \zeta)\zeta \tag{2.78}$$

$$\varphi_9 = 4\xi\zeta \tag{2.79}$$

$$\varphi_{10} = 4\eta\zeta. \tag{2.80}$$

The 10-node tetrahedral element is a very flexible element due to its shape. Furthermore, automatic reliable tetrahedral meshing routines have been developed that are able to cope with nearly any structure (George and Borouchaki 1998), (Freitag and Knupp 2002). The quality of the 10-node element is comparable to the 20-node brick element. Disadvantages are the enormous amount of elements generated by automatic meshing routines and the nontrivial quality check of the mesh. Indeed, owing to the irregular shape of the tetrahedra, a visual check is nearly impossible and one has to rely on mathematical measures such as the dihedral angle.

2.2.5 The 6-node wedge element

For this element type, the local coordinates are such that (Figure 2.5)

$$0 \leq \xi, \eta \leq 1, \quad -1 \leq \zeta \leq 1 \tag{2.81}$$

$$\xi + \eta \leq 1. \tag{2.82}$$

For $\zeta = -1$, one obtains the lower triangle $1 - 2 - 3$. The values of ξ and η of a point A are given by the surface ratio of triangle $1 - A - 3$ and triangle $1 - A - 2$ with respect to triangle $1 - 2 - 3$ respectively. The 6-node wedge element is linear, that is, the connection of the nodes in Figure 2.5 is straight. Its shape functions take the form

$$\varphi_1 = (1 - \xi - \eta)(1 - \zeta)/2 \tag{2.83}$$

$$\varphi_2 = \xi(1 - \zeta)/2 \tag{2.84}$$

$$\varphi_3 = \eta(1 - \zeta)/2 \tag{2.85}$$

$$\varphi_4 = (1 - \xi - \eta)(1 + \zeta)/2 \tag{2.86}$$

$$\varphi_5 = \xi(1 + \zeta)/2 \tag{2.87}$$

$$\varphi_6 = \eta(1 + \zeta)/2. \tag{2.88}$$

2.2.6 The 15-node wedge element

The 15-node wedge element is the quadratic version of the 6-node wedge element (see Figure 2.6). Equations (2.81) and (2.82) also apply here. The shape functions take the form

$$\varphi_1 = -(1 - \xi - \eta)(1 - \zeta)(2\xi + 2\eta + \zeta)/2 \tag{2.89}$$

$$\varphi_2 = \xi(1 - \zeta)(2\xi - \zeta - 2)/2 \tag{2.90}$$

$$\varphi_3 = \eta(1 - \zeta)(2\eta - \zeta - 2)/2 \tag{2.91}$$

$$\varphi_4 = -(1-\xi-\eta)(1+\zeta)(2\xi+2\eta-\zeta)/2 \tag{2.92}$$

$$\varphi_5 = \xi(1+\zeta)(2\xi+\zeta-2)/2 \tag{2.93}$$

$$\varphi_6 = \eta(1+\zeta)(2\eta+\zeta-2)/2 \tag{2.94}$$

$$\varphi_7 = 2\xi(1-\xi-\eta)(1-\zeta) \tag{2.95}$$

$$\varphi_8 = 2\xi\eta(1-\zeta) \tag{2.96}$$

$$\varphi_9 = 2\eta(1-\xi-\eta)(1-\zeta) \tag{2.97}$$

$$\varphi_{10} = 2\xi(1-\xi-\eta)(1+\zeta) \tag{2.98}$$

$$\varphi_{11} = 2\xi\eta(1+\zeta) \tag{2.99}$$

$$\varphi_{12} = 2\eta(1-\xi-\eta)(1+\zeta) \tag{2.100}$$

$$\varphi_{13} = (1-\xi-\eta)(1-\zeta^2) \tag{2.101}$$

$$\varphi_{14} = \xi(1-\zeta^2) \tag{2.102}$$

$$\varphi_{15} = \eta(1-\zeta^2). \tag{2.103}$$

Wedge elements are often used as fill-in elements by automatic hexahedral meshing codes. Their quality is comparable to the 20-node brick and 10-node tetrahedral element.

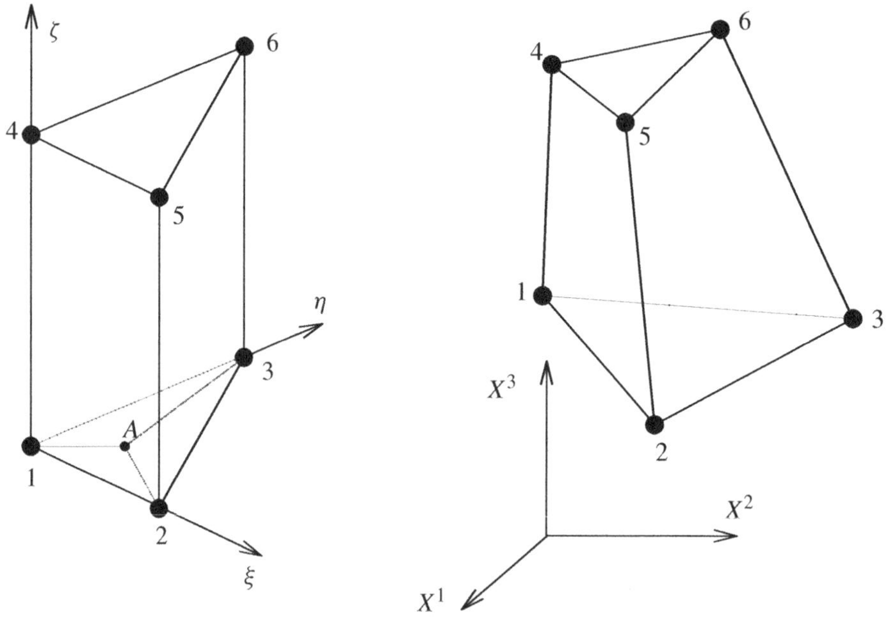

Figure 2.5 6-node wedge element

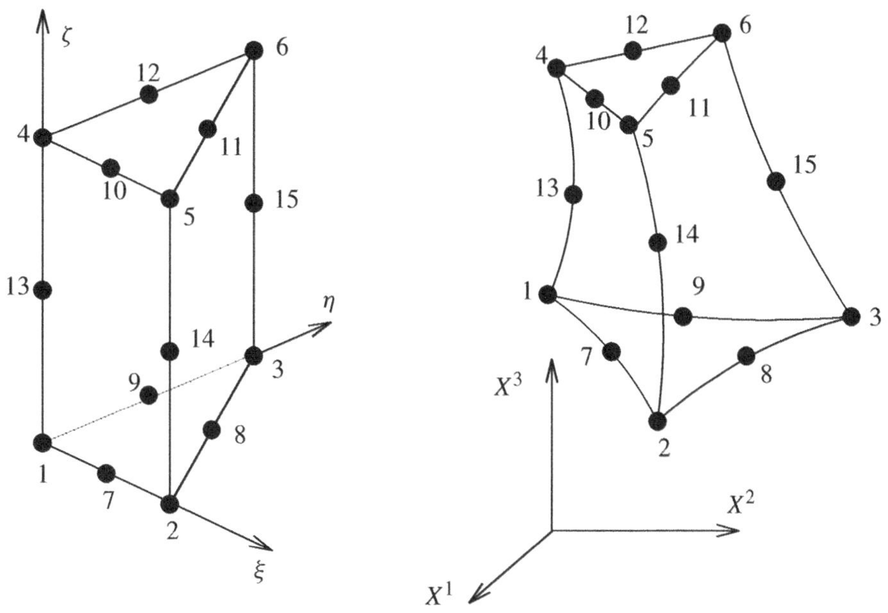

Figure 2.6 15-node wedge element

2.3 Numerical Integration

To obtain the force, the stiffness matrix and the mass matrix (see Equations (2.22)–(2.24)), integration over the element is required. To calculate an integral of the form

$$I = \int_{V_{0e}} f(\boldsymbol{X})\,\mathrm{d}V \tag{2.104}$$

numerical integration is used. Indeed, $f(\boldsymbol{X})$ is frequently a complicated function of space (it usually contains material properties depending on the temperature, which is itself a function of space) and the shape of an element can be quite irregular. Consequently, analytical integration is not feasible. Before applying numerical integration to Equation (2.104) the integration domain is transformed from the global to the local element coordinate system:

$$I = \int_{V_{0eL}} f[\boldsymbol{X}(\xi,\eta,\zeta)]J^*(\xi,\eta,\zeta)\,\mathrm{d}\xi\,\mathrm{d}\eta\,\mathrm{d}\zeta =: \int_{V_{0eL}} g(\xi,\eta,\zeta)\,\mathrm{d}\xi\,\mathrm{d}\eta\,\mathrm{d}\zeta \tag{2.105}$$

where $J^*(\xi,\eta,\zeta)$ is the Jacobian determinant of the transformation $\boldsymbol{X}(\boldsymbol{\gamma})$ (see Equation (2.17)). In this way, the integration domain is identical for all elements belonging to the same type, for example, brick elements. Now, the analytical integration is approximated by a numerical integration scheme (Stroud 1971). This basically means that the integral in Equation (2.105) is replaced by a linear combination of function values at specific locations, the so-called integration points, within the domain of integration:

$$\int_{V_{0eL}} g(\xi,\eta,\zeta)\,\mathrm{d}\xi\,\mathrm{d}\eta\,\mathrm{d}\zeta \approx \sum_{i=1}^{N} g(\xi_i,\eta_i,\zeta_i)w_i. \tag{2.106}$$

The value of the weights, the location of the integration points and their number constitute together an integration scheme. Different schemes lead to different calculational expenditure and different accuracy. For finite element calculations, the Gauss schemes are very popular, because of their high accuracy compared to the numerical expenditure. Since the integration schemes essentially depend on the shape of the domain, a distinction is made between hexahedral, tetrahedral and wedge elements.

2.3.1 Hexahedral elements

The domain in local coordinates for a hexahedral element is a cube extending from -1 to $+1$ ($-1 \leq \xi, \eta, \zeta \leq 1$) along each coordinate axis. The integration schemes are symmetric in each direction. The lowest scheme has one integration point in each direction ($1 \times 1 \times 1 = 1$, Figure 2.7), the next ones have two ($2 \times 2 \times 2 = 8$, Figure 2.8) or three ($3 \times 3 \times 3 = 27$, Figure 2.9) points in each direction. The location of the integration points and their weights are summarized in Table 2.1.

The $1 \times 1 \times 1$ scheme, the $2 \times 2 \times 2$ scheme and the $3 \times 3 \times 3$ scheme are exact for a constant function, a trilinear function and a triquadratic function respectively. Therefore, the $2 \times 2 \times 2$ scheme represents full integration for a linear element (8–node brick) and the $3 \times 3 \times 3$ scheme stands for full integration in a quadratic element (20–node brick). The term *reduced integration* is used if one selects the next coarser scheme: $1 \times 1 \times 1$ for linear elements and $2 \times 2 \times 2$ for quadratic elements. Reduced integration frequently has a beneficial effect: it produces less shear locking and less volumetric locking; therefore, it is ideal for plates, shells and incompressible materials (rubber, plasticity in metals). Furthermore, Barlow has shown ((Barlow 1976), see also (Mackinnon and Carey 1989) and

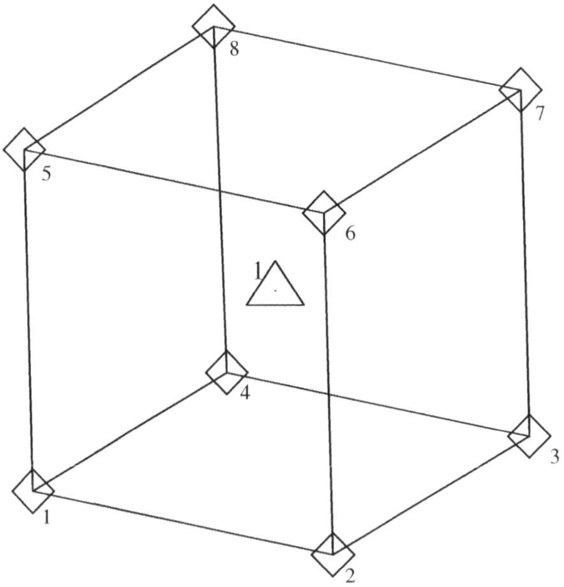

Figure 2.7 Hexahedral element: $1 \times 1 \times 1$ scheme

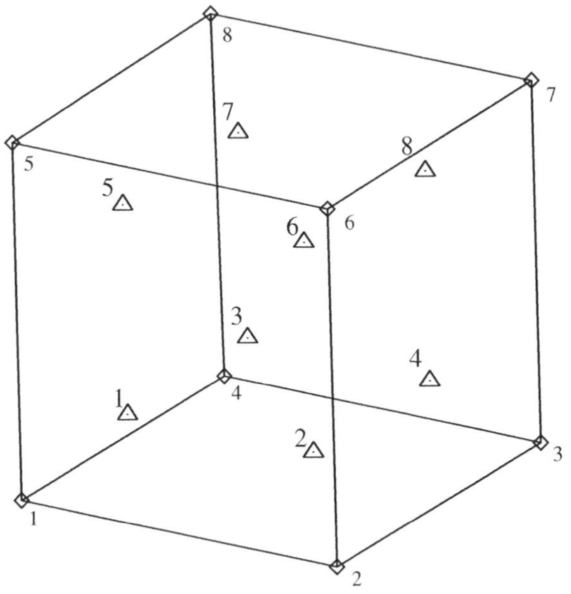

Figure 2.8 Hexahedral element: 2 × 2 × 2 scheme

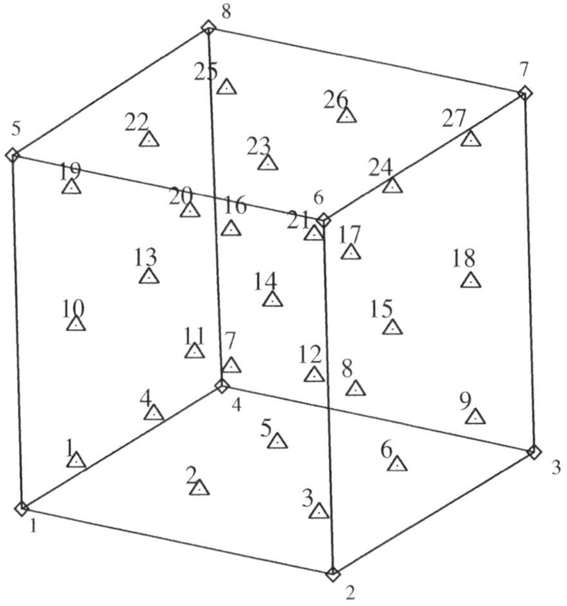

Figure 2.9 Hexahedral element: 3 × 3 × 3 scheme

Table 2.1 Location of the integration points in hexahedral elements.

Scheme	Location (ξ_i, η_i, ζ_i) and any perturbation	Number	Weight
$1 \times 1 \times 1$	$(0, 0, 0)$	1	8
$2 \times 2 \times 2$	$\left(\pm\frac{1}{\sqrt{3}}, \pm\frac{1}{\sqrt{3}}, \pm\frac{1}{\sqrt{3}}\right)$	8	1
$3 \times 3 \times 3$	$\left(\pm\sqrt{\frac{3}{5}}, \pm\sqrt{\frac{3}{5}}, \pm\sqrt{\frac{3}{5}}\right)$	8	$\left(\frac{5}{9}\right)^3$
	$\left(0, \pm\sqrt{\frac{3}{5}}, \pm\sqrt{\frac{3}{5}}\right)$	12	$\left(\frac{8}{9}\right)\left(\frac{5}{9}\right)^2$
	$\left(0, 0, \pm\sqrt{\frac{3}{5}}\right)$	6	$\left(\frac{8}{9}\right)^2\left(\frac{5}{9}\right)$
	$(0, 0, 0)$	1	$\left(\frac{8}{9}\right)^3$

(Liew and Rajendran 2002)) that the reduced integration points are the so-called superconvergent points, in which the stress is one order more accurate than in any other point. However, because of reduced integration, so-called zero-energy modes can arise, leading to hourglassing. Shear locking, volumetric locking and hourglassing are discussed in Section 2.5.

2.3.2 Tetrahedral elements

For tetrahedral elements, the integration domain in local coordinates is depicted in Figure 2.10. The most frequently used Gauss integration schemes are summarized in Table 2.2. Linear tetrahedral elements are usually integrated with one integration point, quadratic elements with four. Figure 2.10 visualizes the scheme with four integration points. The scheme with 15 integration points improves the condition of the consistent mass matrix (cf Section 2.11.6). For tetrahedral elements, the term *reduced integration* is not used.

2.3.3 Wedge elements

The integration domain for a wedge element consists of a prism with triangular lower and upper surface (Figure 2.11). Linear wedge elements are usually integrated with a 2-point scheme, quadratic wedges with a 9-point scheme. The 18-point scheme is used for the integration of the consistent mass matrix (cf Section 2.11.6). The integration schemes are summarized in Table 2.3.

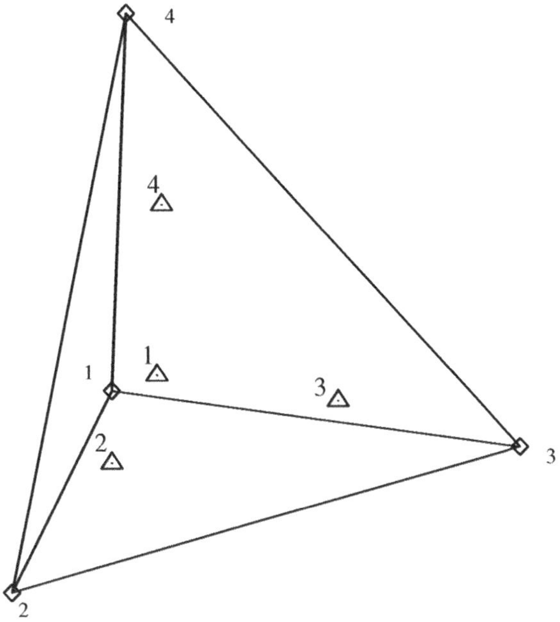

Figure 2.10 Tetrahedral element: 4 integration points

Table 2.2 Location of the integration points in tetrahedral elements.

Total number of integration points	Location $(\xi_i, \eta_i, \zeta_i, 1-\xi_i-\eta_i-\zeta_i)$ and any perturbation	Number	Weight
1	$\left(\frac{1}{4}, \frac{1}{4}, \frac{1}{4}, \frac{1}{4}\right)$	1	$\frac{1}{6}$
4	$\left(\frac{5-\sqrt{5}}{20}, \frac{5-\sqrt{5}}{20}, \frac{5-\sqrt{5}}{20}, \frac{5+3\sqrt{5}}{20}\right)$	4	$\frac{1}{24}$
15	$\left(\frac{1}{4}, \frac{1}{4}, \frac{1}{4}, \frac{1}{4}\right)$	1	$\frac{16}{810}$
	$\left(\frac{7-\sqrt{15}}{34}, \frac{7-\sqrt{15}}{34}, \frac{7-\sqrt{15}}{34}, \frac{13+3\sqrt{15}}{34}\right)$	4	$\frac{2665+14\sqrt{15}}{226\,800}$
	$\left(\frac{7+\sqrt{15}}{34}, \frac{7+\sqrt{15}}{34}, \frac{7+\sqrt{15}}{34}, \frac{13-3\sqrt{15}}{34}\right)$	4	$\frac{2665-14\sqrt{15}}{226\,800}$
	$\left(\frac{10-2\sqrt{15}}{40}, \frac{10-2\sqrt{15}}{40}, \frac{10+2\sqrt{15}}{40}, \frac{10+2\sqrt{15}}{40}\right)$	6	$\frac{20}{2268}$

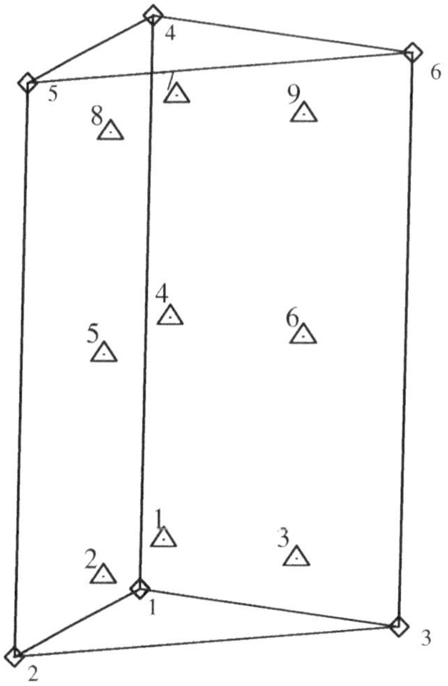

Figure 2.11 Wedge element: 9 integration points

Table 2.3 Location of the integration points in wedge elements.

Total number of integration points	Location $(\xi_i, \eta_i, 1-\xi_i-\eta_i-\zeta_i, \zeta_i)$ and any perturbation	Number	Weight
2	$\left(\frac{1}{3}, \frac{1}{3}, \frac{1}{3}; \pm\frac{1}{\sqrt{3}}\right)$	2	$\frac{1}{2}$
9	$\left(\frac{1}{6}, \frac{1}{6}, \frac{4}{6}; \pm\sqrt{\frac{3}{5}}\right)$	6	$\frac{5}{54}$
	$\left(\frac{1}{6}, \frac{1}{6}, \frac{4}{6}; 0\right)$	3	$\frac{8}{54}$
18	$\left(\frac{1}{6}, \frac{1}{6}, \frac{4}{6}; \pm\sqrt{\frac{3}{5}}\right)$	6	$\frac{1}{12}$
	$\left(\frac{1}{6}, \frac{1}{6}, \frac{4}{6}; 0\right)$	3	$\frac{2}{15}$
	$\left(\frac{1}{2}, \frac{1}{2}, 0; \pm\sqrt{\frac{3}{5}}\right)$	6	$\frac{1}{108}$
	$\left(\frac{1}{2}, \frac{1}{2}, 0; 0\right)$	3	$\frac{2}{135}$

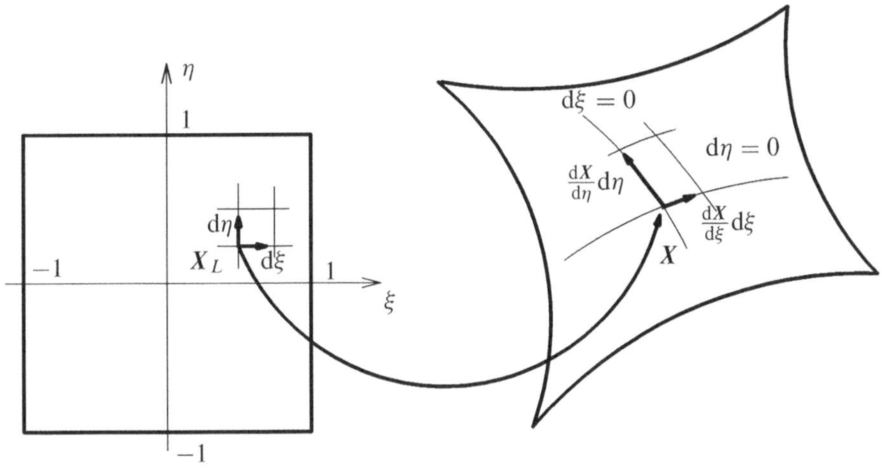

Figure 2.12 Mapping a surface element in local coordinates onto global coordinates

2.3.4 Integration over a surface in three-dimensional space

Occasionally, the domain of an integral is a surface. For instance, for a distributed pressure, the first term in Equation (2.23) takes the form

$$-\int_{A_{0e}} pN^K \varphi_i \, \mathrm{d}A_e = -\int_{A_{0e}} p\varphi_i \, \mathrm{d}A_e^K \tag{2.107}$$

or, generically,

$$\boldsymbol{I} = \int_{A_{0e}} f(\boldsymbol{X}) \, \mathrm{d}\boldsymbol{A}. \tag{2.108}$$

On the left-hand side of Figure 2.12, the surface is shown in local coordinates, on the right-hand side in global coordinates. The infinitesimal surface $\mathrm{d}\boldsymbol{A}$ satisfies

$$\mathrm{d}\boldsymbol{A} = \frac{\partial \boldsymbol{X}}{\partial \xi}\mathrm{d}\xi \times \frac{\partial \boldsymbol{X}}{\partial \eta}\mathrm{d}\eta \tag{2.109}$$

where "×" is the vector product. Consequently, Equation (2.108) can be replaced by

$$\boldsymbol{I} = \int_{A_{eL}} f[\boldsymbol{X}(\xi, \eta)]\boldsymbol{J}^* \, \mathrm{d}\xi \, \mathrm{d}\eta \tag{2.110}$$

where

$$\boldsymbol{J}^* := \frac{\partial \boldsymbol{X}}{\partial \xi} \times \frac{\partial \boldsymbol{X}}{\partial \eta} = \frac{\partial X^K}{\partial \xi}\frac{\partial X^L}{\partial \eta}\boldsymbol{G}_K \times \boldsymbol{G}_L \tag{2.111}$$

can be considered as a Jacobian vector. Since

$$\boldsymbol{G}_K \times \boldsymbol{G}_L = e_{KLM}\boldsymbol{G}^M \sqrt{\det \boldsymbol{G}^\flat} \tag{2.112}$$

Equation (2.111) can also be written as

$$J^* = e_{KLM} \frac{\partial X^K}{\partial \xi} \frac{\partial X^L}{\partial \eta} G^M \sqrt{\det G^\flat} \tag{2.113}$$

or

$$J^* = \begin{vmatrix} G^1 & G^2 & G^3 \\ \dfrac{\partial X^1}{\partial \xi} & \dfrac{\partial X^2}{\partial \xi} & \dfrac{\partial X^3}{\partial \xi} \\ \dfrac{\partial X^1}{\partial \eta} & \dfrac{\partial X^2}{\partial \eta} & \dfrac{\partial X^3}{\partial \eta} \end{vmatrix} \sqrt{\det G^\flat} \tag{2.114}$$

where the vertical lines denote the determinant. Notice that the vector product of two vectors yields a one-form. This confirms the one-form nature of a differential surface element.

Equation (2.107) is the expression for the force in direction K in local node i due to a distributed pressure on A_{0e}. Focusing on a hexahedral element type and assuming that $x = \xi$, $y = \eta$, $z = 0$ and that the pressure is constant, the force F_{zi} in z-direction in local node i takes the form

$$F_{zi} = -p \int_{-1}^{1} \int_{-1}^{1} \varphi_i(\xi, \eta)\, \mathrm{d}\xi\, \mathrm{d}\eta. \tag{2.115}$$

The total force F_z on the surface amounts to

$$F_z = -4p. \tag{2.116}$$

Accordingly, the relative force in local node i satisfies

$$F_{zi}/F_z = \frac{1}{4} \int_{-1}^{1} \int_{-1}^{1} \varphi_i(\xi, \eta)\, \mathrm{d}\xi\, \mathrm{d}\eta. \tag{2.117}$$

The shape functions in the face are the three-dimensional shape functions from Section 2.2 for which one local coordinate is kept constant. Performing the integration in Equation (2.117) and similarly for the other element types leads to the force distributions in Figure 2.13. One notices that for linear elements each node takes the same amount of force. This is not the case for quadratic elements. For 10-node tetrahedral elements, the vertex nodes do not take any force at all and the middle nodes carry the complete force. For 20-node brick elements, the middle nodes carry even more than the complete force resulting in tensile forces in the vertex nodes. This substantially complicates the detection of contact conditions in quadratic elements.

2.4 Extrapolation of Integration Point Values to the Nodes

Solution of the governing finite element equations (2.27) yields the displacements at all nodes. These can be used to calculate the strains (apply Equations (2.14) and (2.4)) and the stresses (through Equation (2.2)) throughout each element. Because of the numerical integration, the strains and stresses are more accurate at the integration points than anywhere else. Therefore, the field variables are usually evaluated at the integration points and, if

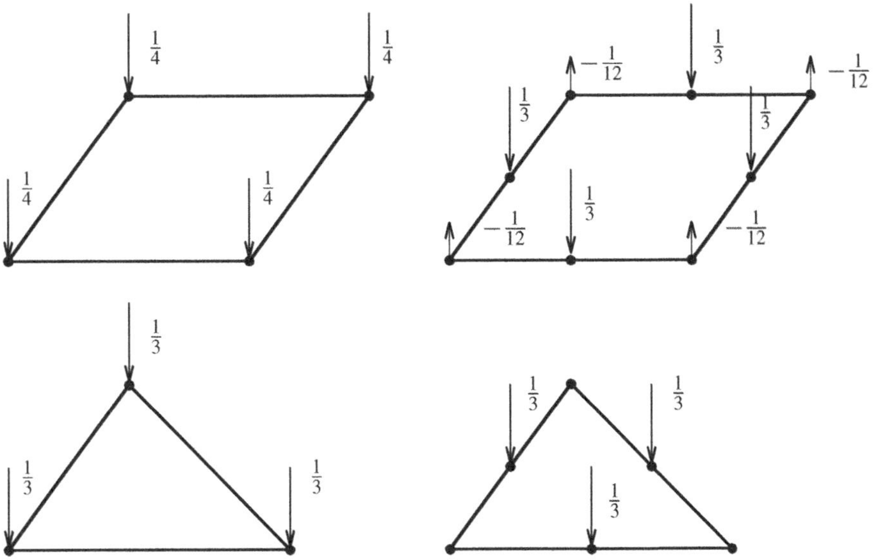

Figure 2.13 Relative nodal forces due to constant pressure

needed, extrapolated to the nodes. This extrapolation is done on an element basis, that is, one obtains for a given node as many values as the number of elements it belongs to. These values are usually discontinuous at the element borders. This is taken care of by calculating the mean value over all elements the node belongs to.

Extrapolation toward the nodes is sometimes replaced by interpolation within patches of elements. This is closely related to the very important topic of error estimation. For further information the reader is referred to (Zienkiewicz and Zhu 1992a), (Zienkiewicz and Zhu 1992b), (Gabaldón and Goicolea 2002) and (Prudhomme *et al.* 2003).

Now, extrapolation schemes will be presented for the three-dimensional elements introduced in the previous sections.

2.4.1 The 8-node hexahedral element

The extrapolation of the field variables in the integration points toward the nodes for the fully integrated linear hexahedral element is usually trilinear, that is, the shape functions that are used for the displacements are also used for the stresses, strains and any other dependent fields. Assume that the field variables are known in the nodes. Then, the integration point values are obtained by (the stress σ_{xx} stands for any field variable)

$$\sigma_{xxj} = \sum_{i=1}^{8} \varphi_i(\xi_j, \eta_j, \zeta_j)\,\sigma_{xxi} \tag{2.118}$$

(i are the nodes, j are the eight integration points) or in matrix form

$$\left\{\sigma_{xx}\right\}_{\text{integration points}} = \left[A\right]\left\{\sigma_{xx}\right\}_{\text{nodes}}. \tag{2.119}$$

Consequently, the nodal values are found by inverting Equation (2.119)

$$\{\sigma_{xx}\}_{\text{nodes}} = [A]^{-1}\{\sigma_{xx}\}_{\text{integration points}}. \tag{2.120}$$

$[A]$ is an 8×8 matrix and can be evaluated explicitly since both the shape functions (Section 2.2.1) and the location of the integration points (Section 2.3.1) are known. Consequently, the inverse matrix can also be coded explicitly into the finite element program. For the node and integration point numbering of Figure 2.8 the matrix $[A]^{-1}$ satisfies

$$[A]^{-1} =$$

$$\begin{bmatrix} +2.549 & -0.683 & -0.683 & +0.183 & -0.683 & +0.183 & +0.183 & -0.049 \\ -0.683 & +2.549 & +0.183 & -0.683 & +0.183 & -0.683 & -0.049 & +0.183 \\ +0.183 & -0.683 & -0.683 & +2.549 & -0.049 & +0.183 & +0.183 & -0.683 \\ -0.683 & +0.183 & +2.549 & -0.683 & +0.183 & -0.049 & -0.683 & +0.183 \\ -0.683 & +0.183 & +0.183 & -0.049 & +2.549 & -0.683 & -0.683 & +0.183 \\ 0.183 & -0.683 & -0.049 & +0.183 & -0.683 & +2.549 & +0.183 & -0.683 \\ -0.049 & +0.183 & +0.183 & -0.683 & +0.183 & -0.683 & -0.683 & +2.549 \\ +0.183 & -0.049 & -0.683 & +0.183 & -0.683 & +0.183 & +2.549 & -0.683 \end{bmatrix}. \tag{2.121}$$

Notice that Equation (2.120) defines the nodal values as a linear combination of the integration point values. For the reduced integration 8-node element, there is only 1 integration point, yielding one field value per element. This value is copied to the nodes and corresponds to a constant function extrapolation.

2.4.2 The 20-node hexahedral element

For the fully integrated 20-node element, a similar scheme as for the fully integrated 8-node element is proposed: the field variables are interpolated using the shape functions of the element:

$$\sigma_{xxj} = \sum_{i=1}^{20} \varphi_i(\xi_j, \eta_j, \zeta_j)\sigma_{xxi}, \quad j = 1, \dots, 27 \tag{2.122}$$

(i are the nodes, j are the integration points). This, however, leads to 27 equations in 20 unknowns (the nodal values σ_{xxi}) and cannot be inverted: the system is overdetermined. A standard procedure to solve overdetermined systems is the least-squares method. Writing Equation (2.122) as

$$b_j = \sum_{i=1}^{20} a_{ji} x_i, \quad j = 1, \dots, 27 \tag{2.123}$$

corresponds to minimizing

$$I := \sum_{j=1}^{27} \left(\sum_{i=1}^{20} a_{ji} x_i - b_j \right)^2 . \tag{2.124}$$

The solution can be found by differentiation:

$$\frac{\partial I}{\partial x_k} = 2 \sum_{j=1}^{27} \left[\left(\sum_{i=1}^{20} a_{ji} x_i - b_j \right) a_{jk} \right] = 0, \quad k = 1, \ldots, 20 \tag{2.125}$$

is equivalent to

$$\sum_{i=1}^{20} \left[\left(\sum_{j=1}^{27} a_{ji} a_{jk} \right) x_i \right] = \sum_{j=1}^{27} a_{jk} b_j, \quad k = 1, \ldots, 20 \tag{2.126}$$

or

$$\sum_{i=1}^{20} c_{ki} x_i = d_k, \quad k = 1, \ldots, 20 \tag{2.127}$$

where

$$c_{ki} = \sum_{j=1}^{27} a_{jk} a_{ji}, \quad k = 1, \ldots, 20 \tag{2.128}$$

$$d_k = \sum_{j=1}^{27} a_{jk} b_j, \quad k = 1, \ldots, 20. \tag{2.129}$$

Equation (2.127) is a system of 20 equations in 20 unknowns. Let $\{b_1\}$ be a unit vector with a unit value in its first row. Then, the solution $\{x_1\}$ contains the nodal values for a unit value in the first integration point and zero in all other integration points. This can be repeated for all other integration points. One finally obtains the 20×27 matrix $[B]$ in the equation

$$\{\sigma_{xx}\}_{\text{nodes}} = [B] \{\sigma_{xx}\}_{\text{integration points}} . \tag{2.130}$$

It takes the form

$$[B] = [\{x_1\} \{x_2\} \ldots \{x_{27}\}] . \tag{2.131}$$

The numerical values can be found in the CalculiX® code (CalculiX 2003).

For the reduced integration element, there are only 8 integration point values. The same scheme as for the fully integrated 8-node element is used to obtain the vertex nodal values. The values of the middle nodes are obtained by taking the mean of the neighboring vertex nodal values.

2.4.3 The tetrahedral elements

For the linear tetrahedral element, there is only 1 integration point and its value is simply copied to the nodes.

The quadratic tetrahedral element contains 4 vertex nodes and 4 integration points. Consequently, exactly the same procedure can be used as for the reduced integrated 20-node elements: one takes the displacement shape functions of the corresponding linear element – the linear tetrahedron – and writes Equations (2.118) to (2.119). The inversion of $[A]$ yields the vertex nodal values by Equation (2.120). Here, $[A]^{-1}$ takes the form

$$[A]^{-1} = \begin{bmatrix} +1.92705 & -0.30902 & -0.30902 & -0.30902 \\ -0.30902 & +1.92705 & -0.30902 & -0.30902 \\ -0.30902 & -0.30902 & +1.92705 & -0.30902 \\ -0.30902 & -0.30902 & -0.30902 & +1.92705 \end{bmatrix} \tag{2.132}$$

for the node and integration point numbering of Figure 2.10. The values in the middle nodes are obtained by taking the mean of the neighboring vertex nodal values.

2.4.4 The wedge elements

For the linear wedge element, the values in the two integration points are linearly extrapolated toward the nodes in the upper and lower triangle.

The quadratic wedge element has 15 nodes and 9 integration points. It is underdetermined. However, there are only 6 vertex nodes. Consequently, one can apply a least-squares scheme for the vertex nodes. It results in the following $6 \times 9\,[B]$ matrix (cf Equation (2.130)):

$$[B] = \begin{bmatrix} +1.6314 & -0.3263 & -0.3263 & +0.5556 & -0.1111 & -0.1111 & -0.5203 & +0.1041 & +0.1041 \\ -0.3263 & +1.6314 & -0.3263 & -0.1111 & +0.5556 & -0.1111 & +0.1041 & -0.5203 & +0.1041 \\ -0.3263 & -0.3263 & +1.6314 & -0.1111 & -0.1111 & +0.5556 & +0.1041 & +0.1041 & -0.5203 \\ -0.5203 & +0.1041 & +0.1041 & +0.5556 & -0.1111 & -0.1111 & +1.6314 & -0.3263 & -0.3263 \\ +0.1041 & -0.5203 & +0.1041 & -0.1111 & +0.5556 & -0.1111 & -0.3263 & +1.6314 & -0.3263 \\ +0.1041 & +0.1041 & -0.5203 & -0.1111 & -0.1111 & +0.5556 & -0.3263 & -0.3263 & +1.6314 \end{bmatrix} \tag{2.133}$$

for the node and integration points numbering of Figure 2.11. The midnode values are obtained by taking the mean of the neighboring vertex nodal values.

2.5 Problematic Element Behavior

Some of the elements discussed earlier exhibit anomalies under certain conditions. This invariably results from the approximations in the finite element formulation. These are twofold: the real displacement field is approximated by the shape functions and the continuous integration is replaced by a sum in discrete points. The most important anomalies that can occur in the volume elements considered here are shear locking, volumetric locking and hourglassing.

2.5.1 Shear locking

Shear locking predominantly occurs in linear elements with full integration (8-node brick). It results in a deformation behavior that is too stiff, that is, the displacements are too small. This can best be explained by looking at a two-dimensional view of a beam subjected to pure bending in Figure 2.14.

The shear force is zero and the shearing strain should everywhere be zero. However, the linear brick element cannot model the curvature appropriately and will approximate the deformed shape by a piecewise-linear curve (recall that the edges of an 8-node brick element are straight). Whereas in the real deformation cross sections remain perpendicular to the beam axis (Figure 2.14(b)), this is not necessarily the case in the finite element approximation (Figure 2.14(c)). Figures 2.14(d) and 2.14(e) show just one element from Figure 2.14(c). If full integration is used ($2 \times 2 \times 2$ integration points) as shown in Figure 2.14(d), the shear strain at the integration points is not zero and a considerable amount of energy is absorbed by the fake shearing phenomenon, not leaving enough energy for bending: the displacements are too small. The problem can be alleviated by using reduced integration ($1 \times 1 \times 1$ integration point) as shown in Figure 2.14(e): the shear strain at the integration point is zero and the correct displacements result.

2.5.2 Volumetric locking

The problem of volumetric locking occurs for incompressible or nearly incompressible behavior. It can be explained using the example in Figure 2.15. Element 1 is fixed alongside 1–2 and 1–4. It is a standard two-dimensional quadrilateral element. The material is assumed to be incompressible. Accordingly, $J = 1$ everywhere. Since $u_1 = u_2 = u_4 = v_1 = v_2 = v_4 = 0$, the displacements in the element amount to (reduce the three-dimensional shape function in Equation (2.39) to the present two-dimensional case)

$$\begin{aligned} u &= \frac{1}{4}(1+\xi)(1+\eta)u_3 \\ v &= \frac{1}{4}(1+\xi)(1+\eta)v_3. \end{aligned} \tag{2.134}$$

For simplicity, the local and global coordinates are assumed to coincide, that is, $x_{,\xi} = y_{,\eta} = 1$, $x_{,\eta} = y_{,\xi} = 0$. Hence,

$$\begin{aligned} u_{,\xi} &= \frac{1}{4}(1+\eta)u_3 \\ u_{,\eta} &= \frac{1}{4}(1+\xi)u_3 \\ v_{,\xi} &= \frac{1}{4}(1+\eta)v_3 \\ v_{,\eta} &= \frac{1}{4}(1+\xi)v_3. \end{aligned} \tag{2.135}$$

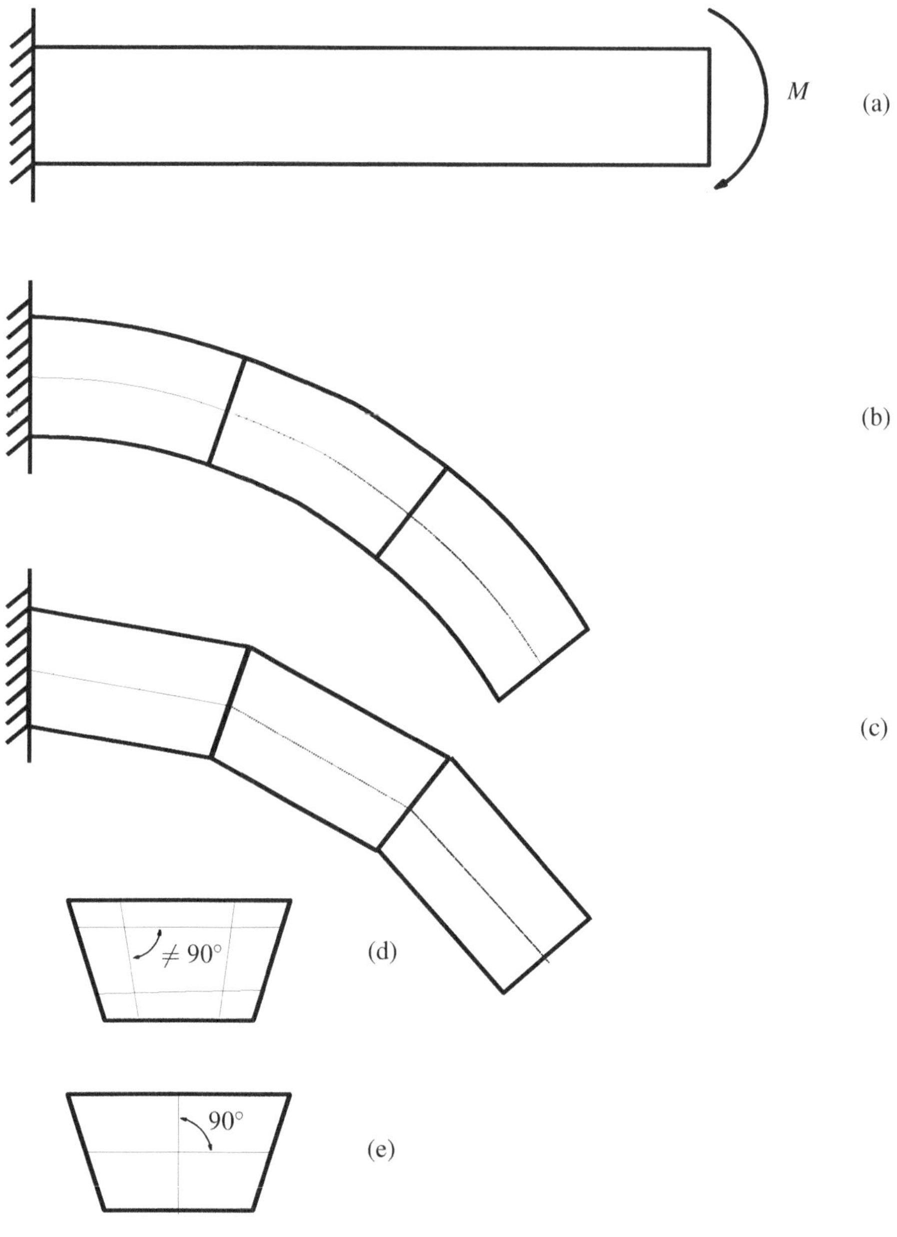

Figure 2.14 The shear locking phenomenon

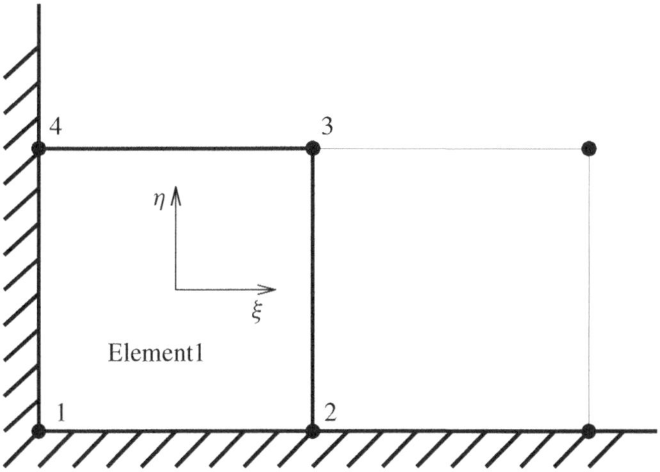

Figure 2.15 Locking behavior in corner elements

Since

$$x^k_{,K} = (X^L \delta^k{}_L + u^k)_{,K} = \delta^k{}_K + u^k_{,K} \tag{2.136}$$

and

$$J = \det(x^k_{,K}) \tag{2.137}$$

one finds

$$J = 1 + \frac{1}{4}(1+\eta)u_3 + \frac{1}{4}(1+\xi)v_3. \tag{2.138}$$

The requirement $J = 1$ (incompressibility) amounts to

$$\frac{1}{4}(1+\eta)u_3 + \frac{1}{4}(1+\xi)v_3 = 0. \tag{2.139}$$

If we take full integration, Equation (2.139) has to be satisfied at $\xi, \eta = \pm 0.57$:

$$\begin{aligned} 0.39u_3 + 0.39v_3 &= 0 \\ 0.11u_3 + 0.39v_3 &= 0 \\ 0.39u_3 + 0.11v_3 &= 0 \\ 0.11u_3 + 0.11v_3 &= 0 \end{aligned} \tag{2.140}$$

which can only be satisfied if $u_3 = v_3 = 0$. This results in a zero-deformation field for the complete element: the element locks. The same argument can be repeated for the neighboring elements (dashed line in Figure 2.15). If, on the other hand, reduced integration is applied, there is only one integration point at $\xi = \eta = 0$. Now, there is only one equation to satisfy:

$$u_3 + v_3 = 0 \tag{2.141}$$

and no locking occurs. Therefore, it is often advantageous to use reduced integration to avoid locking.

Another option is to use hybrid elements (Zienkiewicz and Taylor 1989), in which the pressure is considered as an additional independent variable. Indeed, the problem is that for truly incompressible behavior the pressure cannot be derived from the displacement field, since increasing the hydrostatic pressure does not lead to a change in the displacement field. The resulting hybrid elements are a special case of what are now called *assumed stress* and *assumed strain* elements. In these elements, the stresses and/or strains are interpolated independent of the displacements. They require the application of multifield variational principles such as the Hu–Washizu weak form. Interested readers are referred to (Belytschko *et al.* 2000).

2.5.3 Hourglassing

Hourglassing implies the existence of zero-energy modes: these are displacement modes that do not lead to any strain or stress at the integration points. Since the field values at the integration points are the only ones entering the integration scheme (Equation (2.106)), hourglass modes can be added at will without disturbing the equilibrium condition (Equation (2.1)). This usually results in wildly varying displacement fields but correct stress and strain fields. Reducing the number of integration points naturally increases the number of hourglassing modes. A linear brick element with reduced integration has one integration point where six strain components prevail. However, the same element has 8 nodes leading to 24 degrees of freedom. Accordingly, 18 undetermined modes are left of which 6 are rigid body modes leading to 12 hourglass modes in total. Figure 2.16 shows how hourglassing for a beam under bending might look like.

To get rid of hourglassing, several stabilization methods such as the introduction of artificial stiffness and the enhanced strain method have been proposed. For details, the reader is referred to (Belytschko *et al.* 2000), (Belytschko and Bindeman 1993), (Puso 2000), (Reese and Wriggers 2000) and (Reese 2003a). Notice that the solution in Figure 2.14(e) to eliminate shear locking works because the shear deformation is an hourglass mode for this one integration point. This shows that the locking phenomena and hourglassing are intimately related, (see also (Reese 2002) for more information).

For the 20-node brick element with reduced integration, there are only (3 degrees of freedom) × (20 nodes) − (8 integration points) × (6 strain components) − (6 rigid body modes) = 6 hourglass modes. Because of the quadratic shape of the element sides, these modes cannot propagate through the mesh. Consequently, hourglassing in these elements is rare. In fully integrated brick elements, in tetrahedral elements and wedge elements, hourglassing cannot occur.

Figure 2.16 Hourglassing in a cantilever beam

2.6 Linear Constraints

In addition to the equilibrium equations expressed by Equation (2.27), the solution of a field problem requires the formulation of boundary conditions. One distinguishes geometric boundary conditions, which contain the independent variables such as displacements and temperatures, from natural boundary conditions, which are formulated in terms of the dependent variables such as stress and heat flux. In this section, the focus is on geometric boundary conditions. If only one degree of freedom is involved, the constraint is called a *single point constraint*, else it is a *multiple point constraint*. Constraints can be linear or nonlinear. Examples of truly linear multiple point constraints are the constraint of a degree of freedom in nonrectangular coordinates, cyclic symmetry conditions and equations connecting dissimilar meshes. Examples for nonlinear multiple point constraints are constraints involving finite rotations such as rigid body motions and incompressibility conditions.

2.6.1 Inclusion in the global system of equations

The multiple point constraint concept is extremely powerful. Therefore, a truly efficient way must be found to deal with it numerically. One option is to augment the matrix with the additional equations using Lagrangian multipliers. This, however, leads to larger systems of equations, increased computational times and a nonsymmetric system. This can be avoided by eliminating one degree of freedom per single point constraint or per multiple point constraint during the construction of the global stiffness matrix. How this is done will be explained in the present section.

Assume that our global system of equations contains N degrees of freedom. Equation l reads

$$\sum_{j=1}^{N} a_{lj} u_j = b_l, \quad l = 1, \dots, N. \tag{2.142}$$

Assume we have M additional multiple point constraints of the form

$$\sum_{k=1}^{N} e_{ik} u_k = f_i, \quad i = 1, \dots, M. \tag{2.143}$$

In each of these equations i, we choose one degree of freedom k_i, which is suited to be eliminated and which will be called a *dependent degree of freedom*. All dependent degrees of freedom $k_i, i = 1, \dots, M$ must be distinct. Now, Equation (2.143) can be rearranged by collecting all dependent degrees of freedom on the left-hand side:

$$\sum_{j=1}^{M} e_{ik_j} u_{k_j} = f_i - \sum_{\substack{k=1 \\ k \notin \{k_1, \dots, k_M\}}}^{N} e_{ik} u_k, \quad i = 1, \dots, M. \tag{2.144}$$

This is a set of M equations in M unknowns and can be solved for the dependent degrees of freedom provided that the equations on the left-hand side are linearly independent. It results in equations of the form

$$u_{k_j} = \sum_{\substack{k=1 \\ k \notin \{k_1, \dots, k_M\}}}^{N} c_{k_j k} u_k + d_{k_j}. \tag{2.145}$$

Now, assume that there is only one multiple point constraint. Equation (2.145) reduces to

$$u_i = \sum_{\substack{k=1 \\ k \neq i}}^{N} c_{ik} u_k + d_i \tag{2.146}$$

which allows us to eliminate u_i. Equation (2.146) is of a most general form including single point constraints (all $c_{ik} = 0$). Substituting Equation (2.146) into Equation (2.142) yields (no implicit summation in the section)

$$\sum_{\substack{j=1 \\ j \neq i}}^{N} (a_{lj} + a_{li} c_{ij}) u_j = b_l - a_{li} d_i, \quad l = 1, \dots, N. \tag{2.147}$$

The new coefficient a_{lj}° in the global matrix at position (l, j) now reads

$$a_{lj}^{\circ} = a_{lj} + a_{li} c_{ij}, \quad j, l = 1, \dots, N; j \neq i. \tag{2.148}$$

The global stiffness matrix is Hermitian (complex matrices arise because of cyclic symmetry conditions, cf Section 2.10; if the matrix is real, "Hermitian" can be replaced by "symmetric") and consequently,

$$\overline{a_{lj}} = a_{jl}, \quad j, l = 1, \ldots, N \tag{2.149}$$

where $\overline{a}$ stands for the complex conjugate of a. However, for the new coefficient we have

$$\overline{a^{\circ}_{jl}} = \overline{a_{jl}} + \overline{a_{ji}}\ \overline{c_{il}} \neq a^{\circ}_{lj}, \quad j, l = 1, \ldots, N;\ j \neq i. \tag{2.150}$$

The Hermitian structure is destroyed! This is a serious drawback since it means that computational advantages due to the Hermitian structure are lost. The Hermitian structure, however, can be restored by multiplying row i, which reads

$$\sum_{\substack{j=1\\ j\neq i}}^{N} (a_{ij} + a_{ii}c_{ij})u_j = b_i - a_{ii}d_i \tag{2.151}$$

by $\overline{c_{im}}$ and adding it to row m, $m = 1, \ldots, N, m \neq i$. Now, coefficient a^*_{lj} at position (l, j) reads

$$a^*_{lj} = a_{lj} + a_{li}c_{ij} + a_{ij}\overline{c_{il}} + a_{ii}c_{ij}\overline{c_{il}}, \quad j, l = 1, \ldots, N;\ j, l \neq i \tag{2.152}$$

which coincides with the complex conjugate of

$$a^*_{jl} = a_{jl} + a_{ji}c_{il} + a_{il}\overline{c_{ij}} + a_{ii}c_{il}\overline{c_{ij}}, \quad j, l = 1, \ldots, N;\ j, l \neq i \tag{2.153}$$

which reads

$$\overline{a^*_{jl}} = \overline{a_{jl}} + \overline{a_{ji}}\ \overline{c_{il}} + \overline{a_{il}}c_{ij} + \overline{a_{ii}}\ \overline{c_{il}}c_{ij}, \quad j, l = 1, \ldots, N;\ j, l \neq i. \tag{2.154}$$

The right-hand side coefficient b^*_l satisfies

$$b^*_l = b_l - a_{li}d_i + b_i\overline{c_{il}} - a_{ii}d_i\overline{c_{il}}, \quad l = 1, \ldots, N;\ l \neq i. \tag{2.155}$$

Row i and column i are dropped altogether from the set.

What happens if two multiple point constraints apply? Let us say that there is a second multiple point constraint of the form

$$u_m = \sum_{\substack{k=1\\ k\neq m}}^{N} c_{mk}u_k + d_m, \quad m \neq i. \tag{2.156}$$

We assume that both constraints were brought in the form of Equation (2.145) such that $c_{mi} = c_{im} = 0$. Consequently,

$$u_i = \sum_{\substack{k=1\\ k\neq i, k\neq m}}^{N} c_{ik}u_k + d_i \tag{2.157}$$

$$u_m = \sum_{\substack{k=1\\ k\neq m, k\neq i}}^{N} c_{mk}u_k + d_m. \tag{2.158}$$

Applying Equation (2.152) twice, first to eliminate multiple point constraint m, and then to eliminate multiple point constraint i, one obtains

$$a_{lj}^{**} = a_{lj}^{*} + a_{lm}^{*} c_{mj} + a_{mj}^{*} \overline{c_{ml}} + a_{mm}^{*} c_{mj} \overline{c_{ml}} \tag{2.159}$$

$$\begin{aligned} &= a_{lj} + a_{li} c_{ij} + a_{ij} \overline{c_{il}} + a_{ii} c_{ij} \overline{c_{il}} \\ &+ (a_{lm} + a_{li} c_{im} + a_{im} \overline{c_{il}} + a_{ii} c_{im} \overline{c_{il}}) c_{mj} \\ &+ (a_{mj} + a_{mi} c_{ij} + a_{ij} \overline{c_{im}} + a_{ii} c_{ij} \overline{c_{im}}) \overline{c_{ml}} \\ &+ (a_{mm} + a_{mi} c_{im} + a_{im} \overline{c_{im}} + a_{ii} c_{im} \overline{c_{im}}) c_{mj} \overline{c_{ml}} \end{aligned} \tag{2.160}$$

$$j, l = 1, \ldots, N; \; j, l \neq i; \; j, l \neq m;$$

or taking into account that $c_{im} = 0$

$$\begin{aligned} a_{lj}^{**} &= a_{lj} + (a_{li} c_{ij} + a_{lm} c_{mj}) \\ &+ (a_{ij} \overline{c_{il}} + a_{mj} \overline{c_{ml}}) \\ &+ (a_{im} c_{mj} \overline{c_{il}} + a_{mi} c_{ij} \overline{c_{ml}}) \\ &+ (a_{ii} c_{ij} \overline{c_{il}} + a_{mm} c_{mj} \overline{c_{ml}}). \end{aligned} \tag{2.161}$$

$$j, l = 1, \ldots, N; \; j, l \neq i; \; j, l \neq m.$$

In a similar way one obtains for the right-hand side

$$b_{l}^{**} = b_{l}^{*} - a_{lm}^{*} d_{m} + b_{m}^{*} \overline{c_{ml}} - a_{mm}^{*} d_{m} \overline{c_{ml}} \tag{2.162}$$

$$\begin{aligned} &= b_{l} - a_{li} d_{i} + b_{i} \overline{c_{il}} - a_{ii} d_{i} \overline{c_{il}} \\ &- (a_{lm} + a_{li} c_{im} + a_{im} \overline{c_{il}} + a_{ii} c_{im} \overline{c_{il}})\, d_{m} \\ &+ (b_{m} - a_{mi} d_{i} + b_{i} \overline{c_{im}} - a_{ii} d_{i} \overline{c_{im}}) \overline{c_{ml}} \\ &- (a_{mm} + a_{mi} c_{im} + a_{im} \overline{c_{im}} + a_{ii} c_{im} \overline{c_{im}})\, d_{m} \overline{c_{ml}} \end{aligned} \tag{2.163}$$

$$l = 1, \ldots, N; \; l \neq i, m;$$

or, taking into account that $c_{im} = 0$

$$\begin{aligned} b_{l}^{**} &= b_{l} + (b_{i} \overline{c_{il}} + b_{m} \overline{c_{ml}}) \\ &- (a_{ii} d_{i} \overline{c_{il}} + a_{mm} d_{m} \overline{c_{ml}}) \\ &- (a_{li} d_{i} + a_{lm} d_{m} + a_{im} \overline{c_{il}} d_{m} + a_{mi} \overline{c_{ml}} d_{i}). \end{aligned} \tag{2.164}$$

$$l = 1, \ldots, N; \; l \neq i, m.$$

Equations (2.161) and (2.164) cover all possibilities for coefficients a and b. Indeed, in a finite element code the element matrices are calculated first, Equations (2.21) to (2.24). Then, these matrices are transferred into the global matrix. This operation is symbolized by Equations (2.28) to (2.30) and will be looked at in more detail now.

Suppose an entry a in the local matrix corresponds to global degrees of freedom p and q (row and column). Now, the following possibilities arise:

1. p and q are independent degrees of freedom. Then

$$a_{pq} + = a \tag{2.165}$$

where the C-notation $+=$ was used to indicate that the global matrix entry a_{pq} is to be augmented by a.

2. p is a dependent degree of freedom, q is not. Accordingly, row p in the global matrix is eliminated, column q is not , and a has a similar status as a_{ij} in the term $a_{ij}\overline{c_{il}}$ in Equation (2.152). Since row p is eliminated, the contribution of a is transferred to whatever position a_{ij} in Equation (2.152) has gone to, now substituting p for i and q for j

$$a_{lq} + = a\overline{c_{pl}}. \tag{2.166}$$

This applies to all degrees of freedom l for which $\overline{c_{pl}} \neq 0$. Because of Equation (2.145) these are only independent degrees of freedom.

3. q is a dependent degree of freedom, p is not. This contribution is comparable to $a_{li}c_{ij}$ in Equation (2.152) and $-a_{li}d_i$ in Equation (2.155). Now we have

$$a_{pj} + = ac_{qj} \quad \forall j \text{ such that } c_{qj} \neq 0 \tag{2.167}$$

and

$$b_p + = -ad_q. \tag{2.168}$$

4. p and q are dependent degrees of freedom, $p \neq q$, cf $a_{im}\overline{c_{il}}c_{mj}$ in Equation (2.161) and $a_{im}\overline{c_{il}}d_m$ in Equation (2.164):

$$a_{lj} + = a\overline{c_{pl}}c_{qj} \quad \forall l, j \text{ such that } \overline{c_{pl}} \neq 0 \text{ and } c_{qj} \neq 0 \tag{2.169}$$

and

$$b_l + = -a\overline{c_{pl}}d_q \quad \forall l \text{ such that } \overline{c_{pl}} \neq 0. \tag{2.170}$$

5. p and q are dependent degrees of freedom, $p = q$, cf $a_{ii}c_{ij}\overline{c_{il}}$ in Equation (2.152) and $-a_{ii}d_i\overline{c_{il}}$ in Equation (2.155):

$$a_{lj} + = ac_{pj}\overline{c_{pl}} \quad \forall l, j \text{ such that } \overline{c_{pl}} \neq 0, c_{pj} \neq 0 \tag{2.171}$$

$$b_l + = -ad_p\overline{c_{pl}} \quad \forall l \text{ such that } \overline{c_{pl}} \neq 0. \tag{2.172}$$

For an entry b in the local right-hand side corresponding to global degree of freedom p, there are only two possibilities:

1. p is an independent degree of freedom

$$b_p + = b. \tag{2.173}$$

2. p is a dependent degree of freedom, cf $b_i \overline{c_{il}}$ in Equation (2.155):

$$b_l + = b\overline{c_{pl}} \quad \forall l \text{ such that } \overline{c_{pl}} \neq 0. \tag{2.174}$$

In this way, all entries in the local matrices are transferred to independent degrees of freedom in the global matrices. Consequently, the elimination of degrees of freedom involved in multiple point constraints is taken care of implicitly while transferring the local matrices into the global ones.

Notice that to each entry a in row i_p and column i_q in the element stiffness matrix corresponding to global degrees of freedom p and q, respectively, there is a symmetric entry with the same value a in row i_q and column i_p. If p is a dependent degree of freedom and q is an independent degree of freedom, Equation (2.166) applies to entry a in row i_p and column i_q:

$$a_{lq} + = a\,\overline{c_{pl}} \tag{2.175}$$

and Equation (2.167) applies to entry a in row i_q and column i_p:

$$a_{ql} + = a c_{pl} \tag{2.176}$$

which keeps the Hermitian structure of the global matrix. In practice, only half of the Hermitian matrix is calculated and stored.

2.6.2 Forces induced by linear constraints

The introduction of multiple point constraints induces forces. Indeed, imagine a constraint of the form

$$u_i = u_j. \tag{2.177}$$

Then both degrees of freedom are coupled and behave as if a rigid bar connects both: degree of freedom i will experience a force F, degree of freedom j will experience the inverse force. How does this translate to general multiple point constraints of the form in Equation (2.146)? Recall that the global system contains N degrees of freedom leading to an $N \times N$ stiffness matrix. Adding Equation (2.146) leads to $N + 1$ equations in N unknowns. After substitution into the global set, column i was eliminated leading to N equations in $N - 1$ unknowns. This is still an overdetermined system and generally has no solution. This problem was solved by adding multiples of row i to the other rows and deleting row i afterward. Consequently, row i is not being satisfied. Indeed, the residual of row i is exactly the multiple point constraint force we are looking for (no implicit summation in this section):

$$\sum_{\substack{j=1 \\ j\neq i}}^{N} (a_{ij} + a_{ii}c_{ij})u_j - (b_i - a_{ii}d_i) = F_i. \tag{2.178}$$

Accordingly, the addition of multiples of row i to other rows is equivalent to the addition of multiples of F_i. Indeed, row l reads

$$\sum_{\substack{j=1 \\ j\neq i}}^{N} (a_{lj} + a_{li}c_{ij})u_j + \left[\sum_{\substack{j=1 \\ j\neq i}}^{N} (a_{ij} + a_{ii}c_{ij})u_j\right]\overline{c_{il}} = (b_l - a_{li}d_i) + (b_i - a_{ii}d_i)\overline{c_{il}} \tag{2.179}$$

or

$$\sum_{\substack{j=1 \\ j\neq i}}^{N} (a_{lj} + a_{li}c_{ij})u_j - (b_l - a_{li}d_i) = -F_i\overline{c_{il}} \tag{2.180}$$

and degree of freedom l experiences the force

$$F_l = -F_i\overline{c_{il}}. \tag{2.181}$$

Notice that the force is proportional to the conjugate coefficient in the multiple point constraint. For instance, if the multiple point constraint reads

$$u_i = 2u_j \tag{2.182}$$

degree of freedom j experiences a force that is twice the force acting on degree of freedom i.

2.7 Transformations

Transformations are an important tool for the finite element practitioner. For instance, if a structure exhibits cylindrical symmetry, boundary conditions are more easily formulated in a cylindrical coordinate system than in the global rectangular system. Another important application is the definition of anisotropic material properties in cases in which the material axes do not coincide with the global axes. In all these instances, it is advantageous to introduce a local coordinate system. Here, we will concentrate on local rectangular and local cylindrical systems. Both systems are orthogonal, that is, the covariant and contravariant unit vectors coincide. If

$$\boldsymbol{I}_{I'} := \boldsymbol{G}_{\underline{I'}}/\sqrt{G_{\underline{I'I'}}} \tag{2.183}$$

and similarly for the contravariant base vectors, one can write

$$\boldsymbol{I}_{I'} = \boldsymbol{I}^{I'}, \quad I' = 1, 2, 3 \tag{2.184}$$

$$\boldsymbol{G}_{I'} \cdot \boldsymbol{G}_{J'} = \boldsymbol{I}_{I'} \cdot \boldsymbol{I}_{J'} = 0, \quad I' \neq J'. \tag{2.185}$$

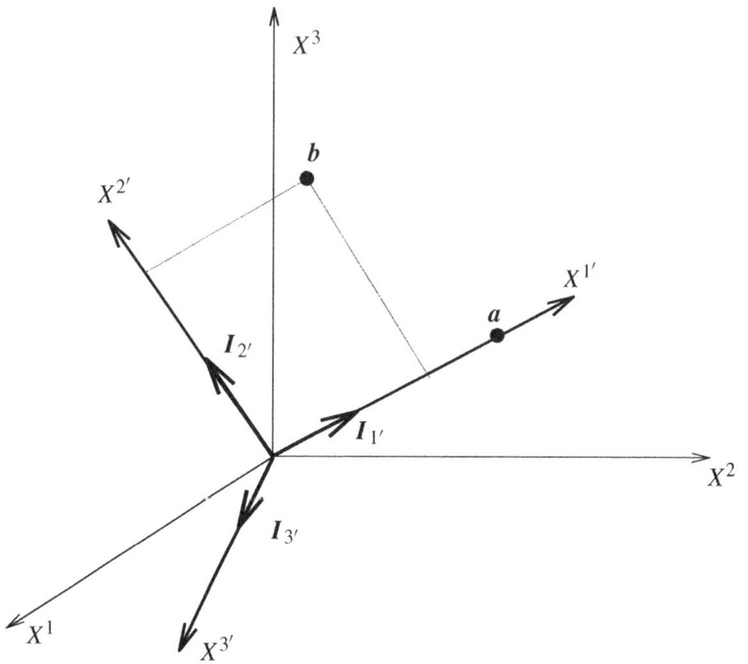

Figure 2.17 Local rectangular system

Let us characterize the global rectangular coordinate system by unit vectors $\boldsymbol{I}_1$, $\boldsymbol{I}_2$, $\boldsymbol{I}_3$ and coordinates X^1, X^2 and X^3.

A local rectangular coordinate system $X^{1'}$-$X^{2'}$-$X^{3'}$ can be defined by a point $\boldsymbol{a}$ on the $X^{1'}$-axis and a second point $\boldsymbol{b}$ within the $X^{1'}$-$X^{2'}$ plane excluding the $X^{1'}$-axis (Figure 2.17). For transformation purposes, it is important to determine unit base vectors in the local coordinate system. The unit vector along the $X^{1'}$-axis is easily determined

$$\boldsymbol{I}_{1'} = \frac{\boldsymbol{a}}{\|\boldsymbol{a}\|}. \tag{2.186}$$

A vector on the $X^{2'}$-axis can be found by moving $\boldsymbol{b}$ in direction $\boldsymbol{I}_{1'}$ such that the resulting vector is orthogonal to $\boldsymbol{I}_{1'}$:

$$(\boldsymbol{b} + \lambda \boldsymbol{I}_{1'}) \perp \boldsymbol{I}_{1'} \tag{2.187}$$

or

$$(\boldsymbol{b} + \lambda \boldsymbol{I}_{1'}) \cdot \boldsymbol{I}_{1'} = 0 \Rightarrow \lambda = -\boldsymbol{b} \cdot \boldsymbol{I}_{1'}. \tag{2.188}$$

Consequently,

$$\boldsymbol{I}_{2'} = \frac{\boldsymbol{b} - (\boldsymbol{b} \cdot \boldsymbol{I}_{1'})\boldsymbol{I}_{1'}}{\|\boldsymbol{b} - (\boldsymbol{b} \cdot \boldsymbol{I}_{1'})\boldsymbol{I}_{1'}\|}. \tag{2.189}$$

Finally,

$$I_{3'} = I_{1'} \times I_{2'} \tag{2.190}$$

where $\times$ symbolizes the vector product.

A local cylindrical coordinate system can be defined by two points on the cylindrical axis (Figure 2.18). A local cylindrical system is also orthogonal, that is, the three unit vectors are perpendicular to each other. However, the orientation of the local unit vectors varies in space. In the finite element code CalculiX® (CalculiX 2003), the first unit vector is in radial direction, the second in tangential direction and the third in axial direction. Let us determine a set of unit vectors in point $\boldsymbol{p}$. From Figure 2.18 we have

$$\boldsymbol{I}_{3'} = \frac{\boldsymbol{b} - \boldsymbol{a}}{\|\boldsymbol{b} - \boldsymbol{a}\|}. \tag{2.191}$$

Point $\boldsymbol{q}$ is a point on the axis such that

$$(\boldsymbol{p} - \boldsymbol{q}) \perp (\boldsymbol{b} - \boldsymbol{a}) \tag{2.192}$$

or, since a point on the axis can be written as $\boldsymbol{a} + \lambda \boldsymbol{I}_{3'}$, $\lambda \in R$

$$(\boldsymbol{p} - \boldsymbol{a} - \lambda \boldsymbol{I}_{3'}) \cdot \boldsymbol{I}_{3'} = 0 \tag{2.193}$$

from which

$$\lambda = (\boldsymbol{p} - \boldsymbol{a}) \cdot \boldsymbol{I}_{3'}. \tag{2.194}$$

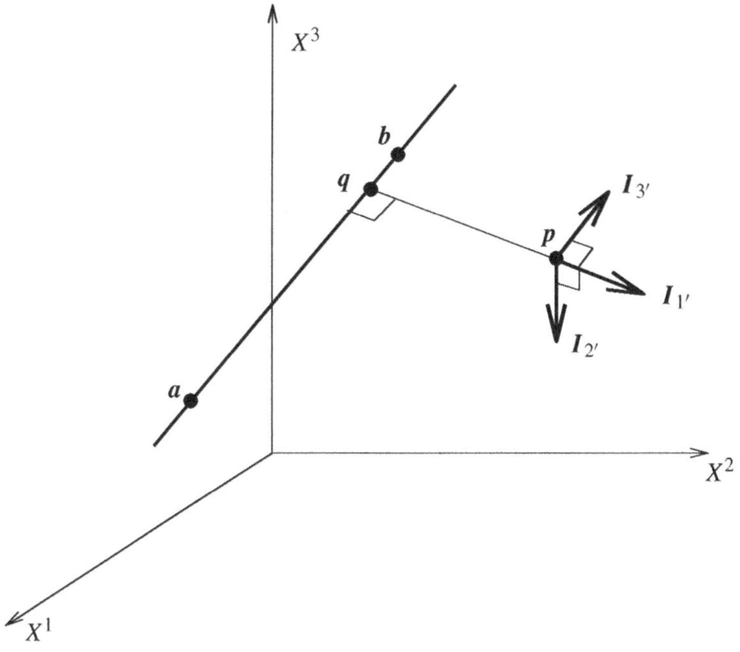

Figure 2.18 Local cylindrical system

Accordingly,

$$\boldsymbol{p} - \boldsymbol{q} = (\boldsymbol{p} - \boldsymbol{a}) - [(\boldsymbol{p} - \boldsymbol{a}) \cdot \boldsymbol{I}_{3'}]\boldsymbol{I}_{3'} \tag{2.195}$$

and

$$\boldsymbol{I}_{1'} = \frac{\boldsymbol{p} - \boldsymbol{q}}{\|\boldsymbol{p} - \boldsymbol{q}\|}. \tag{2.196}$$

If $\boldsymbol{p}$ is on the axis, $\|\boldsymbol{p} - \boldsymbol{q}\| = 0$ and Equation (2.196) cannot be applied. In that case, any direction perpendicular to $\boldsymbol{I}_{3'}$ can be taken for $\boldsymbol{I}_{1'}$. Finally,

$$\boldsymbol{I}_{2'} = \boldsymbol{I}_{3'} \times \boldsymbol{I}_{1'}. \tag{2.197}$$

This concludes the determination of local unit vectors for rectangular and cylindrical systems.

An arbitrary vector $\boldsymbol{p}$ can be expressed as a function of $\boldsymbol{I}_1$, $\boldsymbol{I}_2$ and $\boldsymbol{I}_3$ or $\boldsymbol{I}_{1'}$, $\boldsymbol{I}_{2'}$ and $\boldsymbol{I}_{3'}$:

$$\boldsymbol{p} = X^1\boldsymbol{I}_1 + X^2\boldsymbol{I}_2 + X^3\boldsymbol{I}_3 \tag{2.198}$$

$$= X^{1'}\boldsymbol{I}_{1'} + X^{2'}\boldsymbol{I}_{2'} + X^{3'}\boldsymbol{I}_{3'}. \tag{2.199}$$

Taking the scalar product of Equation (2.198) with $\boldsymbol{I}^{1'}$ we arrive at

$$\boldsymbol{p} \cdot \boldsymbol{I}^{1'} = X^1(\boldsymbol{I}_1 \cdot \boldsymbol{I}^{1'}) + X^2(\boldsymbol{I}_2 \cdot \boldsymbol{I}^{1'}) + X^3(\boldsymbol{I}_3 \cdot \boldsymbol{I}^{1'}) = X^{1'} \tag{2.200}$$

and similarly,

$$X^1(\boldsymbol{I}_1 \cdot \boldsymbol{I}^{2'}) + X^2(\boldsymbol{I}_2 \cdot \boldsymbol{I}^{2'}) + X^3(\boldsymbol{I}_3 \cdot \boldsymbol{I}^{2'}) = X^{2'} \tag{2.201}$$

$$X^1(\boldsymbol{I}_1 \cdot \boldsymbol{I}^{3'}) + X^2(\boldsymbol{I}_2 \cdot \boldsymbol{I}^{3'}) + X^3(\boldsymbol{I}_3 \cdot \boldsymbol{I}^{3'}) = X^{3'}. \tag{2.202}$$

Notice that we have multiplied $\boldsymbol{p}$ by the contravariant unit vectors in the local coordinate system, which, for rectangular and cylindrical coordinate systems happen to coincide with the covariant unit vectors. Equations (2.200) to (2.202) can also be written as

$$X^{K'} = Q^{K'}_{\ L}X^L \tag{2.203}$$

where

$$Q^{K'}_{\ L} = \boldsymbol{I}^{K'} \cdot \boldsymbol{I}_L. \tag{2.204}$$

In a completely similar way, one arrives at

$$X^K = T^K_{\ L'}X^{L'} \tag{2.205}$$

where

$$T^K_{\ L'} = \boldsymbol{I}^K \cdot \boldsymbol{I}_{L'}. \tag{2.206}$$

$\boldsymbol{T}$ is the inverse of $\boldsymbol{Q}$, that is, $\left[Q^{K'}_{\ L}\right]^{-1} = \left[T^{L}_{\ K'}\right]$. For orthogonal systems, where covariant and contravariant unit vectors coincide, one can write

$$Q_{K'L} = \boldsymbol{I}_{K'} \cdot \boldsymbol{I}_L \tag{2.207}$$

$$T_{LK'} = \boldsymbol{I}_L \cdot \boldsymbol{I}_{K'} \tag{2.208}$$

which, in terms of matrix operations, means

$$\boldsymbol{T} = \boldsymbol{Q}^{\mathrm{T}}. \tag{2.209}$$

Consequently,

$$\boldsymbol{Q}^{-1} = \boldsymbol{Q}^{\mathrm{T}} \tag{2.210}$$

that is, $\boldsymbol{Q}$ is an orthogonal matrix. Contravariant vectors satisfy

$$\begin{aligned} \boldsymbol{U} &= U^L \boldsymbol{I}_L \\ &= U^L \frac{\partial X^{K'}}{\partial X^L} \boldsymbol{G}_{K'} \\ &= \sum_{K'=1}^{3} U^L \frac{\partial X^{\underline{K'}}}{\partial X^L} \sqrt{G_{\underline{K'}\underline{K'}}}\, \boldsymbol{I}_{\underline{K'}} \\ &= U^{K'} \boldsymbol{I}_{K'} \end{aligned} \tag{2.211}$$

from which one obtains

$$U^{K'} = U^L \frac{\partial X^{\underline{K'}}}{\partial X^L} \sqrt{G_{\underline{K'}\underline{K'}}}. \tag{2.212}$$

Since (Equation (2.203))

$$U^{K'} = Q^{K'}_{\ L} U^L \tag{2.213}$$

one finds

$$Q^{K'}_{\ L} = \frac{\partial X^{\underline{K'}}}{\partial X^L} \sqrt{G_{\underline{K'}\underline{K'}}}. \tag{2.214}$$

For covariant tensors, we have

$$C_{K'L'} = C_{MN} \frac{\partial X^M}{\partial X^{\underline{K'}}} \frac{\partial X^N}{\partial X^{\underline{L'}}} \sqrt{G^{\underline{K'}\underline{K'}}} \sqrt{G^{\underline{L'}\underline{L'}}} \tag{2.215}$$

$$= C_{MN} T^{M}_{\ K'} T^{N}_{\ L'}. \tag{2.216}$$

Notice that $\boldsymbol{Q}$ and $\boldsymbol{T}$ are not symmetric.

If boundary conditions or material orientations are expressed in local coordinate systems, they have to be transformed into the global system used to formulate Equation (2.27), usually a global rectangular system. In practice, the following situations occur:

1. Single point constraints are formulated in local coordinates:

$$U^{K'} = a \tag{2.217}$$

This is equivalent to

$$Q^{K'}_{\ L} U^L = a \tag{2.218}$$

that is, an inhomogeneous single point constraint in local coordinates is transformed into an inhomogeneous multiple point constraint in global coordinates.

2. Multiple point constraints are formulated in local coordinates. The same procedure applies as under item 1: a homogeneous multiple point constraint in local coordinates is transformed in a (usually longer) homogeneous multiple point constraint in global coordinates, an inhomogeneous multiple point constraint in local coordinates is expanded into an inhomogeneous multiple point constraint in global coordinates.

3. Forces are given in local coordinates:

$$F^{K'} = f. \tag{2.219}$$

This transforms into

$$Q^{K'}_{\ L} F^L = f \tag{2.220}$$

or

$$F^L = T^L_{\ K'} f. \tag{2.221}$$

Consequently, a force with one nonzero local component in direction K' generally results in three nonzero force components L in global coordinates.

4. The material orientation is given in local coordinates. The tangent stiffness matrix, which is a generalization of the elasticity matrix, satisfies

$$\mathrm{d}S^{KL} = \Sigma^{KLMN}\,\mathrm{d}E_{MN}. \tag{2.222}$$

Since

$$\mathrm{d}S^{P'Q'} = \mathrm{d}S^{KL} Q^{P'}_{\ K} Q^{Q'}_{\ L} \tag{2.223}$$

$$\mathrm{d}E_{MN} = \mathrm{d}E_{R'S'} Q^{R'}_{\ M} Q^{S'}_{\ N}. \tag{2.224}$$

Equation (2.222) can be transformed into

$$\mathrm{d}S^{P'Q'} = \Sigma^{KLMN} Q^{P'}_{\ K} Q^{Q'}_{\ L} Q^{R'}_{\ M} Q^{S'}_{\ N}\,\mathrm{d}E_{R'S'}. \tag{2.225}$$

Hence,

$$\Sigma^{P'Q'R'S'} = \Sigma^{KLMN} Q^{P'}_{\ K} Q^{Q'}_{\ L} Q^{R'}_{\ M} Q^{S'}_{\ N}. \tag{2.226}$$

Similar relationships apply to other tensors such as the matrix of expansion coefficients.

In the CalculiX® code, (CalculiX 2003) all quantities expressed in local coordinates are internally transformed into global coordinates using the previous relationships.

2.8 Loading

The loading essentially consists of the terms in Equation (2.23). Here, we focus on centrifugal loading and temperature loading.

2.8.1 Centrifugal loading

Centrifugal loading is a body force that is activated when a body rotates at an angular speed ω about an axis. The force in a point $\boldsymbol{q}$ is proportional to the distance from the axis and the square of the angular speed and is directed away from and orthogonal to the axis (Figure 2.19).

Consider two points on the axis $\boldsymbol{p}_1$ and $\boldsymbol{p}_2$. Let

$$\boldsymbol{e} := \frac{\boldsymbol{p}_2 - \boldsymbol{p}_1}{\|\boldsymbol{p}_2 - \boldsymbol{p}_1\|} \tag{2.227}$$

be a unit vector on the axis. Then, the point $\boldsymbol{p}$ obtained by dropping $\boldsymbol{q}$ orthogonally on the axis satisfies

$$\boldsymbol{p} = \boldsymbol{p}_1 + [(\boldsymbol{q} - \boldsymbol{p}_1) \cdot \boldsymbol{e}]\boldsymbol{e}. \tag{2.228}$$

Accordingly, the centrifugal force in $\boldsymbol{q}$ satisfies

$$\boldsymbol{f} = (\boldsymbol{q} - \boldsymbol{p})\omega^2 \tag{2.229}$$

$$= \{(\boldsymbol{q} - \boldsymbol{p}_1) - [(\boldsymbol{q} - \boldsymbol{p}_1) \cdot \boldsymbol{e}]\boldsymbol{e}\}\,\omega^2. \tag{2.230}$$

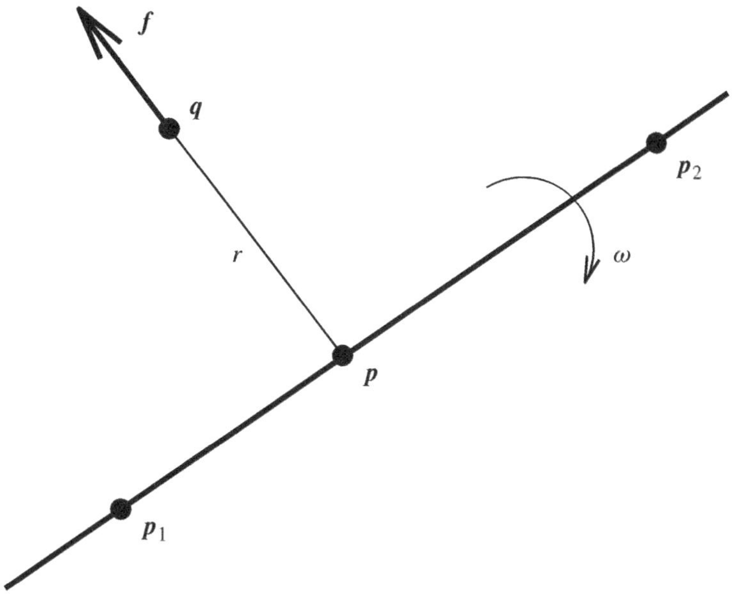

Figure 2.19 Definition of the centrifugal axis

The components of f in the reference configuration, satisfying

$$f = f^K \boldsymbol{G}_K, \tag{2.231}$$

can be directly substituted in Equation (2.23).

2.8.2 Temperature loading

Temperature loading acts as residual stress and corresponds to the term $\beta^{KL}(\theta)T$ in Equation (2.23). Indeed, imagine you heat a sphere while completely suppressing the expansion: a compressive stress builds up, which will lead to the expansion of the sphere if you relax the constraint. Accordingly, the residual stress $-\beta^{KL}(\theta)T$ is related to the expansion of the material. Indeed, for isotropic linear elastic materials, it was shown in Section 1.14.3 that (Equation (1.447))

$$\beta^{KL}(\theta) = [3\lambda(\theta) + 2\mu(\theta)]\alpha(\theta)G^{KL}. \tag{2.232}$$

More generally, defining the anisotropic expansion coefficient by

$$E_{KL} = \alpha_{KL}(\theta)T \tag{2.233}$$

the corresponding stress needed to avoid this expansion for a linear material yields (Equation (1.420))

$$S^{KL} = -\Sigma^{KLMN}(\theta)\alpha_{MN}(\theta)T \tag{2.234}$$

leading to

$$\beta^{KL}(\theta) = \Sigma^{KLMN}(\theta)\alpha_{MN}(\theta). \tag{2.235}$$

In Equation (2.23), thermal loading is the integral of the negative thermal stress. Indeed, an increase in temperature has the same effect as a pulling force (assuming that the body expands as the temperature increases). Numerical integration requires the knowledge of the temperature at the integration points. If the temperatures are given in the nodes, an interpolation has to be performed to obtain the integration point values. Usually, the shape functions that are used to interpolate the displacements are also used to interpolate the temperature (Equation (2.11)):

$$T(\boldsymbol{X}) = \sum_{i=1}^{N} \varphi_i(\xi, \eta, \zeta)T(\boldsymbol{X}_i). \tag{2.236}$$

Because of Equation (2.234), and assuming that the material properties do not vary wildly within an element, the same interpolation is used for the thermal stresses too. However, because of the fact that the strains are obtained through differentiation of the displacements (Equation (2.4)), the degree of the mechanical stress interpolation pattern is one less than for the thermal stress interpolation. This leads to numerical problems unless reduced integration is used for the interpolation of the temperatures. Indeed, the number of reduced integration points is such that interpolating polynomials have a degree that is one less than the degree of the shape functions: linear for quadratic brick elements and constant for linear elements.

Accordingly, the interpolated temperature and the thermal stress is at most trilinear in quadratic elements with reduced integration and constant in linear elements with reduced integration.

For fully integrated bricks, the temperature integration has to be reduced. This is now illustrated for the 20-node brick.

First, the temperature is calculated at the reduced integration points:

$$T_j^* = \sum_{k=1}^{20} \varphi_{kj}^R T_k, \quad j = 1, \ldots, 8. \tag{2.237}$$

This temperature is linearly extrapolated to the nodes of the element:

$$T_i^{**} = \sum_{j=1}^{8} a_{ij} T_j^*, \quad i = 1, \ldots, 20 \tag{2.238}$$

and finally the linearly extrapolated temperature is interpolated at the full integration points:

$$T_l^{***} = \sum_{i=1}^{20} \varphi_{il} T_i^{**}, \quad l = 1, \ldots, 27. \tag{2.239}$$

Substituting Equation (2.237) and Equation (2.238) into Equation (2.239) yields a linear relationship:

$$T_l^{***} = \sum_{k=1}^{20} c_{kl} T_k, \quad l = 1, \ldots, 27 \tag{2.240}$$

where

$$c_{kl} = \sum_{i=1}^{20} \sum_{j=1}^{8} \varphi_{il} a_{ij} \varphi_{kj}^R, \quad \begin{array}{l} k = 1, \ldots, 20 \\ l = 1, \ldots, 27. \end{array} \tag{2.241}$$

φ_{il} and φ_{kj}^R are the values of the shape functions for the 20-node brick element at the full and reduced integration points, respectively, and a_{ij} are the trilinear functions, which are also used for the 8-node brick element ($[A]^{-1}$ in Equation (2.121)). Equation (2.240) replaces Equation (2.236) for the temperature interpolation in the 20-node brick element with full integration. For the concrete coefficients, the reader is referred to the CalculiX® code (CalculiX 2003).

For the fully integrated linear element, the reduced temperature interpolation leads to a constant temperature at the full integration points, which is equal to the mean of the temperature at the nodes. Thus, we get in this case for Equation (2.240)

$$T_l^{***} = \frac{1}{8} \sum_{k=1}^{8} T_k \quad l = 1, \ldots, 8. \tag{2.242}$$

The reduced integration for the temperature in fully integrated elements ensures that the thermal stress and the mechanical stress are modeled with interpolation functions of the same degree.

2.9 Modal Analysis

Mechanical structures exhibit eigenmodes. These are oscillating homogeneous solutions of the linear (or linearized) governing equations. Their amplitude can be freely scaled. The corresponding frequency of the oscillation is called the *eigenfrequency*. Modal analysis, that is, the determination of the eigenfrequencies and eigenmodes is important in structural analysis since the eigenmodes are the preferred shapes a structure will assume when subject to loading. Indeed, one way of calculating dynamic response is by assuming that it is a linear combination of the lowest eigenfrequencies (Meirovitch 1967).

2.9.1 Frequency calculation

A frequency analysis starts from the governing equation (2.27) in homogeneous form

$$\left[M\right]\left\{\ddot{U}\right\}+\left[K\right]\left\{U\right\}=\left\{0\right\} \tag{2.243}$$

with initial conditions

$$\left\{U\right\}_{t=t_0}=\left\{U_0\right\} \tag{2.244}$$

$$\left\{\dot{U}\right\}_{t=t_0}=\left\{V_0\right\}. \tag{2.245}$$

To obtain the eigenmodes, a solution in the form

$$\left\{U\right\}=\left\{U_j\right\}\mathrm{e}^{\mathrm{i}\omega_j t} \tag{2.246}$$

is proposed (separation of the space and time variables). Consequently, Equation (2.243) yields

$$\left[K\right]\left\{U_j\right\}=\omega_j^2\left[M\right]\left\{U_j\right\}. \tag{2.247}$$

This is a classical generalized eigenvalue problem with well-known properties. Since $\left[K\right]$ is symmetric and $\left[M\right]$ is symmetric and positive-definite, the eigenvalues are real and the eigenmodes are orthogonal with respect to $\left[M\right]$. Indeed, suppose that $\lambda_j := \omega_j^2$ is complex with eigenvector $\left\{U_j\right\}$, then, taking the complex conjugate of Equation (2.247)

$$\left[K\right]\left\{\overline{U_j}\right\}=\overline{\lambda_j}\left[M\right]\left\{\overline{U_j}\right\} \tag{2.248}$$

reveals that $\overline{\lambda_j}$ must also be an eigenvalue with eigenvector $\left\{\overline{U_j}\right\}$. Premultiplying Equation (2.248) by $\left\{U_j\right\}^{\mathrm{T}}$ and Equation (2.247) by $\left\{\overline{U_j}\right\}^{\mathrm{T}}$ yields

$$\left\{U_j\right\}^{\mathrm{T}}\left[K\right]\left\{\overline{U_j}\right\}=\overline{\lambda_j}\left\{U_j\right\}^{\mathrm{T}}\left[M\right]\left\{\overline{U_j}\right\} \tag{2.249}$$

$$\left\{\overline{U_j}\right\}^{\mathrm{T}}\left[K\right]\left\{U_j\right\}=\lambda_j\left\{\overline{U_j}\right\}^{\mathrm{T}}\left[M\right]\left\{U_j\right\}. \tag{2.250}$$

Taking the transpose of Equation (2.249) and subtracting the results from Equation (2.250) leads to

$$0=(\lambda_j-\overline{\lambda_j})\left\{\overline{U_j}\right\}^{\mathrm{T}}\left[M\right]\left\{U_j\right\}. \tag{2.251}$$

Since $\left[M\right]$ is positive definite, we have (Greenberg 1978)

$$\left\{\overline{U_j}\right\}^{\mathrm{T}}\left[M\right]\left\{U_j\right\} > 0 \quad \text{if}\left\{U_j\right\} \neq 0. \tag{2.252}$$

Accordingly,

$$\lambda_j = \overline{\lambda_j} \tag{2.253}$$

and $\lambda_j = \omega_j^2$ is real, and so are the corresponding eigenmodes.

Now, let $\left\{U_i\right\}$ and $\left\{U_j\right\}$ be two different solutions:

$$\left[K\right]\left\{U_i\right\} = \lambda_i\left[M\right]\left\{U_j\right\} \tag{2.254}$$

$$\left[K\right]\left\{U_j\right\} = \lambda_j\left[M\right]\left\{U_j\right\}. \tag{2.255}$$

Multiplying Equation (2.254) by $\left\{U_j\right\}^{\mathrm{T}}$ and Equation (2.255) by $\left\{U_i\right\}^{\mathrm{T}}$, taking the transpose of Equation (2.255), subtracting both and taking the symmetry of $\left[K\right]$ and $\left[M\right]$ into account, one obtains

$$(\lambda_i - \lambda_j)\left\{U_j\right\}^{\mathrm{T}}\left[M\right]\left\{U_i\right\} = 0. \tag{2.256}$$

For $\lambda_i \neq \lambda_j$ one has

$$\left\{U_j\right\}^{\mathrm{T}}\left[M\right]\left\{U_i\right\} = 0 \tag{2.257}$$

which shows that the eigenmodes are orthogonal indeed. They are generally normed such that

$$\left\{U_j\right\}^{\mathrm{T}}\left[M\right]\left\{U_j\right\} = 1. \tag{2.258}$$

Premultiplying Equation (2.247) by $\left\{U_j\right\}^{\mathrm{T}}$ yields

$$\left\{U_j\right\}^{\mathrm{T}}\left[K\right]\left\{U_j\right\} = \lambda_j\left\{U_j\right\}^{\mathrm{T}}\left[M\right]\left\{U_j\right\}. \tag{2.259}$$

The matrix $\left[M\right]$ is positive definite. If $\left[K\right]$ is positive definite as well, $\lambda_j = \omega_j^2$ is not only real but also strictly positive. This implies that for each eigenvalue λ_j there are two real eigenfrequencies: ω_j and $-\omega_j$. They correspond to the solutions $\left\{U_j(\boldsymbol{X})\right\}\mathrm{e}^{\mathrm{i}\omega_j t}$ and $\left\{U_j(\boldsymbol{X})\right\}\mathrm{e}^{-\mathrm{i}\omega_j t}$, or, alternatively, to $\left\{U_j(\boldsymbol{X})\right\}\cos(\omega_j t)$ and $\left\{U_j(\boldsymbol{X})\right\}\sin(\omega_j t)$. Notice that these homogeneous solutions are bounded by $\|U_j(\boldsymbol{X})\|$.

If $\lambda_j = 0$, then $\omega_j = 0$ is a double root. The solution of Equation (2.243) now amounts to $\left\{U_j(\boldsymbol{X})\right\}$ and $\left\{U_j(\boldsymbol{X})\right\}t$. For $\lambda_j < 0$ the eigenfrequencies are imaginary: $\omega_j = \pm\mathrm{i}\sqrt{(-\lambda_j)}$ leading to the solutions $\left\{U_j(\boldsymbol{X})\right\}\mathrm{e}^{-\sqrt{(-\lambda_j t)}}$ and $\left\{U_j(\boldsymbol{X})\right\}\mathrm{e}^{\sqrt{(-\lambda_j t)}}$. Accordingly, for $\lambda_j \leq 0$, at least one of the solutions is not bounded.

Eigenvalue problems such as Equation (2.247) are usually solved with dedicated numerical packages such as ARPACK (Lehoucq *et al.* 1998). A continuous system has infinitely many eigenmodes. Usually, only the lowest ones (10 up to maybe 100) are practically important. In what follows, ω_j will be assumed to be positive.

2.9.2 Linear dynamic analysis

A general linear dynamic analysis starts from Equation (2.27):

$$[M]\{\ddot{U}\}+[K]\{U\}=\{F\}. \tag{2.260}$$

Frequently, a damping term linear in the velocity $\{\dot{U}\}$ is added

$$[M]\{\ddot{U}\}+[C]\{\dot{U}\}+[K]\{U\}=\{F\}. \tag{2.261}$$

If the damping is of the Rayleigh type, $[C]$ is defined as a linear combination of $[M]$ and $[K]$:

$$[C]=\alpha[M]+\beta[K]. \tag{2.262}$$

The quintessence of modal dynamic analysis is the fact that the response $\{U\}$, solution of Equation (2.261), can be written as a linear combination of the eigenmodes $\{U_i\}$, which is the solution of Equation (2.247). This relates to the fact that the eigenmodes constitute an orthogonal basis for the solution space of Equation (2.260). Accordingly,

$$\{U(t)\}=\sum_i b_i(t)\{U_i\}. \tag{2.263}$$

Notice that only the coefficients $b_i(t)$ are a function of time, the eigenmodes $\{U_i\}$ are not. In reality, only a finite number of eigenmodes is calculated and the series in Equation (2.263) is truncated. The truncated series is an approximation of $\{U(t)\}$. The quality of the approximation depends on the number of eigenmodes and the frequency content of the loading. Substituting Equation (2.263) into Equation (2.261), one obtains

$$\sum_i[M]\{U_i\}\ddot{b}_i(t)+\sum_i[C]\{U_i\}\dot{b}_i(t)+\sum_i[K]\{U_i\}b_i(t)=\{F(t)\}. \tag{2.264}$$

Premultiplying by $\{U_j\}$ and using Equations (2.247) and (2.262) yields

$$\sum_i\{U_j\}^{\mathrm{T}}[M]\{U_i\}\left[\ddot{b}_i(t)+(\alpha+\beta\omega_i^2)\dot{b}_i(t)+\omega_i^2 b_i(t)\right]=\{U_j\}^{\mathrm{T}}\{F(t)\} \tag{2.265}$$

and because of the orthogonality condition, Equation (2.257), and norming condition, Equation (2.258),

$$\ddot{b}_j(t)+(\alpha+\beta\omega_j^2)\dot{b}_j(t)+\omega_j^2 b_j(t)=\{U_j\}^{\mathrm{T}}\{F(t)\}. \tag{2.266}$$

Equation (2.266) is the central equation for modal dynamics. It can be written for each mode and constitutes a linear inhomogeneous second-order ordinary differential equation with constant coefficients. The key point is that due to the choice of $[C]$, the modes are independent of each other. The differential equations have to be complemented by the initial conditions $b_j(0)$ and $\dot{b}_j(0)$ obtained from $\{U_0\} := \{U(t=0)\}$ and $\{V_0\} := \{\dot{U}(t=0)\}$ (Equation (2.263))

$$\sum_i b_i(0)\{U_i\} = \{U_0\} \Rightarrow b_j(0) = \{U_j\}^{\mathrm{T}}[M]\{U_0\} \tag{2.267}$$

and

$$\sum_i \dot{b}_i(0)\{U_i\} = \{V_0\} \Rightarrow \dot{b}_j(0) = \{U_j\}^{\mathrm{T}}[M]\{V_0\}. \tag{2.268}$$

Equation (2.266) is frequently written as

$$\ddot{b}_j(t) + 2\zeta_j\omega_j\dot{b}_j(t) + \omega_j^2 b_j(t) = \{U_j\}^{\mathrm{T}}\{F(t)\} \tag{2.269}$$

where ζ_j is the friction coefficient defined by

$$\zeta_j := \frac{\alpha + \beta\omega_j^2}{2\omega_j}. \tag{2.270}$$

Notice that because of Equation (2.270), the friction coefficient depends on the eigenvalues. A large α-coefficient leads to low frequency damping and a large β-coefficient to high-frequency damping.

The solution of Equation (2.269) basically depends on the character of the discriminant, defined by

$$\omega_{jd} := \omega_j\sqrt{1-\zeta_j^2}. \tag{2.271}$$

It arises in the solution of the quadratic equation obtained by substituting $\mathrm{e}^{\lambda t}$ in the homogeneous differential equation. One obtains the following cases:

1. $\omega_{jd} \in \mathbb{R}^+$

$$\begin{aligned} b_j(t) = \frac{1}{\omega_{jd}}\int_0^t \{U_j\}^{\mathrm{T}}\{F(\tau)\}\,\mathrm{e}^{-\zeta_j\omega_j(t-\tau)}\sin[\omega_{jd}(t-\tau)]\,\mathrm{d}\tau \\ + \mathrm{e}^{-\zeta_j\omega_j t}\left[\cos[\omega_{jd}t] + \frac{\zeta_j}{\sqrt{1-\zeta_j^2}}\sin[\omega_{jd}t]\right] b_j(0) \\ + \left[\frac{1}{\omega_{jd}}\mathrm{e}^{-\zeta_j\omega_j t}\sin[\omega_{jd}t]\right]\dot{b}_j(0). \end{aligned} \tag{2.272}$$

To obtain Equation (2.272), formula 2.663.1 and 2.667.5 in (Gradshteyn and Ryzhik 1980) were used. The solution is called *subcritical* and consists of oscillatory functions.

2. $\omega_{jd} = 0$

$$b_j(t) = \int_0^t \{U_j\}^{\mathrm{T}} \{F(\tau)\} \mathrm{e}^{-\zeta_j\omega_j(t-\tau)}(t-\tau)\,\mathrm{d}\tau$$
$$+ \mathrm{e}^{-\zeta_j\omega_j t}[1 + \zeta_j\omega_j t]b_j(0) + t\mathrm{e}^{-\zeta_j\omega_j t}\dot{b}_j(0). \quad (2.273)$$

To obtain Equation (2.273), formula 2.322.1 in (Gradshteyn and Ryzhik 1980) was used. The solution is called *critical* and exhibits an exponential nonoscillatory behavior.

3. $\omega_{jd} = \mathrm{i}\omega^*_{jd},\ \omega^*_{jd} \in \mathbb{R}^+$

$$b_j(t) = \frac{1}{\omega^*_{jd}} \int_0^t \{U_j\}^{\mathrm{T}} \{F(\tau)\} \mathrm{e}^{-\zeta_j\omega_j(t-\tau)} \sinh[\omega^*_{jd}(t-\tau)]\,\mathrm{d}\tau$$
$$+ \mathrm{e}^{-\zeta_j\omega_j t}\left[\cosh[\omega^*_{jd}t] + \frac{\zeta_j}{\sqrt{\zeta_j^2 - 1}}\sinh[\omega^*_{jd}t]\right]b_j(0)$$
$$+ \left[\frac{1}{\omega^*_{jd}}\mathrm{e}^{-\zeta_j\omega_j t}\sinh[\omega^*_{jd}t]\right]\dot{b}_j(0). \quad (2.274)$$

The solution is supercritical and exhibits an exponential nonoscillatory behavior. It can also be written as

$$b_j(t) = \frac{1}{2\omega^*_{jd}} \int_0^t \{U_j\}^{\mathrm{T}} \{F(\tau)\} \left[\mathrm{e}^{\omega^-(t-\tau)} - \mathrm{e}^{-\omega^+(t-\tau)}\right] \mathrm{d}\tau$$
$$+ \frac{1}{2}\left[\mathrm{e}^{\omega^- t} + \mathrm{e}^{-\omega^+ t}\right] b_j(0)$$
$$+ \left(\frac{\zeta_j}{2\sqrt{\zeta_j^2 - 1}} b_j(0) + \frac{1}{2\omega^*_{jd}}\dot{b}_j(0)\right)\left[\mathrm{e}^{\omega^- t} - \mathrm{e}^{-\omega^+ t}\right] \quad (2.275)$$

where

$$\omega^- := \omega^*_{jd} - \zeta_j\omega_j \quad (2.276)$$

$$\omega^+ := \omega^*_{jd} + \zeta_j\omega_j. \quad (2.277)$$

In Equations (2.272), (2.273) and (2.275), the right-hand side loading is written inside an integral sign. For pointwise linear loading, the integral can be evaluated exactly. Indeed, let the interval $[0, t]$ be split in subintervals $[t_{i-1}, t_i]$ in which the loading is linear in time

$$\{U_j\}^{\mathrm{T}}\{F(\tau)\} = a_{ij} + b_{ij}\tau \quad \text{for } \tau \in [t_{i-1}, t_i] \tag{2.278}$$

with $t_0 = 0$ and $t_n = t$, and let

$$\sigma := t - \tau \tag{2.279}$$

then (use formulas 2.663.1 and 2.667.5 in (Gradshteyn and Ryzhik 1980))

$$\begin{aligned}
&\int_0^t \{U_j\}^{\mathrm{T}}\{F(\tau)\}\,\mathrm{e}^{-\zeta_j\omega_j(\sigma)}\sin\omega_{jd}(\sigma)\,\mathrm{d}\tau \\
&\quad = \sum_{i=1}^{n}\left\{[a_{ij} + b_{ij}t]\left[\frac{\mathrm{e}^{-\zeta_j\omega_j\sigma}(-\zeta_j\omega_j\sin[\omega_{jd}\sigma] - \omega_{jd}\cos[\omega_{jd}\sigma])}{\zeta_j^2\omega_j^2 + \omega_{jd}^2}\right]_{t-t_i}^{t-t_{i-1}}\right. \\
&\qquad -b_j\left[\frac{\mathrm{e}^{-\zeta_j\omega_j\sigma}}{\zeta_j^2\omega_j^2 + \omega_{jd}^2}\left(\left(-\zeta_j\omega_j\sigma - \frac{(\zeta_j^2\omega_j^2 - \omega_{jd}^2)}{(\zeta_j^2\omega_j^2 + \omega_{jd}^2)}\right)\sin[\omega_{jd}\sigma]\right.\right. \\
&\qquad\qquad \left.\left.\left. - \left(\omega_{jd}\sigma + \frac{2\zeta_j\omega_j\omega_{jd}}{(\zeta_j^2\omega_j^2 + \omega_{jd}^2)}\right)\cos[\omega_{jd}\sigma]\right)\right]_{t-t_i}^{t-t_{i-1}}\right\}
\end{aligned} \tag{2.280}$$

in Equation (2.272), (use formulas 2.322.1 and 2.322.2 in (Gradshteyn and Ryzhik 1980))

$$\begin{aligned}
&\int_0^t \{U_j\}^{\mathrm{T}}\{F(\tau)\}\,\mathrm{e}^{-\zeta_j\omega_j(\sigma)}(\sigma)\,\mathrm{d}\tau \\
&\quad = \sum_{i=1}^{n}\left\{[a_{ij} + b_{ij}t]\left[\mathrm{e}^{-\zeta_j\omega_j\sigma}\left(-\frac{\sigma}{\zeta_j\omega_j} - \frac{1}{\zeta_j^2\omega_j^2}\right)\right]_{t-t_{i-1}}^{t-t_i}\right. \\
&\qquad \left. -b_j\left[\mathrm{e}^{-\zeta_j\omega_j\sigma}\left(-\frac{\sigma^2}{\zeta_j\omega_j} - \frac{2\sigma}{\zeta_j^2\omega_j^2} - \frac{2}{\zeta_j^3\omega_j^3}\right)\right]_{t-t_{i-1}}^{t-t_i}\right\}
\end{aligned} \tag{2.281}$$

in Equation (2.273) and (use formulas 2.311 and 2.322.1 in (Gradshteyn and Ryzhik 1980))

$$\begin{aligned}
&\int_0^t \{U_j\}^{\mathrm{T}}\{F(\tau)\}\left[\mathrm{e}^{\omega^-(\sigma)} - \mathrm{e}^{-\omega^+(\sigma)}\right]\mathrm{d}\tau \\
&\quad = \sum_{i=1}^{n}\left\{[a_{ij} + b_{ij}t]\left[\frac{\mathrm{e}^{\omega^-\sigma}}{\omega^-} + \frac{\mathrm{e}^{-\omega^+\sigma}}{\omega^+}\right]_{t-t_{i-1}}^{t-t_i}\right. \\
&\qquad \left. -b_j\left[\mathrm{e}^{\omega^-\sigma}\left(\frac{\sigma}{\omega^-} - \frac{1}{(\omega^-)^2}\right) + \mathrm{e}^{-\omega^+\sigma}\left(\frac{\sigma}{\omega^+} + \frac{1}{(\omega^+)^2}\right)\right]_{t-t_{i-1}}^{t-t_i}\right\}
\end{aligned} \tag{2.282}$$

in Equation (2.275). Consequently, the solution for piecewise-linear loading can be written down explicitly.

A special case of loading is the harmonic excitation satisfying

$$\{F(t)\} = \{F_R + \mathrm{i}F_I\}\,\mathrm{e}^{\mathrm{i}\Omega t}. \tag{2.283}$$

F_R and F_I are the in-phase and out-of-phase amplitude, respectively, and Ω is the frequency of the excitation. Now, Equation (2.266) reads

$$\ddot{b}_j(t) + (\alpha + \beta\omega_j^2)\dot{b}_j(t) + \omega_j^2 b_j(t) = \{U_j\}^{\mathrm{T}}\{F_R + \mathrm{i}F_I\}\,\mathrm{e}^{\mathrm{i}\Omega t}. \tag{2.284}$$

Inspired by the form of the right-hand side, we assume a complex solution in the form

$$b_j(t) = (b_{jR} + \mathrm{i}b_{jI})\,\mathrm{e}^{\mathrm{i}\Omega t} \tag{2.285}$$

the derivatives of which yield

$$\dot{b}_j(t) = \mathrm{i}\Omega(b_{jR} + \mathrm{i}b_{jI})\,\mathrm{e}^{\mathrm{i}\Omega t} \tag{2.286}$$

$$\ddot{b}_j(t) = -\Omega^2(b_{jR} + \mathrm{i}b_{jI})\,\mathrm{e}^{\mathrm{i}\Omega t}. \tag{2.287}$$

Substitution into Equation (2.284) leads to

$$-\Omega^2(b_{jR} + \mathrm{i}b_{jI}) + \mathrm{i}(\alpha + \beta\omega_j^2)\,\Omega(b_{jR} + \mathrm{i}b_{jI}) + \omega^2(b_{jR} + \mathrm{i}b_{jI}) = \{U_j\}^{\mathrm{T}}\{F_R + \mathrm{i}F_I\}. \tag{2.288}$$

Separating the real and imaginary parts of the equation yields two real equations:

$$\begin{cases} -\Omega^2 b_{jR} - (\alpha + \beta\omega_j^2)\,\Omega b_{jI} + \omega_j^2 b_{jR} = \{U_j\}^{\mathrm{T}}\{F_R\} \\ -\Omega^2 b_{jI} + (\alpha + \beta\omega_j^2)\,\Omega b_{jR} + \omega_j^2 b_{jI} = \{U_j\}^{\mathrm{T}}\{F_I\} \end{cases} \tag{2.289}$$

which is equivalent to

$$\begin{bmatrix} (\omega_j^2 - \Omega^2) & -(\alpha + \beta\omega_j^2)\,\Omega \\ (\alpha + \beta\omega_j^2)\,\Omega & (\omega_j^2 - \Omega^2) \end{bmatrix} \begin{Bmatrix} b_{jR} \\ b_{jI} \end{Bmatrix} = \begin{Bmatrix} \{U_j\}^{\mathrm{T}}\{F_R\} \\ \{U_j\}^{\mathrm{T}}\{F_I\} \end{Bmatrix} \tag{2.290}$$

the solution of which reads

$$b_{jR} = \frac{\{U_j\}^{\mathrm{T}}\{F_R\}(\omega_j^2 - \Omega^2) + \{U_j\}^{\mathrm{T}}\{F_I\}(\alpha + \beta\omega_j^2)\,\Omega}{(\omega_j^2 - \Omega^2)^2 + (\alpha + \beta\omega_j^2)^2\,\Omega^2} \tag{2.291}$$

$$b_{jI} = \frac{\{U_j\}^{\mathrm{T}}\{F_I\}(\omega_j^2 - \Omega^2) - \{U_j\}^{\mathrm{T}}\{F_R\}(\alpha + \beta\omega_j^2)\Omega}{(\omega_j^2 - \Omega^2)^2 + (\alpha + \beta\omega_j^2)^2\Omega^2}. \tag{2.292}$$

2.9.3 Buckling

Buckling calculations are a special case of frequency calculations with preload. In Equation (2.243), it was emphasized that frequency calculations are essentially homogeneous. However, the eigenfrequencies of a structure do depend on the loading. For instance, a rotating blade has other eigenfrequencies in comparison to a static blade. This effect manifests itself through a modified stiffness of the structure due to the stresses and displacements. This is explained in Chapter 3 where the following modified stiffness matrix is defined (Equation (3.17)):

$$\left[K\right]_{\text{mod}} = \left[K\right]_{\text{LE}} + \left[K\right]_{\text{ST}} + \left[K\right]_{\text{LD}}. \tag{2.293}$$

Here, $\left[K\right]_{\text{LE}}$ is the linear elastic stiffness matrix, $\left[K\right]_{\text{ST}}$ is the stress stiffness contribution and $\left[K\right]_{\text{LD}}$ is the large deformation stiffness. By replacing $\left[K\right]$ in Equation (2.247) by $\left[K\right]_{\text{mod}}$, the loading is taken into account in the frequency calculation. The buckling load can be defined as the load for which the lowest eigenfrequency reaches zero. Then, a small perturbation will lead to buckling. Indeed, in Section 2.9.1 it was shown that a zero eigenvalue, or equivalently a zero eigenfrequency, leads to an unbounded homogeneous solution: the system is unstable.

As an example, look at the beam in Figure 2.20, loaded by a point force at its end. Figure 2.21 shows the lowest eigenfrequency ω of the beam. It corresponds with a bending mode with zero displacements at its fixed end. As the tensile force decreases, the lowest eigenvalue ω^2 decreases until it is zero and buckling occurs. Notice that as the eigenvalue ω^2 becomes zero, the eigenfrequency ω is zero too and the solution is unbounded (Section 2.9.1).

Suppose that the structure is loaded by a static force system 1 and a buckling load system 2. The static load system is defined as a system that is permanently acting and the magnitude of which is not changing. The buckling load system varies in magnitude and the basic question is at what value of the buckling load system will the collapse occur. To this end, the buckling load system is scaled with a factor λ and the problem is reduced to the question: at what value of λ is the lowest eigenvalue of the system zero? Equation (2.247) is now equivalent to

$$\left[\left[K\right]_{\text{LE}} + \left[K\right]_{\text{ST1}} + \left[K\right]_{\text{LD1}} + \left[K\right]_{\text{ST2}\lambda} + \left[K\right]_{\text{LD2}\lambda}\right]\left\{U\right\} = 0. \tag{2.294}$$

Index 1 stands for load system 1, index 2λ for λ times load system 2. $\left\{U\right\}$ is only nonzero if the total stiffness matrix is singular which also implies that the matrix is not positive-definite. Equation (3.17) reveals that $\left[K\right]_{\text{ST}}$ is linear in the load but $\left[K\right]_{\text{LD}}$ is not (notice the quadratic term in the displacement). Accordingly,

$$\left[K\right]_{\text{ST}2\lambda} = \lambda\left[K\right]_{\text{ST2}} \tag{2.295}$$

$$\left[K\right]_{\text{LD}2\lambda} \neq \lambda\left[K\right]_{\text{LD}2\lambda}. \tag{2.296}$$

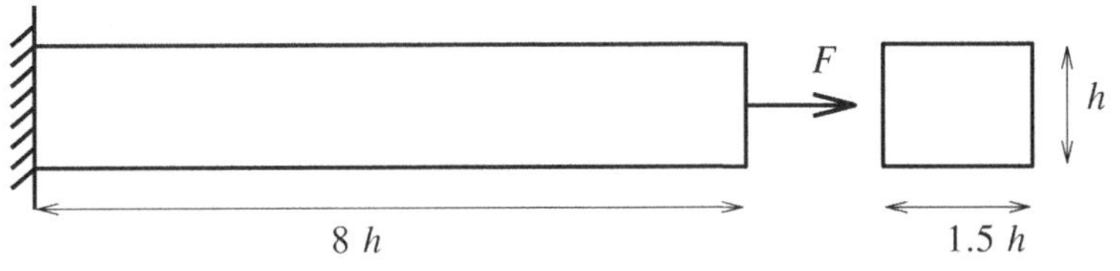

Figure 2.20 Geometry of the beam

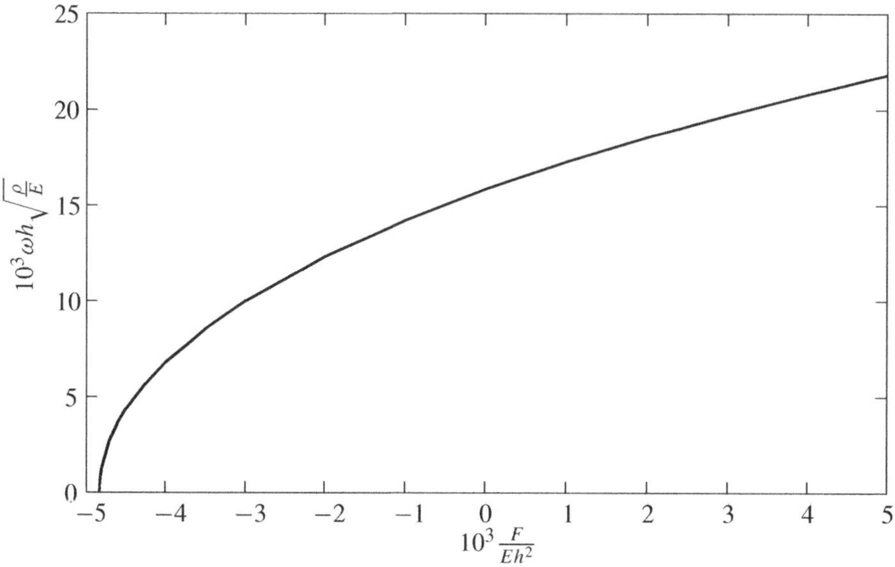

Figure 2.21 Lowest eigenfrequency for a beam under tension

Therefore, for linear buckling calculations, $[K]_{LD2\lambda}$ will be neglected leading to the following eigenvalue problem:

$$[[K]_{LE} + [K]_{ST1} + [K]_{LD1}]\{U\} = -\lambda [K]_{ST2}\{U\}. \tag{2.297}$$

This is again a generalized eigenvalue problem with symmetric matrices similar to Equation (2.247) except that $[K]_{ST2}$ is not positive-definite. It is the governing buckling equation and can be solved using the ARPACK package (Lehoucq *et al.* 1998).

2.10 Cyclic Symmetry

Cyclic symmetry is an important issue in rotating structures such as disks (Ramamurti and Seshu 1990). It basically enables you to calculate eigenmodes for a complete disk by modeling a segment only. Look at the deformed disk in Figure 2.22. It exhibits an eigenmode with a nodal diameter of two. This means that there are two diameters for which the displacements are zero. This corresponds to four zero crossings or to two complete waves along the circumference of the disk. In general, a nodal diameter N corresponds to N waves along the circumference and $2N$ zero crossings. Suppose only a segment extending over an angle Φ_S is modeled. For the disk in Figure 2.23, Φ_S can take any value smaller or equal to 2π. For practical models, the value of Φ_S depends on the size of substructures such as blades. In general, if there are M identical sectors along the circumference, Φ_S must be a multiple of $2\pi/M$. For instance, for the structure in Figure 2.24, there are four identical sectors and Φ_S must be a multiple of $\pi/2$.

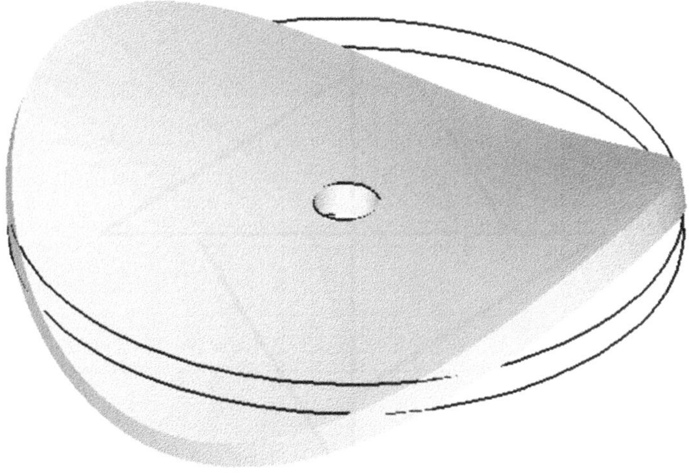

Figure 2.22 Eigenmode of a disk with nodal diameter two

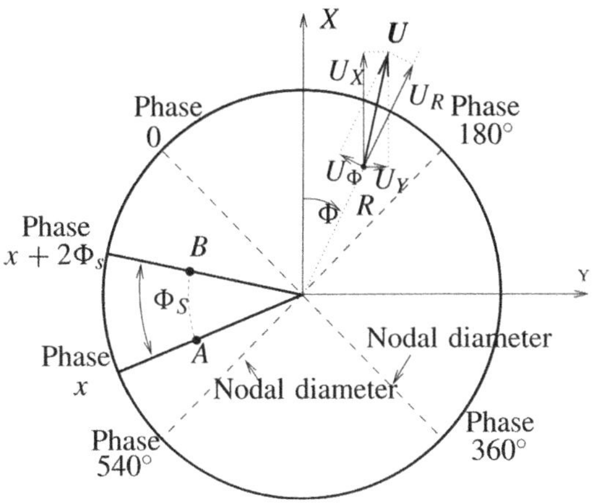

Figure 2.23 Phase along the boundary of the disk and modeled segment

Figure 2.23 shows that for the mode shown in Figure 2.22, the displacements in cylindrical coordinates in any node B on the "clockwise" side are phase shifted with respect to those of the corresponding node A on the "counter clockwise" side by $2\Phi_S$. For a mode with nodal diameter N, this shift takes the value $N\phi_S$. Taking for Φ_S the smallest possible value $\Phi_S = 2\pi/M$, one arrives at

$$\{U\}_{\text{cyl},B} = \{U\}_{\text{cyl},A}\, e^{i\frac{2\pi N}{M}}. \tag{2.298}$$

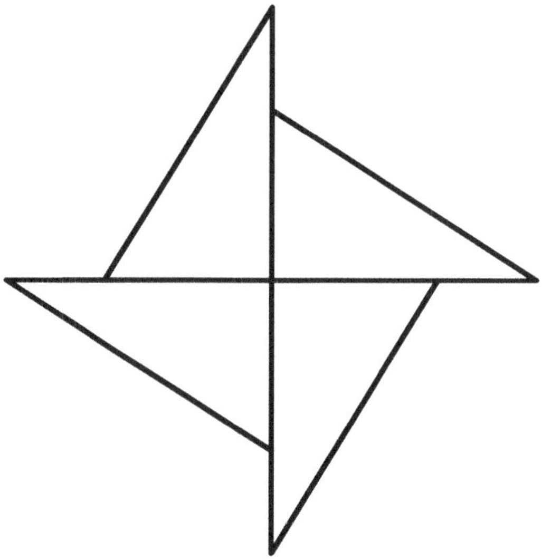

Figure 2.24 Structure consisting of four identical sectors

This is the central equation of cyclic symmetry. For modal analysis no other equation is needed. This basically means the following:

1. The governing equations are not modified but only the boundary conditions are changed.

2. Because of the complex nature of the boundary conditions, the resulting eigenvalue problem is a generalized complex eigenvalue problem. It will be shown that it can be reduced to a generalized real eigenvalue problem twice the size.

3. Because of the presence of N in Equation (2.298), the eigenvalue system is different for a different nodal diameter. For a given nodal diameter N, the solution of the eigenvalue system yields all modes having $2N$ zeros along the circumference. Since cosine and sine are periodic functions with period 2π, one can write

$$e^{i\frac{2\pi N}{M}} = e^{i\frac{2\pi}{M}[N+kM]} \tag{2.299}$$

   where k is an integer. This means that the application of Equation (2.298) will yield all modes with nodal diameter $N + kM$. For other nodal diameters, Equation (2.298) is different and another eigenvalue system results. Accordingly, a cyclic symmetry calculation takes longer, but not as long as when the complete disk is modeled.

4. Equation (2.299) shows that it is sufficient to perform calculations for nodal diameters $0, 1, \ldots, M/2$ if M is even and up to $(M-1)/2$ for M odd. For instance, if M is odd, calculations for $N = (M-1)/2$ also yield the eigenmodes for $N = |(M-1)/2 - M| = (M+1)/2$.

Using the cylindrical coordinate system in Figure 2.23, Equation (2.298) is equivalent to

$$\begin{cases} U_{R,B} = U_{R,A}\mathrm{e}^{\mathrm{i}N\Phi_S} \\ U_{\Phi,B} = U_{\Phi,A}\mathrm{e}^{\mathrm{i}N\Phi_S} \\ U_{Z,B} = U_{Z,A}\mathrm{e}^{\mathrm{i}N\Phi_S}. \end{cases} \tag{2.300}$$

The cylindrical and rectangular coordinates in Figure 2.23 are related by

$$\begin{cases} X = R\cos\Phi \\ Y = R\sin\Phi \\ Z = Z \end{cases} \tag{2.301}$$

or

$$\begin{cases} R = \sqrt{X^2 + Y^2} \\ \Phi = \tan^{-1}\frac{Y}{X} \\ Z = Z \end{cases} \tag{2.302}$$

leading to (see Equation (2.212)),

$$\begin{cases} U_R = U_X\cos\Phi + U_Y\sin\Phi \\ U_\Phi = -U_X\sin\Phi + U_Y\cos\Phi \\ U_Z = U_Z \end{cases} \tag{2.303}$$

since (Equation (1.7))

$$\begin{cases} \boldsymbol{G}_{1'} = \cos\Phi\boldsymbol{I}_1 + \sin\Phi\boldsymbol{I}_2 \\ \boldsymbol{G}_{2'} = -r\sin\Phi\boldsymbol{I}_1 + r\cos\Phi\boldsymbol{I}_2 \\ \boldsymbol{G}_{3'} = \boldsymbol{I}_3 \end{cases} \tag{2.304}$$

and accordingly,

$$G_{1'1'} = G_{3'3'} = 1,\ G_{2'2'} = r^2. \tag{2.305}$$

Inverting Equation (2.303) yields

$$\begin{cases} U_X = U_R\cos\Phi - U_\Phi\sin\Phi \\ U_Y = U_R\sin\Phi + U_\Phi\cos\Phi \\ U_Z = U_Z. \end{cases} \tag{2.306}$$

Now, Equations (2.300) lead in rectangular coordinates to the following linear complex equations:

$$\begin{cases} U_{X,B}\cos\Phi_B + U_{Y,B}\sin\Phi_B = \left[U_{X,A}\cos\Phi_A + U_{Y,A}\sin\Phi_A\right]\mathrm{e}^{\mathrm{i}N\Phi_S} \\ -U_{X,B}\sin\Phi_B + U_{Y,B}\cos\Phi_B = \left[-U_{X,A}\sin\Phi_A + U_{Y,A}\cos\Phi_A\right]\mathrm{e}^{\mathrm{i}N\Phi_S} \\ U_{Z,B} = U_{Z,A}\mathrm{e}^{\mathrm{i}N\Phi_S}. \end{cases} \tag{2.307}$$

Solving for $U_{X,B}$, $U_{Y,B}$ and $U_{Z,B}$, we get three linear multiple point constraints with $U_{X,B}$, $U_{Y,B}$ and $U_{Z,B}$ as dependent variables. Notice that Equation (2.300) and Equation (2.306) lead to

$$\begin{aligned} U_{X,B} &= U_{R,B}\cos\Phi_B - U_{\Phi,B}\sin\Phi_B \\ &= (U_{R,A}\cos\Phi_B - U_{\Phi,A}\sin\Phi_B)\mathrm{e}^{\mathrm{i}N\Phi_S} \\ &\neq U_{X,A}\mathrm{e}^{\mathrm{i}N\Phi_S}, \end{aligned} \tag{2.308}$$

that is, Equations (2.300) do not apply in rectangular coordinates.

The resulting complex eigenvalue system

$$\left[K_{\mathcal{R}} + \mathrm{i}K_{\mathcal{I}}\right]\left\{U_{\mathcal{R}} + \mathrm{i}U_{\mathcal{I}}\right\} = \omega^2\left[M\right]\left\{U_{\mathcal{R}} + \mathrm{i}U_{\mathcal{I}}\right\} \tag{2.309}$$

where the index $\mathcal{R}$ denotes the real part and $\mathcal{I}$ the imaginary part is equivalent to

$$\begin{bmatrix} K_{\mathcal{R}} & -K_{\mathcal{I}} \\ K_{\mathcal{I}} & K_{\mathcal{R}} \end{bmatrix}\begin{Bmatrix} U_{\mathcal{R}} \\ U_{\mathcal{I}} \end{Bmatrix} = \omega^2\begin{bmatrix} M & 0 \\ 0 & M \end{bmatrix}\begin{Bmatrix} U_{\mathcal{R}} \\ U_{\mathcal{I}} \end{Bmatrix}. \tag{2.310}$$

Since the basic equilibrium equations lead to a real symmetric and consequently a Hermitian stiffness matrix, and the treatment of the boundary conditions discussed in Section 2.6 conserves the Hermitian character, $K_{\mathcal{R}} + \mathrm{i}K_{\mathcal{I}}$ is Hermitian. Accordingly,

$$\begin{gathered} \overline{K_{\mathcal{R}} + \mathrm{i}K_{\mathcal{I}}} = (K_{\mathcal{R}} + \mathrm{i}K_{\mathcal{I}})^{\mathrm{T}} \\ \Downarrow \\ K_{\mathcal{R}} - \mathrm{i}K_{\mathcal{I}} = K_{\mathcal{R}}^{\mathrm{T}} + \mathrm{i}K_{\mathcal{I}}^{\mathrm{T}} \\ \Downarrow \\ \begin{cases} K_{\mathcal{R}} = K_{\mathcal{R}}^{T} \\ -K_{\mathcal{I}} = K_{\mathcal{I}}^{T} \end{cases} \end{gathered} \tag{2.311}$$

which shows that Equation (2.310) is a symmetric eigenvalue problem.
Solving Equation (2.310), we get each eigenvalue twice. Indeed, one can check that if

$$\begin{Bmatrix} U_{\mathcal{R}} \\ U_{\mathcal{I}} \end{Bmatrix} \tag{2.312}$$

is a solution,

$$\begin{Bmatrix} -U_{\mathcal{I}} \\ U_{\mathcal{R}} \end{Bmatrix} \tag{2.313}$$

is a solution too with the same eigenfrequency. Recomposing the complex form, the first solution corresponds to $\{U_1\} = \{U_{\mathcal{R}} + \mathrm{i}U_{\mathcal{I}}\}$, the second to $\{U_2\} = \{-U_{\mathcal{I}} + \mathrm{i}U_{\mathcal{R}}\} = \{U_{\mathcal{R}} + \mathrm{i}U_{\mathcal{I}}\}\,\mathrm{e}^{\mathrm{i}\pi/2}$ which shows that the difference between both is a phase shift of 90°. The resulting solution amounts to (Equation (2.246))

$$\{U\} = \{U_{\mathcal{R}} + \mathrm{i}U_{\mathcal{I}}\}\,\mathrm{e}^{\mathrm{i}\omega t} \tag{2.314}$$

or

$$\{U\} = [\{U_{\mathcal{R}}\} \cos \omega t - \{U_{\mathcal{I}}\} \sin \omega t] + \mathrm{i}\,[\{U_{\mathcal{R}}\} \sin \omega t + \{U_{\mathcal{I}}\} \cos \omega t]. \qquad (2.315)$$

Since the governing equation is linear, both the real and imaginary part are a solution. Taking the real part, one arrives at

$$\{U\} = \{U_{\mathcal{R}}\} \cos \omega t - \{U_{\mathcal{I}}\} \sin \omega t. \qquad (2.316)$$

This is not a standing wave. However, if ω is a solution, so is $-\omega$ and accordingly

$$\{U\} = \{U_{\mathcal{R}}\} \cos \omega t + \{U_{\mathcal{I}}\} \sin \omega t \qquad (2.317)$$

is a solution too and any linear combination of Equation (2.316) and Equation (2.317) as well. Consequently, half the sum and half the difference, which are both standing waves, are also solutions:

$$\{U\} = \{U_{\mathcal{R}}\} \cos \omega t \qquad (2.318)$$

$$\{U\} = \{U_{\mathcal{I}}\} \sin \omega t. \qquad (2.319)$$

How do we arrive at the solution in the other sectors? Let the solution in a point P in the primary sector be $\boldsymbol{U}$. The solution in a point Q exactly K sectors ahead satisfies (Equation (2.300))

$$\begin{cases} U_{R,Q} = U_{R,P} e^{\mathrm{i} K N \Phi_S} \\ U_{\Phi,Q} = U_{\Phi,P} e^{\mathrm{i} K N \Phi_S} \\ U_{Z,Q} = U_{Z,P} e^{\mathrm{i} K N \Phi_S}. \end{cases} \qquad (2.320)$$

$\{U_{R,P}, U_{\Phi,P}, U_{Z,P}\}$ are related to $\{U_{X,P}, U_{Y,P}, U_{Z,P}\}$ through the relations in Equation (2.303). $\{U_{X,P}, U_{Y,P}, U_{Z,P}\}$ are generally complex and so are $\{U_{R,P}, U_{\Phi,P}, U_{Z,P}\}$. Because the Equations (2.303) are linear, they can be applied to the real and imaginary parts of the solution separately. The first equality in Equation (2.320) now reads

$$(U_{R,Q})_{\mathcal{R}} + \mathrm{i}(U_{R,Q})_{\mathcal{I}} = [(U_{R,P})_{\mathcal{R}} + \mathrm{i}(U_{R,P})_{\mathcal{I}}] e^{\mathrm{i} K N \Phi_S} \qquad (2.321)$$

which leads to

$$\begin{cases} (U_{R,Q})_{\mathcal{R}} = (U_{R,P})_{\mathcal{R}} \cos(K N \Phi_S) - (U_{R,P})_{\mathcal{I}} \sin(K N \Phi_S) \\ (U_{R,Q})_{\mathcal{I}} = (U_{R,P})_{\mathcal{R}} \sin(K N \Phi_S) + (U_{R,P})_{\mathcal{I}} \cos(K N \Phi_S). \end{cases} \qquad (2.322)$$

The rectangular components of the solution in Q are obtained through Equation (2.306). Accordingly, the solution in point Q is obtained by

1. converting the solution in P to cylindrical coordinates (Equation (2.303)),
2. applying the mapping in Equation (2.321) to each component to obtain the solution in Q in cylindrical coordinates,
3. converting this solution into rectangular coordinates (Equation (2.306)).

This also applies to higher-order tensors such as stresses or strains.

Cyclic symmetry properties can also be used in static calculations by expanding the circumferential loading in its Fourier components.

2.11 Dynamics: The α-Method

Instead of using modal dynamics to solve Equation (2.27), a direct solution in the space–time domain is also feasible. Usually, the space domain is meshed with finite elements, whereas the time domain is discretized with finite differences. The α-method, which will be presented here because of its excellent performance, is a further development of the Newmark algorithm (Zienkiewicz and Taylor 1989). Major contributions to the α-method were made in Hilber's Ph.D. Thesis (Hilber 1976) and in (Hilber *et al.* 1977), (Hilber and Hughes 1978), (Hulbert and Hughes 1987) and (Miranda *et al.* 1989). Here, the α-method is introduced in its classical form. For extensions of the α-method and other time-integration schemes see (Hughes 2000), (Muğan and Hulbert 2001a), (Muğan and Hulbert 2001b) and (Chung *et al.* 2003).

Important criteria to evaluate a numerical procedure are accuracy, consistency, stability and high-frequency dissipation. In general, second-order accuracy is strived at. This means that the error in each iteration is $O(\Delta t^2)$ where Δt is the size of a time increment. Accordingly, the error decreases as $\Delta t \to 0$, which also implies consistency. The stability issue is generally linked to the size of Δt. For some algorithms, the solution grows out of bounds for large values of Δt. This is, especially in explicit codes, a matter of concern. If the size of Δt does not matter, the algorithm is called *unconditionally stable*. Stability and consistency together imply convergence. High-frequency dissipation is related to the wish to attenuate high frequencies, which are generally less accurate due to the limited resolution of the finite element mesh.

2.11.1 Implicit formulation

The equation to be solved is Equation (2.27):

$$[K]\{U\} + [M]\{\ddot{U}\} = \{F\}. \tag{2.323}$$

Finite difference discretization in time means that the finite element variables are calculated at discrete times, for example, $t = t_0, t_1, \ldots, t_n, t_{n+1}, \ldots$. For simplicity, let us focus on a material point with displacement $\boldsymbol{u}$, velocity $\boldsymbol{v}$ and acceleration $\boldsymbol{a}$. Integration of

$$\boldsymbol{a} = \dot{\boldsymbol{v}} \tag{2.324}$$

yields

$$\boldsymbol{v}(t) = \boldsymbol{v}_n + \int_{t_n}^{t} \boldsymbol{a}(\xi)\,\mathrm{d}\xi \tag{2.325}$$

or, for $t = t_{n+1}$,

$$\boldsymbol{v}_{n+1} = \boldsymbol{v}_n + \int_{t_n}^{t_{n+1}} \boldsymbol{a}(\boldsymbol{\xi})\,\mathrm{d}\xi. \tag{2.326}$$

The integral on the right-hand side of Equation (2.326) cannot be solved since $\boldsymbol{a}$ is unknown except at $t = t_n$ and $t = t_{n+1}$. However, if we approximate $\boldsymbol{a}(\xi)$ by a linear combination of $\boldsymbol{a}_n$ and $\boldsymbol{a}_{n+1}$,

$$\boldsymbol{a}(t) \sim (1-\gamma)a_n + \gamma a_{n+1}, \quad t \in [t_n, t_{n+1}] \tag{2.327}$$

we can perform the integration leading to

$$\boldsymbol{v}_{n+1} = \boldsymbol{v}_n + \Delta t[(1-\gamma)\boldsymbol{a}_n + \gamma \boldsymbol{a}_{n+1}]. \tag{2.328}$$

Similarly for $\boldsymbol{u}$

$$\dot{\boldsymbol{u}} = \boldsymbol{v}. \tag{2.329}$$

Hence,

$$\boldsymbol{u}(t) = \boldsymbol{u}_n + \int_{t_n}^{t} \boldsymbol{v}(\eta)\,\mathrm{d}\eta. \tag{2.330}$$

Substituting Equation (2.325) into Equation (2.330) yields

$$\boldsymbol{u}(t) = \boldsymbol{u}_n + \boldsymbol{v}_n(t-t_n) + \int_{t_n}^{t}\int_{t_n}^{\eta} \boldsymbol{a}(\xi)\,\mathrm{d}\xi\,\mathrm{d}\eta. \tag{2.331}$$

Assuming $\boldsymbol{a}$ in the interval $[t_n, t_{n+1}]$ to be a linear combination of $\boldsymbol{a}_n$ and $\boldsymbol{a}_{n+1}$ (not necessarily the same as in Equation (2.328), i.e. $2\beta \neq \gamma$ in general),

$$\boldsymbol{a}(t) = (1-2\beta)\boldsymbol{a}_n + 2\beta \boldsymbol{a}_{n+1}, \quad t \in [t_n, t_{n+1}] \tag{2.332}$$

one obtains

$$\boldsymbol{u}(t) = \boldsymbol{u}_n + \boldsymbol{v}_n(t-t_n) + \frac{1}{2}(t-t_n)^2[(1-2\beta)\,\boldsymbol{a}_n + 2\beta \boldsymbol{a}_{n+1}] \tag{2.333}$$

or, for $t = t_{n+1}$,

$$\boldsymbol{u}_{n+1} = \boldsymbol{u}_n + \Delta t \boldsymbol{v}_n + \frac{1}{2}\Delta t^2[(1-2\beta)\,\boldsymbol{a}_n + 2\beta \boldsymbol{a}_{n+1}]. \tag{2.334}$$

Notice that $\boldsymbol{u}_n + \Delta t \boldsymbol{v}_n$ is the displacement which applies if the acceleration is zero. Denoting $\{V\} := \{\dot{U}\}$ and $\{A\} := \{\ddot{U}\}$, and letting

$$\{A\}_{n+1} = \{A\}_n + \{\Delta A\} \tag{2.335}$$

Equations (2.328) and (2.334) can be written for the complete mesh in the form

$$\{V\}_{n+1} = \{V\}_n + \Delta t\left[(1-\gamma)\{A\}_n + \gamma\{A\}_{n+1}\right] \tag{2.336}$$

$$\{U\}_{n+1} = \{U\}_n + \Delta t\{V\}_n + \frac{1}{2}(\Delta t)^2\left[(1-2\beta)\{A\}_n + 2\beta\{A\}_{n+1}\right]. \tag{2.337}$$

Equation (2.323) has to be satisfied at t_{n+1}. Hence,

$$[M]\{A\}_{n+1} + [C]\{V\}_{n+1} + [K]\{U\}_{n+1} = \{F\}_{n+1}^{\text{ext}}. \tag{2.338}$$

Here, a friction term was inserted (cf Section 2.9.2), and the index "ext" stands for external force. Substitution of Equations (2.336) and (2.337) into Equation (2.338) leads to the Newmark algorithm. However, experience has shown that the high-frequency dissipation can be much improved if all terms in Equation (2.338) except the acceleration term are evaluated at an intermediate position between t_n and t_{n+1}:

$$\begin{aligned}[M]\{A\}_{n+1} + (1+\alpha)[C]\{V\}_{n+1} - \alpha[C]\{V\}_n + (1+\alpha)[K]\{U\}_{n+1}\\ - \alpha[K]\{U\}_n = (1+\alpha)\{F\}_{n+1}^{\text{ext}} - \alpha\{F\}_n^{\text{ext}}, \quad -1 \leq \alpha \leq 0. \quad (2.339)\end{aligned}$$

Since the stiffness term and friction term can also be considered as an internal force:

$$\{F\}_{n+1}^{\text{int}} := [C]\{V\}_{n+1} + [K]\{U\}_{n+1} \quad (2.340)$$

Equation (2.339) amounts to

$$[M]\{A\}_{n+1} + (1+\alpha)\{F\}_{n+1}^{\text{int}} - \alpha\{F\}_n^{\text{int}} = (1+\alpha)\{F\}_{n+1}^{\text{ext}} - \alpha\{F\}_n^{\text{ext}}. \quad (2.341)$$

Substitution of Equations (2.336) and (2.337) into Equation (2.340) yields the *α-method* (called after the parameter α). In the next sections, it will be proven that if β and γ satisfy

$$\beta = \frac{1}{4}(1-\alpha)^2 \quad (2.342)$$

$$\gamma = \frac{1}{2} - \alpha \quad (2.343)$$

the algorithm is second-order accurate and unconditionally stable for $\alpha \in [-1/3, 0]$. Maximum high-frequency dissipation is obtained for $\alpha = -1/3$. For $\alpha = 0$, there is no high-frequency dissipation and the α-method reduces to a special case of the Newmark method (also called the *average acceleration method* since $\gamma = 1/2$ and $\beta = 1/4$ corresponds to taking the mean of $\{A\}_n$ and $\{A\}_{n+1}$). Defining

$$\{\tilde{V}\}_{n+1} = \{V\}_n + (1-\gamma)\Delta t\{A\}_n \quad (2.344)$$

$$\{\tilde{U}\}_{n+1} = \{U\}_n + \Delta t\{V\}_n + \frac{1}{2}(\Delta t)^2(1-2\beta)\{A\}_n \quad (2.345)$$

Equations (2.336) and (2.337) yield

$$\{V\}_{n+1} = \{\tilde{V}\}_{n+1} + \gamma\Delta t\{A\}_{n+1} \quad (2.346)$$

$$\{U\}_{n+1} = \{\tilde{U}\}_{n+1} + \beta(\Delta t)^2\{A\}_{n+1}. \quad (2.347)$$

The quantities $\{\tilde{U}\}_{n+1}$ and $\{\tilde{V}\}_{n+1}$ can be considered as predictor values and depend on values at time t_n only. Substitution of Equations (2.346) and (2.347) into Equation (2.339) yields

$$\begin{aligned}\Big[[M] + (1+\alpha)[C]\Delta t\gamma + (1+\alpha)[K](\Delta t)^2\beta\Big]\{A\}_{n+1}\\ + (1+\alpha)[C]\{\tilde{V}\}_{n+1} - \alpha[C]\{V\}_n + (1+\alpha)[K]\{\tilde{U}\}_{n+1}\\ - \alpha[K]\{U\}_n = (1+\alpha)\{F\}_{n+1}^{\text{ext}} - \alpha\{F\}_n^{\text{ext}} \quad (2.348)\end{aligned}$$

which is equivalent to

$$[M^*]\{A\}_{n+1} = \{F\} \tag{2.349}$$

where

$$[M^*] = [M] + (1+\alpha)[C]\gamma\Delta t + (1+\alpha)[K]\beta(\Delta t)^2 \tag{2.350}$$

$$\{F\} = (1+\alpha)\{F\}_{n+1}^{\text{ext}} - \alpha\{F\}_n^{\text{ext}} - (1+\alpha)\{\tilde{F}\}_{n+1}^{\text{int}} + \alpha\{F\}_n^{\text{int}} \tag{2.351}$$

$$\{F\}_n^{\text{int}} = [C]\{V\}_n + [K]\{U\}_n \tag{2.352}$$

$$\{\tilde{F}\}_{n+1}^{\text{int}} = [C]\{\tilde{V}\}_{n+1} + [K]\{\tilde{U}\}_{n+1}. \tag{2.353}$$

After solving for $\{A\}_{n+1}$, $\{V\}_{n+1}$ and $\{U\}_{n+1}$ can be determined using Equations (2.346) and (2.347).

2.11.2 Extension to nonlinear applications

The procedure can also be extended to nonlinear problems, in which $[C]$ and $[K]$ are nonlinear functions of $\{U\}$ and $\{V\}$. Now, Equation (2.339) is replaced by

$$\begin{aligned}[M]\{A\}_{n+1} &+ (1+\alpha)[C(\{U\}_{n+1},\{V\}_{n+1})]\{V\}_{n+1} \\ &- \alpha[C(\{U\}_n,\{V\}_n)]\{V\}_n + (1+\alpha)[K(\{U\}_{n+1},\{V\}_{n+1})]\{U\}_{n+1} \\ &- \alpha[K(\{U\}_n,\{V\}_n)]\{U\}_n = (1+\alpha)\{F\}_{n+1}^{\text{ext}} - \alpha\{F\}_n^{\text{ext}}.\end{aligned} \tag{2.354}$$

Defining

$$[K]_n := [K(\{U\}_n,\{V\}_n)] \tag{2.355}$$

$$[C]_n := [C(\{U\}_n,\{V\}_n)] \tag{2.356}$$

Equations (2.349) and (2.351) still apply but Equations (2.350) and (2.352) to Equation (2.353) now yield

$$[M^*] = [M] + (1+\alpha)[C]_{n+1}\gamma\Delta t + (1+\alpha)[K]_{n+1}\beta(\Delta t)^2 \tag{2.357}$$

$$\{F\}_n^{\text{int}} = [C]_n\{V\}_n + [K]_n\{U\}_n \tag{2.358}$$

$$\{\tilde{F}\}_{n+1}^{\text{int}} = [C]_{n+1}\{\tilde{V}\}_{n+1} + [K]_{n+1}\{\tilde{U}\}_{n+1}. \tag{2.359}$$

$[K]_{n+1}$ and $[C]_{n+1}$ are, however, not *a priori* known. Therefore, they are calculated on the basis of $\{\tilde{U}\}_{n+1}$ and $\{\tilde{V}\}_{n+1}$ (the predictor values) and iterations are run till convergence. Denoting the predictor values by

$$\{V\}_{n+1}^{(1)} = \{\tilde{V}\}_{n+1} = \{V\}_n + (1-\gamma)\Delta t\{A\}_n \tag{2.360}$$

$$\{U\}_{n+1}^{(1)} = \{\tilde{U}\}_{n+1} = \{U\}_n + \Delta t\{V\}_n + \frac{1}{2}(\Delta t)^2(1-2\beta)\{A\}_n \tag{2.361}$$

the force in iteration 1 amounts to

$$\{F\}^{(1)} = (1+\alpha)\{F\}_{n+1}^{\text{ext}} - \alpha\{F\}_n^{\text{ext}} - (1+\alpha)\{F\}_{n+1}^{(1)\text{int}} + \alpha\{F\}_n^{\text{int}} \tag{2.362}$$

where

$$\{F\}_n^{\text{int}} = [C]_n\{V\}_n + [K]_n\{U\}_n \tag{2.363}$$

$$\{F\}_{n+1}^{(1)\text{int}} = [C]_{n+1}^{(1)}\{V\}_{n+1}^{(1)} + [K]_{n+1}^{(1)}\{U\}_{n+1}^{(1)}. \tag{2.364}$$

In Equation (2.364) use was made of the following abbreviations:

$$[C]_{n+1}^{(i)} := \left[C(\{U\}_{n+1}^{(i)}, \{V\}_{n+1}^{(i)})\right] \tag{2.365}$$

and

$$[K]_{n+1}^{(i)} := \left[K(\{U\}_{n+1}^{(i)}, \{V\}_{n+1}^{(i)})\right] \tag{2.366}$$

Equation (2.349) now amounts to

$$[M^*]^{(1)}\{A\}_{n+1}^{(2)} = \{F\}^{(1)} \tag{2.367}$$

where

$$[M^*]^{(1)} = [M] + (1+\alpha)[C]_{n+1}^{(1)}\gamma\Delta t + (1+\alpha)[K]_{n+1}^{(1)}\beta(\Delta t)^2. \tag{2.368}$$

The corrected velocity and displacement satisfy

$$\{V\}_{n+1}^{(2)} = \{V\}_{n+1}^{(1)} + \gamma\Delta t\{A\}_{n+1}^{(2)} \tag{2.369}$$

$$\{U\}_{n+1}^{(2)} = \{U\}_{n+1}^{(1)} + \beta(\Delta t)^2\{A\}_{n+1}^{(2)}. \tag{2.370}$$

Now one can use the corrected displacements and velocities to calculate new values for $[M^*]$ and $\{F\}$ and iterate until $\{A\}_{n+1}$ has converged. For iteration i, one obtains

$$\{F\}^{(i)} = (1+\alpha)\{F\}_{n+1}^{\text{ext}} - \alpha\{F\}_n^{\text{ext}} - (1+\alpha)\{F\}_{n+1}^{(i)\text{int}} + \alpha\{F\}_n^{\text{int}} \tag{2.371}$$

$$\{F\}_{n+1}^{(i)\text{int}} = [C]_{n+1}^{(i)}\{V\}_{n+1}^{(1)} + [K]_{n+1}^{(i)}\{U\}_{n+1}^{(1)} \tag{2.372}$$

$$[M^*]^{(i)} = [M] + (1+\alpha)[C]_{n+1}^{(i)}\gamma\Delta t + (1+\alpha)[K]_{n+1}^{(i)}\beta(\Delta t)^2 \tag{2.373}$$

$$[M^*]^{(i)}\{A\}_{n+1}^{(i+1)} = \{F\}^{(i)} \tag{2.374}$$

$$\{V\}_{n+1}^{(i+1)} = \{V\}_{n+1}^{(1)} + \gamma\Delta t\{A\}_{n+1}^{(i+1)} \tag{2.375}$$

$$\{U\}_{n+1}^{(i+1)} = \{U\}_{n+1}^{(1)} + \beta(\Delta t)^2\{A\}_{n+1}^{(i+1)}. \tag{2.376}$$

This scheme has the disadvantage that $\{V\}_{n+1}^{(1)}$ and $\{U\}_{n+1}^{(1)}$ have to be stored. This can be avoided by writing Equations (2.375) and (2.376) for i and solving for $\{U\}_{n+1}^{(1)}$ and $\{V\}_{n+1}^{(1)}$:

$$\{V\}_{n+1}^{(1)} = \{V\}_{n+1}^{(i)} - \Delta t\gamma\{A\}_{n+1}^{(i)} \tag{2.377}$$

$$\{U\}_{n+1}^{(1)} = \{U\}_{n+1}^{(i)} - (\Delta t)^2\beta\{A\}_{n+1}^{(i)}. \tag{2.378}$$

Substituting these equations into Equations (2.371) to (2.374) one obtains

$$\begin{aligned}\left\{[M] + (1+\alpha)[C]_{n+1}^{(i)}\gamma\Delta t + (1+\alpha)[K]_{n+1}^{(i)}\beta(\Delta t)^2\right\}\{A\}_{n+1}^{(i+1)} \\ - \left\{(1+\alpha)[C]_{n+1}^{(i)}\gamma\Delta t + (1+\alpha)[K]_{n+1}^{(i)}\beta(\Delta t)^2\right\}\{A\}_{n+1}^{(i)} \\ = (1+\alpha)\{F\}_{n+1}^{\text{ext}} - \alpha\{F\}_n^{\text{ext}} + \alpha\{F\}_n^{\text{int}} \\ - (1+\alpha)\left\{[C]_{n+1}^{(i)}\{V\}_{n+1}^{(i)} + [K]_{n+1}^{(i)}\{U\}_{n+1}^{(i)}\right\}. \quad (2.379)\end{aligned}$$

By defining

$$\{\Delta A\}^{(i)} = \{A\}_{n+1}^{(i+1)} - \{A\}_{n+1}^{(i)} \quad (2.380)$$

$$\{A\}_{n+1}^{(1)} = \{0\} \quad (2.381)$$

and

$$\{R\}^{(i)} = \{F\}^{(i)*} - [M]\{A\}_{n+1}^{(i)} \quad (2.382)$$

where

$$\{F\}^{(i)*} = (1+\alpha)\{F\}_{n+1}^{\text{ext}} - \alpha\{F\}_n^{\text{ext}} - (1+\alpha)\{F\}_{n+1}^{(i)\text{int}*} + \alpha\{F\}_n^{\text{int}} \quad (2.383)$$

and

$$\{F\}_{n+1}^{(i)\text{int}*} = [C]_{n+1}^{(i)}\{V\}_{n+1}^{(i)} + [K]_{n+1}^{(i)}\{U\}_{n+1}^{(i)} \quad (2.384)$$

Equation (2.379) can be reduced to

$$[M^*]^{(i)}\{\Delta A\}^{(i)} = \{R\}^{(i)}. \quad (2.385)$$

Furthermore, Equations (2.375) to (2.378) yield

$$\{V\}_{n+1}^{(i+1)} = \{V\}_{n+1}^{(i)} + \gamma\Delta t\{\Delta A\}^{(i)} \quad (2.386)$$

$$\{U\}_{n+1}^{(i+1)} = \{U\}_{n+1}^{(i)} + \beta(\Delta t)^2\{\Delta A\}^{(i)}. \quad (2.387)$$

$\{R\}^{(i)}$ is the residual in iteration i and its size can be used as a criterion to exit the loop. Summarizing,

1. Calculate the predictor values: Equations (2.360), (2.361)
2. Loop i, $i = 1, 2, \ldots$: multicorrector step
   (a) Calculate $\{F\}^{(i)*}$ (Equation (2.383)), $[M^*]^{(i)}$ (Equation (2.373)), $\{R\}^{(i)}$ (Equation (2.382)). If $\|\{R\}^{(i)}\| < \epsilon$: exit.
   (b) Calculate $\{\Delta A\}^{(i)}$ (Equation (2.385).
   (c) Update $\{U\}_{n+1}$, $\{V\}_{n+1}$ and $\{A\}_{n+1}$ (Equations (2.386), (2.387), (2.380)).

Those degrees of freedom for which boundary values are given are not a part of the solution system. For these degrees of freedom, either $\{U\}_{n+1}$, $\{V\}_{n+1}$ or $\{A\}_{n+1}$ is given. The predictor step yields $\{U\}_{n+1}^{(1)}$, $\{V\}_{n+1}^{(1)}$ and $\{A\}_{n+1}^{(1)}$ (Equations (2.360), (2.361) and (2.381)) as usual. However, now only one corrector step is necessary. We set

$$\{U\}_{n+1}^{(2)} = \{U\}_{n+1} \tag{2.388}$$

or

$$\{V\}_{n+1}^{(2)} = \{V\}_{n+1} \tag{2.389}$$

or

$$\{A\}_{n+1}^{(2)} = \{A\}_{n+1} . \tag{2.390}$$

Then, Equations (2.387), (2.386) and (2.380) yield for $i = 1$

$$\{\Delta A\}^{(1)} = \frac{1}{\beta(\Delta t)^2}\left[\{U\}_{n+1} - \{U\}_{n+1}^{(1)}\right] \tag{2.391}$$

$$\{\Delta A\}^{(1)} = \frac{1}{\gamma \Delta t}\left[\{V\}_{n+1} - \{V\}_{n+1}^{(1)}\right] \tag{2.392}$$

or

$$\{\Delta A\}^{(1)} = \{A\}_{n+1} . \tag{2.393}$$

Substituting $\{\Delta A\}^{(1)}$ into Equation (2.386) or Equation (2.387) yields the remaining unknowns (displacement if the velocity is given, velocity if the displacement is given and displacement and velocity if the acceleration is given).

The assumption that $\boldsymbol{a}$ in $[t_n, t_{n+1}]$ is a linear combination of $\boldsymbol{a}_n$ and $\boldsymbol{a}_{n+1}$ works well unless $\boldsymbol{a}$ changes discontinuously. For instance, if the external force jumps at $t = t_n^+$, the acceleration a_n^+ has to be adjusted accordingly to get accurate results. The correct acceleration is obtained by using Equation (2.338):

$$\begin{aligned}[M]\{A\}_{n+} &= \{F\}_{n+}^{\text{ext}} - \{F\}_n^{\text{int}} \\ &= \{F\}_n^{\text{ext}} + \{\Delta F\} - \{F\}_n^{\text{int}}\end{aligned} \tag{2.394}$$

where $\{\Delta F\}$ is the force jump. Consequently, the acceleration jump amounts to

$$[M]\ \{A\}_{n+} - \{A\}_n\big) = \{\Delta F\} . \tag{2.395}$$

2.11.3 Consistency and accuracy of the implicit formulation

In order to examine the consistency, accuracy and stability of the implicit scheme, Equation (2.349) together with Equations (2.344) to (2.345) and Equations (2.350) to (2.353) are written for a homogeneous (no external force) single degree of freedom system:

$$
\begin{aligned}
&[m + (1+\alpha)c\gamma\Delta t + (1+\alpha)k\beta(\Delta t)^2]a_{n+1} \\
&\qquad = \alpha c v_n + \alpha k u_n - (1+\alpha)c[v_n + (1-\gamma)\Delta t a_n] \\
&\qquad\qquad - (1+\alpha)k[u_n + \Delta t v_n + \frac{1}{2}(\Delta t)^2(1-2\beta)a_n]
\end{aligned} \tag{2.396}
$$

or

$$
\begin{aligned}
&[m + (1+\alpha)c\gamma\Delta t + (1+\alpha)k\beta(\Delta t)^2]a_{n+1} \\
&\qquad = -(1+\alpha)[c(1-\gamma)\Delta t + \frac{1}{2}k(\Delta t)^2(1-2\beta)]a_n \\
&\qquad\qquad - [c + (1+\alpha)k\Delta t]v_n - k u_n.
\end{aligned} \tag{2.397}
$$

Defining the frequency Ω and the friction coefficient ξ by

$$\Omega^2 := \frac{k}{m}(\Delta t)^2 \tag{2.398}$$

$$\xi := \frac{c}{2m\Omega}\Delta t \tag{2.399}$$

which is equivalent to

$$k = \frac{m\Omega^2}{(\Delta t)^2} \tag{2.400}$$

$$c = \frac{2m\Omega\xi}{\Delta t} \tag{2.401}$$

one can replace k and c in Equation (2.397), yielding

$$
\begin{aligned}
&[1 + (1+\alpha)2\Omega\xi\gamma + (1+\alpha)\Omega^2\beta]a_{n+1}(\Delta t)^2 \\
&\qquad = -(1+\alpha)[(1-\gamma)2\Omega\xi + \tfrac{1}{2}(1-2\beta)\Omega^2]a_n(\Delta t)^2 \\
&\qquad\qquad - [2\Omega\xi + (1+\alpha)\Omega^2]v_n\Delta t - \Omega^2 u_n.
\end{aligned} \tag{2.402}
$$

The one-dimensional equivalent of Equations (2.344) to (2.347) yields

$$\Delta t v_{n+1} = \Delta t v_n + (1-\gamma)(\Delta t)^2 a_n + \gamma(\Delta t)^2 a_{n+1} \tag{2.403}$$

$$u_{n+1} = u_n + \Delta t v_n + \frac{1}{2}(1-2\beta)(\Delta t)^2 a_n + \beta(\Delta t)^2 a_{n+1}. \tag{2.404}$$

After substitution of Equation (2.402) into the right-hand side of Equations (2.403) and (2.404), these three equations form a system expressing u_{n+1}, $\Delta t v_{n+1}$ and $(\Delta t)^2 a_{n+1}$ in terms of u_n, $\Delta t v_n$ and $(\Delta t)^2 a_n$:

$$\begin{Bmatrix} u_{n+1} \\ (\Delta t)v_{n+1} \\ (\Delta t)^2 a_{n+1} \end{Bmatrix} = [A] \begin{Bmatrix} u_n \\ (\Delta t)v_n \\ (\Delta t)^2 a_n \end{Bmatrix} \tag{2.405}$$

where

$$[A] = \begin{bmatrix} 1+\beta P & 1+\beta Q & \frac{1}{2}+\beta(R-1) \\ \gamma P & 1+\gamma Q & 1+\gamma(R-1) \\ P & Q & R \end{bmatrix} \tag{2.406}$$

and

$$P = -\Omega^2/d \tag{2.407}$$

$$Q = -[2\xi\Omega + (1+\alpha)\Omega^2]/d \tag{2.408}$$

$$R = -(1+\alpha)[(1-\gamma)2\Omega\xi + \tfrac{1}{2}(1-2\beta)\Omega^2]/d \tag{2.409}$$

$$d = 1 + (1+\alpha)2\Omega\xi\gamma + (1+\alpha)\Omega^2\beta. \tag{2.410}$$

Equation (2.405) is a set of three homogeneous finite difference equations of the first order with constant coefficients. Solutions are obtained by setting

$$\begin{Bmatrix} u_n \\ (\Delta t)v_n \\ (\Delta t)^2 a_n \end{Bmatrix} = \lambda^n \{X\} \tag{2.411}$$

leading to

$$[A]\{X\} = \lambda\{X\} \tag{2.412}$$

which is a classical eigenvalue problem. Solutions exist if the characteristic equation is satisfied

$$\lambda^3 - I_1\lambda^2 + I_2\lambda - I_3 = 0 \tag{2.413}$$

where I_1, I_2 and I_3 are the invariants of A (the elements of A are denoted by $A_{11}, \ldots$) :

$$I_1 = \operatorname{tr} A = A_{11} + A_{22} + A_{33} \tag{2.414}$$

$$I_2 = \begin{vmatrix} A_{22} & A_{23} \\ A_{32} & A_{33} \end{vmatrix} + \begin{vmatrix} A_{11} & A_{13} \\ A_{31} & A_{33} \end{vmatrix} + \begin{vmatrix} A_{11} & A_{12} \\ A_{21} & A_{22} \end{vmatrix} \tag{2.415}$$

$$I_3 = \det A. \tag{2.416}$$

Equation (2.413) is also the characteristic equation of the following third-order equation:

$$D = u_{n+1} - I_1 u_n + I_2 u_{n-1} - I_3 u_{n-2} = 0 \tag{2.417}$$

which is equivalent to Equation (2.405). After some algebra, one obtains the invariants of Equation (2.406):

$$I_1 = 2 - \left\{2\xi\Omega\left[1+\alpha(1-\gamma)\right] + \Omega^2\left[(1+\alpha)(\gamma+\tfrac{1}{2}) - \beta\alpha\right]\right\}/d \tag{2.418}$$

$$I_2 = 1 - \left\{2\xi\Omega\left[1+2\alpha(1-\gamma)\right] + \Omega^2\left[\gamma - 1/2 + 2\alpha(\gamma-\beta)\right]\right\}/d \tag{2.419}$$

$$I_3 = -\left[2\xi\Omega\alpha(1-\gamma) + \Omega^2\alpha(\gamma - \beta - \tfrac{1}{2}\right]/d. \tag{2.420}$$

To examine the consistency and accuracy, Equation (2.417) is expanded by Taylor series about $u_n = u$:

$$u_{n+1} = u_n + \dot{u}\Delta t + \ddot{u}\frac{(\Delta t)^2}{2!} + \dddot{u}\frac{(\Delta t)^3}{3!} + \cdots \tag{2.421}$$

and similarly for u_{n-1} and u_{n-2}. Collecting terms, one obtains

$$\begin{aligned} D = {} & (1 - I_1 + I_2 - I_3)u + (1 - I_2 + 2I_3)\dot{u}\Delta t \\ & + (1 + I_2 - 4I_3)\ddot{u}\frac{(\Delta t)^2}{2} + (1 - I_2 + 8I_3)\dddot{u}\frac{(\Delta t)^3}{3!} + \cdots \end{aligned} \tag{2.422}$$

Defining

$$\overline{I}_i := I_i(d = 1) \tag{2.423}$$

one obtains after some algebra

$$1 - \overline{I}_1 + \overline{I}_2 - \overline{I}_3 = \Omega^2 \tag{2.424}$$

$$1 - \overline{I}_2 + 2\overline{I}_3 = \tfrac{\Omega}{2}[4\xi + (-1 + 2\alpha + 2\gamma)\Omega] \tag{2.425}$$

$$1 + \overline{I}_2 - 4\overline{I}_3 = \tfrac{1}{2}\left\{4 - 4[1 + 2\alpha(\gamma - 1)]\xi\Omega + (1 - 4\alpha - 4\alpha\beta - 2\gamma + 4\alpha\gamma)\Omega^2\right\} \tag{2.426}$$

$$1 - \overline{I}_2 + 8\overline{I}_3 = -\tfrac{1}{2}\Omega\{-4[1 + 6\alpha(\gamma - 1)]\xi + [1 - 4\alpha(2 + 3\beta - 3\gamma) - 2\gamma]\Omega\} \tag{2.427}$$

and

$$1 - I_1 + I_2 - I_3 = (1 - \overline{I}_1 + \overline{I}_2 - \overline{I}_3)/d \tag{2.428}$$

$$1 - I_2 + 2I_3 = (1 - \overline{I}_2 + 2\overline{I}_3)/d \tag{2.429}$$

$$1 + I_2 - 4I_3 = [2(d - 1) + 1 + \overline{I}_2 - 4\overline{I}_3]/d \tag{2.430}$$

$$1 - I_2 + 8I_3 = (1 - \overline{I}_2 + 8\overline{I}_3)/d. \tag{2.431}$$

Now, we will assume there is no friction: $\xi = 0$. Defining ω by

$$\Omega := \omega\Delta t \tag{2.432}$$

collecting terms in Equation (2.422) yields

$$D = (\Delta t)^2(\omega^2 u + \ddot{u}) + (\Delta t)^3\left[\tfrac{1}{2}(-1 + 2\alpha + 2\gamma)\omega^2\dot{u}\right] + O(\Delta t)^4 \tag{2.433}$$

or

$$\ddot{u} + \omega^2 u = \frac{D}{(\Delta t)^2} - \Delta t\left[(\alpha + \gamma - \tfrac{1}{2})\omega^2\dot{u}\right] + O(\Delta t)^2. \tag{2.434}$$

The left-hand side is the governing equation without friction, the first term on the right-hand side is its approximation. For $\Delta t \to 0$ both are identical. Consequently, the numerical scheme is consistent. The accuracy is given by the power of the subsequent Δt terms on the right-hand side. In general, the scheme is first order. However, if

$$\gamma = \tfrac{1}{2} - \alpha \tag{2.435}$$

the scheme is second order. This will be assumed to be the case in the following section.

2.11.4 Stability of the implicit scheme

A numerical scheme is stable if its amplification matrix $[A]$, Equation (2.406), has no eigenvalues with size greater than one. Indeed, if $|\lambda| > 1$, the homogeneous solution, Equation (2.411) diverges. Since a particular solution (i.e. a solution of the inhomogeneous problem) augmented by a linear combination of the homogeneous solutions is a particular solution as well, any instable homogeneous solution will lead to an instability of the inhomogeneous scheme. Accordingly, we have to solve Equation (2.413) for λ and check that $|\lambda| \leq 1$.

Since the coefficients I_1, I_2 and I_3 are a function of Ω, Equations (2.418) to (2.420), so is its solution λ. Recall that the parameter γ is defined by Equation (2.435), α and β are still adjustable. The strategy that will be adopted here follows the treatise in (Hilber 1976): first, we check the value of $|\lambda|$ for $\Omega = 0$, then we examine $\Omega \to \infty$ and finally look at the values in-between.

1. For $\Omega = 0$ we have $I_1 = 2$, $I_2 = 1$ and $I_3 = 0$. The roots are $\lambda = 0$ and $\lambda = 1$, the latter is a double root. The condition $|\lambda| \leq 1$ is satisfied.

2. For $\Omega \to \infty$ I_1, I_2 and I_3 reduce to

$$I_1 \sim 2 - \frac{(1+\alpha)(\gamma+\frac{1}{2}) - \beta\alpha}{(1+\alpha)\beta}, \qquad \Omega \to \infty \tag{2.436}$$

$$I_2 \sim 1 - \frac{\gamma - \frac{1}{2} + 2\alpha(\gamma-\beta)}{(1+\alpha)\beta}, \qquad \Omega \to \infty \tag{2.437}$$

$$I_3 \sim \frac{\alpha(\gamma - \beta - \frac{1}{2})}{(1+\alpha)\beta}, \qquad \Omega \to \infty. \tag{2.438}$$

Notice that ξ drops out for $\Omega \to \infty$. Substituting $\gamma = \frac{1}{2} - \alpha$ and rearranging, Equation (2.413) yields

$$\begin{aligned}\alpha(\alpha+\beta) &- (2\alpha^2 + \beta(1+\alpha) + 2\alpha\beta)\lambda \\ &+ [2\beta(1+\alpha) + \alpha\beta - 1 + \alpha^2]\lambda^2 - \lambda^3(1+\alpha)\beta = 0. \end{aligned}\tag{2.439}$$

Since the term $(1+\alpha)/\beta$ that is multiplying λ^3 originates from the denominator of Equations (2.436) to (2.438), $\alpha = -1$ and $\beta = 0$ must be excluded. The solution of Equation (2.439) is (by inspection or by using a symbolic mathematical program)

$$\lambda_3 = \frac{\alpha}{1+\alpha} \tag{2.440}$$

$$\lambda_{1,2} = 1 - \frac{1-\alpha}{2\beta} \pm \frac{1}{2\beta}\sqrt{(1-\alpha)^2 - 4\beta}. \tag{2.441}$$

Let us first have a closer look at λ_1 and λ_2, Figures 2.25 and 2.26.

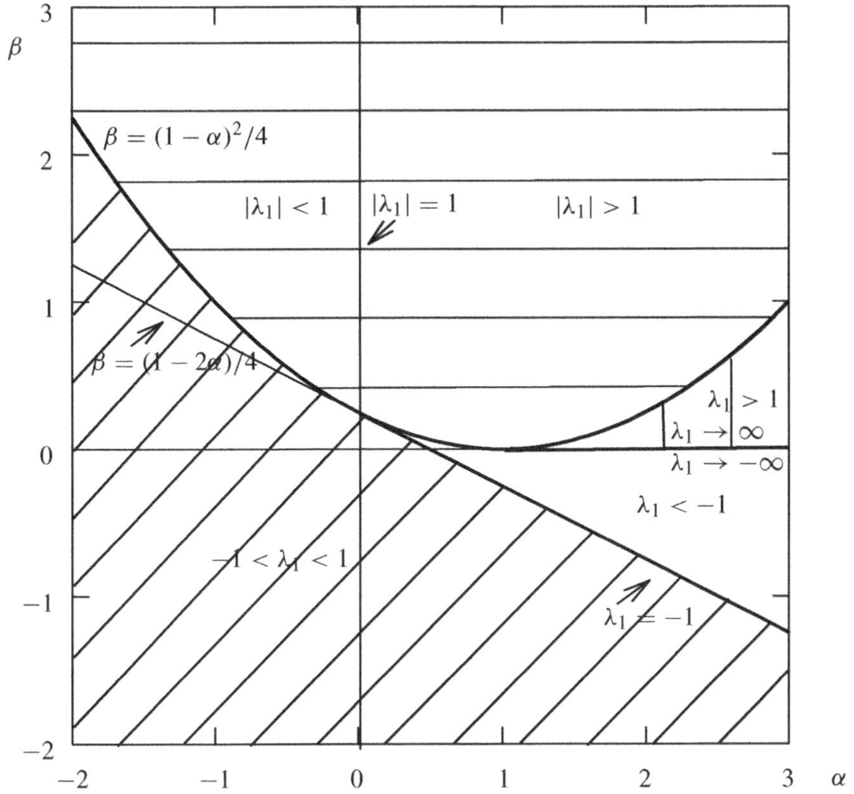

Figure 2.25 Evaluation of λ_1

(a) For $(1-\alpha)^2 < 4\beta$, λ_1 and λ_2 are complex and

$$|\lambda_1| = |\lambda_2| = \sqrt{\frac{\alpha+\beta}{\beta}}. \tag{2.442}$$

$|\lambda_1| = |\lambda_2| = 1$ for $\alpha = 0$, to the left of the β-axis we have $|\lambda_1| < 1$, $|\lambda_2| < 1$, to the right we obtain $|\lambda_1| > 1$, $|\lambda_2| > 1$.

(b) On the parabola $(1-\alpha)^2 = 4\beta$ and one finds

$$\lambda_1 = \lambda_2 = -\frac{1+\alpha}{1-\alpha} \tag{2.443}$$

leading to

$$\alpha < 0 \Rightarrow -1 < \lambda_{1,2} < 1$$

$$\alpha = 0 \Rightarrow \lambda_{1,2} = -1$$

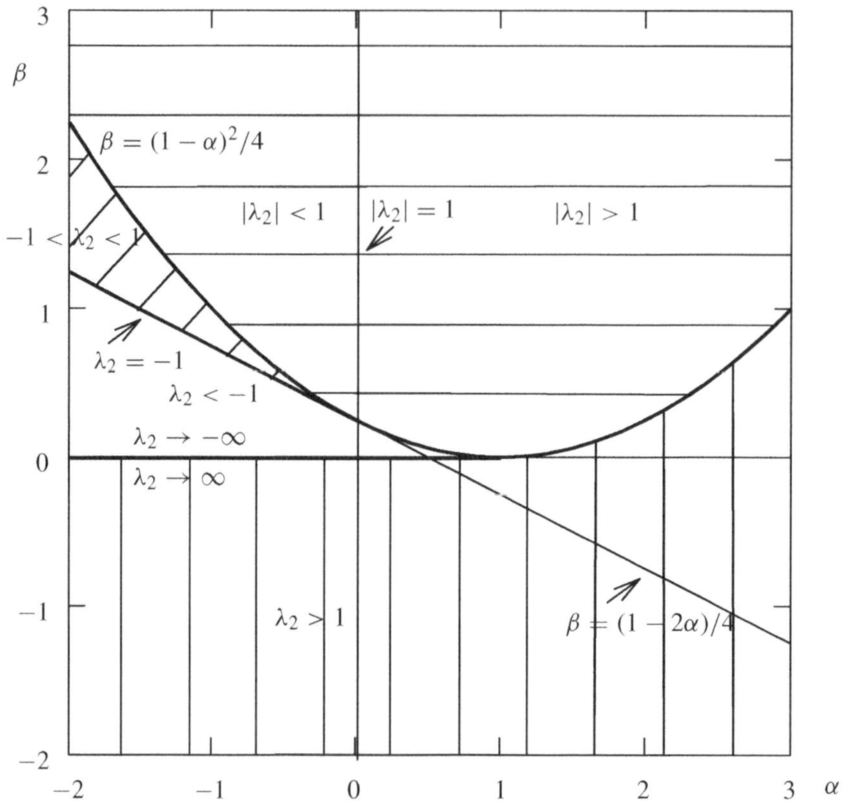

Figure 2.26 Evaluation of λ_2

$$0 < \alpha < 1 \Rightarrow \lambda_{1,2} < -1$$

$$\alpha > 1 \Rightarrow \lambda_{1,2} > 1. \tag{2.444}$$

(c) If $(1-\alpha)^2 \geq 4\beta$, λ_1 and λ_2 are both real. Closer examination reveals

$$\lambda_1 \neq 1, \lambda_2 \neq 1 \tag{2.445}$$

$$\lambda_1 = -1 \Leftrightarrow \beta = \frac{1-2\alpha}{4}, \quad \alpha \geq 0 \tag{2.446}$$

$$\lambda_2 = -1 \Leftrightarrow \beta = \frac{1-2\alpha}{4}, \quad \alpha \leq 0. \tag{2.447}$$

The straight line $\beta = (1-2\alpha)/4$ is tangent to the parabola at $\alpha = 0$. For $\beta \to 0$, $\lambda_{1,2}$ is not properly defined by Equation (2.441) and an asymptotic expansion must be developed

$$\lambda_{1,2} = 1 - \frac{1-\alpha}{2\beta} \pm \frac{|1-\alpha|}{2\beta}\sqrt{1 - \frac{4\beta}{(1-\alpha)^2}}$$

$$= 1 - \frac{1-\alpha}{2\beta} \pm \frac{|1-\alpha|}{2\beta}\left[1 - \frac{2\beta}{(1-\alpha)^2} + O(\beta^2)\right], \quad \beta \to 0$$

$$= 1 - \frac{1-\alpha}{2\beta} \pm \frac{|1-\alpha|}{2\beta} \mp \frac{|1-\alpha|}{(1-\alpha)^2} + O(\beta), \quad \beta \to 0. \tag{2.448}$$

This yields for λ_1

$$\begin{aligned} \lambda_1 &= 1 - \frac{1}{1-\alpha} + O(\beta), \quad \beta \to 0, \alpha < 1 \\ \lambda_1 &= \frac{\alpha - 1}{\beta} + O(1), \quad \beta \to 0, \alpha > 1 \end{aligned} \tag{2.449}$$

and for λ_2

$$\begin{aligned} \lambda_2 &= -\frac{1-\alpha}{\beta} + O(1), \quad \beta \to 0, \alpha < 1 \\ \lambda_2 &= 1 + \frac{1}{\alpha - 1} + O(\beta), \quad \beta \to 0, \alpha > 1 \end{aligned} . \tag{2.450}$$

Summarizing, the straight line $\beta = (1-2\alpha)/4$ divides the region under the parabola into three zones. Only for $\alpha \leq 0$ and $\beta \geq (1-2\alpha)/4$ we have $|\lambda_1| \leq 1$ and $|\lambda_2| \leq 1$. λ_1 and λ_2 are sometimes called the *principal roots*.

For the third eigenvalue (sometimes called the *spurious root*), one obtains

$$|\lambda_3| = 1 \Leftrightarrow \lambda_3 = -1 \Leftrightarrow \alpha = -\frac{1}{2}. \tag{2.451}$$

However (see Figure 2.27),

$$\begin{aligned} |\lambda_3| \leq 1 &\Leftrightarrow \alpha \geq -\frac{1}{2} \\ |\lambda_3| > 1 &\Leftrightarrow \alpha < -\frac{1}{2}. \end{aligned} \tag{2.452}$$

Accordingly, the implicit scheme is unconditionally stable (i.e. $|\lambda_1| \leq 1$, $|\lambda_2| \leq 1$ and $|\lambda_3| \leq 1$) at high frequencies only if

$$\begin{aligned} -\frac{1}{2} &\leq \alpha \leq 0 \\ \beta &\geq \frac{1-2\alpha}{4}. \end{aligned} \tag{2.453}$$

Now, for $\Omega \to \infty$, we would like to maximize the dissipation to get rid of spurious high-frequency effects, that is, we seek to minimize $\max(|\lambda_1|, |\lambda_2|, |\lambda_3|)$. How does β affect $\max(|\lambda_1|, |\lambda_2|, |\lambda_3|)$? Since λ_3 is not a function of β, we focus on λ_1 and λ_2. One can prove that for fixed $\alpha \geq -\frac{1}{2}$, $\max(|\lambda_1|, |\lambda_2|)$ attains a minimum for

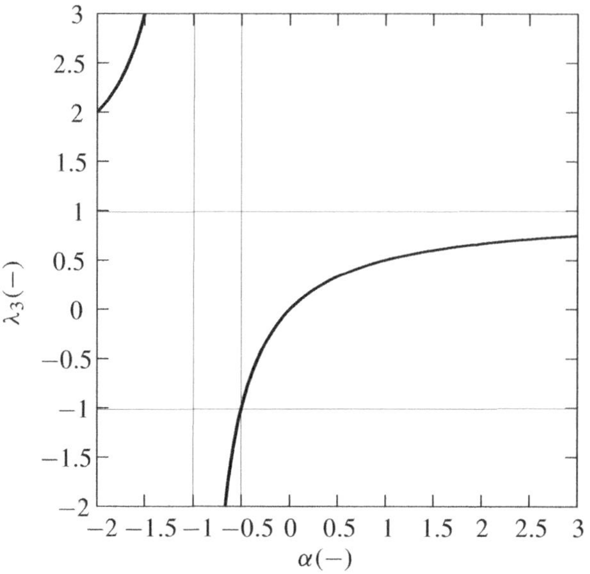

Figure 2.27 Evaluation of λ_3

$\beta = (1-\alpha)^2/4$, that is, on the parabola, satisfying

$$|\lambda_1| = |\lambda_2| = \frac{1+\alpha}{1-\alpha}. \tag{2.454}$$

Indeed, for $\beta > (1-\alpha)^2/4$ we deduced Equation (2.442) which is a monotonic increasing function of β toward 1 for $\beta \to \infty$. The derivative of Equation (2.441) with respect to β satisfies

$$\frac{\partial \lambda_{1,2}}{\partial \beta} = \frac{1-\alpha}{2\beta^2} \mp \frac{1}{2\beta^2}\sqrt{(1-\alpha)^2+4\beta} \mp \frac{1}{\beta\sqrt{(1-\alpha)^2+4\beta}}. \tag{2.455}$$

For $\beta = (1-\alpha)^2/4$ one gets

$$\lambda_1 = \lambda_2 = -\frac{1+\alpha}{1-\alpha} < 0 \quad \text{for } \alpha > -1. \tag{2.456}$$

Now, $\frac{\partial \lambda_2}{\partial \beta} > 0$ (Equation (2.455)), hence, λ_2 decreases with decreasing β. Accordingly, $|\lambda_2|$ and $\max(|\lambda_1|, |\lambda_2|)$ increase with decreasing β in a neighborhood of $\beta = (1-\alpha)^2/4$. This completes the proof.

Consequently, we maximize the high-frequency dissipation if we take

$$\beta = \frac{(1-\alpha)^2}{4} \tag{2.457}$$

and the only parameter left is α. Summarizing, for $-\frac{1}{2} \leq \alpha \leq 0$:

$$|\lambda_1| = |\lambda_2| = \frac{1+\alpha}{1-\alpha} \leq 1 \tag{2.458}$$

$$|\lambda_3| = \frac{-\alpha}{1+\alpha} \leq 1 \tag{2.459}$$

$|\lambda_1| = |\lambda_2| = |\lambda_3| = \frac{1}{2}$ for $\alpha = -\frac{1}{3}$. $|\lambda_1| = |\lambda_2|$ is a monotonic decreasing function of α and $|\lambda_3|$ is monotonic increasing. Consequently, $\max(|\lambda_1|, |\lambda_2|, |\lambda_3|)$ has a minimum at $\lambda = -1/3$. The complete range $[-\frac{1}{2}, 0]$ for $\max(|\lambda_1|, |\lambda_2|, |\lambda_3|)$ is covered if one takes $\alpha \in [-\frac{1}{2}, 0]$. This concludes the treatment for $\Omega \to \infty$.

3. To assure that $\max(|\lambda_1|, |\lambda_2|, |\lambda_3|) \leq 1$ for $0 < \Omega < \infty$, the feasibility of $|\lambda| = 1$ as solution of Equation (2.413) will be checked. In general, the solution can be complex. Substituting $\lambda = e^{i\varphi}$ in Equation (2.413) and separating the real and imaginary part of the equation yields

$$\cos 3\varphi - I_1 \cos 2\varphi + I_2 \cos\varphi - I_3 = 0 \tag{2.460}$$

$$\sin 3\varphi - I_1 \sin 2\varphi + I_2 \sin\varphi = 0. \tag{2.461}$$

Both equations must be satisfied. Expanding $\sin 2\varphi = 2\sin\varphi\cos\varphi$ and $\sin 3\varphi = -3\cos\varphi + 4\cos^3\varphi$ in Equation (2.461) yields

$$\sin\varphi(-1 + 4\cos^2\varphi - 2I_1\cos\varphi + I_2) = 0 \tag{2.462}$$

which implies

$$\sin\varphi = 0 \Leftrightarrow \varphi = 0, \pi \tag{2.463}$$

or

$$4\cos^2\varphi - 2I_1\cos\varphi + I_2 = 1. \tag{2.464}$$

(a) For $\varphi = 0$, Equation (2.460) yields

$$1 - I_1 + I_2 - I_3 = 0 \tag{2.465}$$

(b) for $\varphi = \pi$ one obtains

$$1 + I_1 + I_2 + I_3 = 0. \tag{2.466}$$

(c) If we expand the terms $\cos 2\varphi = \cos^2\varphi - \sin^2\varphi$ and $\cos 3\varphi = -3\cos\varphi + 4\cos^3\varphi$ in Equation (2.460), we get

$$\cos\varphi(-3 + 4\cos^2\varphi - 2I_1\cos\varphi + I_2) + I_1 - I_3 = 0 \tag{2.467}$$

and Equation (2.467) can finally be transformed into

$$\cos\varphi = \frac{I_1 - I_3}{2}. \tag{2.468}$$

Substitution of Equation (2.468) into Equation (2.464) finally yields

$$I_3(I_3 - I_1) + I_2 - 1 = 0. \tag{2.469}$$

Equations (2.465), (2.466) and (2.469) cover all cases for which $|\lambda| = 1$.

(a) Equation (2.465) leads to (Equation (2.428) and Equation (2.424))

$$\Omega^2 = 0 \tag{2.470}$$

which yields a double root for $\Omega = 0$. This was already covered previously.

(b) Substituting $\gamma = \frac{1}{2} - \alpha$ and $\beta = (1-\alpha)^2/4$ in Equations (2.418) to (2.420) and using these in Equation (2.466) yields

$$\alpha^2(1+2\alpha)\Omega^2 - 8\alpha(1+2\alpha)\xi\Omega + 4 = 0 \tag{2.471}$$

with roots

$$\Omega_{1,2} = \frac{2\xi}{\alpha}\left[1 \pm \sqrt{1 - \frac{1}{4\xi^2(1+2\alpha)}}\right]. \tag{2.472}$$

For $4\xi^2(1+2\alpha) < 1$, there are no real solutions, for $4\xi^2(1+2\alpha) \geq 1$ the solutions are both negative since $\alpha < 0$ (for $\alpha = 0$, Equation (2.471) has no solution either). Accordingly, there are no positive real solutions of Equation (2.471).

(c) Finally, Equation (2.469) yields

$$\Omega[8\xi + 8\xi^2\Omega + 2\xi(1+\alpha^2)\Omega^2 - \alpha(1+\alpha)^2\Omega^3] = 0. \tag{2.473}$$

$\Omega = 0$ was already covered. For $\xi \neq 0$ and $\alpha \neq 0$ all the coefficients in Equation (2.473) are strictly positive ($\xi > 0$, $-\frac{1}{2} \leq \alpha < 0$) and $\Omega = 0$ is the only solution. If one of them is zero but not both, the same reasoning applies to the nonzero terms. If $\xi = \alpha = 0$ the equation is satisfied for all Ω. This corresponds to the classical Newmark algorithm.

Summarizing, for the parameter combinations $\gamma = \frac{1}{2} - \alpha$, $\beta = (1-\alpha)^2/4$, $\alpha \in [-\frac{1}{3}, 0]$, the implicit scheme is unconditionally stable and second-order accurate. The spectral radius for different values of α and $\xi = 0$ is shown in Figure 2.28, whereas the effect of ξ for $\alpha = -\frac{1}{3}$ is plotted in Figure 2.29. The figure shows that there is nearly no numerical dissipation for small Ω-values. For increasing values of Ω the dissipation gradually increases. In the previous derivation, it was shown that the solutions of the characteristic equation can be complex. This leads to oscillatory damping and a corresponding (small) period error of the solution. For more information the reader is referred to (Miranda *et al.* 1989), (Hilber and Hughes 1978) and (Bathe 1995).

2.11.5 Explicit formulation

The method in Section 2.11.1 is essentially implicit owing to the formulation of $[M^*]$ in Equation (2.350) and Equation (2.357). If $[M^*]$ is diagonal, the method is explicit. The mass matrix can be made diagonal by lumping. The problem is the damping matrix $[C]$ and the stiffness matrix $[K]$, which are usually not diagonal. In the explicit predictor–corrector

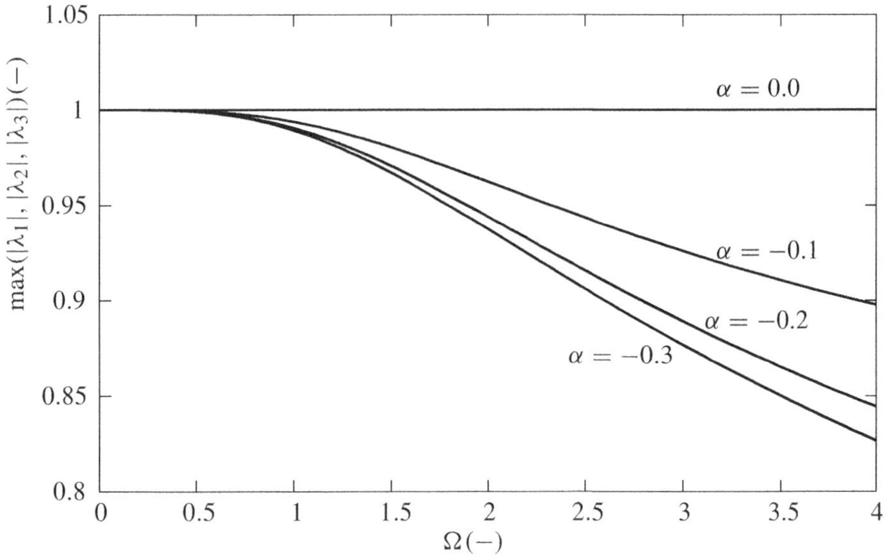

Figure 2.28 Spectral radius for $\xi = 0$

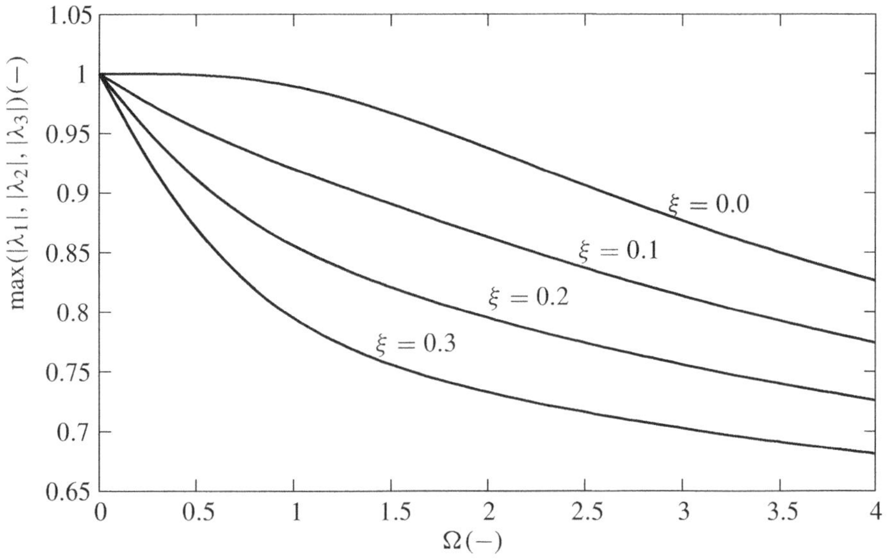

Figure 2.29 Spectral radius for $\alpha = -0.3$

procedure introduced here (Miranda *et al.* 1989), the diagonalization of $\left[M^*\right]$ is achieved by simply dropping the last two terms in Equation (2.350), that is, by setting

$$\left[M^*\right] = \left[M\right] \tag{2.474}$$

and lumping $\left[M\right]$. This really amounts to replacing Equation (2.339) by

$$\begin{aligned}\left[M\right]\left\{A\right\}_{n+1} + (1+\alpha)\left[C\right]\left\{\tilde{V}\right\}_{n+1} - \alpha\left[C\right]\left\{V\right\}_n + (1+\alpha)\left[K\right]\left\{\tilde{U}\right\}_{n+1}\\ - \alpha\left[K\right]\left\{U\right\}_n = (1+\alpha)\left\{F\right\}^{\text{ext}}_{n+1} - \alpha\left\{F\right\}^{\text{ext}}_n, \quad -1 \leq \alpha \leq 0.\end{aligned} \tag{2.475}$$

It can be shown that this iterative scheme is second-order accurate if Equation (2.342) and Equation (2.343) are satisfied. Furthermore, high-frequency dissipation is achieved for $\alpha < 0$. However, the explicit scheme is not unconditionally stable. Indeed, the one-dimensional equivalent of the explicit scheme corresponds to the equivalent model of the implicit scheme in which the parameter d defined in Equation (2.410) is replaced by 1. Equation (2.433) still applies and the explicit scheme is consistent and second-order accurate for $\xi = 0$ if $\gamma = \frac{1}{2} - \alpha$. However, the explicit scheme is not stable for $\Omega \to \infty$. To check stability, Equation (2.465), (2.466) and (2.469), which still apply, are analyzed. $\beta = (1-\alpha)^2/4$ is assumed throughout. Equation (2.465) reduces to $\Omega^2 = 0$ and deserves no further attention. Equation (2.466) now yields

$$(1 - \alpha - 2\alpha^2 - \alpha^3)\Omega^2 + 4(1 + \alpha + 2\alpha^2)\xi\Omega - 4 = 0 \tag{2.476}$$

leading to

$$\Omega_{1,2} = \frac{2(1+\alpha+2\alpha^2)}{1-\alpha-2\alpha^2-\alpha^3}\left[\pm\sqrt{\xi^2 + \frac{1-\alpha-2\alpha^2-\alpha^3}{(1+\alpha+2\alpha^2)^2}} - \xi\right]. \tag{2.477}$$

Since $1 - \alpha - 2\alpha^2 - \alpha^3 > 0$ and $1 + \alpha + 2\alpha^2 > 0$ for $-\frac{1}{2} \leq \alpha \leq 0$, the positive root in Equation (2.477) marks a relevant crossing of the $|\lambda| = 1$ line. Figure 2.30 shows Ω_1 as a function of α for different ξ values.

Equation (2.469) reduces to

$$\Omega\{-8\xi - 8\alpha(1+2\alpha)\xi^2\Omega + 2\xi[-\alpha(1+2\alpha) + \alpha^3]\Omega^2 + \alpha(1+\alpha)^2\Omega^3\} = 0. \tag{2.478}$$

For $\xi = 0$ the only solution is $\Omega = 0$. A numerical analysis shows that also for $\xi > 0$ there are no positive real roots of Equation (2.478) smaller than Ω_1 in Equation (2.477). Summarizing, the stable regime is limited by a critical "frequency" value given by Ω_1 in Equation (2.477) and plotted in Figure 2.30.

2.11.6 The consistent mass matrix

The consistent mass matrix is obtained by evaluating Equation (2.24). This is usually performed by one of the integration schemes from Section 2.3. In the previous sections, it was explained that a force jump leads to a jump in the acceleration. To this end Equation (2.395) has to be evaluated. This is a system of equations with the mass matrix on the left-hand side.

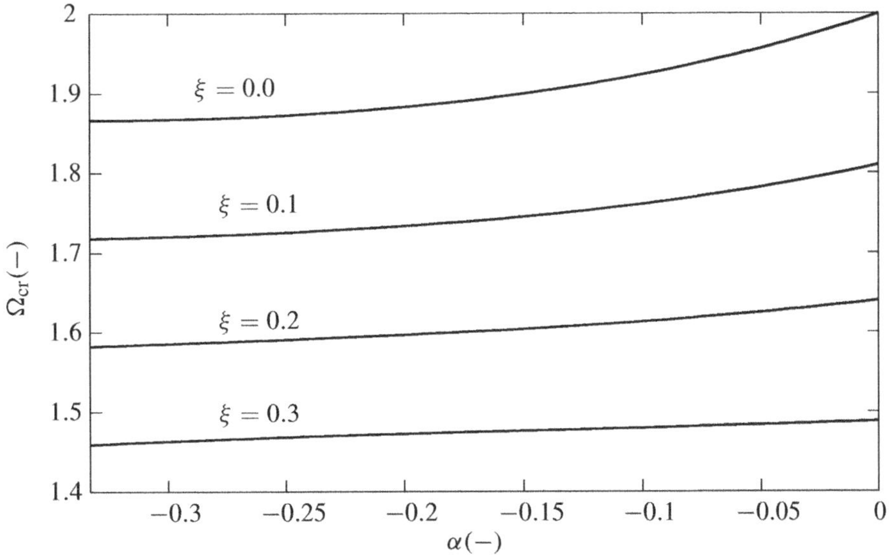

Figure 2.30 Critical frequency for the explicit scheme

Equation (2.395) cannot be solved if the mass matrix is singular. Physically, this situation cannot arise since the mass matrix is positive-definite. Indeed, the kinetic energy satisfies

$$\mathcal{K} = \frac{1}{2}\left\{V\right\}^{\mathrm{T}}\left[M\right]\left\{V\right\} = 0 \Leftrightarrow \left\{V\right\} = 0. \tag{2.479}$$

Consequently,

$$\left[M\right]\left\{V\right\} = 0 \Rightarrow \left\{V\right\} = 0 \tag{2.480}$$

and $\left[M\right]$ is regular. However, $\left[M\right]$ can become singular because of the numerical integration. To realize this, consider just one finite element with constant initial density and recall that the numerical integration of Equation (2.24) amounts to

$$M_{ij} \approx \rho_0 \sum_{k=1}^{N} w_k \varphi_{ik} \varphi_{jk} J_k^* \tag{2.481}$$

where M_{ij} denotes the entry in row i and column j of the matrix $\left[M\right]$, w_k are the weighting functions, J_k^* is the Jacobian determinant of the global–local coordinate transformation and the indices K and M in Equation (2.24) were dropped since the mass matrix does not depend on them (the mass matrix really consists of three identical submatrices, one for each coordinate direction). The size of the matrix in Equation (2.481) is equal to the number of nodes in the element. A matrix $\left[a\right]$ is singular if its determinant vanishes.

Suppose there is only one integration point. This is the case for the 8-node brick element with reduced integration and the four-node tetrahedral element with standard integration. Accordingly, Equation (2.481) reduces to

$$M_{ij} \approx \rho_0 w_1 \varphi_{i1} \varphi_{j1} J_1^*. \tag{2.482}$$

Furthermore, the shape functions and the location of the integration point for these elements are such that

$$\varphi_{i1} = \varphi_{j1} =: \varphi_1 \quad \forall i, j \tag{2.483}$$

and we get

$$M_{ij} \approx \rho_0 w_1 \varphi_1 \varphi_1 J_1^*. \tag{2.484}$$

All entries in the matrix are identical: the matrix is singular. Also, 20-node brick elements with reduced integration frequently lead to badly conditioned mass matrices. Therefore, it is advisable to use the higher-order schemes to integrate the mass matrix.

2.11.7 Lumped mass matrix

In the explicit formulation the mass matrix is reduced to a diagonal form. This can be performed in several ways (Zienkiewicz and Taylor 1989). Here, only one method will be discussed, which is used in the CalculiX® code (CalculiX 2003). In this method, the lumped mass matrix is obtained by scaling the diagonal terms of the consistent mass matrix such that the total mass is recovered. Denoting the consistent element mass matrix by $\left[M_{Cij}\right]$ and the lumped element mass matrix by $\left[M_{Lij}\right]$ one finds

$$M_{Lii} = M_{Cii} \frac{M_e}{\sum_{j=1}^{n} M_{Cjj}} \tag{2.485}$$

where M_e is the total mass of the matrix, that is,

$$M_e = \sum_{i=1}^{n} \sum_{j=1}^{n} M_{Cij} \tag{2.486}$$

and n is the number of nodes in the element. This rule is applied to linear elements. For quadratic elements, a distinction is made between vertex node contributions and midside node contributions. Denoting the set of vertex nodes by VN and the set of midside nodes by MN, we define

$$\alpha := \frac{\sum_{i \in \mathrm{VN}} \int_{V_e} \varphi_i^2 \, \mathrm{d}V}{\sum_{j \in \mathrm{MN}} \int_{V_e} \varphi_i^2 \, \mathrm{d}V}. \tag{2.487}$$

α is a measure for the mass concentrated in the vertex nodes relative to the mass in the midside nodes. The integration in Equation (2.487) is performed in local coordinates. The lumped mass entries are now obtained by

$$M_{Lii} = M_{Cii} \left(\frac{M_e}{\sum_{j \in \mathrm{VN}} M_{Cjj}} \right) \left(\frac{\alpha}{1+\alpha} \right), \quad i \in \mathrm{VN} \tag{2.488}$$

$$M_{Lii} = M_{Cii} \left(\frac{M_e}{\sum_{j \in \mathrm{MN}} M_{Cjj}} \right) \left(\frac{1}{1+\alpha} \right), \quad i \in \mathrm{MN}. \tag{2.489}$$

Summing the masses in Equation (2.488) and Equation (2.489) readily shows that the total element mass is correctly reproduced. The factor α for some widely used quadratic elements is listed in Table 2.4.

Table 2.4 The lumping factor α for several element types.

Element type	α
20-node brick element	0.2917
10-node tetrahedral element	0.1203
15-node wedge element	0.2141

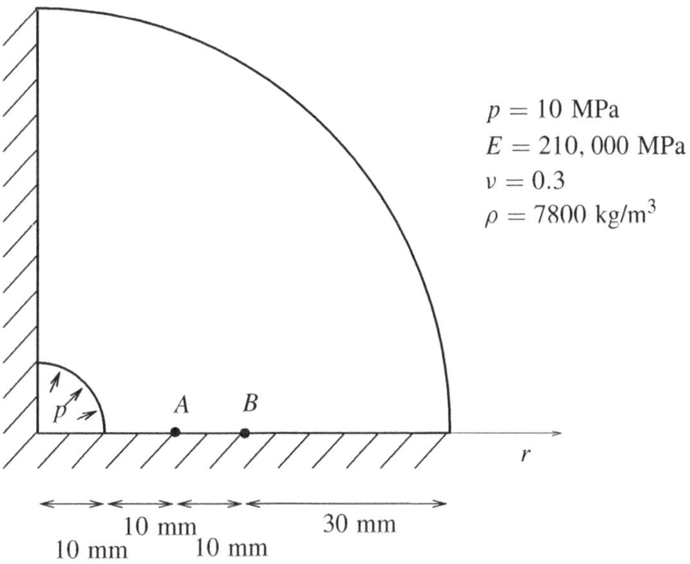

Figure 2.31 Geometry of the spherical shell and material data

2.11.8 Spherical shell subject to a suddenly applied uniform pressure

Consider the thick spherical shell in Figure 2.31 (only one-eighth is shown). At $t = 0$, a pressure p is applied and we are interested in the radial stresses as a function of time. It is known that the ensuing pressure waves travel at a speed c_1 satisfying (Graff 1975)

$$c_1 = \sqrt{\frac{\lambda + 2\mu}{\rho}} = 6.0202 \times 10^6 \text{ mm/s} \tag{2.490}$$

where λ and μ are Lamé's constants and ρ is the density of the material (λ and μ can be calculated from Young's modulus E and the Poisson coefficient ν in Figure (2.31) by use of Equations (1.450) and (1.451)). This means that they reach the outer surface of the shell after 8.3×10^{-6} s. One-eighth of the shell is meshed by 10 rows of 20-node brick elements with reduced integration across the thickness and 75 elements in circumferential direction, resulting in 750 elements.

Figure 2.32 shows the radial stress in points A and B by using the implicit α-method with $\alpha = -0.05$ and compares these results with the analytical solution (dashed lines,

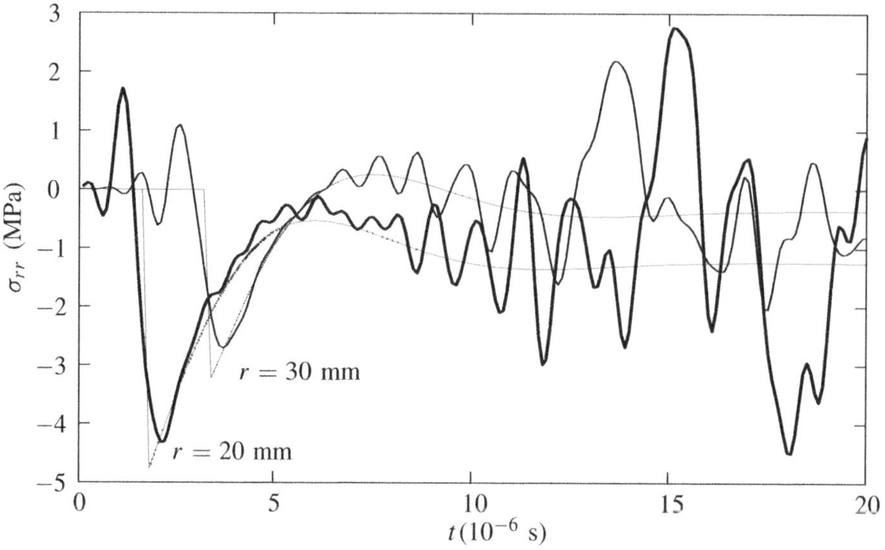

Figure 2.32 Radial stress after pressure surge

(Eringen 1980)). The analytical solution applies to a hole in infinite space, and therefore no reflection takes place. Since the radial wave equation does not exhibit dispersion (Graff 1975) the signal form is kept during propagation, although its amplitude changes. The numerical solution hits the extremal values of the analytical solution and the time at which they occur well. The finite element results are nodal values and therefore they are smeared out due to the extrapolation within the element (cf Section 2.4). For times exceeding the transversal time of the wall thickness, the wave is reflected leading to maxima at $t = 1.4 \times 10^{-5}$ s in B and $t = 1.5 \times 10^{-5}$ s in A. The subsequent maximum in A (after two reflections) takes place at $t = 1.82 \times 10^{-5}$ s. The results also show that quadratic elements tend to lead to oscillatory solutions for short-time calculations.

3

Geometric Nonlinear Effects

3.1 General Equations

Nonlinearities are involved in a lot of applications. Either the strains and/or rotations are large, such that the Lagrangian strain cannot be approximated by the infinitesimal strain, or there are discontinuities such as in contact phenomena. Another frequent source of nonlinearities is nonlinear material behavior. Although this chapter focuses on geometric nonlinearities, the present section treats both geometric and material nonlinearities.

Nonlinear problems are usually broken down into a repetition of linear ones. This can best be illustrated by a one-dimensional nonlinear problem. Consider the nonlinear equation

$$f(x) = F. \tag{3.1}$$

Both the left-hand side and the right-hand side are plotted in Figure 3.1 as a function of x. Suppose we know a starting value x_0, which is reasonably close to the solution of our equation (or close to "a" solution, since a nonlinear equation can have multiple solutions).

To find the solution, the function $f(x)$ is locally linearized at $x = x_0$ by replacing it by its tangent line. Accordingly, Equation (3.1) now reads

$$f(x_0) + (x - x_0)f'(x_0) = F \tag{3.2}$$

which can be solved using a linear equation solver. This yields a first approximation of the solution, which we will call x_1. Now the same procedure can be repeated until the relative difference between two subsequent solutions is smaller than a specified value ϵ:

$$\left| \frac{x_i - x_{i-1}}{x_{i-1}} \right| \leq \epsilon. \tag{3.3}$$

For solutions close to zero, one sometimes has to resort to the absolute difference. This is called the *Newton–Raphson method*. If the true tangent is taken, it exhibits a quadratic rate of convergence. However, whether it converges at all largely depends on the following:

The Finite Element Method for Three-dimensional Thermomechanical Applications Guido Dhondt
 ISBN: 0-470-85752-8

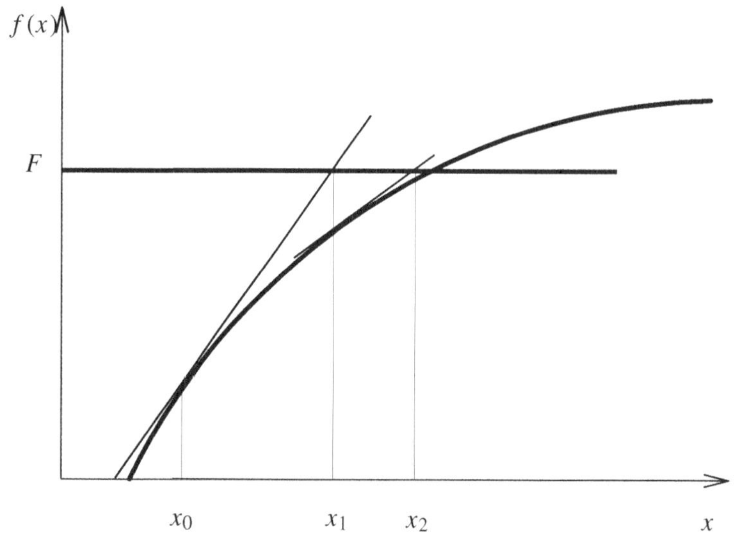

Figure 3.1 The Newton–Raphson method

1. How close the starting solution is to the final solution. The local maximum between the starting guess and the true solution in Figure 3.2 leads to no convergence.
2. The smoothness of the nonlinear function. Because of the jump in Figure 3.3, the Newton–Raphson procedure does not converge.

For our applications, the Newton–Raphson method will be used throughout. For other solution methods, the reader is referred to (Zienkiewicz and Taylor 1989) and (Matthies and Strang 1979). A nice treatise on the computability of nonlinear problems is given in (Belytschko and Mish 2001).

How can the Newton–Raphson method be applied to the governing finite element equations? The major equation for mechanical problems is Equation (2.1). The nonlinearities arise twofold in the term $S^{KL}\delta E_{KL}$ on the left-hand side:

1. For materials of mechanical grade 1 and thermal grade 1, the second Piola–Kirchhoff stress $\boldsymbol{S}$ is generally a nonlinear function of $\boldsymbol{E}$ and its time derivatives (Equation (1.382)).
2. The Lagrange strain $\boldsymbol{E}$ is a nonlinear function of $\boldsymbol{U}$ (Equation (1.84)), in rectangular coordinates:

$$2E_{KL} = U_{K,L} + U_{L,K} + U^{M}_{\ ,K} U_{M,L}. \tag{3.4}$$

For the material nonlinearity, the Newton–Raphson method is applied in a straightforward manner. Assume that we find an intermediate solution $\boldsymbol{E}^0$ with corresponding stress $\boldsymbol{S}_0(\boldsymbol{E}^0)$. Linearizing $\boldsymbol{S}$ at $\boldsymbol{E}^0$ yields

$$S^{KL} \approx S_0^{KL} + \left.\frac{\partial S^{KL}}{\partial E_{MN}}\right|_0 (E_{MN} - E^0_{MN}). \tag{3.5}$$

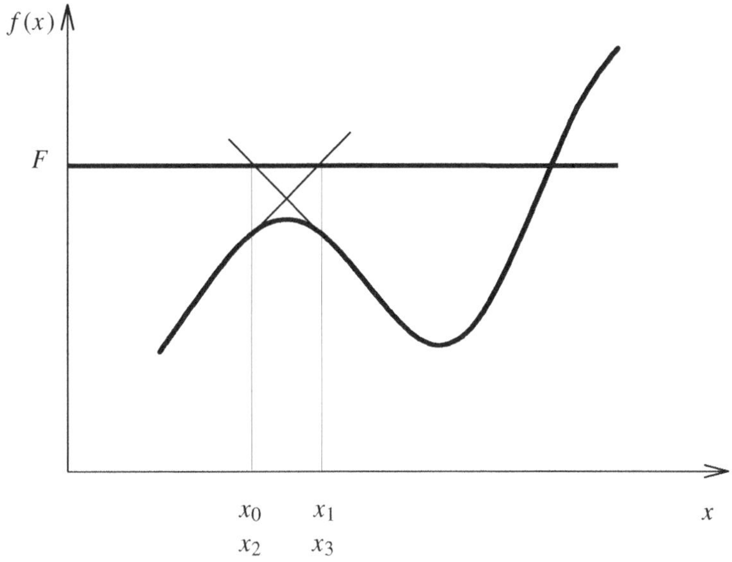

Figure 3.2 Local maximum

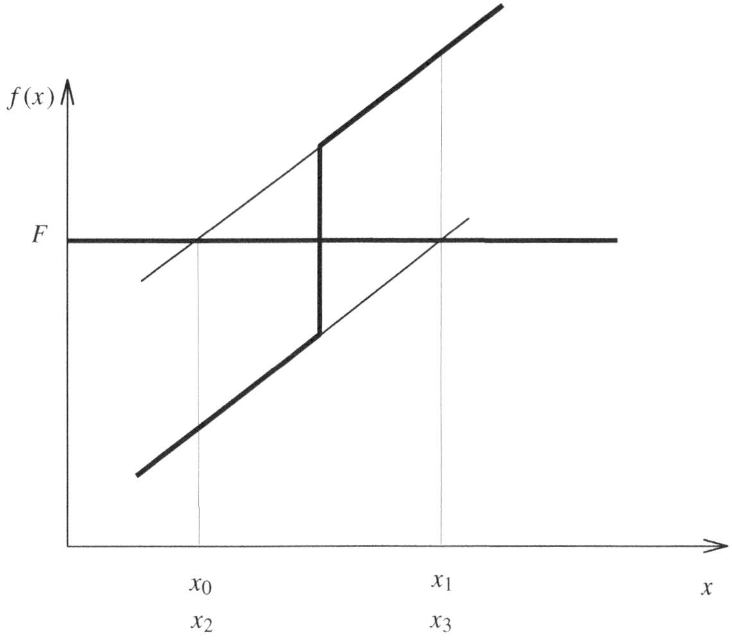

Figure 3.3 Discontinuous function

Denoting

$$\Sigma_0^{KLMN} := \left.\frac{\partial S^{KL}}{\partial E_{MN}}\right|_0 \tag{3.6}$$

Equation (3.5) yields

$$S^{KL} \approx S_0^{KL} + \Sigma_0^{KLMN}(E_{MN} - E_{MN}^0). \tag{3.7}$$

Differentiating Equation (3.4) yields an expression for the infinitesimal perturbation δE_{KL}:

$$\delta E_{KL} = \tfrac{1}{2}(\delta U_{K,L} + \delta U_{L,K} + U^M{}_{,K}\delta U_{M,L} + U_{M,L}\delta U^M{}_{,K}). \tag{3.8}$$

Accordingly,

$$\begin{aligned} &S^{KL}\delta E_{KL} \\ &= \left\{S_0^{KL} + \tfrac{1}{2}\Sigma_0^{KLMN}\left[(U_{M,N} + U_{N,M} + U^R{}_{,M}U_{R,N}) - (V_{M,N} + V_{N,M} + V^R{}_{,M}V_{R,N})\right]\right\} \\ &\qquad \cdot \tfrac{1}{2}\cdot(\delta U_{K,L} + \delta U_{L,K} + U^P{}_{,K}\delta U_{P,L} + U_{P,L}\delta U^P{}_{,K}) \end{aligned} \tag{3.9}$$

where $\boldsymbol{V}$ is the displacement corresponding to $\boldsymbol{E}^0$. Defining the new displacement increment $\boldsymbol{W}$ (to reduce the length of the equations $\boldsymbol{V}$ and $\boldsymbol{W}$ are used instead of the more intuitive notation $\boldsymbol{U}_0$ and $\Delta\boldsymbol{U}$ respectively)

$$\boldsymbol{W} := \boldsymbol{U} - \boldsymbol{V} \tag{3.10}$$

and replacing $\boldsymbol{U}$ in Equation (3.9) by $\boldsymbol{V} + \boldsymbol{W}$ leads to

$$\begin{aligned} &S^{KL}\delta E_{KL} \\ &= \left[S_0^{KL} + \tfrac{1}{2}\Sigma_0^{KLMN}(W_{M,N} + W_{N,M} + V^R{}_{,M}W_{R,N} + W^R{}_{,M}V_{R,N} + W^R{}_{,M}W_{R,N})\right] \\ &\qquad \cdot \tfrac{1}{2}\cdot\left[\delta W_{K,L} + \delta W_{L,K} + (V^P{}_{,K} + W^P{}_{,K})\delta W_{P,L} + (V_{P,L} + W_{P,L})\delta W^P{}_{,K}\right]. \end{aligned} \tag{3.11}$$

In the above equations, $\boldsymbol{V}$ is the displacement calculated thus far and known. The unknown is the incremental displacement $\boldsymbol{W}$. In Equation (3.11), the terms linear in $\boldsymbol{W}$ are force contributions, the quadratic terms contribute to the stiffness and the higher-order terms are neglected. Consequently, Equation (3.11) yields

$$\begin{aligned} S^{KL}\delta E_{KL} \approx{}& \tfrac{1}{2}S_0^{KL}\ \ \delta W_{K,L} + \delta W_{L,K} + V^P{}_{,K}\delta W_{P,L} + V_{P,L}\delta W^P{}_{,K}\Big) \\ &+ \tfrac{1}{2}S_0^{KL}\ \ W^M{}_{,K}\delta W_{M,L} + W_{M,L}\delta W^M{}_{,K}\Big) \\ &+ \tfrac{1}{4}\Sigma_0^{KLMN}\ \ W_{M,N} + W_{N,M} + V^R{}_{,M}W_{R,N} + W^R{}_{,M}V_{R,N}\Big) \\ &\cdot\ \ \delta W_{K,L} + \delta W_{L,K} + V^P{}_{,K}\delta W_{P,L} + V_{P,L}\delta W^P{}_{,K}\Big). \end{aligned} \tag{3.12}$$

Because of the symmetries in S_0^{KL} and Σ_0^{KLMN} ($S_0^{KL} = S_0^{LK}$ and $\Sigma_0^{KLMN} = \Sigma_0^{LKMN} = \Sigma_0^{KLNM}$), Equation (3.12) further reduces to

$$S^{KL}\delta E_{KL} \approx S_0^{KL} \quad \delta W_{K,L} + V^P{}_{,K}\delta W_{P,L}\Big) + S_0^{KL} W^P{}_{,K}\delta W_{P,L}$$
$$+\,\Sigma_0^{KLMN} \quad W_{M,N} + V^R{}_{,M} W_{R,N}\Big) \cdot \quad \delta W_{K,L} + V^P{}_{,K}\delta W_{P,L}\Big). \tag{3.13}$$

This equation applies to linear as well as to nonlinear materials. The only difference is that for linear materials Σ_0^{KLMN} is constant, for nonlinear materials it is a function of E_{KL}. By substituting Equation (3.13) into Equation (2.1), one obtains, instead of Equation (2.6),

$$\int_{V_0} W_{M,N}\Sigma_0^{KLMN}\delta W_{K,L}\,\mathrm{d}V + \int_{V_0} S_0^{KL} W^P{}_{,K}\delta W_{P,L}\,\mathrm{d}V$$
$$+\int_{V_0} \Sigma_0^{KLMN} \quad V^R{}_{,M} W_{R,N}\delta W_{K,L} + V^P{}_{,K} W_{M,N}\delta W_{P,L} + V^R{}_{,M} V^P{}_{,K} W_{R,N}\delta W_{P,L}\Big)\,\mathrm{d}V$$
$$= \int_{A_{0t}} \overline{T}^K_{(N)}\delta W_K\,\mathrm{d}A + \int_{V_0} \rho_0 f^K \delta W_K\,\mathrm{d}V + \int_{V_0} [\beta^{KL}(\theta)T - \gamma^{KL}]\delta U_{K,L}\,\mathrm{d}V$$
$$- \int_{V_0} S_0^{KL}(\delta W_{K,L} + V^P{}_{,K}\delta W_{P,L})\,\mathrm{d}V - \rho_0 \int_{V_0} \frac{D^2 V^K}{Dt^2}\delta W_K\,\mathrm{d}V$$
$$- \rho_0 \int_{V_0} \frac{D^2 W^K}{Dt^2}\delta W_K\,\mathrm{d}V. \tag{3.14}$$

The first term on the left-hand side is the traditional (linear) stiffness term, the second is the stress stiffness and the third is the large deformation stiffness. The last term on the right-hand side is the mass term. By renaming indices, one can also write for Equation (3.14),

$$\int_{V_0} \quad \Sigma_0^{KLMN} + S_0^{NL} G^{MK} + \Sigma_0^{KLRN} V^M{}_{,R} + \Sigma_0^{SLMN} V^K{}_{,S}$$
$$+\Sigma_0^{SLRN} V^M{}_{,R} V^K{}_{,S}\Big) W_{M,N}\delta W_{K,L}\,\mathrm{d}V$$
$$= \int_{A_{0t}} \overline{T}^K_{(N)}\delta W_K\,\mathrm{d}A + \int_{V_0} \rho_0 f^K \delta W_K\,\mathrm{d}V + \int_{V_0} [\beta^{KL}(\theta)T - \gamma^{KL}]\delta U_{K,L}\,\mathrm{d}V$$
$$- \int_{V_0} \quad S_0^{KL} + S_0^{ML} V^K{}_{,M}\Big)\,\delta W_{K,L}\,\mathrm{d}V - \rho_0 \int_{V_0} \frac{D^2 V^K}{Dt^2}\delta W_K\,\mathrm{d}V$$
$$- \rho_0 \int_{V_0} \frac{D^2 W^K}{Dt^2}\delta W_K\,\mathrm{d}V \tag{3.15}$$

which has a completely similar form to Equation (2.6). Accordingly, Equation (2.27)

$$[K]\{W\} + [M]\frac{D^2}{Dt^2}\{W\} = \{F\} \tag{3.16}$$

also applies here together with Equations (2.28) to (2.30), where now

$$[K]_{\mathrm{e}(iK)(jM)} = \int_{V_{0\mathrm{e}}} \varphi_{i,L}\varphi_{j,N} \quad \Sigma_0^{KLMN} + S_0^{NL} G^{MK}$$

$$+ \Sigma^{KLRN} V^M_{,R} + \Sigma^{SLMN} V^K_{,S} + \Sigma^{SLRN} V^M_{,R} V^K_{,S}) \, \mathrm{d}V_\mathrm{e} \quad (3.17)$$

$$[M]_{\mathrm{e}(iK)(jM)} = \rho_0 \int_{V_{0\mathrm{e}}} \varphi_i \varphi_j \, \mathrm{d}V_\mathrm{e} \quad (3.18)$$

$$\{F\}_{\mathrm{e}(iK)} = \{F\}^{\mathrm{ext}}_{\mathrm{e}(iK)} - \{F\}^{\mathrm{int}}_{\mathrm{e}(iK)} - \int_{V_{0\mathrm{e}}} \rho_0 \frac{D^2 V^K}{Dt^2} \varphi_i \, \mathrm{d}V_\mathrm{e} \quad (3.19)$$

$$\{F\}^{\mathrm{ext}}_{\mathrm{e}(iK)} = \int_{A_{t0\mathrm{e}}} \overline{T}^K_{(N)} \varphi_i \, \mathrm{d}A_\mathrm{e} + \int_{V_{0\mathrm{e}}} \rho_0 f^K \varphi_i \, \mathrm{d}V_\mathrm{e}$$

$$+ \int_{V_{0\mathrm{e}}} [\beta^{KL}(\theta) T - \gamma^{KL}] \varphi_{i,L} \, \mathrm{d}V_\mathrm{e}. \quad (3.20)$$

$$\{F\}^{\mathrm{int}}_{\mathrm{e}(iK)} = \int_{V_{0\mathrm{e}}} S_0^{KL} + S_0^{ML} V^K_{,M}) \, \varphi_{i,L} \, \mathrm{d}V_\mathrm{e}. \quad (3.21)$$

Consequently, each iteration (Figure 3.1) in a nonlinear calculation leads to a linear set of equations and the same solvers can be used as in the linear case.

3.2 Application to a Snapping-through Plate

Prediction and modeling of local instabilities is an important issue in engineering problems. These phenomena are characterized by a local or temporal decrease of the load-carrying capacity. This means that the load cannot be used as a time parameter since it is not monotonically increasing. In general, powerful techniques such as the Riks method (Riks 1987) (Crisfield 1983), which use the path length as the time parameter, have to be followed. However, in some applications, such as the one discussed in this section, other more simple time parameters can be selected.

Consider the bent plate in Figure 3.4 loaded by a force in the center. As the force increases, the plane bends until it snaps through. The snapping is an instability accompanied by a complete loss of force-carrying capacity. Therefore, if the force is increased with time (or pseudo-time), equilibrium is lost at the onset of instability. The time increments are decreased, but the Newton–Raphson procedure fails to find a solution. This problem can be solved by taking the displacement u of the loading point in the direction of the force as a parameter since it is monotonically increasing with time. Figure 3.5 shows the force-displacement curve for the loading point. Before the onset of instability, marked by the force maximum, the force steadily increases. During snapping-through, the force crosses the zero-axis (unstable equilibrium, characterized by a negative force-displacement slope) while

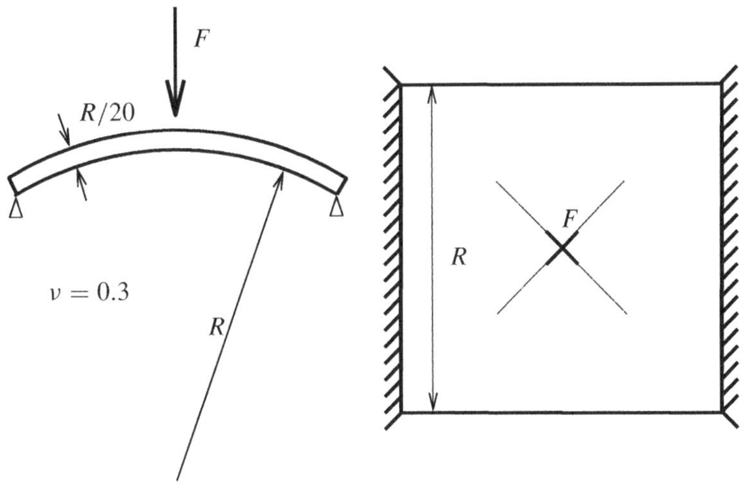

Figure 3.4 Bent plate

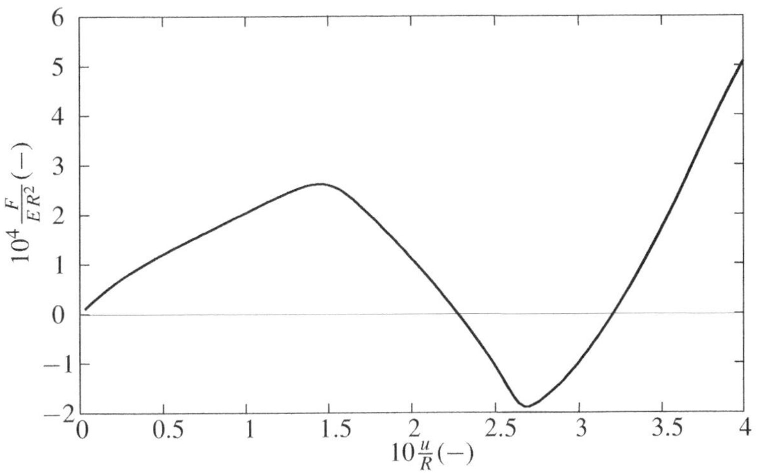

Figure 3.5 Force-displacement curve for the bent plate

decreasing steadily, reaches a minimum and increases again until a new stable configuration is found (stable equilibrium, characterized by a positive force-displacement slope). Notice that at times an upward force must be exerted to keep the plate in its position. In the new stable configuration, the force is zero. Increasing the force again now leads to a monotonic force-displacement curve. This is an example of a strongly nonlinear behavior. It also shows how a diligent choice of the loading parameter can lead to convergent solutions even in the presence of instabilities. Other applications of the instability theory are treated in (Mang *et al.* 2001) and (Kim *et al.* 2003).

3.3 Solution-dependent Loading

In the previous section, the loading terms on the right-hand side of Equation (2.1) were assumed to be independent of the displacements. This, however, is not necessarily the case. The effect of the surface traction depends on the size and orientation of the surface it acts on. Because of the deformation, both the size and the orientation can change. The body force term too can depend on the displacements. For instance, the centrifugal force depends on the distance from the rotation axis. This distance can change because of the deformation of the structure.

3.3.1 Centrifugal forces

In general, the body forces $\boldsymbol{f}$ can be linearly approximated at $\boldsymbol{U} = \boldsymbol{V}$ by

$$\boldsymbol{f}(\boldsymbol{X} + \boldsymbol{V} + \boldsymbol{W}) = \boldsymbol{f}(\boldsymbol{X} + \boldsymbol{V}) + \left.\frac{\partial \boldsymbol{f}}{\partial \boldsymbol{U}}\right|_{\boldsymbol{U}=\boldsymbol{V}} \cdot \boldsymbol{W}. \tag{3.22}$$

The centrifugal body forces $\boldsymbol{f}$ take the form (Equation (2.230))

$$\boldsymbol{f} = \{(\boldsymbol{q} - \boldsymbol{p}_1) - [(\boldsymbol{q} - \boldsymbol{p}_1) \cdot \boldsymbol{e}]\boldsymbol{e}\}\omega^2. \tag{3.23}$$

Now, we assume that the location of the rotation axis does not change because of the deformation, that is,

$$\boldsymbol{p}_1 = \boldsymbol{P}_1 \tag{3.24}$$

$$\boldsymbol{e} = \boldsymbol{E} \tag{3.25}$$

whereas

$$\boldsymbol{q} = \boldsymbol{Q} + \boldsymbol{U}. \tag{3.26}$$

$\boldsymbol{Q}$ is the original position of $\boldsymbol{q}$. Accordingly, Equation (3.22) now reads

$$\boldsymbol{f}(\boldsymbol{Q} + \boldsymbol{V} + \boldsymbol{W}) = \{(\boldsymbol{Q} + \boldsymbol{V} - \boldsymbol{P}_1) - [(\boldsymbol{Q} + \boldsymbol{V} - \boldsymbol{P}_1) \cdot \boldsymbol{E}]\boldsymbol{E}\}\omega^2 + [\boldsymbol{W} - (\boldsymbol{W} \cdot \boldsymbol{E})\boldsymbol{E}]\omega^2. \tag{3.27}$$

Notice that $\boldsymbol{f}$ is linear in $\boldsymbol{W}$. By comparison with Equation (3.22), one observes

$$\boldsymbol{f}(\boldsymbol{Q} + \boldsymbol{V}) = \{(\boldsymbol{Q} + \boldsymbol{V} - \boldsymbol{P}_1) - [(\boldsymbol{Q} + \boldsymbol{V} - \boldsymbol{P}_1) \cdot \boldsymbol{E}]\boldsymbol{E}\}\omega^2 \tag{3.28}$$

and

$$\begin{aligned}
\left.\frac{\partial \boldsymbol{f}}{\partial \boldsymbol{U}}\right|_{\boldsymbol{U}=\boldsymbol{V}} \cdot \boldsymbol{W} &= [\boldsymbol{W} - (\boldsymbol{W} \cdot \boldsymbol{E}^\flat)\boldsymbol{E}^\sharp]\omega^2 \\
&= [\boldsymbol{W} - \boldsymbol{E}^\sharp(\boldsymbol{E}^\flat \cdot \boldsymbol{W})]\omega^2 \\
&= [\boldsymbol{W} - (\boldsymbol{E}^\sharp \otimes \boldsymbol{E}^\flat) \cdot \boldsymbol{W}]\omega^2 \\
&= [\boldsymbol{I} - \boldsymbol{E}^\sharp \otimes \boldsymbol{E}^\flat] \cdot \boldsymbol{W}\omega^2.
\end{aligned} \tag{3.29}$$

Consequently, the centrifugal term in Equation (3.15) amounts to

$$\int_{V_0} \rho_0 f^K \delta W_K \,\mathrm{d}V = \int_{V_0} \rho_0 f_0^K \delta W_K \,\mathrm{d}V + \int_{V_0} \rho_0 \left[W^K - (W^L E_L) E^K \right] \omega^2 \delta W_K \,\mathrm{d}V \tag{3.30}$$

where

$$f_0^K = \{(\boldsymbol{Q} + \boldsymbol{V} - \boldsymbol{P}_1) - [(\boldsymbol{Q} + \boldsymbol{V} - \boldsymbol{P}_1) \cdot \boldsymbol{E}]\boldsymbol{E}\}\,\omega^2 \cdot \boldsymbol{G}^K \tag{3.31}$$

does not depend on the deformation. The first term on the right-hand side of Equation (3.30) is the instantaneous force contribution, which has already been taken into account in Equation (3.15). The second term, however, is new and contributes to the stiffness matrix. Indeed, writing

$$W^K = \sum_i \varphi_j W_j^K \tag{3.32}$$

and similar for δW_K leads to

$$\sum_{\mathrm{e}} \int_{V_{0\mathrm{e}}} \rho_0 \left(\sum_j \varphi_j W_j^{\ K} - \sum_j \varphi_j W_j^{\ L} E_L E^K \right) \left(\sum_i \varphi_i \delta W_{iK} \right) \omega^2 \,\mathrm{d}V \tag{3.33}$$

or

$$\sum_{\mathrm{e}} \sum_i \sum_j \int_{V_{0\mathrm{e}}} \rho_0 \varphi_i \varphi_j \,\mathrm{d}V \quad \delta_L^{\ K} - E_L E^K \big) \, W_j^{\ L} \delta W_{iK} \omega^2. \tag{3.34}$$

The contribution to the stiffness matrix amounts to

$$[K]_{(iK)(jL)} = - \int_{V_{0\mathrm{e}}} \rho_0 \varphi_i \varphi_j \,\mathrm{d}V \quad \delta_L^{\ K} - E_L E^K \big) \, \omega^2. \tag{3.35}$$

The minus sign results from bringing the stiffness contribution to the left-hand side. Notice that, because of the direction of the rotation axis, the contribution to the stiffness matrix is anisotropic.

3.3.2 Traction forces

The traction term in Equation (3.15) amounts to

$$I = \int_{A_{0t}} \overline{T}^K_{(N)} \delta W_K \,\mathrm{d}A. \tag{3.36}$$

Here, $\overline{T}^K_{(N)}$ is a function of the deformation. Recall that it is defined by (Equation (1.263)):

$$\overline{\boldsymbol{T}}_{(N)} = \overline{\boldsymbol{t}}_{(\boldsymbol{n})} \left(\frac{\mathrm{d}a}{\mathrm{d}A} \right). \tag{3.37}$$

For a uniform pressure $\boldsymbol{\sigma} = -p\boldsymbol{g}^{\sharp}$, one arrives at

$$
\begin{aligned}
\overline{\boldsymbol{T}}_{(N)} &= \boldsymbol{\sigma} \cdot \boldsymbol{n} \left(\frac{\mathrm{d}a}{\mathrm{d}A} \right) \\
&= -p\boldsymbol{g}^{\sharp} \cdot \boldsymbol{n} \left(\frac{\mathrm{d}a}{\mathrm{d}A} \right).
\end{aligned} \tag{3.38}
$$

Hence,

$$
\begin{aligned}
\overline{T}^{K}_{(N)} &= \overline{\boldsymbol{T}}_{(N)} \cdot \boldsymbol{G}^{K} \\
&= -p\boldsymbol{G}^{K} \cdot \boldsymbol{g}^{\sharp} \cdot \boldsymbol{n} \left(\frac{\mathrm{d}a}{\mathrm{d}A} \right) \\
&= -p g^{K}{}_{l} g^{lk} n_k \left(\frac{\mathrm{d}a}{\mathrm{d}A} \right) \\
&= -p g^{Kk} \left(\frac{\mathrm{d}a_k}{\mathrm{d}A} \right).
\end{aligned} \tag{3.39}
$$

Recall that (Equation (1.66))

$$
\mathrm{d}a_k = J X^{L}{}_{,k} \, \mathrm{d}A_L. \tag{3.40}
$$

Accordingly,

$$
\overline{T}^{K}_{(N)} = -p g^{Kk} (J X^{L}{}_{,k}) N_L \tag{3.41}
$$

where

$$
N_L = \frac{\mathrm{d}A_L}{\mathrm{d}A}. \tag{3.42}
$$

Consequently, Equation (3.36) now reads

$$
I = -\int_{A_{0t}} p g^{Kk} (J X^{L}{}_{,k}) \delta W_K \, \mathrm{d}A_L. \tag{3.43}
$$

$\left[X^{K}{}_{,k}\right]$ is the inverse of $\left[x^{k}{}_{,K}\right]$. The inverse of a matrix is the transpose of the matrix of its cofactors divided by its determinant:

$$
X^{L}{}_{,k} = \frac{1}{2J} e_{knm} \, \mathrm{e}^{LNM} x^{n}_{,N} x^{m}_{,M}. \tag{3.44}
$$

Assuming that p does not vary over the surface, substitution of Equation (3.44) into Equation (3.43) yields

$$
I = -\frac{p}{2} g^{Kk} e_{knm} \, \mathrm{e}^{LNM} \int_{A_{0t}} x^{n}_{,N} x^{m}_{,M} \delta W_K \, \mathrm{d}A_L. \tag{3.45}
$$

Since

$$x^m = X^M g^m{}_M + U^m \tag{3.46}$$

$$= X^M g^m{}_M + V^m + W^m \tag{3.47}$$

$$=: \overline{x}^m + W^m \tag{3.48}$$

where Equation (3.48) defines $\overline{x}^m$, and using the shape functions

$$W^m = \sum_j \varphi_j W_j{}^m \tag{3.49}$$

$$\delta W^K = \sum_i \varphi_i \delta W_{iK} \tag{3.50}$$

one obtains

$$I = -\frac{p}{2} g^{Kk} e_{knm} \, \mathrm{e}^{LNM} \left(\sum_i \int_{A_{0t}} \overline{x}^n_{,N} \overline{x}^m_{,M} \varphi_i \, \mathrm{d}A_L \delta W_{iK} \right.$$

$$+ \sum_i \sum_j \int_{A_{0t}} \varphi_{j,N} \overline{x}^m_{,M} \varphi_i \, \mathrm{d}A_L W_j{}^n \delta W_{iK}$$

$$\left. + \sum_i \sum_j \int_{A_{0t}} \overline{x}^n_{,N} \varphi_{j,M} \varphi_i \, \mathrm{d}A_L W_j{}^m \delta W_{ik} \right) + O(\|\boldsymbol{W}\|^3), \quad \|\boldsymbol{W}\| \to 0 \tag{3.51}$$

$$\approx - \sum_i \int_{A_{0t}} p g^{Kk} \varphi_i \, \mathrm{d}a_k \delta W_{iK}$$

$$- \frac{p}{2} g^{Kk} e_{knm} \, \mathrm{e}^{LNM} \sum_i \sum_j \int_{A_{0t}} \left(\overline{x}^n_{,N} \varphi_{j,M} - \varphi_{j,N} \overline{x}^n_{,M} \right) \varphi_i \, \mathrm{d}A_L W_j{}^m \delta W_{iK}. \tag{3.52}$$

The first term in Equation (3.52) is a force term already encountered in Section 3.1, the second term yields a stiffness contribution:

$$\left[K\right]_{(iK)(jM)} = \frac{p}{2} g^{Kk} g^{mM} e_{knm} \, \mathrm{e}^{LNP} \int_{A_{0t}} \left(\overline{x}^n_{,N} \varphi_{j,P} - \varphi_{j,N} \overline{x}^n_{,P} \right) \varphi_i \, \mathrm{d}A_L \tag{3.53}$$

or, interchanging m, M with l, L,

$$\left[K\right]_{(iK)(jL)} = \frac{p}{2} g^{Kk} g^{lL} e_{knl} \, \mathrm{e}^{MNP} \int_{A_{0t}} \left(\overline{x}^n_{,N} \varphi_{j,P} - \varphi_{j,N} \overline{x}^n_{,P} \right) \varphi_i \, \mathrm{d}A_M. \tag{3.54}$$

This stiffness contribution is not symmetric. To reduce the computational costs, a symmetrization can be performed by replacing $\left[K\right]$ by $\frac{1}{2} \left(\left[K\right] + \left[K\right]^T \right)$.

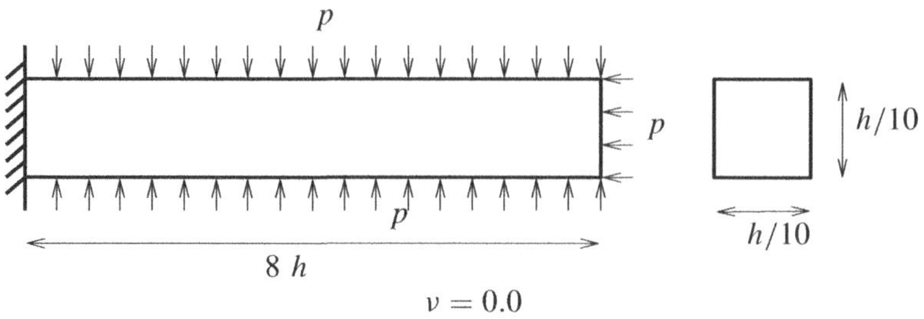

Figure 3.6 Slender beam under hydrostatic pressure

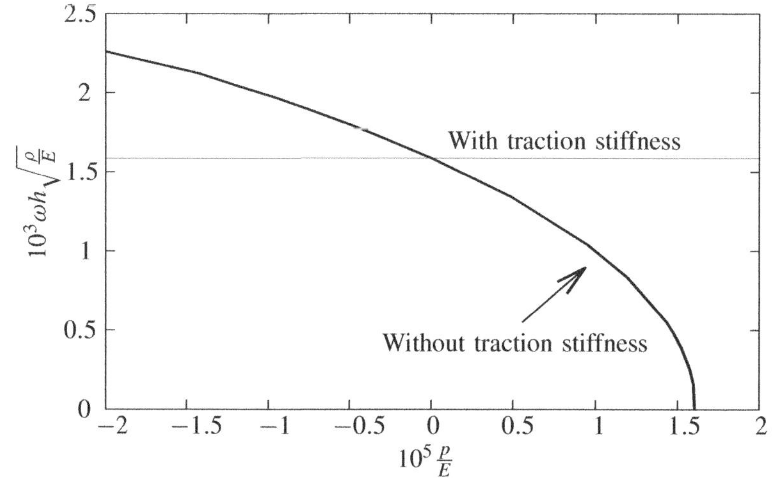

Figure 3.7 Lowest eigenfrequency of the beam

3.3.3 Example: a beam subject to hydrostatic pressure

A slender beam (Figure 3.6) is dropped in the ocean. As it sinks, the pressure steadily increases and the question arises whether the beam will buckle. Therefore, the eigenfrequencies are calculated (cf. Section 2.9.3) since buckling will occur at any zero-crossing of the lowest eigenfrequency. Applying the stress stiffness and large deformation stiffness leads to the solid curve in Figure 3.7. Buckling occurs for a large enough pressure. However, taking the traction stiffness also into account yields the dashed curve: no buckling takes place! Intuitively, as soon as the beam tends to buckle, the deformation-induced traction forces along the sides of the beam stabilize its state. Other applications can be found in (Rumpel and Schweizerhof 2003).

3.4 Nonlinear Multiple Point Constraints

Sometimes there are extra constraints that are not covered by the constitutive equations. The simplest ones are single point constraints, expressing that a degree of freedom has to assume a specific value. These are simple boundary conditions. In other cases, a relationship is established among several degrees of freedom. These are called *multiple point constraints* (MPC). They can be linear or nonlinear. Linear multiple point constraints were encountered in Chapter 2, for instance, in Section 2.10 on cyclic symmetry. Examples of nonlinear equations are given in the following sections and include rigid body motion, incompressible behavior and others. In Section 2.6, it was shown that a linear multiple point constraint can be taken care of right away at the creation time of the stiffness matrix by expressing the dependent degree of freedom as a function of the independent degrees of freedom. A nonlinear multiple point constraint can be treated in the same way after linearization.

The linearization follows exactly the scheme sketched in Section 3.1. Let

$$\boldsymbol{U} := \left\{u_{i_1}, u_{i_2}, \ldots, u_{i_n}\right\} \tag{3.55}$$

be the degrees of freedom involved in the nonlinear multiple point constraint $f(\boldsymbol{U}) = F$. Then, a linearization at $\boldsymbol{U} = \boldsymbol{U}_0$ yields

$$f(\boldsymbol{U}_0) + \nabla f_{\boldsymbol{U}}(\boldsymbol{U}_0) \cdot (\boldsymbol{U} - \boldsymbol{U}_0) = F \tag{3.56}$$

or

$$\nabla f_{\boldsymbol{U}}(\boldsymbol{U}_0) \cdot \Delta\boldsymbol{U} = F - f(\boldsymbol{U}_0) \tag{3.57}$$

where

$$\Delta\boldsymbol{U} := \boldsymbol{U} - \boldsymbol{U}_0. \tag{3.58}$$

This equation is updated as soon as a new solution $\boldsymbol{U}_0$ is obtained. Notice that not only can the coefficients of a linearized multiple point constraint change from iteration to iteration, but also the degrees of freedom involved. This can lead to a change of the dependent degrees of freedom as the calculation proceeds.

Accordingly, a stream chart of a nonlinear solution procedure that includes nonlinear multiple point constraints looks like the one shown in Figure 3.8. The box "update MPC" not only stands for the update of the multiple point constraints but also for the update of any solution dependent boundary conditions such as contact areas or radiation heat flux rates.

3.5 Rigid Body Motion

A first example of nonlinear multiple point constraints constitutes rigid body motion. Here, nonlinearity arises because of large rotations. In what follows, rectangular coordinates are assumed and the spatial frame coincides with the material frame.

3.5.1 Large rotations

Consider a vector $\boldsymbol{\theta} = \theta\boldsymbol{n}$ along an axis AB (Figure 3.9), and a vector $\boldsymbol{r}_0$. Now, the vector $\boldsymbol{r}_0$ is rotated about the axis AB until the new vector $\boldsymbol{r}$ includes an angle $\theta = \|\boldsymbol{\theta}\|$ with $\boldsymbol{r}_0$. We would like to find an expression for $\boldsymbol{r}$ as a function of $\boldsymbol{r}_0$ and $\boldsymbol{\theta}$.

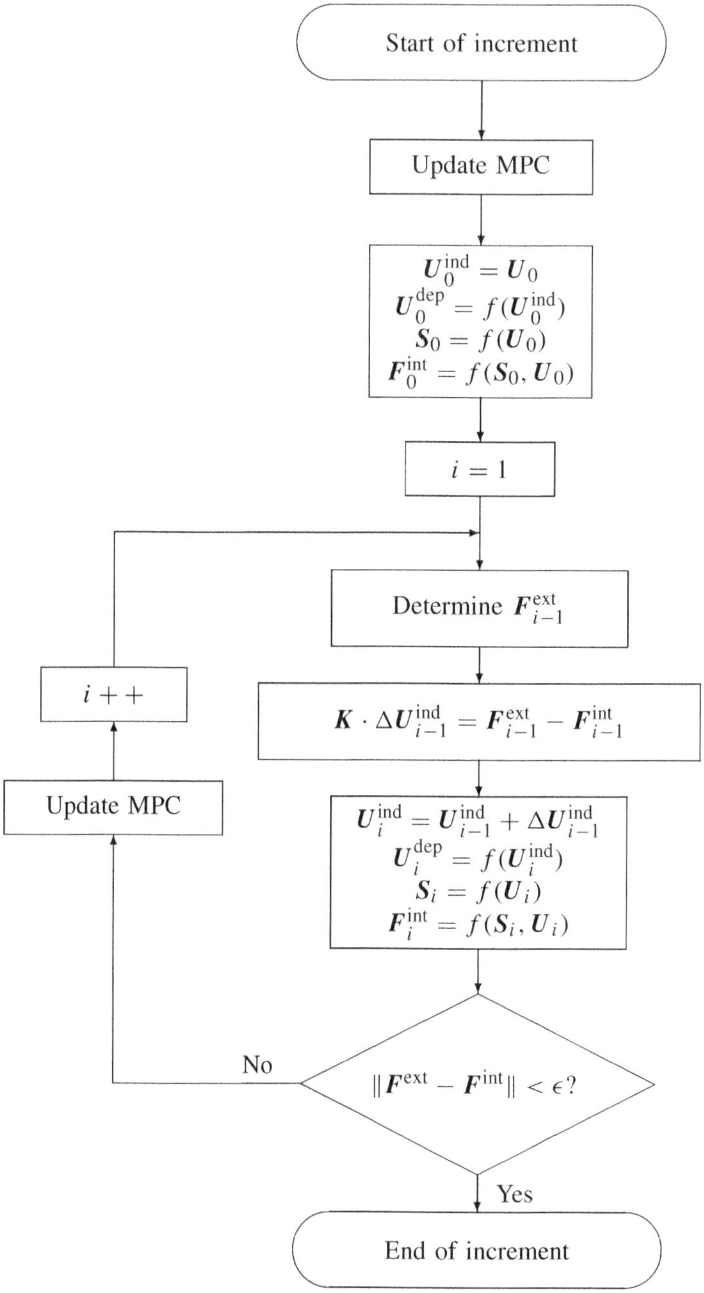

Figure 3.8 Stream chart of the nonlinear solution procedure
"dep" = dependent, "ind" = independent, "int" = internal, "ext" = external

For an infinitesimal angle $\mathrm{d}\theta$, the change $\mathrm{d}\boldsymbol{r}$ of $\boldsymbol{r}$ is perpendicular to $\boldsymbol{r}$ and satisfies

$$\mathrm{d}\boldsymbol{r} = \mathrm{d}\theta(\boldsymbol{n} \times \boldsymbol{r}) \tag{3.59}$$

in component notation:

$$\mathrm{d}r_i = \mathrm{d}\theta e_{ijk} n_j r_k. \tag{3.60}$$

Defining the matrix $\boldsymbol{S}$ by

$$S_{ik} := e_{ijk} n_j \tag{3.61}$$

one finds

$$\mathrm{d}\boldsymbol{r} = \mathrm{d}\theta \boldsymbol{S} \cdot \boldsymbol{r} \tag{3.62}$$

or

$$\frac{\mathrm{d}\boldsymbol{r}}{\mathrm{d}\theta} = \boldsymbol{S} \cdot \boldsymbol{r}. \tag{3.63}$$

This is a linear homogeneous vector differential equation with the solution

$$\boldsymbol{r} = \mathrm{e}^{S\theta} \cdot \boldsymbol{r}_0 \tag{3.64}$$

satisfying the initial condition $\boldsymbol{r}(0) = \boldsymbol{r}_0$. Equation (3.64) can be expanded into

$$\boldsymbol{r} = \left(\boldsymbol{I} + \theta \boldsymbol{S} + \frac{1}{2!}\theta^2 \boldsymbol{S}^2 + \frac{1}{3!}\theta^3 \boldsymbol{S}^3 + \cdots\right) \cdot \boldsymbol{r}_0. \tag{3.65}$$

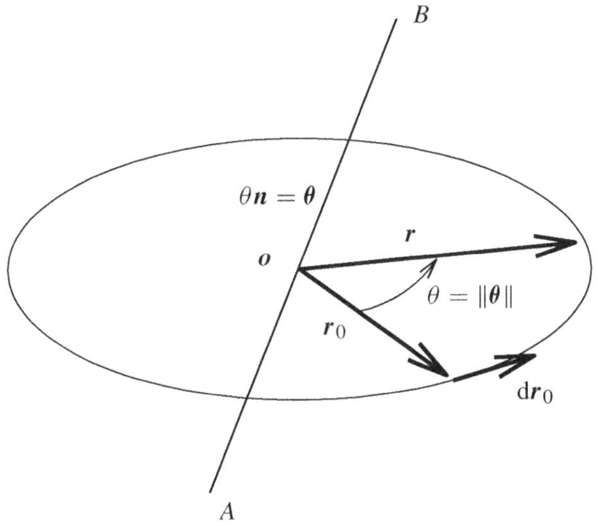

Figure 3.9 Large rotation about the axis AB

Since $\boldsymbol{S} \cdot \boldsymbol{r} = \boldsymbol{n} \times \boldsymbol{r}$ (Equations (3.59) and (3.62)) and $\boldsymbol{a} \times (\boldsymbol{b} \times \boldsymbol{c}) = (\boldsymbol{a} \cdot \boldsymbol{c})\boldsymbol{b} - (\boldsymbol{a} \cdot \boldsymbol{b})\boldsymbol{c}$, one finds

$$\boldsymbol{S}^2 \cdot \boldsymbol{r} = \boldsymbol{S} \cdot (\boldsymbol{S} \cdot \boldsymbol{r}) = \boldsymbol{n} \times (\boldsymbol{n} \times \boldsymbol{r}) = (\boldsymbol{n} \cdot \boldsymbol{r})\boldsymbol{n} - \boldsymbol{r} \tag{3.66}$$

$$\boldsymbol{S}^3 \cdot \boldsymbol{r} = \boldsymbol{S} \cdot (\boldsymbol{S}^2 \cdot \boldsymbol{r}) = \boldsymbol{n} \times [(\boldsymbol{n} \cdot \boldsymbol{r})\boldsymbol{n} - \boldsymbol{r}] = -\boldsymbol{n} \times \boldsymbol{r} = -\boldsymbol{S} \cdot \boldsymbol{r} \tag{3.67}$$

from which one finds

$$\boldsymbol{S}^3 = -\boldsymbol{S}. \tag{3.68}$$

Accordingly, all powers of $\boldsymbol{S}$ exceeding 2 can be reduced to $\pm\boldsymbol{S}$ or $\pm\boldsymbol{S}^2$. Consequently,

$$\begin{aligned} e^{S\theta} &= \boldsymbol{I} + \boldsymbol{S}\left(\theta - \frac{1}{3!}\theta^3 + \frac{1}{5!}\theta^5 - \cdots\right) \\ &\quad + \boldsymbol{S}^2\left(\frac{1}{2!}\theta^2 - \frac{1}{4!}\theta^4 + \frac{1}{6!}\theta^6 - \cdots\right) \\ &= \boldsymbol{I} + \sin\theta\boldsymbol{S} + (1 - \cos\theta)\boldsymbol{S}^2. \end{aligned} \tag{3.69}$$

Hence,

$$\boldsymbol{r} = \left. \boldsymbol{I} + \sin\theta\boldsymbol{S} + (1 - \cos\theta)\boldsymbol{S}^2\right) \cdot \boldsymbol{r}_0. \tag{3.70}$$

Since

$$\boldsymbol{S}^2 = \boldsymbol{n} \otimes \boldsymbol{n} - \boldsymbol{I} \tag{3.71}$$

this also reduces to

$$\boldsymbol{r} = [\cos\theta\boldsymbol{I} + \sin\theta\boldsymbol{S} + (1 - \cos\theta)\boldsymbol{n} \otimes \boldsymbol{n}] \cdot \boldsymbol{r}_0. \tag{3.72}$$

Defining

$$\hat{\boldsymbol{\theta}} = \theta\boldsymbol{S} \tag{3.73}$$

finally yields

$$\boldsymbol{r} = \boldsymbol{C} \cdot \boldsymbol{r}_0 \tag{3.74}$$

where

$$\boldsymbol{C} = \left[\cos\theta\boldsymbol{I} + \frac{\sin\theta}{\theta}\hat{\boldsymbol{\theta}} + (1 - \cos\theta)\frac{\boldsymbol{\theta} \otimes \boldsymbol{\theta}}{\theta^2}\right] \tag{3.75}$$

or in component notation,

$$C_{ij} = \delta_{ij}\cos\theta + \frac{\sin\theta}{\theta}e_{ikj}\theta_k + \left(\frac{1 - \cos\theta}{\theta^2}\right)\theta_i\theta_j. \tag{3.76}$$

Notice that this is a nonlinear relation in $\boldsymbol{\theta}$. Therefore, only a truly nonlinear calculation can take large rotations into account. In simple linear calculations, Equation (3.59) is sometimes used for finite rotations, yielding

$$\boldsymbol{r} = \boldsymbol{r}_0 + \theta(\boldsymbol{n} \times \boldsymbol{r}_0). \tag{3.77}$$

Using this relation amounts to the motion in Figure 3.10 and is only feasible for a small θ. The true angle α satisfies

$$\alpha = \arctan\theta \approx \theta - \frac{\theta^3}{3} + \cdots \tag{3.78}$$

and $\|\boldsymbol{r}\|$ satisfies

$$\|\boldsymbol{r}\| = r_0\sqrt{\theta^2 + 1} \approx r_0\left(1 + \frac{\theta^2}{2}\right). \tag{3.79}$$

3.5.2 Rigid body formulation

Defining a set of nodes to behave like a rigid body means that all degrees of freedom of the set are reduced to six degrees of freedom: three translations $\boldsymbol{w}$ of a point A and three rotations $\boldsymbol{\theta}$ about point A. Point A can be the center of gravity of the node set, but this does not have to be. Any point will do. Usually, we take an existing node belonging to the rigid node set to be point A. However, we can also generate an additional fictitious node to be point A. Hence, the motion $\boldsymbol{u}$ of a node at location $\boldsymbol{p}$ can be described as (Figure 3.11)

$$\boldsymbol{u} = \boldsymbol{w} + [\boldsymbol{C}(\boldsymbol{\theta}) - \boldsymbol{I}] \cdot (\boldsymbol{p} - \boldsymbol{q}) \tag{3.80}$$

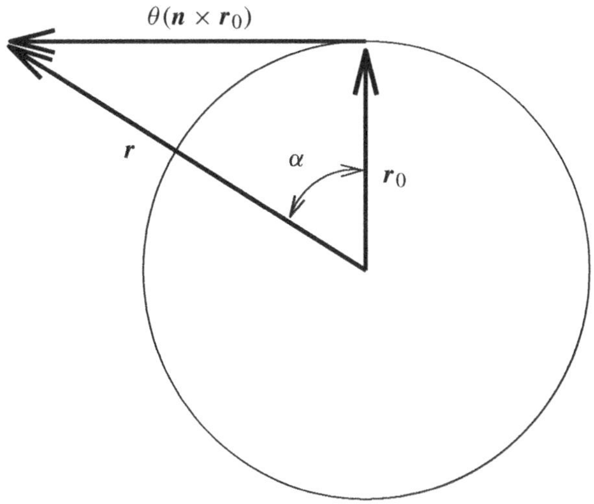

Figure 3.10 Linearized rotation

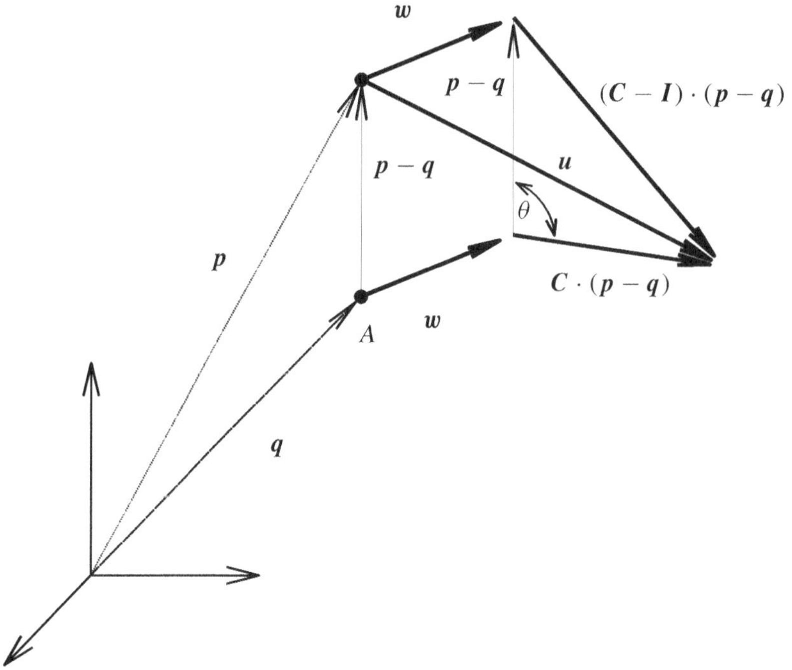

Figure 3.11 Rigid body motion of $\boldsymbol{p}$ about A

where $\boldsymbol{w}$ represents the motion of point A and $\boldsymbol{q}$ its location. The first term on the right-hand side represents the translation and the second represents the rotation. Equation (3.80) is a nonlinear relationship since $\boldsymbol{C}(\boldsymbol{\theta})$ is nonlinear (Equation (3.75)). Linearizing at $(\boldsymbol{w}_0, \boldsymbol{\theta}_0)$, as described in Section 3.4, yields

$$\begin{aligned}\boldsymbol{u}_0 + \boldsymbol{I} \cdot (\boldsymbol{u} - \boldsymbol{u}_0) = \boldsymbol{w}_0 + \boldsymbol{I} \cdot (\boldsymbol{w} - \boldsymbol{w}_0) + [\boldsymbol{C}(\boldsymbol{\theta}_0) - \boldsymbol{I}] \cdot (\boldsymbol{p} - \boldsymbol{q}) \\ + \left[\frac{\partial \boldsymbol{C}}{\partial \boldsymbol{\theta}}(\boldsymbol{\theta}_0) \cdot (\boldsymbol{\theta} - \boldsymbol{\theta}_0) \right] \cdot (\boldsymbol{p} - \boldsymbol{q})\end{aligned} \tag{3.81}$$

or

$$\begin{aligned}(\boldsymbol{u} - \boldsymbol{u}_0) = \boldsymbol{I} \cdot (\boldsymbol{w} - \boldsymbol{w}_0) + \left[\frac{\partial \boldsymbol{C}}{\partial \boldsymbol{\theta}}(\boldsymbol{\theta}_0) \cdot (\boldsymbol{\theta} - \boldsymbol{\theta}_0) \right] \cdot (\boldsymbol{p} - \boldsymbol{q}) \\ + \boldsymbol{w}_0 + [\boldsymbol{C}(\boldsymbol{\theta}_0) - \boldsymbol{I}] \cdot (\boldsymbol{p} - \boldsymbol{q}) - \boldsymbol{u}_0.\end{aligned} \tag{3.82}$$

In component notation, this reads

$$\begin{aligned}u_i - u_{0i} = w_i - w_{0i} + \left(\frac{\partial \boldsymbol{C}}{\partial \boldsymbol{\theta}}(\boldsymbol{\theta}_0) \right)_{ijl} (\theta_l - \theta_{0l})(\boldsymbol{p} - \boldsymbol{q})_j \\ + w_{0i} + \left[C_{ij}(\boldsymbol{\theta}_0) - \delta_{ij} \right] (\boldsymbol{p} - \boldsymbol{q})_j - u_{0i}\end{aligned} \tag{3.83}$$

where, differentiating Equation (3.76),

$$\left(\frac{\partial \boldsymbol{C}}{\partial \boldsymbol{\theta}}(\boldsymbol{\theta}_0)\right)_{ijl} = \frac{\partial C_{ij}}{\partial \theta_l} \tag{3.84}$$

$$\begin{aligned} &= \frac{\partial \cos\theta}{\partial \theta_l}\delta_{ij} + \frac{\partial}{\partial \theta_l}\left(\frac{\sin\theta}{\theta}\right) e_{ikj}\theta_k \\ &+ \left(\frac{\sin\theta}{\theta}\right) e_{ilj} + \frac{\partial}{\partial \theta_l}\left(\frac{1-\cos\theta}{\theta^2}\right)\theta_i\theta_j \\ &+ \left(\frac{1-\cos\theta}{\theta^2}\right)\left(\delta_{il}\theta_j + \theta_i\delta_{jl}\right). \end{aligned} \tag{3.85}$$

The first term in Equation (3.83) is linear in the translations w_i, the second term is linear in the rotations θ_l and the third term is constant. The derivatives in Equation (3.85) are easily determined

$$\frac{\partial \cos\theta}{\partial \theta_l} = -\theta_l \frac{\sin\theta}{\theta} \tag{3.86}$$

$$\frac{\partial}{\partial \theta_l}\left(\frac{\sin\theta}{\theta}\right) = \frac{\theta_l}{\theta^3}(\theta\cos\theta - \sin\theta) \tag{3.87}$$

$$\frac{\partial}{\partial \theta_l}\left(\frac{1-\cos\theta}{\theta}\right) = \frac{\theta_l}{\theta^4}(\theta\sin\theta - 2 + 2\cos\theta). \tag{3.88}$$

For small values of θ, these expressions are undetermined and the limit must be taken

$$\lim_{\theta\to 0}\frac{\sin\theta}{\theta} = 1 \tag{3.89}$$

$$\lim_{\theta\to 0}\frac{1-\cos\theta}{\theta^2} = \frac{1}{2} \tag{3.90}$$

$$\lim_{\theta\to 0}\frac{\theta\cos\theta - \sin\theta}{\theta^3} = -\frac{1}{3} \tag{3.91}$$

$$\lim_{\theta\to 0}\frac{\theta\sin\theta - 2 + 2\cos\theta}{\theta^4} = -\frac{1}{12}. \tag{3.92}$$

Equations (3.83) are the linearized rigid body multiple point constraints at $(\boldsymbol{w}_0, \boldsymbol{\theta}_0)$. Whereas the translational degrees of freedom can be associated with an existing node, this is not the case for the rotational degrees of freedom. The easiest solution is to assign them to a new fictitious node, that is, the translational degrees of freedom of the new node are interpreted as the rotational degrees of freedom of the rigid body.

The above procedure assumes that there is a one-to-one relationship between the motion of the body and the translation and rotation expression given by $(\boldsymbol{w}, \boldsymbol{\theta})$. If this is not the case, additional measures must be taken. For instance, if the body consists of points lying on a straight line, the rotation about this line is not uniquely determined. In that case, the rotation about the line must be explicitly assigned. Assume that $\boldsymbol{a}$ is a unit vector along the line, then, setting the rotation about the line to zero amounts to the linear multiple point constraint $\boldsymbol{a} \cdot \boldsymbol{\theta} = 0$.

3.5.3 Beam and shell elements

The present section looks into a three-dimensional expansion theory of beam and shell elements. Beam and shell structures are characterized by small dimensions across their thickness. Therefore, simplified assumptions can be applied in the thickness direction, leading to different formulations. In the simplest forms, straight fibers orthogonal to the midplane in plates and shells and to the midline in beams are assumed to stay straight and orthogonal during deformation. This leads to the *Kirchhoff theory for plates* and the *Bernoulli–Euler theory for beams*. If the fibers do remain straight during deformation but not necessarily orthogonal to the midplane/midline, the formulation is called the *Mindlin theory for plates* and the *Timoshenko theory for beams*, see also (Zienkiewicz and Taylor 1989), (Graff 1975) and (Meirovitch 1967). The assumptions regarding the displacement field across the thickness have the advantage that only the middle plane (midline) needs to be modeled, while the changes across the thickness are covered by the introduction of additional rotational degrees of freedom in the nodes. Accordingly, modeling needs are basically reduced to the creation of a two-dimensional mesh of the (curved) surface (for shells/plates) or a one-dimensional mesh of the beam axis. The price to be paid is the need for the derivation of the material stiffness matrix specifically for shell and/or beam elements, due to the special formulation in terms of rotational degrees of freedom. Therefore, the idea of hybrid shell-solid and even pure-solid formulations has come up in different forms in recent years, (Bischoff and Ramm 1999), (Flores and Oñate 2001), (Wriggers *et al.* 1996), (Düster *et al.* 2001) and (Sze *et al.* 2002).

In the present derivation, a new, pure-solid way is selected. The ease of modeling is kept by reducing the shells and beams to their midplane and centerline respectively. However, instead of introducing rotational degrees of freedom, 8-node quadratic shell or plate elements and 3-node quadratic beam elements are expanded into 1 layer of 20-node brick elements (with full or reduced integration). Quadratic elements are chosen because of their intrinsically good properties: they are known to behave well for slender structures and rarely exhibit locking or hourglassing. The way of expansion is shown in Figures 3.12 and 3.13.

As long as the plate, shell or beam is smooth, the expansion results in a three-dimensional connected continuum model. However, problems arise as soon as sharp kinks need to be modeled, in areas where several shells and beams cross or the thickness of the shells or beams changes discontinuously. At such locations, all nodes expanded from one and the same node are considered to behave like a rigid body, and will be called a *knot*. At a knot, the degrees of freedom are reduced to three translational and three rotational degrees. All participating structures are expanded as stand-alone parts. Figure 3.14 shows the expansion at a knot between shells and Figure 3.15 at a knot between beams. The structures partially overlap.

A knot is also introduced between beams with a different offset and/or with different cross section (Figure 3.16) and in composed shells and beams. The I-cross section in Figure 3.17 consists of three beam elements with exactly the same nodes, but with different cross section and different offset. Since the cross section is defined as a *rigid body*, it will remain plane and no warping will occur. However, shear deformation is possible since the cross section does not have to remain orthogonal to the central axis. The expanded structure is a volume model and has no rotational

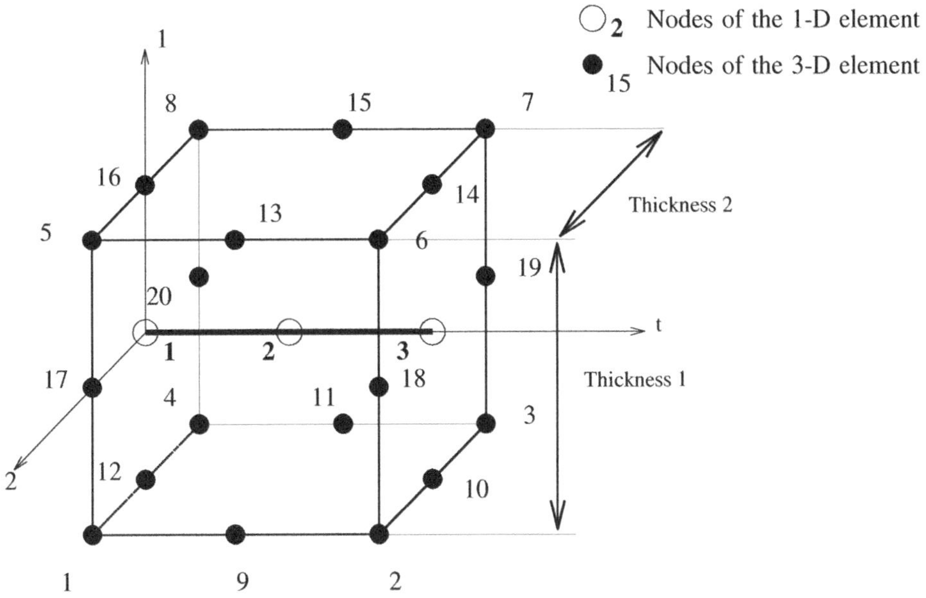

Figure 3.12 Expansion of the one-dimensional element

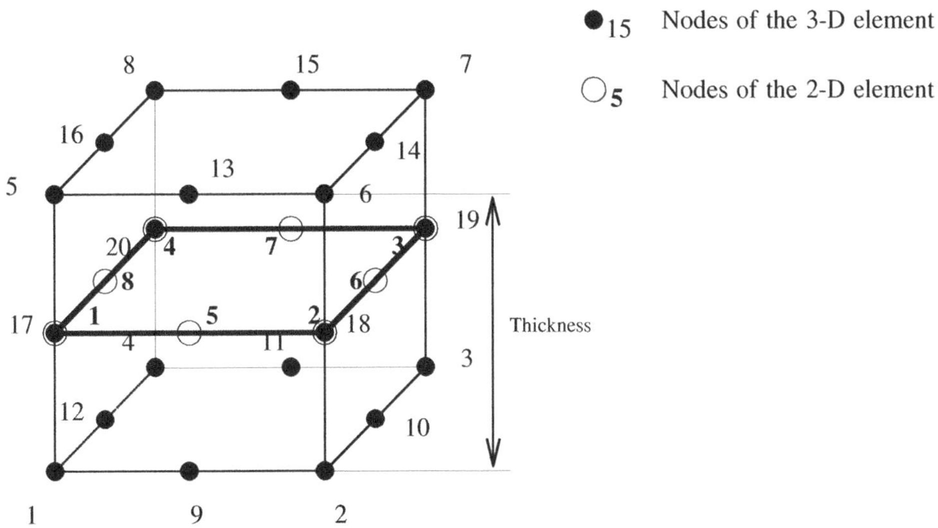

Figure 3.13 Expansion of the two-dimensional element

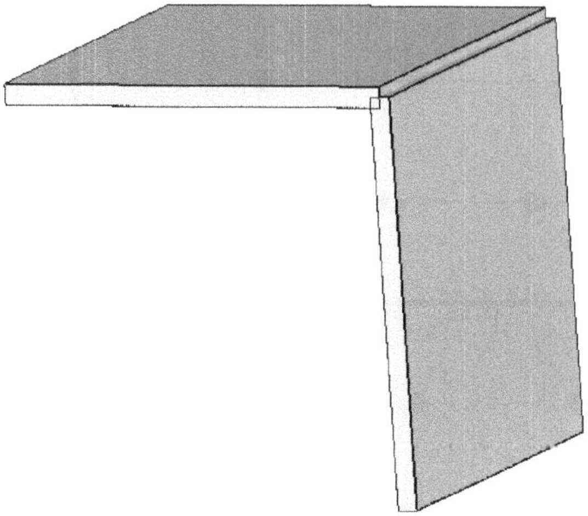

Figure 3.14 Knot between shells

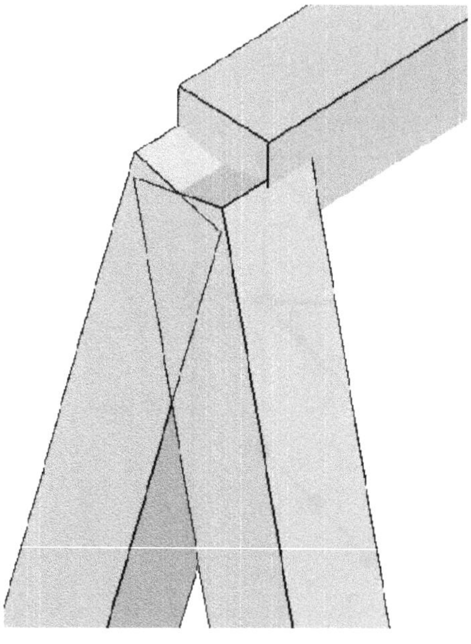

Figure 3.15 Knot between beams

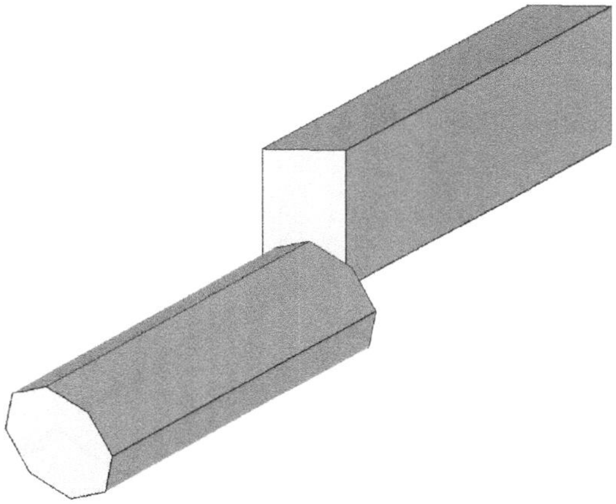

Figure 3.16 Knot between beams with different offset and different cross section

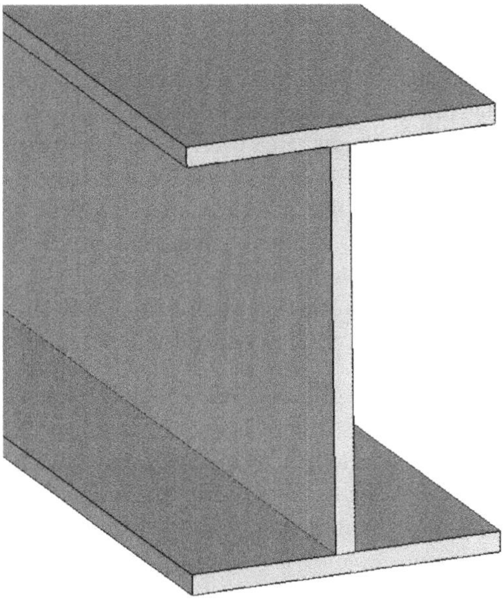

Figure 3.17 I-cross section composed of three simple beam elements each with a different offset

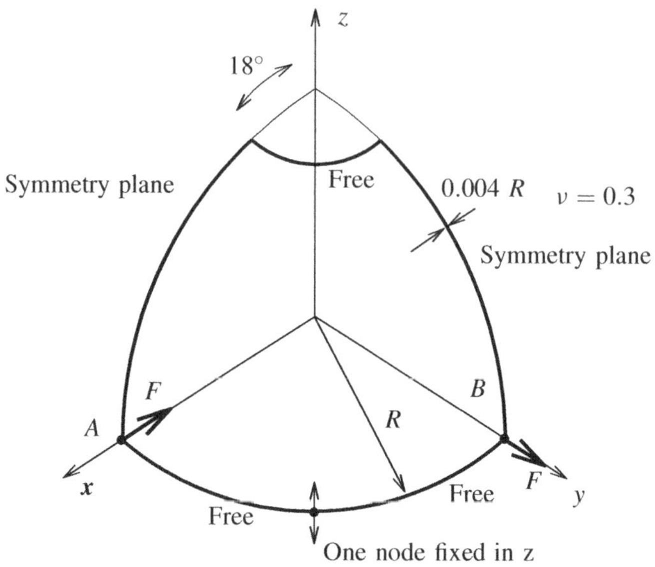

Figure 3.18 Hemispherical shell loaded by concentrated forces

degrees of freedom. Therefore, knots are also introduced at nodes where the user has defined rotations.

The foregoing expansion can also be applied to plane stress, plane strain and axisymmetric elements. Any mixing of these element types among each other or with beams and shells is also taken care of by the knots. However, in the case of plane stress, plane strain or axisymmetric elements, the rigid body definition is restricted to the nodes in the midplane or along the centerline. Indeed, the off-center nodes in plane stress, plane strain and axisymmetric elements are subject to additional conditions due to the z-symmetry or axisymmetry, which would collide with the rigid body definition.

As an example, consider the thin hemispherical shell with a hole at the top and loaded by concentrated forces (Figure 3.18). The shell is meshed in three different ways:

1. As a three-dimensional structure using genuine 20-node brick elements with full integration. The 8×10 element mesh contains one element over the thickness (1872 degrees of freedom in total). The length to the thickness ratio of the elements is about 40. All nodes in the $x - z$ plane are fixed in the y-direction, and the nodes in the $y - z$ plane are fixed in the x-direction. This description contains translational degrees of freedom only.

2. As a three-dimensional structure using genuine 20-node brick elements with reduced integration. The same comments as under 1 also apply here.

3. As a shell structure meshed by 8×10 quadratic shell elements with reduced integration. In the $x - z$ plane, the translational degrees of freedom in the y-direction and the rotational degrees of freedom about the x-axis and z-axis are fixed, in the $y - z$ plane the translational degrees of freedom in the x-direction and the rotational

Table 3.1 Displacements of nodes A and B.

Load $10^9 \frac{F}{ER^2}$	ABAQUS® 4-node shell		20-node brick full integration		20-node brick red. integration		8-node shell red. integration	
	$\frac{1}{R}u_{x,A}$	$\frac{1}{R}u_{y,B}$	$\frac{1}{R}u_{x,A}$	$\frac{1}{R}u_{y,B}$	$\frac{1}{R}u_{x,A}$	$\frac{1}{R}u_{y,B}$	$\frac{1}{R}u_{x,A}$	$\frac{1}{R}u_{y,B}$
5.86	−0.326	0.232	−0.138	0.114	−0.329	0.227	−0.324	0.222
8.79	−0.434	0.282	−0.191	0.147	−0.447	0.277	−0.442	0.272
14.65	−0.590	0.341	−0.271	0.190	−0.618	0.334	−0.610	0.328

degrees of freedom about the y-axis and z-axis are fixed. The shell elements are internally automatically expanded into 20-node brick elements with reduced integration. Along $x = 0$ and $y = 0$, rigid knots are introduced to take care of the rotational degrees of freedom.

The displacements of nodes A and B in x- and y-direction respectively, are listed in Table 3.1 and compared with ABAQUS® reference results. The 20-node brick elements with full integration are clearly too stiff. However, the elements with reduced integration show good agreement with the reference results even for highly nonlinear deformations.

3.6 Mean Rotation

Sometimes a rigid body motion is just too restrictive. Consider a beam with square cross section, fixed at one end. The other end is twisted by an angle γ. It is not known what motion each node on the twisted surface makes, only the mean rotation γ is known. Hence, the twisted surface can expand, contract, warp, and so on, which violates a rigid surface condition.

To formulate an appropriate multiple point constraint, consider the motion of the nodes on the twisted surface as a translation of the center of gravity of this set of nodes, followed by a motion about it. The location of the center of gravity $\boldsymbol{p}_{\text{cg}}$ of a set of N nodes at locations $\boldsymbol{p}_i$ satisfies

$$\boldsymbol{p}_{\text{cg}} = \frac{1}{N}\sum_{j=1}^{N} \boldsymbol{p}_j. \tag{3.93}$$

The relative location $\boldsymbol{p}'_i$ of node i is

$$\boldsymbol{p}'_i = \boldsymbol{p}_i - \frac{1}{N}\sum_{j=1}^{N} \boldsymbol{p}_j. \tag{3.94}$$

The translation of the center of gravity is given by the mean of the displacements $\boldsymbol{u}_i$:

$$\boldsymbol{u}_{\text{cg}} = \frac{1}{N}\sum_{j=1}^{N} \boldsymbol{u}_j \tag{3.95}$$

and the relative displacement $\boldsymbol{u}'_i$ of each nodes i satisfies

$$\boldsymbol{u}'_i = \boldsymbol{u}_i - \frac{1}{N}\sum_{j=1}^{N}\boldsymbol{u}_j. \tag{3.96}$$

The rotation of each node i about the center of gravity is expressed by the angle α_i in Figure 3.19. This angle satisfies

$$|\sin\alpha_i| = \frac{\|\boldsymbol{p}'_{\underline{i}} \times (\boldsymbol{p}'_{\underline{i}} + \boldsymbol{u}'_{\underline{i}})\|}{\|\boldsymbol{p}'_{\underline{i}}\| \cdot \|\boldsymbol{p}'_{\underline{i}} + \boldsymbol{u}'_{\underline{i}}\|} \tag{3.97}$$

(the underscore removes implicit summation). However, the plane defined by $\boldsymbol{p}'_i$, $\boldsymbol{u}'_i$ in Figure 3.19 will generally be different for each node i. Generally, we are interested in the rotation about an axis. Let this axis be defined by a unit vector $\boldsymbol{a}$. Then, the rotation γ_i of node $\boldsymbol{p}_i$ about this axis can be expressed as

$$\gamma_i = \arcsin\frac{\boldsymbol{a} \cdot [\boldsymbol{p}'_{\underline{i}} \times (\boldsymbol{p}'_{\underline{i}} + \boldsymbol{u}'_{\underline{i}})]}{\|\boldsymbol{p}'_{\underline{i}}\| \cdot \|\boldsymbol{p}'_{\underline{i}} + \boldsymbol{u}'_{\underline{i}}\|} \tag{3.98}$$

and expressing that the mean angle amounts to γ leads to

$$\frac{1}{N}\sum_i \gamma_i = \gamma. \tag{3.99}$$

Because of Equation (3.98), this is a nonlinear equation in the displacements. To linearize this equation, we first focus on Equation (3.98) and define

$$\lambda_i := \sin\gamma_i \tag{3.100}$$

and use component notation in a rectangular coordinate system yielding

$$\lambda_i = \frac{e_{knj} a_k p'_{\underline{i}n}(p'_{\underline{i}j} + u'_{\underline{i}j})}{\|\boldsymbol{p}'_{\underline{i}}\| \cdot \|\boldsymbol{p}'_{\underline{i}} + \boldsymbol{u}'_{\underline{i}}\|} \tag{3.101}$$

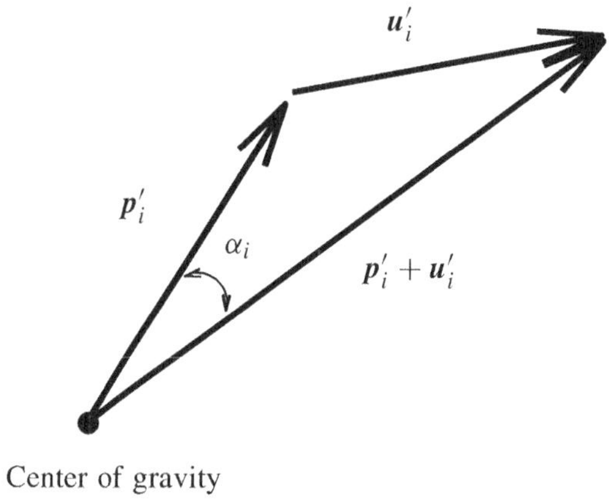

Figure 3.19 Rotation about the center of gravity

where p'_{ij} and u'_{ij} are the j components of $\boldsymbol{p}_i$ and $\boldsymbol{u}_i$, respectively. The only terms in Equation (3.101) depending on $\boldsymbol{u}_k, k = 1, \ldots, N$ are u'_{ij} and $\boldsymbol{u}'_i$ through Equation (3.96). Because of the term $\|\boldsymbol{p}'_i + \boldsymbol{u}'_i\|$ in the denominator of Equation (3.101), λ_i is nonlinear in $\boldsymbol{u}_k$. To linearize $\lambda_i(\boldsymbol{u}_k)$, we first focus on the derivative of some simpler expressions:

$$\frac{\partial u'_{ij}}{\partial u_{pq}} = \delta_{ip}\delta_{jq} - \frac{1}{N}\sum_{k=1}^{N}\delta_{kp}\delta_{jq}$$
$$= \delta_{jq}(\delta_{ip} - \tfrac{1}{N}). \tag{3.102}$$

Now,

$$\|\boldsymbol{p}'_i + \boldsymbol{u}'_i\|^2 = p'_{\underline{i}j}p'_{\underline{i}j} + 2u'_{\underline{i}j}p'_{\underline{i}j} + u'_{\underline{i}j}u'_{\underline{i}j}. \tag{3.103}$$

Hence,

$$\frac{\partial\|\boldsymbol{p}'_i + \boldsymbol{u}'_i\|^2}{\partial u_{pq}} = 2(\delta_{\underline{i}p} - \tfrac{1}{N})(p'_{\underline{i}q} + u'_{\underline{i}q}) \tag{3.104}$$

and

$$\frac{\partial\|\boldsymbol{p}'_i + \boldsymbol{u}'_i\|}{\partial u_{pq}} = \frac{(\delta_{\underline{i}p} - \frac{1}{N})(p'_{\underline{i}q} + u'_{\underline{i}q})}{\|\boldsymbol{p}'_{\underline{i}} + \boldsymbol{u}'_{\underline{i}}\|}. \tag{3.105}$$

Using Equations (3.102) to (3.105), one obtains for the derivative of γ_i:

$$\frac{\partial\gamma_i}{\partial u_{pq}} = \frac{1}{\sqrt{1-\lambda_{\underline{i}}^2}}\frac{e_{knj}a_k p'_{\underline{i}n}(\delta_{\underline{i}p} - \frac{1}{N})}{\|\boldsymbol{p}'_{\underline{i}}\|\cdot\|\boldsymbol{p}'_{\underline{i}} + \boldsymbol{u}'_{\underline{i}}\|^3}\left[\delta_{jq}\|\boldsymbol{p}'_{\underline{i}} + \boldsymbol{u}'_{\underline{i}}\|^2 - (p'_{\underline{i}q} + u'_{\underline{i}q})(p'_{\underline{i}j} + u'_{\underline{i}j})\right]. \tag{3.106}$$

Defining

$$\boldsymbol{\xi}_i := \frac{(\boldsymbol{p}'_{\underline{i}} + \boldsymbol{u}'_{\underline{i}})}{\|\boldsymbol{p}'_{\underline{i}} + \boldsymbol{u}'_{\underline{i}}\|} \tag{3.107}$$

and

$$\eta_i := \frac{\boldsymbol{p}'_{\underline{i}}}{\|\boldsymbol{p}'_{\underline{i}}\|} \tag{3.108}$$

Equation (3.106) can be transformed into

$$\frac{\partial\gamma_i}{\partial u_{pq}} = \frac{1}{\sqrt{1-\lambda_{\underline{i}}^2}}\frac{\delta_{\underline{i}p} - \frac{1}{N})}{\|\boldsymbol{p}'_{\underline{i}} + \boldsymbol{u}'_{\underline{i}}\|}[e_{knq}a_k\eta_{\underline{i}n} - \lambda_{\underline{i}}\xi_{\underline{i}q}] \tag{3.109}$$

where

$$\lambda_i = e_{knj}a_k\eta_{\underline{i}n}\xi_{\underline{i}j}. \tag{3.110}$$

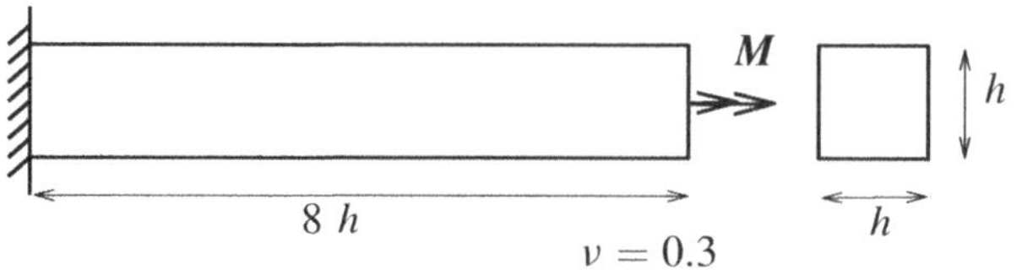

Figure 3.20 Cantilever beam with square cross section subject to torsion

The governing nonlinear equation, Equation (3.99), can finally be linearized at position 0 yielding

$$\sum_{i=1}^{N} \left\{ \gamma_i|_0 + \left.\frac{\partial \gamma_i}{\partial u_{pq}}\right|_0 (u_{pq} - u_{pq}|_0) \right\} = N\gamma \tag{3.111}$$

or

$$\left(\sum_{i=1}^{N} \left.\frac{\partial \gamma_i}{\partial u_{pq}}\right|_0 \right) (u_{pq} - u_{pq}|_0) = N\gamma - \sum_{i=1}^{N} \gamma_i|_0 . \tag{3.112}$$

This is a linear scalar equation in the unknowns u_{pq}, $p = 1, \ldots, N, q = 1, \ldots, 3$. Notice that the coefficients of the linear terms can at times be zero. Since the dependent term in an equation must have a nonzero coefficient, the selection of the dependent variable may have to change from one iteration to the next.

The mean-rotation concept only makes sense if more than one node is involved. If one of the nodes k happens to coincide with the center of gravity of the node set, the angle γ_k is not determinate since $\boldsymbol{p}'_k = 0$ and the contribution $\gamma_k|_0$ and $\left.\frac{\partial \gamma_k}{\partial u_{pq}}\right|_0$ are left out in the sums in Equation (3.112). Equation (3.112) is less restrictive than a rigid body motion. Accordingly, less energy is needed for applying a mean rotation than for a rigid body motion.

This can be nicely illustrated by the cantilever beam in Figure 3.20. A torque is applied at the free end such that a rotation of 45° results. Three conditions are examined here:

1. The beam theory is applied and the torque is determined analytically by (Popov 1968)

$$M = \frac{0.141\varphi(bc)^3 G}{L} \tag{3.113}$$

   where, in the actual example, $\varphi = \pi/4$, $b = c = h$ and $L = 8\ h$.

2. The cross section at the free end of the beam is considered as a rigid body.

3. The mean-rotation condition is applied.

The torque required for each of these conditions is listed in Table 3.2. The analytical result is close to the rigid body condition. The mean-rotation condition requires a torque that is 10% less due to the relaxed constraints.

Table 3.2 Torque needed for a twist of 45°.

Condition	$\frac{M}{Gh^3}$
Beam theory	0.0138
Rigid body	0.0141
Mean rotation	0.0126

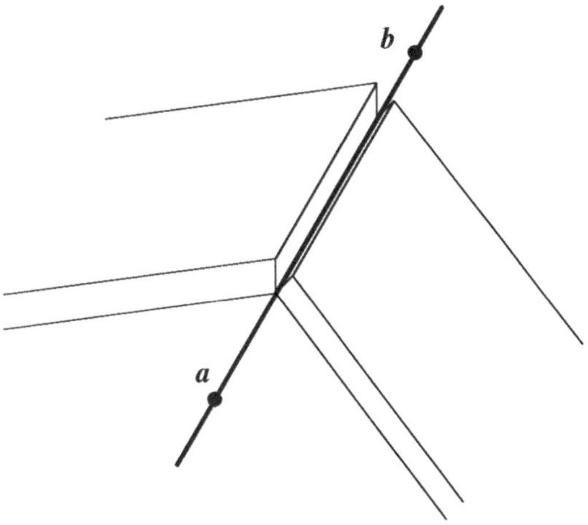

Figure 3.21 A straight-line kinematic constraint

3.7 Kinematic Constraints

As in the previous sections, rectangular coordinates are assumed throughout and the spatial reference system coincides with the material reference system.

3.7.1 Points on a straight line

Occasionally, one comes across the condition that points must stay on a straight line. An example of such a case is a hinge consisting of nodes on a line (Figure 3.21). The line itself can move in space. A node $\boldsymbol{p}$ lies on the straight line defined by distinct nodes $\boldsymbol{a}$ and $\boldsymbol{b}$ if

$$\boldsymbol{p} = \boldsymbol{a} + \lambda(\boldsymbol{b} - \boldsymbol{a}), \quad \lambda \in \mathbb{R} \tag{3.114}$$

which is equivalent to

$$x_p - x_a = \lambda(x_b - x_a) \tag{3.115}$$

$$y_p - y_a = \lambda(y_b - y_a) \tag{3.116}$$

$$z_p - z_a = \lambda(z_b - z_a) \tag{3.117}$$

where (x, y, z) are coordinates in the deformed configuration. Since $\boldsymbol{a}$ and $\boldsymbol{b}$ do not coincide, at least one of the right-hand sides in Equations (3.115) to (3.117) is nonzero. Let $x_b \neq x_a$ and $\boldsymbol{a}$ and $\boldsymbol{b}$ be such that $x_a = x_b$ is highly improbable throughout the complete deformation, then, one obtains by solving for λ in Equation (3.115) and substituting into Equation (3.116) and (3.117),

$$(y_p - y_a)(x_b - x_a) = (x_p - x_a)(y_b - y_a) \tag{3.118}$$

$$(z_p - z_a)(x_b - x_a) = (x_p - x_a)(z_b - z_a). \tag{3.119}$$

Since

$$x_a = X_a + u_a \tag{3.120}$$

where X_a is the x-coordinate of node $\boldsymbol{a}$ in the undeformed configuration and u_a is its displacement in x-direction and similarly for the other coordinates,

$$y_a = Y_a + v_a \tag{3.121}$$

$$z_a = Z_a + w_a. \tag{3.122}$$

Equations (3.118) and (3.119) are a set of two nonlinear equations in $u_a, v_a, w_a, u_b, v_b, w_b$ and u_p, v_p, w_p. Denoting Equation (3.118) by

$$f(v_p, u_p, v_a, u_a, v_b, u_b) = 0 \tag{3.123}$$

and since (Equation (3.120))

$$\frac{\partial f}{\partial u_a} = \frac{\partial f}{\partial x_a}\frac{\partial x_a}{\partial u_a} = \frac{\partial f}{\partial x_a} \tag{3.124}$$

and similarly for the other variables, linearization of Equation (3.123) at $(v_p^0, u_p^0, v_a^0, u_a^0, v_b^0, u_b^0)$ yields

$$f(v_p^0, u_p^0, v_a^0, u_a^0, v_b^0, u_b^0) + \left.\frac{\partial f}{\partial v_p}\right|_0 (v_p - v_p^0) + \left.\frac{\partial f}{\partial u_p}\right|_0 (u_p - u_p^0) + \left.\frac{\partial f}{\partial v_a}\right|_0 (v_a - v_p^0)$$
$$+ \left.\frac{\partial f}{\partial u_a}\right|_0 (u_a - u_p^0) + \left.\frac{\partial f}{\partial v_b}\right|_0 (v_b - v_p^0) + \left.\frac{\partial f}{\partial u_b}\right|_0 (u_b - u_p^0) \approx 0 \tag{3.125}$$

where

$$\left.\frac{\partial f}{\partial v_p}\right|_0 = x_b^0 - x_a^0 \tag{3.126}$$

$$\left.\frac{\partial f}{\partial u_p}\right|_0 = -(y_b^0 - y_a^0) \tag{3.127}$$

$$\left.\frac{\partial f}{\partial v_a}\right|_0 = -(x_b^0 - x_p^0) \tag{3.128}$$

$$\left.\frac{\partial f}{\partial u_a}\right|_0 = -(y_p^0 - y_b^0) \tag{3.129}$$

$$\left.\frac{\partial f}{\partial v_b}\right|_0 = -(x_p^0 - x_a^0) \tag{3.130}$$

$$\left.\frac{\partial f}{\partial u_b}\right|_0 = y_p^0 - y_a^0 \tag{3.131}$$

and

$$x_a^0 := X_a + u_a^0 \tag{3.132}$$

(similarly for the other coordinates). An analogous procedure can be applied to Equation (3.119). In the present case, v_p and w_p are suitable selections for the dependent variables since $x_b^0 \neq x_a^0$ is assumed. It is advantageous to select $\boldsymbol{a}$ and $\boldsymbol{b}$ at an appreciable distance from each other in order to improve the accuracy. Each node $\boldsymbol{p}$ constrained to lie on the line defined by the nodes $\boldsymbol{a}$ and $\boldsymbol{b}$ will lead to two of the above equations.

3.7.2 Points in a plane

The treatment of points constrained to lie in a plane is somewhat similar to the derivation in the previous section. Let the plane α be defined by three nodes $\boldsymbol{a}$, $\boldsymbol{b}$ and $\boldsymbol{c}$, which are not colinear, that is,

$$\boldsymbol{m} := (\boldsymbol{b} - \boldsymbol{c}) \times (\boldsymbol{a} - \boldsymbol{c}) \neq \boldsymbol{0}. \tag{3.133}$$

A node $\boldsymbol{p}$ lies in the plane if

$$\boldsymbol{m} \cdot (\boldsymbol{p} - \boldsymbol{c}) = 0. \tag{3.134}$$

Introducing spatial coordinates (x, y, z), Equation (3.134) is equivalent to

$$f = \begin{vmatrix} x_p - x_c & y_p - y_c & z_p - z_c \\ x_a - x_c & y_a - y_c & z_a - z_c \\ x_b - x_c & y_b - y_c & z_b - z_c \end{vmatrix} = 0. \tag{3.135}$$

The vertical lines denote the determinant of the 3×3 matrix. f is a nonlinear equation in $u_a, v_a, w_a, u_b, v_b, w_b, u_c, v_c, w_c, u_p, v_p$ and w_p since

$$x_a = X_a + u_a \tag{3.136}$$

and similarly for the other coordinates. The derivatives of f at $(u_a^0, v_a^0, w_a^0, u_b^0, \ldots, w_p^0)$ with respect to $u_a, v_a, w_a, u_b, v_b, w_b, u_p, v_p$ and w_p are the corresponding cofactors, that is,

$$\left.\frac{\partial f}{\partial u_a}\right|_0 = -\begin{vmatrix} y_p^0 - y_c^0 & z_p^0 - z_c^0 \\ y_b^0 - y_c^0 & z_b^0 - z_c^0 \end{vmatrix} \tag{3.137}$$

since (Equation (3.136))

$$\left.\frac{\partial f}{\partial u_a}\right|_0 = \left.\frac{\partial f}{\partial x_a}\right|_0 \tag{3.138}$$

and the derivatives with respect to u_c, v_c and w_c are sums of cofactors,

$$\left.\frac{\partial f}{\partial u_c}\right|_0 = -\begin{vmatrix} y_a^0 - y_c^0 & z_a^0 - z_c^0 \\ y_b^0 - y_c^0 & z_b^0 - z_c^0 \end{vmatrix} + \begin{vmatrix} y_p^0 - y_c^0 & z_p^0 - z_c^0 \\ y_b^0 - y_c^0 & z_b^0 - z_c^0 \end{vmatrix} - \begin{vmatrix} y_p^0 - y_c^0 & z_p^0 - z_c^0 \\ y_a^0 - y_c^0 & z_a^0 - z_c^0 \end{vmatrix}. \tag{3.139}$$

Consequently, denoting the elements of the matrix at $(u_a^0, v_a^0, \ldots, w_p^0)$ by $a_{11}, a_{12}, \ldots, a_{33}$ and the corresponding cofactors by $A_{11}, A_{12}, \ldots, A_{33}$, the linearization of Equation (3.135) yields

$$\begin{aligned} f^0 &+ A_{11}(u_p - u_p^0) + A_{12}(v_p - v_p^0) + A_{13}(w_p - w_p^0) \\ &+ A_{21}(u_a - u_a^0) + A_{22}(v_a - v_a^0) + A_{23}(w_a - w_a^0) + A_{31}(u_b - u_b^0) \\ &+ A_{32}(v_b - v_b^0) + A_{33}(w_b - w_b^0) - (A_{11} + A_{21} + A_{31})(u_c - u_c^0) \\ &- (A_{12} + A_{22} + A_{32})(v_c - v_c^0) - (A_{13} + A_{23} + A_{33})(w_c - w_c^0) \approx 0. \end{aligned} \tag{3.140}$$

Since $\boldsymbol{m} \neq 0$, A_{11}, A_{12} and A_{13} cannot all be zero. The variable with the largest coefficient in size should be taken as the dependent degree of freedom. Accordingly, if

$$|A_{12}| \geq |A_{13}| \geq |A_{11}| \tag{3.141}$$

take v_p as the dependent degree of freedom, unless it is already used in another multiple point constraint. Notice that the nonlinearity only arises because of the fact that the plane defined by $\boldsymbol{a}$, $\boldsymbol{b}$ and $\boldsymbol{c}$ is not fixed in space. If the plane is fixed, $x_a, y_a, z_a, x_b, \ldots, z_c$ are constants and Equation (3.135) reduces to a linear equation in x_p, y_p, and z_p.

3.8 Incompressibility Constraint

Many materials such as rubber or organic tissue are either incompressible or can be viewed as such. In Chapter 1, it was shown that this condition is equivalent to $J = 1$. Denoting the undeformed position of $\boldsymbol{X}$ by the rectangular coordinates (X, Y, Z), the deformed position by (x, y, z) and the displacements by (u, v, w), this condition is equivalent to

$$J = \begin{vmatrix} x_{,X} & x_{,Y} & x_{,Z} \\ y_{,X} & y_{,Y} & y_{,Z} \\ z_{,X} & z_{,Y} & z_{,Z} \end{vmatrix} = 1 \tag{3.142}$$

or, using the local coordinates $\boldsymbol{\gamma}(\xi, \eta, \zeta)$,

$$J = \begin{vmatrix} x_{,\xi} & x_{,\eta} & x_{,\zeta} \\ y_{,\xi} & y_{,\eta} & y_{,\zeta} \\ z_{,\xi} & z_{,\eta} & z_{,\zeta} \end{vmatrix} \cdot \begin{vmatrix} \xi_{,X} & \xi_{,Y} & \xi_{,Z} \\ \eta_{,X} & \eta_{,Y} & \eta_{,Z} \\ \zeta_{,X} & \zeta_{,Y} & \zeta_{,Z} \end{vmatrix} = 1. \tag{3.143}$$

This is a function of the displacement components of all nodes belonging to the element at stake. Indeed (cf Equation (2.9) and (2.10)),

$$x = \sum_{i=1}^{N} \varphi_i(\xi, \eta, \zeta) x_i = \sum_{i=1}^{N} \varphi_i(\xi, \eta, \zeta)(X_i + u_i) \tag{3.144}$$

$$y = \sum_{i=1}^{N} \varphi_i(\xi, \eta, \zeta)(Y_i + v_i) \tag{3.145}$$

$$z = \sum_{i=1}^{N} \varphi_i(\xi, \eta, \zeta)(Z_i + w_i). \tag{3.146}$$

Notice that Equations (3.144) to (3.146) only apply if the formulation is isoparametric, that is, the undeformed position and the displacements are interpolated in the same way. Accordingly, one finds

$$x_{,\xi} = \sum_{i=1}^{N} \frac{\partial \varphi_i}{\partial \xi}(\xi, \eta, \zeta)(X_i + u_i) \tag{3.147}$$

and similarly for the other terms. If we write Equation (3.143) as

$$f(u_1, v_1, w_1, u_2, v_2, w_2, \ldots, u_N, v_N, w_N) = 0 \tag{3.148}$$

the linearization yields

$$f(u_1^0, v_1^0, w_1^0, u_2^0, v_2^0, w_2^0, \cdots, u_N^0, v_N^0, w_N^0) + \sum_i \left[\left.\frac{\partial f}{\partial u_i}\right|_0 (u_i - u_i^0) + \left.\frac{\partial f}{\partial v_i}\right|_0 (v_i - v_i^0) + \left.\frac{\partial f}{\partial w_i}\right|_0 (w_i - w_i^0) \right] \approx 0. \tag{3.149}$$

Substitution of Equations (3.147) into Equation (3.143) reveals that f is a linear function of u_i if all v_i and w_i are kept constant, that is, $v_i = v_i^0$ and $w_i = w_i^0$, $\forall i$. Accordingly,

$$\frac{\partial f}{\partial u_i} = \begin{vmatrix} \frac{\partial \varphi_i}{\partial \xi} & \frac{\partial \varphi_i}{\partial \eta} & \frac{\partial \varphi_i}{\partial \zeta} \\ y_{,\xi} & y_{,\eta} & y_{,\zeta} \\ z_{,\xi} & z_{,\eta} & z_{,\zeta} \end{vmatrix} \cdot \begin{vmatrix} \xi_{,X} & \xi_{,Y} & \xi_{,Z} \\ \eta_{,X} & \eta_{,Y} & \eta_{,Z} \\ \zeta_{,X} & \zeta_{,Y} & \zeta_{,Z} \end{vmatrix}. \tag{3.150}$$

Equation (3.149) can be applied at any internal point of the element and leads to one equation in all the degrees of freedom belonging to the element (e.g. 60 degrees of freedom for the 20-node brick element). If it is applied to the points on the border, the degrees of freedom of the adjoining elements must be considered too. In that case, it sounds feasible to require that the mean of the Jacobian determined for each of the adjoining elements separately, must be 1.

The question remains, at what points should Equation (3.142) be applied to yield valid results. Application to too many points leads to volumetric locking of the element. Taking hybrid elements as reference, where the pressure is usually interpolated with a lower degree than the displacements, it is proposed to apply the incompressibility condition to the corner nodes for quadratic elements, and to the center of the element for linear type elements.

4

Hyperelastic Materials

In this chapter, hyperelastic materials will be discussed. They are defined as materials for which a free energy function

$$\Sigma(\boldsymbol{C}, \theta, \boldsymbol{X}) \tag{4.1}$$

exists such that Equations (1.393) and (1.394) apply. The function Σ is sometimes called the *stored-energy function* (Ciarlet 1993), (Simo and Hughes 1997). Because of the functional dependence in Equation (4.1), hyperelastic materials have no memory (Figure 4.1). After unloading, they return without time delay to their starting position. The determination of the second Piola–Kirchhoff stress is straightforward through Equation (1.393):

$$\boldsymbol{S} = 2\frac{\partial \Sigma}{\partial \boldsymbol{C}}. \tag{4.2}$$

The question naturally arises whether the function Σ can be freely chosen or whether physical considerations impose any restrictions. This is treated in the first section. Then, a few popular isotropic models are discussed and applied to simulate a shear test and the inflation of a balloon. Finally, the theory is extended to anisotropic materials such as fiber-reinforced tissues. For further reading, the reader is particularly referred to (Holzapfel 2000) and (Bonet and Wood 1997).

4.1 Polyconvexity of the Stored-energy Function

4.1.1 Physical requirements

Basic physical considerations imply that extreme strains must lead to infinite stress (Antman 1983). The word "extreme" applies equally well to large compressions as well as to large expansions. If the material is extremely compressed such that it is on the verge of being annihilated, $J \rightarrow 0$, large stresses should result. Large stresses should equally well be required to expand a material beyond bounds ($J \rightarrow \infty$). The notion of "extreme strains"

The Finite Element Method for Three-dimensional Thermomechanical Applications Guido Dhondt
 ISBN: 0-470-85752-8

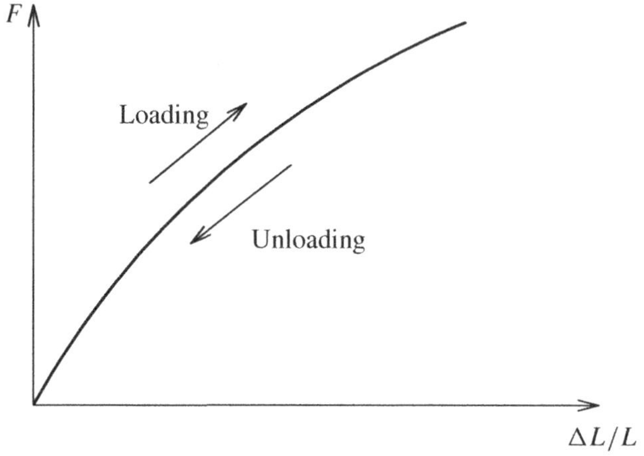

Figure 4.1 Force-stretch diagram for a hyperelastic material in a uniaxial test

can be further concretized by looking at the invariants of $\boldsymbol{C}$ in terms of the principal values Λ_1, Λ_2 and Λ_3 (cf Equations (1.121)–(1.123)):

$$I_1 = \Lambda_1 + \Lambda_2 + \Lambda_3 \tag{4.3}$$

$$I_2 = \Lambda_1\Lambda_2 + \Lambda_1\Lambda_3 + \Lambda_2\Lambda_3 \tag{4.4}$$

$$I_3 = \Lambda_1\Lambda_2\Lambda_3. \tag{4.5}$$

Recall that the eigenvalues of $\boldsymbol{C}$ are the squares of the stretch in the principal direction. Indeed, one finds, using Equation (1.132) and defining the norm for vectors and tensors of rank two by $\|\boldsymbol{N}\| = \sqrt{\boldsymbol{N}\cdot\boldsymbol{N}}$ and $\|\boldsymbol{A}\| = \sqrt{\boldsymbol{A}:\boldsymbol{A}}$ respectively,

$$\Lambda_i = \boldsymbol{N}_{\underline{i}} \cdot \boldsymbol{F}^{\mathrm{T}} \cdot \boldsymbol{F} \cdot \boldsymbol{N}_{\underline{i}} = \|\boldsymbol{F}\cdot\boldsymbol{N}_i\|^2. \tag{4.6}$$

Consequently, $\Lambda_i \geq 0$. Furthermore, the eigenvalues are the solution of the characteristic equation

$$\Lambda^3 - I_1\Lambda^2 + I_2\Lambda - I_3 = 0. \tag{4.7}$$

If at least one $\Lambda_i \to 0$, then $I_3 \to 0$ must apply in order to satisfy the above equation. The other way around, if $\Lambda_1\Lambda_2\Lambda_3 \to 0$, then at least one $\Lambda_i \to 0$. Accordingly, a small value of I_3 is equivalent to small extreme strains. If at least one $\Lambda_i \to \infty$, then $I_1 \to \infty$ since $\Lambda_i \geq 0$. The inverse is also true: if $I_1 \to \infty$, at least one $\Lambda_i \to \infty$. Accordingly, a large value of I_1 is equivalent to large extreme strains. If $I_1 \to \infty$, then $I_1 + I_2 + I_3 \to \infty$ since $I_1, I_2 \geq 0$, and $I_1 + I_2 + I_3$ cannot be large unless at least one $\Lambda_i \to \infty$, which implies that $I_1 \to \infty$. Consequently,

$$I_1 \to \infty \quad \Leftrightarrow \quad I_1 + I_2 + I_3 \to \infty. \tag{4.8}$$

Summarizing,

$$\text{"small" extreme strains} \;\Leftrightarrow\; I_3 \to 0 \tag{4.9}$$

$$\text{"large" extreme strains} \;\Leftrightarrow\; I_1 + I_2 + I_3 \to \infty. \tag{4.10}$$

In treatises on stored-energy functions, the invariants of $\boldsymbol{C}$ are frequently written as a function of $\boldsymbol{F}$. One has

$$\|\boldsymbol{F}\| = \sqrt{\boldsymbol{F}:\boldsymbol{F}} = \sqrt{I_1} \tag{4.11}$$

$$\begin{aligned} \|\mathrm{Cof}\boldsymbol{F}\| &= \sqrt{\mathrm{tr}[(\mathrm{Cof}\boldsymbol{F})^{\mathrm{T}}\cdot(\mathrm{Cof}\boldsymbol{F})]} \\ &= J\sqrt{\mathrm{tr}(\boldsymbol{F}^{-1}\cdot\boldsymbol{F}^{-T})} \\ &= J\sqrt{\mathrm{tr}(\boldsymbol{C}^{-1})} \\ &= J\sqrt{\frac{1}{\Lambda_1}+\frac{1}{\Lambda_2}+\frac{1}{\Lambda_3}} \\ &= J\sqrt{\frac{I_2}{I_3}} \\ &= \sqrt{I_2} \end{aligned} \tag{4.12}$$

$$\det\boldsymbol{F} = J = \sqrt{I_3}. \tag{4.13}$$

Recall that

$$\boldsymbol{F}^{-1} = \frac{(\mathrm{Cof}\boldsymbol{F})^{\mathrm{T}}}{\det\boldsymbol{F}} \tag{4.14}$$

which was used in the derivation of Equation (4.12). Equations (4.9) and (4.10) can now be replaced by

$$\text{“small” extreme strains} \;\Leftrightarrow\; \det\boldsymbol{F}\to 0 \tag{4.15}$$

$$\text{“large” extreme strains} \;\Leftrightarrow\; \|\boldsymbol{F}\| + \|\mathrm{Cof}\boldsymbol{F}\| + \det\boldsymbol{F} \to +\infty. \tag{4.16}$$

“Large” stresses basically mean

$$\left\|\frac{\partial\Sigma}{\partial\boldsymbol{C}}\right\| \to +\infty. \tag{4.17}$$

If $\Sigma(\boldsymbol{C},\boldsymbol{X})$ is continuous on a closed interval $[a,b]$ and differentiable within the open interval (a,b), then the mean value theorem states that

$$\sup_{\boldsymbol{C}\in(\boldsymbol{a},\boldsymbol{b})}\left\|\frac{\partial\Sigma}{\partial\boldsymbol{C}}\right\| \ge \frac{\|\Sigma(\boldsymbol{b})-\Sigma(\boldsymbol{a})\|}{\|\boldsymbol{b}-\boldsymbol{a}\|}. \tag{4.18}$$

This means that $\Sigma\to+\infty$ is sufficient for $\|\frac{\partial\Sigma}{\partial\boldsymbol{C}}\|\to+\infty$. Summarizing, the requirements for Σ are

$$\Sigma(\boldsymbol{C},\boldsymbol{X})\to+\infty \quad \text{if} \quad \det\boldsymbol{F}\to 0^+ \tag{4.19}$$

$$\Sigma(\boldsymbol{C},\boldsymbol{X})\to+\infty \quad \text{if} \quad (\|\boldsymbol{F}\| + \|\mathrm{Cof}\boldsymbol{F}\| + \det\boldsymbol{F})\to+\infty. \tag{4.20}$$

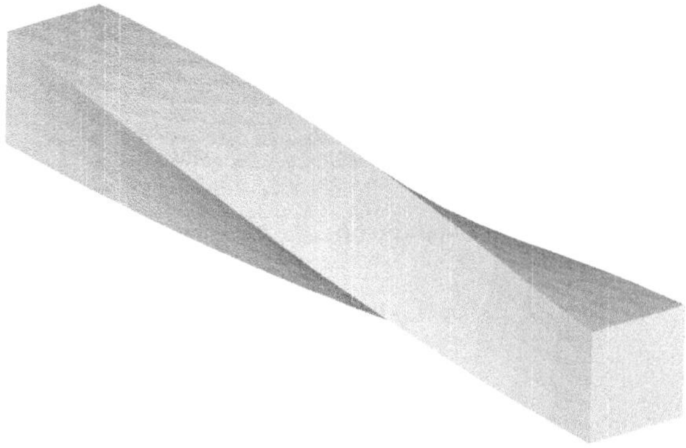

Figure 4.2 Gurtin's experiment

For simplicity, the temperature dependence is dropped from $\Sigma(\boldsymbol{C}, \theta, \boldsymbol{X})$. Recall that the deformation gradient $\boldsymbol{F}$ belongs to the set of 3×3 matrices with a positive determinant, that is,

$$\boldsymbol{F} \in \mathbb{M}^3_+. \tag{4.21}$$

Equation (4.20) is sometimes replaced by the coerciveness inequality, which reads

$$\Sigma(\boldsymbol{C}, \boldsymbol{X}) \geq \alpha \left[\|\boldsymbol{F}\|^p + \|\mathrm{Cof}\boldsymbol{F}\|^q + (\det \boldsymbol{F})^r \right] + \beta,$$
$$\alpha, p, q, r > 0, \boldsymbol{F} \in \mathbb{M}^3_+, \boldsymbol{X} \in V_0. \tag{4.22}$$

This condition plays a major role in proving the existence of a solution. The present section essentially follows (Ciarlet 1993) and is based on the research by John Ball (see, for instance, (Ball 1977)). Here, only the main results will be quoted. For proofs and further reading, the reader is referred to (Ciarlet 1993).

The existence of a solution immediately calls into mind the uniqueness problem. Contrary to linear problems, nonlinear problems can have infinitely many solutions. Merely one example is given here: a beam under torsion fixed at its ends and with stress-free sides (Figure 4.2, Gurtin's experiment). There are infinitely many solutions to this problem, each differing by a torsion angle of a multiple of 2π from the others. Consequently, the solution is physically not unique and accordingly a numerical uniqueness is not desirable either.

4.1.2 Convexity

To proceed, some basic mathematical concepts of convexity have to be explained. Indeed, convexity plays a major role in the derivation of stored-energy functions satisfying Equations (4.19) and (4.22).

Definition 4.1.1 *A subset of a vector space is convex if, for any two elements a and b belonging to the subset, the closed interval $[a, b]$ also belongs to the subset.*

The interval $[a, b]$ consists of all points $a + \lambda(b - a)$, where $\lambda \in [0, 1]$. For example, consider the vector space over $\mathbb{R}$ of all 3×3 matrices $\mathbb{M}^3$. The matrices with positive determinant ($\mathbb{M}^3_+$) form a nonconvex subset. Indeed, $A = \text{Diag}(-3, 2, -1) \in \mathbb{M}^3_+$ and $B = \text{Diag}(2, -3, -1) \in \mathbb{M}^3_+$ but $\text{Diag}(-1, -1, -2) = A + B \notin \mathbb{M}^3_+$. Here, $\text{Diag}(-3, 2, -1)$ is a diagonal 3×3 matrix with elements $-3, 2$ and -1.

Definition 4.1.2 *The closed convex hull* co U *of a subset* U *of a vector space* V *is the smallest closed convex subset of* V *that contains* U.

One can prove (Ciarlet 1993),

$$\text{co } \mathbb{M}^3_+ = \mathbb{M}^3 \tag{4.23}$$

$$\text{co}\{(\boldsymbol{F}, \text{Cof}\boldsymbol{F}, \det \boldsymbol{F}) \in \mathbb{M}^3_+ \times \mathbb{M}^3_+ \times \mathbb{R}_+\} = \mathbb{M}^3 \times \mathbb{M}^3 \times (0, \infty). \tag{4.24}$$

Note that $\text{Cof}\boldsymbol{F} \in \mathbb{M}^3_+$ since

$$\det(\text{Cof}\boldsymbol{F}) = (\det \boldsymbol{F})^2 \tag{4.25}$$

because of Equation (4.14) and the properties of determinants (Gradshteyn and Ryzhik 1980).

Definition 4.1.3 *A function* $f : U \subset V \to \mathbb{R}$ *defined on a convex subset* U *of a vector space* V *is convex on* U *if*

$$\forall a, b \in U, \lambda \in [0, 1] : f[\lambda a + (1 - \lambda)b] \leq \lambda f(a) + (1 - \lambda) f(b). \tag{4.26}$$

The following theorem, formulated here for the special case of an inner product space, can be used to prove the convexity of a function:

Theorem 4.1.4 *Let* $f : U \to \mathbb{R}$ *be a function defined and twice differentiable over a convex subset* U *of an inner product vector space. The function* f *is convex on* U *if and only if*

$$f''(a) \cdot (b - a, b - a) \geq 0, \quad \forall a, b \in U. \tag{4.27}$$

(Ciarlet 1993).

The second derivative is a bilinear mapping of its arguments, that is, it has two arguments and is linear in each of them. In our applications, the bilinear mapping reduces to a classical inner product. As an example, consider

$$f : \boldsymbol{A} \in \mathbb{M}^3 \to \|\boldsymbol{A}\|^2. \tag{4.28}$$

Since $\|\boldsymbol{A}\|^2 = \boldsymbol{A} : \boldsymbol{A} = A_{ij} A_{ij}$ (in rectangular coordinates), one finds

$$\frac{\partial A_{ij} A_{ij}}{\partial A_{kl}} = A_{kl} + A_{kl} = 2A_{kl} \tag{4.29}$$

and for the second derivative

$$2\frac{\partial A_{kl}}{\partial A_{mn}} = 2\delta_{km}\delta_{ln}. \tag{4.30}$$

Accordingly, $\forall \boldsymbol{B}$,

$$f''(\boldsymbol{A})_{klmn} B_{kl} B_{mn} = 2 B_{kl} B_{kl} = 2\|\boldsymbol{B}\|^2. \tag{4.31}$$

Since $\|\boldsymbol{B}\| \geq 0$, f is convex.

On the other hand,

$$f : \boldsymbol{A} \in \mathbb{M}^3 \to \|\mathrm{Cof}\boldsymbol{A}\|^2 \tag{4.32}$$

and

$$g : \boldsymbol{A} \in \mathbb{M}^3 \to \det \boldsymbol{A} = I_{3A} \tag{4.33}$$

are not convex. Indeed,

$$\boldsymbol{A} := \mathrm{Diag}(3, 1, 1) \in \mathbb{M}^3 \tag{4.34}$$

$$\boldsymbol{B} := \mathrm{Diag}(1, 3, 1) \in \mathbb{M}^3 \tag{4.35}$$

$$f(\boldsymbol{A}) = 19 = f(\boldsymbol{B}) \tag{4.36}$$

$$g(\boldsymbol{A}) = 3 = g(\boldsymbol{B}) \tag{4.37}$$

$$\boldsymbol{C} = \lambda \boldsymbol{A} + (1 - \lambda)\boldsymbol{B} \tag{4.38}$$

$$f(\boldsymbol{C}) = 19 + 16\lambda - 32\lambda^3 + 16\lambda^4 \tag{4.39}$$

$$g(\boldsymbol{C}) = 2 + 4\lambda - 4\lambda^2. \tag{4.40}$$

Accordingly,

$$\lambda f(\boldsymbol{A}) + (1 - \lambda) f(\boldsymbol{B}) = 19 \tag{4.41}$$

$$\lambda g(\boldsymbol{A}) + (1 - \lambda) g(\boldsymbol{B}) = 3 \tag{4.42}$$

but

$$f(\boldsymbol{C})|_{\lambda=0.01} > 19 \tag{4.43}$$

$$g(\boldsymbol{C})|_{\lambda=0.01} > 3. \tag{4.44}$$

This concludes the proof.

An important example of a convex function is

$$f : (x, \boldsymbol{A}) \in \mathbb{R}^+ \times \mathbb{M}^3 \to \frac{\|\boldsymbol{A}\|^2}{x^{2/3}}. \tag{4.45}$$

The proof given here goes back to (Hartmann and Neff 2003). First consider

$$g : (x, y) \in \mathbb{R}^+ \times \mathbb{R} \to f(x) \cdot g(y). \tag{4.46}$$

$\mathbb{R}^+ \times \mathbb{R}$ is a convex domain. According to Theorem 4.1.4, g is convex if and only if

$$\begin{Bmatrix} x \\ y \end{Bmatrix}^{\mathrm{T}} \begin{bmatrix} f''(x)g(y) & f'(x)g'(y) \\ f'(x)g'(y) & f(x)g''(y) \end{bmatrix} \begin{Bmatrix} x \\ y \end{Bmatrix} \geq 0, \quad \forall x, y. \tag{4.47}$$

This implies that the 2×2 matrix in Equation (4.47) must be positive semidefinite. A matrix is positive semidefinite if and only if all eigenvalues are not negative. The eigenvalues of a 2×2 symmetric matrix

$$\begin{bmatrix} a_{11} & a_{12} \\ a_{12} & a_{22} \end{bmatrix} \tag{4.48}$$

are

$$\lambda_{1,2} = \frac{1}{2}\left[(a_{11} + a_{22}) \pm \sqrt{(a_{11} + a_{22})^2 - 4(a_{11}a_{22} - a_{12}^2)}\right] \tag{4.49}$$

which are positive if and only if

$$\begin{cases} a_{11}a_{22} - a_{12}^2 \geq 0 \quad \text{and} \\ a_{11} \geq 0 \quad (\text{or } a_{22} \geq 0). \end{cases} \tag{4.50}$$

Accordingly, we require

$$\begin{cases} f''(x)g(y) \geq 0 \\ f''(x)g(y)f(x)g''(x) \geq [f'(x)g'(x)]^2. \end{cases} \tag{4.51}$$

Let

$$\begin{cases} f(x) := x^{-\alpha}, \quad \alpha \geq 0 \\ g(y) := y^p \end{cases} \tag{4.52}$$

then Equations (4.51) are equivalent to

$$\frac{\alpha + 1}{\alpha} \geq \frac{p}{p-1} \Rightarrow \alpha \leq p - 1. \tag{4.53}$$

For instance, for $p = 2$ and $\alpha = 2/3$,

$$g : (x, y) \in \mathbb{R}^+ \times \mathbb{R} \to \frac{y^2}{x^{2/3}} \tag{4.54}$$

is a convex function. Substituting $\|\boldsymbol{A}\|$ for y, one obtains, using the Cauchy–Schwarz condition,

$$\begin{aligned} f[\lambda x_1 + (1-\lambda)x_2, \lambda \boldsymbol{A}_1 + (1-\lambda)\boldsymbol{A}_2] &= \frac{\|\lambda \boldsymbol{A}_1 + (1-\lambda)\boldsymbol{A}_2\|^2}{[\lambda x_1 + (1-\lambda)x_2]^{3/2}} \\ &\leq \frac{[\lambda\|\boldsymbol{A}_1\| + (1-\lambda)\|\boldsymbol{A}_2\|]^2}{[\lambda x_1 + (1-\lambda)x_2]^{3/2}} \\ &\leq \lambda \frac{\|\boldsymbol{A}_1\|^2}{x_1^{3/2}} + (1-\lambda)\frac{\|\boldsymbol{A}_2\|^2}{x_2^{3/2}}. \end{aligned} \tag{4.55}$$

The last step is a consequence of the convexity of g. This concludes the proof that the function f in Equation (4.45) is convex. In a similar way, one can prove that

$$f : (\boldsymbol{x}, \boldsymbol{A}) \in \mathbb{R}^+ \times \mathbb{M}^3 \to \frac{\|\boldsymbol{A}\|^3}{x^2} \tag{4.56}$$

is convex by choosing $p = 3$ and $\alpha = 2$.

The convexity of a function can be extended to nonconvex subsets:

Definition 4.1.5 *A function $f^* : U \to \mathbb{R}$ is convex if there exists a convex function $f : \mathrm{co}\, U \to \mathbb{R}$ such that $f^*(a) = f(a)\ \forall a \in U$.*

Accordingly,

$$f^* : \boldsymbol{F} \in \mathbb{M}^3_+ \to \|\boldsymbol{F}\|^2 = I_1 \tag{4.57}$$

is a convex function since f defined by Equation (4.28) is convex in $\mathbb{M}^3 = \mathrm{co}\, \mathbb{M}^3_+$.

A convex function $f : \mathbb{R}^+ \to \mathbb{R}$ can be extended to a convex function $\overline{f} : \mathbb{R} \to \mathbb{R} \cup \{+\infty\}$ by defining

$$\begin{aligned} \overline{f}(x) &= f(x), \quad x \in \mathbb{R}^+ \\ &= +\infty, \quad x \in \mathbb{R}\backslash\mathbb{R}^+. \end{aligned} \tag{4.58}$$

Convexity is a very nice property and it would be advantageous if we could take simple functions such as in Equation (4.57) to be stored-energy functions. Unfortunately, this is not possible (for a proof, see (Ciarlet 1993)):

Theorem 4.1.6 *Let $\boldsymbol{X} \in V_0$ and $\Sigma : \boldsymbol{F} \in \mathbb{M}^3_+ \to \Sigma(\boldsymbol{X}, \boldsymbol{C}) \in \mathbb{R}$ be convex. Then:*

1. *Equation (4.19) is not satisfied.*
2. *The eigenvalues σ_i of the resulting Cauchy stress satisfy $\sigma_1 + \sigma_2 \geq 0$, $\sigma_1 + \sigma_3 \geq 0$, $\sigma_2 + \sigma_3 \geq 0$ at any $\boldsymbol{X} \in V_0$.*

Accordingly, for a convex function, there is no constraint to prevent the annihilation of material, and some stress states, such as uniform hydrostatic pressure, cannot be simulated. Therefore, convex functions are unsuitable as stored-energy functions.

4.1.3 Polyconvexity

To solve this problem, John Ball (Ball 1977) had the idea of relaxing the convexity requirement to polyconvexity, which is defined as follows:

Definition 4.1.7 *A stored-energy function $\hat{\Sigma} : V_0 \times \mathbb{M}^3_+ \to \mathbb{R}$ is polyconvex, if for each $\boldsymbol{X} \in V_0$ there exists a convex function*

$$\Sigma : V_0 \times \mathbb{M}^3 \times \mathbb{M}^3 \times (0, +\infty) \to \mathbb{R} \tag{4.59}$$

such that

$$\hat{\Sigma}(\boldsymbol{X}, \boldsymbol{F}) = \Sigma(\boldsymbol{X}, \boldsymbol{F}, Cof\boldsymbol{F}, \det \boldsymbol{F}) \quad \forall \boldsymbol{F} \in \mathbb{M}^3_+. \tag{4.60}$$

Using this definition, both $\|\mathrm{Cof}\boldsymbol{F}\|^2$ and $\det \boldsymbol{F}$ are polyconvex (the latter because $f : x \in \mathbb{R}^+ \to x$ is convex). On the basis of Equation (4.45), one also finds that

$$f : \boldsymbol{F} \in \mathbb{M}^3_+ \to \frac{\|\boldsymbol{F}\|^2}{(\det \boldsymbol{F})^{2/3}} \tag{4.61}$$

is polyconvex. Notice that the expression $\|\boldsymbol{F}\|^2/(\det \boldsymbol{F})^{2/3}$ is the first invariant of $\overline{\boldsymbol{C}}$ satisfying

$$\overline{\boldsymbol{C}} = \frac{\boldsymbol{C}}{(\det \boldsymbol{F})^{2/3}} \tag{4.62}$$

and

$$\overline{I}_3 := I_{3\overline{C}} = \det \overline{\boldsymbol{C}} = 1. \tag{4.63}$$

Accordingly, $\overline{\boldsymbol{C}}$ contains the isochoric motion of $\boldsymbol{C}$. In hyperelastic applications and von Mises plasticity, the total motion is frequently split into an isochoric part and a volumetric part. Equation (4.61) is now equivalent to

$$f : \boldsymbol{F} \in \mathbb{M}^3_+ \to I_{1\overline{C}} \tag{4.64}$$

which is a polyconvex function.

Since $f(\boldsymbol{I}) = 3$, the function

$$g : \boldsymbol{F} \in \mathbb{M}^3_+ \to \overline{I}_1 - 3 \tag{4.65}$$

is a convex residual stress-free stored-energy potential. Furthermore, one can prove that $\overline{I}_1 - 3 \geq 0$. Indeed (Schröder and Neff 2001)

$$3I_2 - I_1^2 = (\Lambda_1\Lambda_2 + \Lambda_1\Lambda_3 + \Lambda_2\Lambda_3) - (\Lambda_1 + \Lambda_2 + \Lambda_3)^2 \tag{4.66}$$

$$= \Lambda_1\Lambda_2 + \Lambda_1\Lambda_3 + \Lambda_2\Lambda_3 - \Lambda_1^2 - \Lambda_2^2 - \Lambda_3^2 \tag{4.67}$$

$$= -\tfrac{1}{2}\left[(\Lambda_1 - \Lambda_2)^2 + (\Lambda_2 - \Lambda_3)^2 + (\Lambda_3 - \Lambda_1)^2\right] \leq 0. \tag{4.68}$$

Accordingly,

$$I_1^2 \geq 3I_2. \tag{4.69}$$

Notice that this only applies if the eigenvalues are real, which is guaranteed since $\boldsymbol{C}$ is symmetric. Equation (4.69) can also be obtained by requiring the solution of the characteristic equation to be real (cf the explicit solution of a cubic equation in (Abramowitz and Stegun 1972)).

Equation (4.69) also applies to the inverse of $\boldsymbol{C}$ unless $\det \boldsymbol{C} = 0$:

$$I^2_{1C^{-1}} \geq 3I_{2C^{-1}}. \tag{4.70}$$

Recall that the eigenvalues of the inverse of a matrix are the inverse of the eigenvalues. Accordingly,

$$I_{1C^{-1}} = \frac{I_2}{I_3} \tag{4.71}$$

$$I_{2C^{-1}} = \frac{I_1}{I_3} \tag{4.72}$$

and

$$I_{3C^{-1}} = \frac{1}{I_3}. \tag{4.73}$$

Consequently, Equation (4.70) is equivalent to

$$I_2^2 \geq 3 I_1 I_3. \tag{4.74}$$

Equations (4.66) and (4.74), together with Equations (4.11) to (4.13), yield (recall that all eigenvalues and invariants are strictly positive unless material annihilation is accepted)

$$I_1^4 \geq 27 I_1 I_3 \tag{4.75}$$

$$\Downarrow$$

$$\|\boldsymbol{F}\|^4 \geq 3\sqrt{3}\|\boldsymbol{F}\|(\det \boldsymbol{F}) \tag{4.76}$$

$$\Downarrow$$

$$\overline{I}_1 = \frac{\|\boldsymbol{F}\|^2}{(\det \boldsymbol{F})^{2/3}} \geq 3 \tag{4.77}$$

which completes the proof. Accordingly, the function $g = \overline{I}_1 - 3$ is a positive, polyconvex, residual stress-free stored-energy function.

Now, the following theorem will be used:

Theorem 4.1.8 *If*

$$f : x \in V_0 \to f(x) \in \mathbb{R}^+ \tag{4.78}$$

is convex and

$$g : y \in \mathbb{R}^+ \to g(y) \tag{4.79}$$

is monotonic increasing and convex, then

$$g \circ f : x \in V_0 \to (g \circ f)\,(x) \tag{4.80}$$

is convex.

Proof. f is convex means

$$f[\lambda a + (1-\lambda)b] \leq \lambda f(a) + (1-\lambda) f(b). \tag{4.81}$$

Hence, since g is monotonic increasing,

$$g\{f[\lambda a + (1-\lambda)b]\} \leq g[\lambda f(a) + (1-\lambda) f(b)] \tag{4.82}$$

$$\leq \lambda g[f(a)] + (1-\lambda) g[f(b)] \tag{4.83}$$

due to the convexity of g.

Let us apply this theorem to $f = \overline{I}_1 - 3$. Choosing $g(y) = y^i, i \geq 1$ one finds that

$$h : \boldsymbol{F} \in \mathbb{M}^3_+ \rightarrow (\overline{I}_1 - 3)^i, \quad i \geq 1 \tag{4.84}$$

is polyconvex. This function is frequently used in stored-energy functions for rubber materials.

The second invariant of $\overline{\boldsymbol{C}}$ satisfies

$$\overline{I}_2 = \frac{\|\text{Cof}\boldsymbol{F}\|^2}{(\det \boldsymbol{F})^{4/3}}. \tag{4.85}$$

This function is not polyconvex. However,

$$f : \boldsymbol{F} \in \mathbb{M}^3_+ \rightarrow \frac{\|\text{Cof}\boldsymbol{F}\|^3}{(\det \boldsymbol{F})^2} = \overline{I}_2^{3/2} \tag{4.86}$$

is polyconvex because of Equation (4.56). Equations (4.66) and (4.74) reveal that

$$I_2^4 \geq 27 I_2 I_3^2 \tag{4.87}$$

$$\Downarrow$$

$$\frac{I_2^{3/2}}{I_3} \geq 3\sqrt{3}. \tag{4.88}$$

Using the same reasoning as for $\overline{I}_1$, one finds that

$$h : \boldsymbol{F} \in \mathbb{M}^3_+ \rightarrow \quad \overline{I}_2^{3/2} - 3\sqrt{3}\Big)^i, \quad i \geq 1 \tag{4.89}$$

is polyconvex. Furthermore $h(\boldsymbol{I}) = 0$, since $I_{2I} = 3$, and consequently the initial configuration is stress-free. Terms of the kind in Equation (4.89) are only recently being used in stored-energy functions (see (Hartmann and Neff 2003) and (Düster *et al.* 2003)).

Notice that the basic norm properties in conjunction with Theorem 4.1.8 can be used to prove that $f : \boldsymbol{A} \in \mathbb{M}^3 \rightarrow \|\boldsymbol{A}\|^2$ in Equation (4.28) is convex. Indeed, the norm properties guarantee that

$$\|\lambda \boldsymbol{A}_1 + (1 - \lambda)\boldsymbol{A}_2\| \leq \lambda\|\boldsymbol{A}_1\| + (1 - \lambda)\|\boldsymbol{A}_2\|. \tag{4.90}$$

Consequently, $\|\boldsymbol{A}\|$ is convex and also positive. Application of Theorem 4.1.8 with $g(y) = y^i, i \geq 1$ shows that

$$f : \boldsymbol{A} \in \mathbb{M}^3 \rightarrow \|\boldsymbol{A}\|^i, \quad i \geq 1 \tag{4.91}$$

is convex.

To prove that $\overline{I}_2$ is not polyconvex, the following definitions are introduced:

Definition 4.1.9 *A twice differentiable function* $\Sigma(\boldsymbol{A})$, $\boldsymbol{A} \in \mathbb{M}^3$ *leads to an elliptic system if and only if*

$$\forall \boldsymbol{A} \in \mathbb{M}^3, \forall \boldsymbol{\xi}, \boldsymbol{\eta} \in \mathbb{R}^3 : \Sigma''(\boldsymbol{A}) \cdot (\boldsymbol{\xi} \otimes \boldsymbol{\eta}, \boldsymbol{\xi} \otimes \boldsymbol{\eta}) \geq 0. \tag{4.92}$$

Definition 4.1.10 *A function* $\Sigma(\boldsymbol{A})$, $\boldsymbol{A} \in \mathbb{M}^3$ *is rank-one convex if*

$$f : t \in \mathbb{R} \rightarrow \Sigma(\boldsymbol{A} + t(\boldsymbol{\xi} \otimes \boldsymbol{\eta})) \tag{4.93}$$

is convex $\forall \boldsymbol{A} \in \mathbb{M}^3, \forall \boldsymbol{\xi}, \boldsymbol{\eta} \in \mathbb{R}^3$.

One can prove (Dacorogna 1989),

Theorem 4.1.11 *1. For sufficiently smooth functions* Σ, *one has*

$$\Sigma \text{ leads to an elliptic system}$$

$$\Updownarrow$$

$$\Sigma \text{ is rank-one convex}$$

2. Σ *is polyconvex* $\Rightarrow$ Σ *is rank-one convex*

Notice that for ellipticity, the direction the second derivative is projected on is a rank-one matrix $(\boldsymbol{\xi} \otimes \boldsymbol{\eta})$, whereas for convexity, this direction can have an arbitrary rank (a general matrix $\boldsymbol{A}$, Theorem 4.1.4). Accordingly, convexity implies rank-one convexity. Recall that the rank of a matrix is the dimension of its image. Since

$$(\boldsymbol{\xi} \otimes \boldsymbol{\eta}) \cdot \boldsymbol{\zeta} = \boldsymbol{\xi}(\boldsymbol{\eta} \cdot \boldsymbol{\xi}) \tag{4.94}$$

the rank of $\boldsymbol{\xi} \otimes \boldsymbol{\eta}$ is one. The concept of rank-one convexity is somewhat simpler than convexity. Therefore, invoking Theorem 4.1.11(2), it is mainly used to prove that a function is not polyconvex. Let us apply this to prove that $\overline{I}_2$ in Equation (4.85) is not polyconvex. Rank-one convexity implies that

$$f : t \in \mathbb{R} \rightarrow \frac{\|\text{Cof}[\boldsymbol{A} + t(\boldsymbol{\xi} \otimes \boldsymbol{\eta})]\|^2}{\det[\boldsymbol{A} + t(\boldsymbol{\xi} \otimes \boldsymbol{\eta})]^{4/3}} \tag{4.95}$$

is convex. The expression $\det[\boldsymbol{A} + t(\boldsymbol{\xi} \otimes \boldsymbol{\eta})]$ is linear in t. This can be seen by applying a coordinate rotation (which leaves the determinant unchanged since it is an invariant of its argument) such that $\boldsymbol{\xi}$ coincides with a basis vector. Then the term $t(\boldsymbol{\xi} \otimes \boldsymbol{\eta})$ leads to a linear change of just one row in $\boldsymbol{A}$. Since the cofactors of a matrix are the minor determinants, the numerator is a sum of squares of linear relations. This leads to a quadratic relation with positive coefficients for the quadratic term and the constant term:

$$f(t) = \frac{\lambda_1^2 t^2 + \lambda_2 t + \lambda_3^2}{(\lambda_4 t + \lambda_5)^{4/3}}. \tag{4.96}$$

The second derivative of a function $g = a^{\alpha} b^{\beta}$ has the form

$$g'' = a^{\alpha-2} b^{\beta-2}[\alpha(\alpha-1)(a')^2 b^2 + \alpha a a'' b^2 + \beta(\beta-1)(b')^2 a^2 + \beta b b'' a^2 + 2\alpha a a' \beta b b']. \tag{4.97}$$

Taking

$$\alpha = \gamma/2 \tag{4.98}$$

$$a = \lambda_1^2 t^2 + \lambda_2 t + \lambda_3^2 \tag{4.99}$$

$$b = \lambda_4 t + \lambda_5 \tag{4.100}$$

the term in the square brackets yields

$$\lambda_4^2\lambda_1^4 t^4[(\beta+\gamma)^2-(\beta+\gamma)]. \tag{4.101}$$

This function is only convex for $(\beta+\gamma)^2-(\beta+\gamma)\geq 0$, that is, $\beta+\gamma\leq 0$ or $\beta+\gamma\geq 1$. Since for Equation (4.96), $\gamma=2$ and $\beta=-4/3$, f in Equation (4.95) is not rank-one convex and consequently not polyconvex (Theorem 4.1.11).

Finally, note that all convex functions are polyconvex, but not vice versa.

4.1.4 Suitable stored-energy functions

The polyconvexity concept plays an important role in the existence theorems. Indeed, John Ball proved that a solution exists if

1. the stored-energy function is polyconvex
2. Equation (4.19) applies:

$$\lim_{\det F\to 0^+} \Sigma(C, X)=+\infty \tag{4.102}$$

3. and the coerciveness inequality is satisfied, Equation (4.22).

For details, the reader is referred to (Ciarlet 1993). These are sufficient but not necessary conditions.

An important class of materials satisfying these conditions is evoked by the following theorem:

Theorem 4.1.12 *Let Σ be a stored-energy function of the form*

$$F\in \mathbb{M}_+^3 \to \Sigma(F)=\sum_{i=1}^{M} a_i I_{1(C^{\gamma_i/2})}+\sum_{j=1}^{N} b_j I_{2(C^{\delta_j/2})}+ \quad \sqrt{I_{3C}}) \tag{4.103}$$

where $a_i>0$, $\gamma_i\geq 1$, $b_j>0$, $\delta_j\geq 1$ and $\quad :(0,+\infty)\to\mathbb{R}$ is a convex function, then Σ is polyconvex and satisfies

$$\Sigma(F)\geq \alpha\left\{\|F\|^p+\|\mathrm{Cof}F\|^q\right\}+ \quad \sqrt{I_{3C}}) \tag{4.104}$$

$\forall F\in\mathbb{M}_+^3$ *with* $\alpha>0$, $p=\max_i(\gamma_i)$, $q=\max_j(\delta_j)$.

If, in addition, $\lim_{\delta\to 0^+}\Gamma(\delta)=+\infty$ (Equation (4.19)), the material is called an *Ogden* material. An Ogden material satisfies the conditions in the existence theorem by Ball. Accordingly, a solution exists. Notice that both $a_i>0$ and $b_j>0$ apply. Hence, both I_{1C} and I_{2C} must be present, together with I_{3C} because of Equation (4.19). In Equation (4.103), $I_{1(C^{\gamma_i/2})}$ and $I_{2(C^{\delta_j/2})}$ are defined by

$$I_{1(C^{\gamma_i/2})}:=\sum_j \Lambda_j^{\gamma_i/2} \tag{4.105}$$

$$I_{2(C^{\delta_j/2})}:=\sum_{\substack{k,l\\k\neq l}}(\Lambda_k\Lambda_l)^{\delta_j/2}. \tag{4.106}$$

Notice that Theorem 4.1.12 only gives information on how to construct polyconvex functions that have the desired property. This does not mean that any function not in the form of Equation (4.103) is inappropriate. Yet, in most cases, no existence results will be available.

As an example of a well-known stored-energy function that is not polyconvex, consider the St.Venant–Kirchhoff potential (Equation (1.440)):

$$\Sigma = \tfrac{1}{2}\lambda(\mathrm{tr}\boldsymbol{E})^2 + \mu\mathrm{tr}(\boldsymbol{E}^2). \tag{4.107}$$

For the proof, the reader is referred to (Ciarlet 1993). Using Equations (1.444) and (1.445) together with

$$2I_{1E} = -3 + I_{1C} \tag{4.108}$$

$$4I_{2E} = 3 - 2I_{1C} + I_{2C} \tag{4.109}$$

$$8I_{3E} = -1 + I_{1C} - I_{2C} + I_{3C} \tag{4.110}$$

it is clear that Σ does not depend on I_{3C} and Equation (4.19) cannot be satisfied. Furthermore, the stress obtained by differentiating Equation (4.107)

$$\boldsymbol{S} = \lambda(\mathrm{tr}\boldsymbol{E})\boldsymbol{G}^{\sharp} + 2\mu\boldsymbol{E} \tag{4.111}$$

can be inverted to yield

$$\boldsymbol{E} = \frac{1}{E}[-\nu(\mathrm{tr}\boldsymbol{S})\boldsymbol{G} + (1+\nu)\boldsymbol{S}] \tag{4.112}$$

which implies uniqueness. As explained previously, uniqueness is not desirable for large strains. Although the use of Equation (4.107) will yield better results than the use of infinitesimal strains, it should not be used for large strains. Its field of operation is often called *large deformation–small strains*, which emphasizes the good performance for large rotations (shell applications).

4.2 Isotropic Hyperelastic Materials

In this section, frequently used stored-energy potentials for hyperelastic materials are treated. These include the Arruda–Boyce, the Mooney–Rivlin, the neo-Hooke, the polynomial, the reduced polynomial, the Yeoh and the Ogden model. The preferred form involves a split into an isochoric part and a volumetric part (ABAQUS 1997), (Kaliske and Rothert 1997), (Storåkers 1986). For a treatise on volumetric strain-energy functions, see (Doll and Schweizerhof 2000). This implies that we will use the reduced invariants $\overline{I}_1$ and $\overline{I}_2$ and the reduced principal stretches $\overline{\lambda}_1$, $\overline{\lambda}_2$, and $\overline{\lambda}_3$. Unfortunately, these forms do not fit the generic form of Equation (4.103). However, in some cases we can prove explicitly that John Ball's conditions are satisfied. Notice that the use of the reduced quantities in polynomial-type functions automatically implies that Σ grows beyond bounds as $J \to 0$ (Equation (4.19)).

The split of the stored-energy function in an isochoric part and a volumetric part finds its origin in the near isochoric behavior of most rubber materials. The isochoric coefficients are usually determined by simple tests such as the uniaxial, equibiaxial or planar test (ABAQUS 1997). The compressibility coefficients are derived by volumetric compression tests. Depending on the model, a linear or a nonlinear least-squares procedure is used to find the coefficients (Hartmann 2001a), (Hartmann 2001b). In general, different types of tests are needed for a good description of the material. Even then, extrapolation to stretches significantly exceeding the range of the test data can lead to wildly erroneous behavior. If only one test type is performed (nearly always a uniaxial test), the neo-Hooke and the Arruda–Boyce models seem to perform well because of the physical foundations of these models. More complex phenomenological models such as the Ogden model and the polynomial model with many terms require the availability of different test type data to perform well.

Another issue is the stability of the models. Several criteria exist, such as the Baker–Ericksen inequality and the incremental stability. The Baker–Ericksen inequality states that if a Cauchy principal stress σ_i exceeds another Cauchy principal stress σ_j, the corresponding stretch λ_i should exceed λ_j as well. The incremental stability requires that the incremental power $\dot{\boldsymbol{S}} : \dot{\boldsymbol{E}}$ be positive. For details, the reader is referred to (Hartmann 2003) and (Reese 1994). For most models, stability requirements imply limits on the coefficients.

Taking a stored-energy functional of the form $\Sigma(\boldsymbol{C}, \theta, \boldsymbol{X})$, we are interested in the stress (Equation (4.2))

$$S^{KL} = 2\frac{\partial \Sigma}{\partial C_{KL}} \tag{4.113}$$

and the tangent stiffness (Equation (3.6)):

$$\Sigma^{KLMN} = 2\frac{\partial S^{KL}}{\partial C_{MN}} = 4\frac{\partial^2 \Sigma}{\partial C_{KL} \partial C_{MN}}. \tag{4.114}$$

For isotropic materials, Σ will be of the form $\Sigma(\overline{I}_1, \overline{I}_2, J, \theta, \boldsymbol{X})$ or of the Ogden form $\Sigma(\overline{\lambda}_1, \overline{\lambda}_2, \overline{\lambda}_3, J, \theta, \boldsymbol{X})$, where the dependence on θ and $\boldsymbol{X}$ is hidden in the coefficients of the models.

4.2.1 Polynomial form

The general polynomial stored-energy function takes the form

$$\Sigma = \sum_{i+j=1}^{N} B_{ij}(\overline{I}_1 - 3)^i (\overline{I}_2 - 3)^j + \sum_{i=1}^{N} \frac{1}{D_i}(J^{\text{el}} - 1)^{2i} \tag{4.115}$$

where

$$J^{\text{el}} = \frac{J}{J^{\text{th}}} \tag{4.116}$$

with J^{el} the elastic Jacobian of the deformation, J the total Jacobian and J^{th} the thermal Jacobian,

$$J^{\text{th}} = (1 + \alpha T)^3 \tag{4.117}$$

cf Equation (1.449). Notice that the polynomial form is not polyconvex unless $j = 0$ (since $\overline{I}_2$ is not polyconvex) and $B_{ij}, D_i \geq 0$. Special forms are the Mooney–Rivlin strain-energy potential

$$\Sigma = B_{10}(\overline{I}_1 - 3) + B_{01}(\overline{I}_2 - 3) + \frac{1}{D_1}(J^{\mathrm{el}} - 1)^2 \tag{4.118}$$

the neo-Hooke strain potential

$$\Sigma = B_{10}(\overline{I}_1 - 3) + \frac{1}{D_1}(J^{\mathrm{el}} - 1)^2 \tag{4.119}$$

the Yeoh form

$$\Sigma = B_{10}(\overline{I}_1 - 3) + B_{20}(\overline{I}_1 - 3)^2 + B_{30}(\overline{I}_1 - 3)^3 + \frac{1}{D_1}(J^{\mathrm{el}} - 1)^2 + \frac{1}{D_2}(J^{\mathrm{el}} - 1)^4 + \frac{1}{D_3}(J^{\mathrm{el}} - 1)^6 \tag{4.120}$$

and the reduced polynomial form

$$\Sigma = \sum_{i=1}^{N} B_{i0}(\overline{I}_1 - 3)^i + \sum_{i=1}^{N} \frac{1}{D_i}(J^{\mathrm{el}} - 1)^{2i}. \tag{4.121}$$

Only the neo-Hooke, the Yeoh and the reduced polynomial form are polyconvex because of the absence of $\overline{I}_2$. Since $\overline{I}_2$ is difficult to determine experimentally and its inclusion in the stored-energy function does not necessarily improve its predictive quality (Kaliske and Rothert 1997), it is advisable to start off with a dependence only on $\overline{I}_1$. This especially applies if only uniaxial data are available. Among the models that depend only on $\overline{I}_1$, the neo-Hooke type assumes a special position. Indeed, using Gaussian statistical thermodynamics, its constant can be linked to the molecular chain density of the material (Treloar 1975).

Since Σ is linear in the coefficients B_{ij}, a linear least-squares procedure suffices to determine them. The Baker–Ericksen inequality is assured if all $B_{ij} \geq 0$ (sufficient condition) S^{KL} and Σ^{KLMN} take the form

$$S^{KL} = 2\left\{ \sum_{i+j=1}^{N} B_{ij}\left[i(\overline{I}_1 - 3)^{i-1}(\overline{I}_2 - 3)^j \frac{\partial \overline{I}_1}{\partial C_{kl}} + j(\overline{I}_1 - 3)^i(\overline{I}_2 - 3)^{j-1}\frac{\partial \overline{I}_2}{\partial C_{kl}}\right] + \sum_{i=1}^{N} \frac{2i}{D_i}(J^{\mathrm{el}} - 1)^{2i-1}\frac{\partial J^{\mathrm{el}}}{\partial C_{KL}} \right\} \tag{4.122}$$

$$\Sigma^{KLMN} = 4\left\{\sum_{i+j=1}^{N} B_{ij}(\overline{I}_1-3)^{i-2}(\overline{I}_2-3)^{j-2}\left[i(i-1)(\overline{I}_2-3)^2\frac{\partial\overline{I}_1}{\partial C_{KL}}\frac{\partial\overline{I}_1}{\partial C_{MN}}\right.\right.$$
$$+ij(\overline{I}_1-3)(\overline{I}_2-3)\frac{\partial\overline{I}_1}{\partial C_{KL}}\frac{\partial\overline{I}_2}{\partial C_{MN}}+i(\overline{I}_1-3)(\overline{I}_2-3)^2\frac{\partial^2\overline{I}_1}{\partial C_{KL}\partial C_{MN}}$$
$$+ij(\overline{I}_1-3)(\overline{I}_2-3)\frac{\partial\overline{I}_2}{\partial C_{KL}}\frac{\partial\overline{I}_1}{\partial C_{MN}}+j(j-1)(\overline{I}_1-3)^2\frac{\partial\overline{I}_2}{\partial C_{KL}}\frac{\partial\overline{I}_2}{\partial C_{MN}}$$
$$\left.+j(\overline{I}_1-3)^2(\overline{I}_2-3)\frac{\partial^2\overline{I}_2}{\partial C_{KL}\partial C_{MN}}\right]$$
$$\left.+\sum_{i=1}^{N}\frac{2i(J^{\mathrm{el}}-1)^{2i-2}}{D_i}\left[(2i-1)\frac{\partial J^{\mathrm{el}}}{\partial C_{KL}}\frac{\partial J^{\mathrm{el}}}{\partial C_{MN}}+(J^{\mathrm{el}}-1)\frac{\partial^2 J^{\mathrm{el}}}{\partial C_{KL}\partial C_{MN}}\right]\right\}. \tag{4.123}$$

The derivatives of the invariants with respect to $\boldsymbol{C}$ are treated in Section 4.4.

4.2.2 Arruda–Boyce form

This potential function satisfies

$$\Sigma = \mu\left[\tfrac{1}{2}(\overline{I}_1-3)+\frac{1}{20\lambda_m^2}(\overline{I}_1^2-9)+\frac{11}{1050\lambda_m^4}(\overline{I}_1^3-27)\right.$$
$$\left.+\frac{19}{7000\lambda_m^6}(\overline{I}_1^4-81)+\frac{519}{673\,750\lambda_m^8}(\overline{I}_1^5-243)\right]$$
$$+\frac{1}{D}\left(\frac{(J^{\mathrm{el}})^2-1}{2}-\ln J^{\mathrm{el}}\right),\quad \mu, D\geq 0. \tag{4.124}$$

Notice that all terms are polyconvex and that

$$\lim_{J^{\mathrm{el}}\to 0}\Sigma = +\infty \tag{4.125}$$

$$\lim_{J^{\mathrm{el}}\to\infty}\Sigma = +\infty \tag{4.126}$$

$$\lim_{\substack{I_1\to\infty\\ J^{\mathrm{el}}<M}}\Sigma = +\infty. \tag{4.127}$$

where M is some positive real number. Accordingly, the physical requirements in Section 4.1.1 are fulfilled. The Arruda–Boyce model is based on an 8-chain representation of the macromolecular network of rubber and is extensively described in (Arruda and Boyce 1993). For this model, the Baker–Ericksen inequality is satisfied. The determination of the coefficients requires a nonlinear least-squares procedure. The second Piola–Kirchhoff stress and tangent stiffness satisfy

$$S^{KL} = 2\left[\mu\left(\frac{1}{2} + \frac{1}{10\lambda_m^2}\overline{I}_1 + \frac{33}{1050\lambda_m^4}\overline{I}_1^2 + \frac{76}{7000\lambda_m^6}\overline{I}_1^3 + \frac{2595}{673\,750\lambda_m^8}\overline{I}_1^4\right)\frac{\partial\overline{I}_1}{\partial C_{KL}} + \frac{1}{D}\left(J^{\text{el}} - \frac{1}{J^{\text{el}}}\right)\frac{\partial J^{\text{el}}}{\partial C_{KL}}\right] \quad (4.128)$$

$$\begin{aligned}\Sigma^{KLMN} = 4\Bigg[&\mu\left(\frac{1}{10\lambda_m^2} + \frac{66}{1050\lambda_m^4}\overline{I}_1 + \frac{228}{7000\lambda_m^6}\overline{I}_1^2 + \frac{10\,380}{673\,750\lambda_m^8}\overline{I}_1^3\right)\frac{\partial\overline{I}_1}{\partial C_{KL}}\frac{\partial\overline{I}_1}{\partial C_{MN}} \\ &+\mu\left(\frac{1}{2} + \frac{1}{10\lambda_m^2}\overline{I}_1 + \frac{33}{1050\lambda_m^4}\overline{I}_1^2 + \frac{76}{7000\lambda_m^6}\overline{I}_1^3 + \frac{2595}{673\,750\lambda_m^8}\overline{I}_1^4\right)\frac{\partial^2\overline{I}_1}{\partial C_{KL}\partial C_{MN}} \\ &+\frac{1}{D}\left(1 + \frac{1}{(J^{\text{el}})^2}\right)\frac{\partial J^{\text{el}}}{\partial C_{KL}}\frac{\partial J^{\text{el}}}{\partial C_{MN}} + \frac{1}{D}\left(J^{\text{el}} - \frac{1}{J^{\text{el}}}\right)\frac{\partial^2 J^{\text{el}}}{\partial C_{KL}\partial C_{MN}}\Bigg].\end{aligned} \quad (4.129)$$

4.2.3 The Ogden form

The Ogden form resembles the stored-energy function in Theorem 4.1.12; however, the principal stretches are replaced by their reduced form:

$$\Sigma = \sum_{i=1}^{N}\frac{2\mu_i}{\alpha_i^2}(\overline{\lambda}_1^{\alpha_i} + \overline{\lambda}_2^{\alpha_i} + \overline{\lambda}_3^{\alpha_i} - 3) + \sum_{i=1}^{N}\frac{1}{D_i}(J^{\text{el}} - 1)^{2i} \quad (4.130)$$

where

$$\overline{\lambda}_i := \frac{\lambda_i}{J^{1/3}} = \frac{\lambda_i^{2/3}}{\lambda_j^{1/3}\lambda_k^{1/3}} \quad j, k \neq i. \quad (4.131)$$

For $\alpha_i = 2$, one obtains $\overline{I}_1$, for $\alpha_i = -2$, the invariant $\overline{I}_2$ emerges. Since $\overline{I}_2$ is not polyconvex, Equation (4.130) is not necessarily polyconvex. S^{KL} and Σ^{KLMN} satisfy

$$S^{KL} = 2\left[\sum_{i=1}^{N}\frac{2\mu_i}{\alpha_i}\left(\sum_{k=1}^{3}\overline{\lambda}_k^{\alpha_i - 1}\frac{\partial\overline{\lambda}_k}{\partial C_{KL}}\right) + \sum_{i=1}^{N}\frac{2i}{D_i}(J^{\text{el}} - 1)^{2i-1}\frac{\partial J^{\text{el}}}{\partial C_{KL}}\right] \quad (4.132)$$

$$\begin{aligned}\Sigma^{KLMN} = 4\Bigg\{&\sum_{i=1}^{N}\frac{2\mu_i}{\alpha_i}\left[(\alpha_i - 1)\left(\sum_{k=1}^{3}\overline{\lambda}_k^{\alpha_i - 2}\frac{\partial\overline{\lambda}_k}{\partial C_{KL}}\frac{\partial\overline{\lambda}_k}{\partial C_{MN}}\right) + \sum_{k=1}^{3}\overline{\lambda}_k^{\alpha_i - 1}\frac{\partial^2\overline{\lambda}_k}{\partial C_{KL}\partial C_{MN}}\right] \\ &+\sum_{i=1}^{N}\left[\frac{2i(2i-1)}{D_i}(J^{\text{el}} - 1)^{2i-2}\frac{\partial J^{\text{el}}}{\partial C_{KL}}\frac{\partial J^{\text{el}}}{\partial C_{MN}} + \frac{2i}{D_i}(J^{\text{el}} - 1)^{2i-1}\frac{\partial^2 J^{\text{el}}}{\partial C_{KL}\partial C_{MN}}\right]\Bigg\}.\end{aligned} \quad (4.133)$$

4.2.4 Elastomeric foam behavior

Whereas the potentials in the previous sections are frequently used for materials that are nearly incompressible, such as rubber, elastomeric foams are very compressible. The general form satisfies

$$\Sigma = \sum_{i=1}^{N} \frac{2\mu_i}{\alpha_i^2} \left\{ \hat{\lambda}_1^{\alpha_i} + \hat{\lambda}_2^{\alpha_i} + \hat{\lambda}_3^{\alpha_i} - 3 + \frac{1}{\beta_i} \left[(J^{\mathrm{el}})^{-\alpha_i \beta_i} - 1 \right] \right\} \tag{4.134}$$

where

$$\hat{\lambda}_i := \frac{\lambda_i}{(J^{\mathrm{el}})^{1/3}}. \tag{4.135}$$

This form comes close to the Ogden form defined in Theorem 4.1.12 if $\alpha_i, \mu_i > 0$. Indeed, for $\alpha_i > 0$, the first three terms correspond to $I_{1(C^{\alpha_i/2})}$. For $\alpha_i < 0$, however, they correspond to

$$\frac{I_{2(C^{-\alpha_i/2})}}{I_{3(C^{-\alpha_i/2})}}, \tag{4.136}$$

which is not compatible with the Ogden form. Since the second derivative of the volumetric term satisfies

$$\Sigma''_{\mathrm{vol}} = \sum_{i=1}^{N} \frac{(\alpha_i \beta_i)(\alpha_i \beta_i + 1)}{\beta_i} \frac{2\mu_i}{\alpha_i^2} (J^{\mathrm{el}})^{-\alpha_i \beta_i - 2} \tag{4.137}$$

convexity is guaranteed if

$$\mu_{\underline{i}} \alpha_{\underline{i}} (\alpha_{\underline{i}} \beta_{\underline{i}} + 1) \geq 0 \tag{4.138}$$

(sufficient but not necessary condition). The derivatives yield

$$S^{KL} = 2 \sum_{i=1}^{N} \frac{2\mu_i}{\alpha_i} \left[\sum_{k=1}^{3} \hat{\lambda}_k^{(\alpha_i - 1)} \frac{\partial \hat{\lambda}_k}{\partial C_{KL}} - (J^{\mathrm{el}})^{-\alpha_i \beta_i - 1} \frac{\partial J^{\mathrm{el}}}{\partial C_{KL}} \right] \tag{4.139}$$

$$\begin{aligned} \Sigma^{KLMN} = 8 \sum_{i=1}^{N} \frac{\mu_i}{\alpha_i} \Bigg[& (\alpha_i - 1) \sum_{k=1}^{3} \hat{\lambda}_k^{(\alpha_i - 2)} \frac{\partial \hat{\lambda}_k}{\partial C_{KL}} \frac{\partial \hat{\lambda}_k}{\partial C_{MN}} \\ & + \sum_{k=1}^{3} \hat{\lambda}_k^{(\alpha_i - 1)} \frac{\partial^2 \hat{\lambda}_k}{\partial C_{KL} \partial C_{MN}} \\ & + (\alpha_i \beta_i + 1)(J^{\mathrm{el}})^{-\alpha_i \beta_i - 2} \frac{\partial J^{\mathrm{el}}}{\partial C_{KL}} \frac{\partial J^{\mathrm{el}}}{\partial C_{MN}} - (J^{\mathrm{el}})^{-\alpha_i \beta_i - 1} \frac{\partial^2 J^{\mathrm{el}}}{\partial C_{KL} \partial C_{MN}} \Bigg]. \end{aligned} \tag{4.140}$$

4.3 Nonhomogeneous Shear Experiment

To illustrate the differences between the models, the nonhomogeneous shear experiment investigated in (van den Bogert and de Borst 1994) is discussed. A rubber material is considered and described by the neo-Hooke, Mooney–Rivlin, Yeoh and Arruda–Boyce model. The isochoric constants are taken from (Kaliske and Rothert 1997) and were obtained by fitting tensile test results. The volumetric data are such that $\nu_{\text{eq}} = 0.475$ at zero deformation. They satisfy (coefficients B_{ij} and μ in N/mm^2, D_i in mm^2/N, λ_m is dimensionless)

1. neo-Hooke model

$$B_{10} = 0.525, D_1 = 0.0952 \tag{4.141}$$

2. Mooney–Rivlin model

$$B_{10} = 0.1486, B_{01} = 0.4849, D_1 = 0.0789 \tag{4.142}$$

3. Yeoh model

$$B_{10} = 0.538, B_{20} = -0.0685, B_{30} = 0.0325, \\ D_1 = 0.0929, D_2 = 0.0086, D_3 = 0.0008 \tag{4.143}$$

4. Arruda–Boyce model

$$\mu = 0.71, \lambda_m = 1.7029, D = 0.1408. \tag{4.144}$$

When applied to a $1 \times 1 \times 8$ mm^3 specimen, we get the force versus stretch curves in Figure 4.3. According to (Kaliske and Rothert 1997), the experimental results are best fit by the Yeoh curve exhibiting an S-shape. This typical shape originates from the negative B_{20} coefficient. The neo-Hooke model and the Mooney–Rivlin model are not capable of capturing this effect.

The shear experiment is schematically shown in Figure 4.4. The upper and lower surfaces are rigid. The lower surface cannot translate or rotate, all degrees of freedom are fixed. The upper surface can only translate in x-direction and z-direction. A force is applied in x-direction. A uniform $5 \times 5 \times 10$ 20-node brick element mesh was used with reduced integration.

The displacements in x-direction (Figure 4.5) show similar tendencies as the uniaxial test data. The Yeoh model predicts more hardening than the neo-Hooke and Mooney–Rivlin model. However, up to moderate displacements, all models predict similar results closely fitting the experimental data (overall behavior of the experimental data is symbolized by discrete symbols). Because of the elongation in x-direction, the specimen shrinks in z-direction (Figure 4.6). This is reasonably well modeled by the neo-Hooke, Yeoh and Arruda–Boyce model. The Mooney–Rivlin model, however, shows a completely opposite tendency: the specimen grows thicker. Notice that the Mooney–Rivlin model is the only model including the second invariant. It seems that predictions of models that include the second invariant are not very accurate if model-parameter characterization is based on uniaxial test results only.

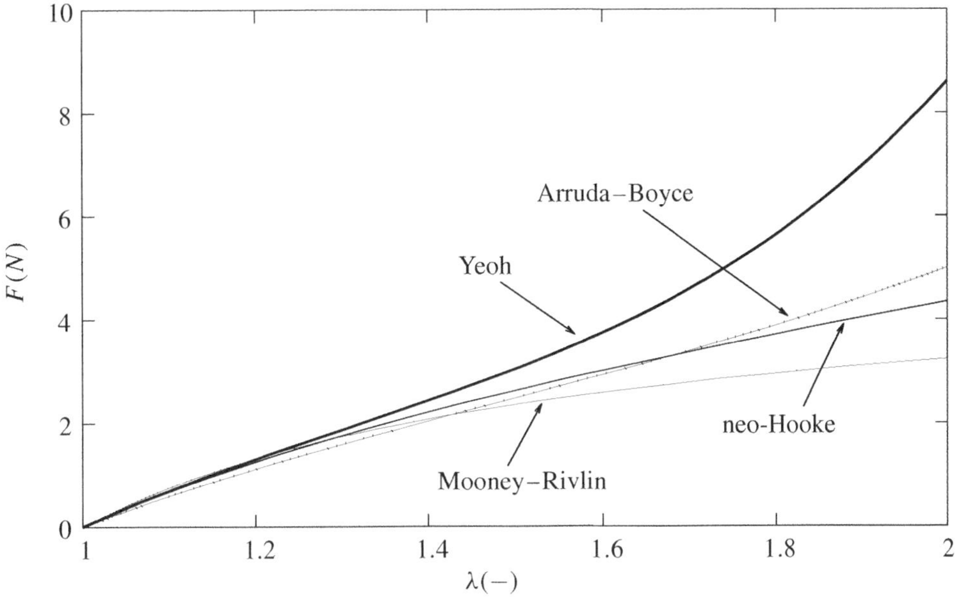

Figure 4.3 Tensile test results

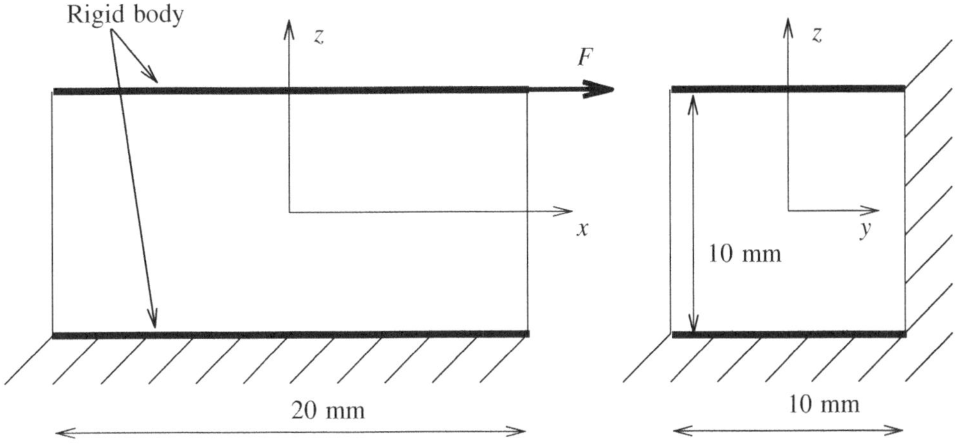

Figure 4.4 Nonhomogeneous shear experiment

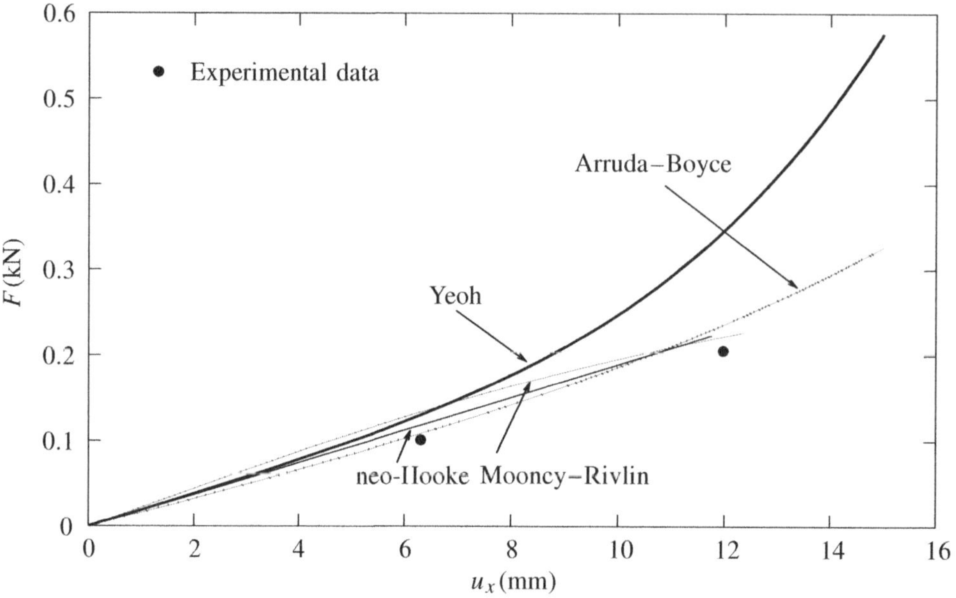

Figure 4.5 Horizontal deformation

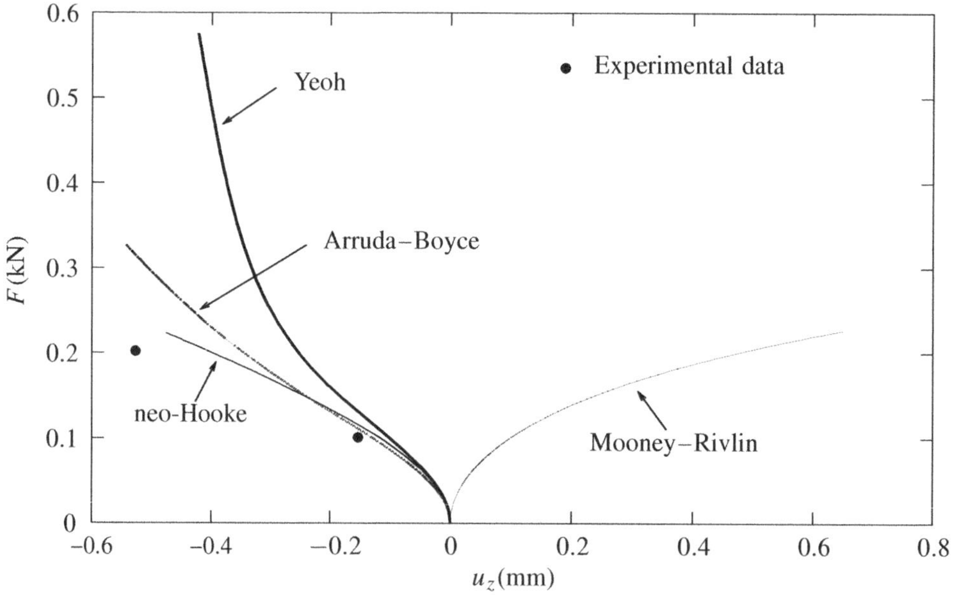

Figure 4.6 Vertical deformation

4.4 Derivatives of Invariants and Principal Stretches

4.4.1 Derivatives of the invariants

In the previous section, the derivatives of the reduced invariants and (reduced) principal stretches with respect to $\boldsymbol{C}$ were used. Recall the definitions of the reduced invariants:

$$\overline{I}_1 = I_3^{-1/3} I_1 \tag{4.145}$$

$$\overline{I}_2 = I_3^{-2/3} I_2 \tag{4.146}$$

$$J^{\text{el}} = I_3^{1/2} / J^{\text{th}}. \tag{4.147}$$

Differentiation yields

$$\frac{\partial \overline{I}_1}{\partial C_{KL}} = -\tfrac{1}{3} I_3^{-4/3} I_1 \frac{\partial I_3}{\partial C_{KL}} + I_3^{-1/3} \frac{\partial I_1}{\partial C_{KL}} \tag{4.148}$$

$$\frac{\partial \overline{I}_2}{\partial C_{KL}} = -\tfrac{2}{3} I_3^{-5/3} I_2 \frac{\partial I_3}{\partial C_{KL}} + I_3^{-2/3} \frac{\partial I_2}{\partial C_{KL}} \tag{4.149}$$

$$\frac{\partial J^{\text{el}}}{\partial C_{KL}} = \tfrac{1}{2} \frac{I_3^{-1/2}}{J^{\text{th}}} \frac{\partial I_3}{\partial C_{KL}} \tag{4.150}$$

and for the second derivatives,

$$\begin{aligned} \frac{\partial^2 \overline{I}_1}{\partial C_{KL} C_{MN}} &= \tfrac{4}{9} I_3^{-7/3} I_1 \frac{\partial I_3}{\partial C_{KL}} \frac{\partial I_3}{\partial C_{MN}} \\ &\quad - \tfrac{1}{3} I_3^{-4/3} \left(\frac{\partial I_1}{\partial C_{MN}} \frac{\partial I_3}{\partial C_{KL}} + \frac{\partial I_1}{\partial C_{KL}} \frac{\partial I_3}{\partial C_{MN}} \right) \\ &\quad - \tfrac{1}{3} I_3^{-4/3} I_1 \frac{\partial^2 I_3}{\partial C_{KL} C_{MN}} + I_3^{-1/3} \frac{\partial^2 I_1}{\partial C_{KL} C_{MN}} \end{aligned} \tag{4.151}$$

$$\begin{aligned} \frac{\partial^2 \overline{I}_2}{\partial C_{KL} C_{MN}} &= \tfrac{10}{9} I_3^{-8/3} I_2 \frac{\partial I_3}{\partial C_{KL}} \frac{\partial I_3}{\partial C_{MN}} \\ &\quad - \tfrac{2}{3} I_3^{-5/3} \left(\frac{\partial I_2}{\partial C_{MN}} \frac{\partial I_3}{\partial C_{KL}} + \frac{\partial I_2}{\partial C_{KL}} \frac{\partial I_3}{\partial C_{MN}} \right) \\ &\quad - \tfrac{2}{3} I_3^{-5/3} I_2 \frac{\partial^2 I_3}{\partial C_{KL} C_{MN}} + I_3^{-2/3} \frac{\partial^2 I_2}{\partial C_{KL} C_{MN}} \end{aligned} \tag{4.152}$$

$$\frac{\partial^2 J^{\text{el}}}{\partial C_{KL} C_{MN}} = -\tfrac{1}{4} \frac{I_3^{-3/2}}{J^{\text{th}}} \frac{\partial I_3}{\partial C_{KL}} \frac{\partial I_3}{\partial C_{MN}} + \tfrac{1}{2} \frac{I_3^{-1/2}}{J^{\text{th}}} \frac{\partial^2 I_3}{\partial C_{KL} C_{MN}}. \tag{4.153}$$

Equations (4.148) to (4.153) yield the derivatives of the reduced invariants as a function of the derivatives of the invariants. The latter yields (Equations (1.507) to (1.509))

$$\frac{\partial I_1}{\partial C_{KL}} = G^{KL} \tag{4.154}$$

$$\frac{\partial I_2}{\partial C_{KL}} = I_1 G^{KL} - C_{PQ} G^{PK} G^{QL} \tag{4.155}$$

$$\frac{\partial I_3}{\partial C_{KL}} = I_3 C^{-1KL} \tag{4.156}$$

$$\frac{\partial^2 I_1}{\partial C_{KL} C_{MN}} = 0 \tag{4.157}$$

$$\begin{aligned}\frac{\partial^2 I_2}{\partial C_{KL} C_{MN}} &= \frac{\partial I_1}{\partial C_{MN}} G^{KL} - \frac{\partial C_{PQ}}{\partial C_{MN}} G^{PK} G^{QL} \\ &= G^{MN} G^{KL} - \tfrac{1}{2}(G^{MK} G^{NL} + G^{ML} G^{NK})\end{aligned} \tag{4.158}$$

$$\frac{\partial^2 I_3}{\partial C_{KL} C_{MN}} = I_3 C^{-1KL} C^{-1MN} + I_3 \frac{\partial C^{-1KL}}{\partial C_{MN}}. \tag{4.159}$$

Since

$$C^{-1KL} C_{LA} = \delta^K_{\ A} \tag{4.160}$$

differentiation yields

$$\frac{\partial C^{-1KL}}{\partial C_{MN}} C_{LA} + C^{-1KL} \frac{\partial C_{LA}}{C_{MN}} = 0 \tag{4.161}$$

which leads to

$$\begin{aligned}\frac{\partial C^{-1KL}}{\partial C_{MN}} C_{LA} &= -\tfrac{1}{2} C^{-1KL} \quad \delta_L^{\ M} \delta_A^{\ N} + \delta_L^{\ N} \delta_A^{\ M}) \\ &= -\tfrac{1}{2} \quad C^{KM} \delta_A^{\ N} + C^{KN} \delta_A^{\ M}).\end{aligned} \tag{4.162}$$

Multiplication of both sides with C^{-1AB} yields

$$\frac{\partial C^{-1KB}}{\partial C_{MN}} = -\tfrac{1}{2} \quad C^{-1KM} C^{-1NB} + C^{-1KN} C^{-1MB}). \tag{4.163}$$

Accordingly, Equation (4.159) can be rewritten as

$$\frac{\partial^2 I_3}{\partial C_{KL} C_{MN}} = I_3 \left[C^{-1KL} C^{-1MN} - \tfrac{1}{2} \quad C^{-1}{}_{KM} C^{-1}{}_{NL} + C^{-1}{}_{KN} C^{-1}{}_{ML}) \right]. \tag{4.164}$$

4.4.2 Derivatives of the principal stretches

The derivatives of the reduced principal stretches can be obtained in a similar way (for an alternative formulation see (Simo and Taylor 1991)). Starting from

$$\bar{\lambda}_i = I_3^{-1/6} \lambda_i = J^{-1/3} \lambda_i \tag{4.165}$$

one obtains

$$\frac{\partial \bar{\lambda}_i}{\partial C_{KL}} = -\tfrac{1}{6} I_3^{-7/6} \lambda_i \frac{\partial I_3}{\partial C_{KL}} + I_3^{-1/6} \frac{\partial \lambda_i}{\partial C_{KL}} \tag{4.166}$$

$$\begin{aligned}\frac{\partial^2 \bar{\lambda}_i}{\partial C_{KL} \partial C_{MN}} &= \tfrac{7}{36} I_3^{-13/6} \lambda_i \frac{\partial I_3}{\partial C_{KL}} \frac{\partial I_3}{\partial C_{MN}} - \tfrac{1}{6} I_3^{-7/6} \frac{\partial \lambda_i}{\partial C_{MN}} \frac{\partial I_3}{\partial C_{KL}} \\ &\quad - \tfrac{1}{6} I_3^{-7/6} \lambda_i \frac{\partial^2 I_3}{\partial C_{KL} \partial C_{MN}} - \tfrac{1}{6} I_3^{-7/6} \frac{\partial \lambda_i}{\partial C_{KL}} \frac{\partial I_3}{\partial C_{MN}} \\ &\quad + I_3^{-1/6} \frac{\partial^2 \lambda_i}{\partial C_{KL} \partial C_{MN}}.\end{aligned} \tag{4.167}$$

To obtain the derivative of the principal stretches with respect to $\boldsymbol{C}$, we start from the characteristic equation

$$\lambda^6 - I_1 \lambda^4 + I_2 \lambda^2 - I_3 = 0. \tag{4.168}$$

Taking the first derivative with respect to $\boldsymbol{C}$, one obtains

$$6\lambda \frac{\partial \lambda}{\partial C_{KL}} - \frac{\partial I_1}{\partial C_{KL}} \lambda^4 - 4 I_1 \lambda^3 \frac{\partial \lambda}{\partial C_{KL}} + \frac{\partial I_2}{\partial C_{KL}} \lambda^2 + 2\lambda I_2 \frac{\partial \lambda}{\partial C_{KL}} - \frac{\partial I_3}{\partial C_{KL}} = 0 \tag{4.169}$$

yielding

$$\frac{\partial \lambda}{\partial C_{KL}} = \left(\lambda^4 \frac{\partial I_1}{\partial C_{KL}} - \lambda^2 \frac{\partial I_2}{\partial C_{KL}} + \frac{\partial I_3}{\partial C_{KL}} \right) \Big/ \left. 6\lambda^5 - 4 I_1 \lambda^3 + 2\lambda I_2 \right). \tag{4.170}$$

Taking the second derivative of Equation (4.168) yields

$$\begin{aligned}\frac{\partial^2 \lambda}{\partial C_{KL} \partial C_{MN}} &= \Bigg[\left. -30\lambda^4 + 12 I_1 \lambda^2 - 2 I_2 \right) \frac{\partial \lambda}{\partial C_{KL}} \frac{\partial \lambda}{\partial C_{MN}} \\ &\quad + 4\lambda^3 \left(\frac{\partial I_1}{\partial C_{KL}} \frac{\partial \lambda}{\partial C_{MN}} + \frac{\partial I_1}{\partial C_{MN}} \frac{\partial \lambda}{\partial C_{KL}} \right) + \lambda^4 \frac{\partial^2 I_1}{\partial C_{KL} \partial C_{MN}} \\ &\quad - \lambda^2 \frac{\partial^2 I_2}{\partial C_{KL} \partial C_{MN}} - 2\lambda \left(\frac{\partial I_2}{\partial C_{KL}} \frac{\partial \lambda}{\partial C_{MN}} + \frac{\partial I_2}{\partial C_{MN}} \frac{\partial \lambda}{\partial C_{KL}} \right) \\ &\quad + \frac{\partial^2 I_3}{\partial C_{KL} \partial C_{MN}} \Bigg] \Big/ \left. 6\lambda^5 - 4 I_1 \lambda^3 + 2\lambda I_2 \right).\end{aligned} \tag{4.171}$$

Equations (4.170) and (4.171) only apply on condition that the denominator is not zero. The denominator is the derivative of the characteristic equation, which can also be written as

$$L = 0 \Leftrightarrow (\lambda^2 - \lambda_1^2)(\lambda^2 - \lambda_2^2)(\lambda^2 - \lambda_3^2) = 0 \tag{4.172}$$

and a zero denominator for λ_i signifies

$$\left. \frac{\partial L}{\partial \lambda} \right|_{\lambda = \lambda_i} = 0 \tag{4.173}$$

which means that λ_i^2 is at least a double root if we exclude $\lambda_i = 0$. To obtain the derivatives for double and triple roots, a different approach has to be taken (Itskov 2001). The eigenvalues of $\boldsymbol{C}$ and its invariants are related by

$$\Lambda_1 + \Lambda_2 + \Lambda_3 = I_1 \tag{4.174}$$

$$\Lambda_1\Lambda_2 + \Lambda_1\Lambda_3 + \Lambda_2\Lambda_3 = I_2 \tag{4.175}$$

$$\Lambda_1\Lambda_2\Lambda_3 = I_3. \tag{4.176}$$

Taking the derivative with respect to $\boldsymbol{C}$, one obtains

$$\begin{bmatrix} 1 & 1 & 1 \\ \Lambda_2 + \Lambda_3 & \Lambda_1 + \Lambda_3 & \Lambda_1 + \Lambda_2 \\ \Lambda_2\Lambda_3 & \Lambda_1\Lambda_3 & \Lambda_1\Lambda_2 \end{bmatrix} \begin{Bmatrix} \Lambda_{1,\boldsymbol{C}} \\ \Lambda_{2,\boldsymbol{C}} \\ \Lambda_{3,\boldsymbol{C}} \end{Bmatrix} = \begin{Bmatrix} I_{1,\boldsymbol{C}} \\ I_{2,\boldsymbol{C}} \\ I_{3,\boldsymbol{C}} \end{Bmatrix}. \tag{4.177}$$

Three cases can be distinguished

1. If $\Lambda_1 \neq \Lambda_2 \neq \Lambda_3 \neq \Lambda_1$, then the solution of Equation (4.177) yields

$$\Lambda_{i,\boldsymbol{C}} = \frac{\Lambda_i^2 I_{1,\boldsymbol{C}} - \Lambda_i I_{2,\boldsymbol{C}} + I_{3,\boldsymbol{C}}}{3\Lambda_i^2 - 2I_1\Lambda_i + I_2} \tag{4.178}$$

which agrees with Equation (4.170) since $\lambda_{i,\boldsymbol{C}} = 2\lambda_i\lambda_{i,\boldsymbol{C}}$. Expanding the derivatives of the invariants (Equations (1.507)–(1.509)),

$$\frac{\partial I_1}{\partial \boldsymbol{C}} = \boldsymbol{G}^\sharp \tag{4.179}$$

$$\frac{\partial I_2}{\partial \boldsymbol{C}} = I_1\boldsymbol{G}^\sharp - \boldsymbol{G}^\sharp \cdot \boldsymbol{C} \cdot \boldsymbol{G}^\sharp \tag{4.180}$$

$$\frac{\partial I_3}{\partial \boldsymbol{C}} = I_3\boldsymbol{C}^{-1} \tag{4.181}$$

and taking Equations (4.174) and (4.175) into account, Equation (4.178) can also be written as

$$\frac{\partial \Lambda_i}{\partial \boldsymbol{C}} = \frac{\Lambda_i(\Lambda_i - I_1)\boldsymbol{G}^\sharp + \Lambda_i\boldsymbol{G}^\sharp \cdot \boldsymbol{C} \cdot \boldsymbol{G}^\sharp + I_3\boldsymbol{C}^{-1}}{(\Lambda_i - \Lambda_j)(\Lambda_i - \Lambda_k)}. \tag{4.182}$$

Since

$$\boldsymbol{G}^\sharp = \boldsymbol{G}^\sharp \cdot \boldsymbol{C} \cdot \boldsymbol{C}^{-1} \tag{4.183}$$

$$\boldsymbol{G}^\sharp \cdot \boldsymbol{C} \cdot \boldsymbol{G}^\sharp = \boldsymbol{G}^\sharp \cdot \boldsymbol{C}^2 \cdot \boldsymbol{C}^{-1} \tag{4.184}$$

and

$$\boldsymbol{C}^{-1} = \boldsymbol{G}^\sharp \cdot \boldsymbol{G} \cdot \boldsymbol{C}^{-1} \tag{4.185}$$

one finds by comparison with Equation (1.126),

$$\frac{\partial \Lambda_i}{\partial \boldsymbol{C}} = \Lambda_i \boldsymbol{G}^\sharp \cdot \boldsymbol{M}^i \cdot \boldsymbol{C}^{-1}. \tag{4.186}$$

Writing $\boldsymbol{C}^{-1}$ in terms of the structural tensors

$$\boldsymbol{C}^{-1} = \sum_j \Lambda_j^{-1} \boldsymbol{N}_j \otimes \boldsymbol{N}_j \tag{4.187}$$

Equation (4.186) can be further simplified to

$$\begin{aligned}
\Lambda_{i,\boldsymbol{C}} &= \Lambda_i \boldsymbol{G}^\sharp \cdot (\boldsymbol{N}^i \otimes \boldsymbol{N}^i) \cdot \sum_j \Lambda_j^{-1} (\boldsymbol{N}_j \otimes \boldsymbol{N}_j) \\
&= \Lambda_i \boldsymbol{G}^\sharp \sum_j \Lambda_j^{-1} (\boldsymbol{N}^i \otimes \boldsymbol{N}_j) \boldsymbol{N}^i \cdot \boldsymbol{N}_j \\
&= \Lambda_i \boldsymbol{G}^\sharp \Lambda_i^{-1} (\boldsymbol{N}^i \otimes \boldsymbol{N}_i) \\
&= \boldsymbol{N}_i \otimes \boldsymbol{N}_i \\
&= \boldsymbol{M}_i.
\end{aligned} \tag{4.188}$$

It is a remarkably simple expression: for three distinct eigenvalues, the derivatives of the eigenvalues are the corresponding contravariant structural tensors. Using Equation (4.188), one also obtains a very elegant expression for the principal stresses in an Ogden material. Indeed,

$$\boldsymbol{S} = 2\frac{\partial \Sigma}{\partial \Lambda_i} \Lambda_{i,\boldsymbol{C}} = 2\frac{\partial \Sigma}{\partial \Lambda_i} \boldsymbol{M}_i. \tag{4.189}$$

Accordingly (Equation (1.132)),

$$\Lambda_{jS} = \boldsymbol{S} : \boldsymbol{M}^j = 2\frac{\partial \Sigma}{\partial \Lambda_j} \tag{4.190}$$

or (Equation (1.428))

$$\lambda_{j\sigma} = \frac{2}{J}\Lambda_j \frac{\partial \Sigma}{\partial \Lambda_j} = \frac{2}{J}\frac{\partial \Sigma}{\partial \ln \Lambda_j} = \frac{1}{J}\frac{\partial \Sigma}{\partial \ln \lambda_j}. \tag{4.191}$$

2. If two eigenvalues are equal, for example, $\Lambda = \Lambda_1 = \Lambda_2 \neq \Lambda_3$ Equation (4.177) reduces to

$$\begin{bmatrix} 1 & 1 & 1 \\ \Lambda + \Lambda_3 & \Lambda + \Lambda_3 & 2\Lambda \\ \Lambda\Lambda_3 & \Lambda\Lambda_3 & \Lambda^2 \end{bmatrix} \begin{Bmatrix} \Lambda_{1,\boldsymbol{C}} \\ \Lambda_{2,\boldsymbol{C}} \\ \Lambda_{3,\boldsymbol{C}} \end{Bmatrix} = \begin{Bmatrix} I_{1,\boldsymbol{C}} \\ I_{2,\boldsymbol{C}} \\ I_{3,\boldsymbol{C}} \end{Bmatrix} \tag{4.192}$$

and column 1 and 2 are identical: the system is singular. It can be reduced to

$$\begin{bmatrix} 1 & 1 \\ \Lambda + \Lambda_3 & 2\Lambda \end{bmatrix} \begin{Bmatrix} \Lambda_{1,C} + \Lambda_{2,C} \\ \Lambda_{3,C} \end{Bmatrix} = \begin{Bmatrix} I_{1,C} \\ I_{2,C} \end{Bmatrix}. \tag{4.193}$$

The solution satisfies

$$\Lambda_{1,C} + \Lambda_{2,C} = \frac{2\Lambda I_{1,C} - I_{2,C}}{\Lambda - \Lambda_3} \tag{4.194}$$

$$\Lambda_{3,C} = \frac{I_{2,C} - (\Lambda + \Lambda_3) I_{1,C}}{\Lambda - \Lambda_3}. \tag{4.195}$$

It is not difficult to prove that

$$\Lambda_{1,C} + \Lambda_{2,C} = \boldsymbol{G}^{\sharp} - \boldsymbol{M}_3 = \boldsymbol{M}_1 + \boldsymbol{M}_2 \tag{4.196}$$

$$\Lambda_{3,C} - \boldsymbol{M}_3 \tag{4.197}$$

where $\boldsymbol{M}_1 + \boldsymbol{M}_2$ and $\boldsymbol{M}_3$ satisfy Equation (1.129) and Equation (1.130).

3. For three equal eigenvalues $\Lambda = \Lambda_1 = \Lambda_2 = \Lambda_3$, Equation (4.193) reduces to one single equation:

$$\Lambda_{1,C} + \Lambda_{2,C} + \Lambda_{3,C} = I_{1,C} = \boldsymbol{G}^{\sharp}. \tag{4.198}$$

Now let us take a look at the second derivatives of λ_i. Instead of using Equation (4.171), one can also express it through the second derivative of $\Lambda_i = \lambda_i^2$:

$$\lambda_{i,CC} = \frac{1}{2\sqrt{\Lambda_i}} \Lambda_{i,CC} - \frac{1}{4\Lambda_i\sqrt{\Lambda_i}} \Lambda_{i,C} \otimes \Lambda_{i,C} \tag{4.199}$$

obtained by differentiating

$$\lambda_{i,C} = \frac{1}{2\sqrt{\Lambda_i}} \Lambda_{i,C} \tag{4.200}$$

with respect to $\boldsymbol{C}$. Again, three cases can be distinguished

1. For $\Lambda_1 \neq \Lambda_2 \neq \Lambda_3 \neq \Lambda_1$ one obtains (Equation (4.188))

$$\Lambda_{i,CC} = \boldsymbol{M}_{i,C}. \tag{4.201}$$

An expression for $\boldsymbol{M}_{i,C}$ is found by differentiating Equation (1.125) leading to (notice that $\boldsymbol{M}_{i,C} = \boldsymbol{G}^{\sharp} \cdot \boldsymbol{M}^{i}_{,C} \cdot \boldsymbol{G}^{\sharp}$)

$$\begin{bmatrix} 1 & 1 & 1 \\ \Lambda_1 & \Lambda_2 & \Lambda_3 \\ \Lambda_1^2 & \Lambda_2^2 & \Lambda_3^2 \end{bmatrix} \begin{Bmatrix} \boldsymbol{M}^{1}_{,C} \\ \boldsymbol{M}^{2}_{,C} \\ \boldsymbol{M}^{3}_{,C} \end{Bmatrix} = \begin{Bmatrix} 0 \\ \boldsymbol{A} \\ \boldsymbol{B} \end{Bmatrix} \tag{4.202}$$

where

$$\boldsymbol{A} = \boldsymbol{C}_{,C} - \sum_i \boldsymbol{M}_i \otimes \boldsymbol{M}^i \tag{4.203}$$

$$\boldsymbol{B} = \boldsymbol{C}^2_{,C} - 2\sum_i \Lambda_i \boldsymbol{M}_i \otimes \boldsymbol{M}^i \tag{4.204}$$

and

$$\boldsymbol{C}_{,C} := \frac{\partial \boldsymbol{C}}{\partial \boldsymbol{C}} \tag{4.205}$$

and similar expressions for the other terms. Straightforward calculation yields for $\boldsymbol{C}_{,C}$ and $\boldsymbol{C}^2_{,C}$,

$$\frac{\partial C_{KL}}{\partial C_{PQ}} = \tfrac{1}{2}(\delta^P{}_K \delta^Q{}_L + \delta^P{}_L \delta^Q{}_K) =: \mathbb{I}_I \tag{4.206}$$

and

$$\frac{\partial C_{KL} C_{NM} G^{LN}}{\partial C_{PQ}} = \tfrac{1}{2}(\delta^P{}_K C^Q{}_M + \delta^Q{}_K C^P{}_M) + \tfrac{1}{2}(\delta^Q{}_M C_K{}^P + \delta^P{}_M C_K{}^Q). \tag{4.207}$$

Notice the following shorthand notation:

$$(\mathbb{I}_I)^{IJ}{}_{KL} := (\mathbb{I}_\delta)^{IJ}{}_{KL} := \tfrac{1}{2}(\delta^I{}_K \delta^J{}_L + \delta^I{}_L \delta^J{}_K) \tag{4.208}$$

$$(\mathbb{I}_G)_{IJKL} := \tfrac{1}{2}(G_{IK} G_{JL} + G_{IL} G_{JK}) \tag{4.209}$$

$$(\mathbb{I}_{G^\sharp})^{IJKL} := \tfrac{1}{2}(G^{IK} G^{JL} + G^{IL} G^{JK}), \tag{4.210}$$

and similarly for other tensor fields. The solution of Equation (4.202) amounts to

$$\boldsymbol{M}^i{}_{,C} = \frac{1}{D_{\underline{i}}}[\boldsymbol{B} - (I_1 - \Lambda_{\underline{i}})\boldsymbol{A}] \tag{4.211}$$

where

$$D_i = (\Lambda_{\underline{i}} - \Lambda_j)(\Lambda_{\underline{i}} - \Lambda_k), \quad i = 1, 2, 3;\ j, k \neq i. \tag{4.212}$$

2. For $\Lambda = \Lambda_1 = \Lambda_2 \neq \Lambda_3$, Equation (4.202) reduces to

$$\begin{bmatrix} 1 & 1 \\ \Lambda & \Lambda_3 \end{bmatrix} \begin{Bmatrix} \boldsymbol{M}^1{}_{,C} + \boldsymbol{M}^2{}_{,C} \\ \boldsymbol{M}^3{}_{,C} \end{Bmatrix} = \begin{Bmatrix} 0 \\ \boldsymbol{A} \end{Bmatrix} \tag{4.213}$$

leading to

$$\boldsymbol{M}^1{}_{,C} + \boldsymbol{M}^2{}_{,C} = -\boldsymbol{A}/(\Lambda_3 - \Lambda) \tag{4.214}$$

$$\boldsymbol{M}^3{}_{,C} = \boldsymbol{A}/(\Lambda_3 - \Lambda). \tag{4.215}$$

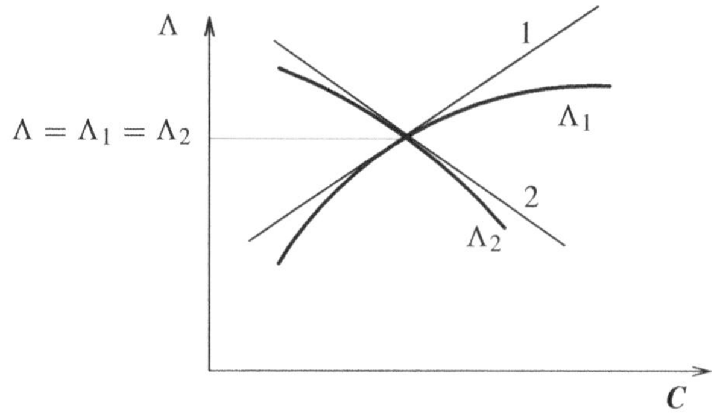

Figure 4.7 Tangent ambiguity for identical eigenvalues

3. For $\Lambda = \Lambda_1 = \Lambda_2 = \Lambda_3$, one obtains in a similar way

$$\boldsymbol{M}^1_{,\boldsymbol{C}} + \boldsymbol{M}^2_{,\boldsymbol{C}} + \boldsymbol{M}^3_{,\boldsymbol{C}} = 0. \tag{4.216}$$

Notice that for double or triple roots, the derivatives of Λ_i (Equation (4.196) and Equation (4.198)) and $\boldsymbol{M}^i$ (Equation (4.214) and Equation (4.216)) are not known separately: only the sum is known. This is not surprising, since double roots cannot be distinguished, and consequently it is not clear whether tangent 1 or tangent 2 applies (Figure 4.7).

4.4.3 Expressions for the stress and stiffness for three equal eigenvalues

In the previous section, it was found that for three equal eigenvalues the derivatives of λ_i are not known separately, only their sum can be calculated. In the present section, it will be shown that this suffices to determine the stress and the stiffness. For $\lambda_1 = \lambda_2 = \lambda_3 = \lambda$, Equation (4.132) and Equation (4.133) reduce to

$$\boldsymbol{S} = 2\left[\sum_{i=1}^{n} \frac{2\mu_i}{\alpha_i} \bar{\lambda}^{\lambda_i - 1} \sum_{k=1}^{3} \bar{\lambda}_{k,\boldsymbol{C}} + \sum_{i=1}^{N} \frac{2i}{D_i}(J^{\mathrm{el}} - 1)^{2i-1} J^{\mathrm{el}}_{,\boldsymbol{C}}\right] \tag{4.217}$$

and

$$\begin{aligned}\boldsymbol{\Sigma} = 4\Bigg\{&\sum_{i=1}^{N} \frac{2\mu_i}{\alpha_i}\left[(\alpha_i - 1)\bar{\lambda}^{\alpha_i - 2}\left(\sum_{k=1}^{3} \bar{\lambda}_{k,\boldsymbol{C}} \otimes \bar{\lambda}_{k,\boldsymbol{C}}\right) + \bar{\lambda}^{\alpha_i - 1} \sum_{k=1}^{3} \bar{\lambda}_{k,\boldsymbol{CC}}\right] \\ &+ \sum_{i=1}^{N}\left[\frac{2i(2i-1)}{D_i}(J^{\mathrm{el}} - 1)^{2i-2} J^{\mathrm{el}}_{,\boldsymbol{C}} \otimes J^{\mathrm{el}}_{,\boldsymbol{C}} + \frac{2i}{D_i}(J^{\mathrm{el}} - 1)^{2i-1} J^{\mathrm{el}}_{,\boldsymbol{CC}}\right]\Bigg\}. \end{aligned} \tag{4.218}$$

Furthermore, Equation (4.166) and Equation (4.167) now lead to

$$\sum_{k=1}^{3} \bar{\lambda}_{k,\boldsymbol{C}} = -\tfrac{1}{2} I_3^{-7/6} \lambda I_{3,\boldsymbol{C}} + I_3^{-1/6} \sum_{k=1}^{3} \lambda_{k,\boldsymbol{C}} \tag{4.219}$$

$$\sum_{k=1}^{3} \bar{\lambda}_{k,\boldsymbol{C}} \otimes \bar{\lambda}_{k,\boldsymbol{C}} = \tfrac{1}{12} I_3^{-7/3} \lambda^2 I_{3,\boldsymbol{C}} \otimes I_{3,\boldsymbol{C}} + I_3^{-1/3} \sum_{k=1}^{3} \lambda_{k,\boldsymbol{C}} \otimes \lambda_{k,\boldsymbol{C}}$$
$$- \tfrac{1}{6} I_3^{-4/3} \lambda I_{3,\boldsymbol{C}} \otimes \sum_{k=1}^{3} \lambda_{k,\boldsymbol{C}} - \tfrac{1}{6} I_3^{-4/3} \lambda \sum_{k=1}^{3} \lambda_{k,\boldsymbol{C}} \otimes I_{3,\boldsymbol{C}} \tag{4.220}$$

$$\sum_{k=1}^{3} \bar{\lambda}_{k,\boldsymbol{CC}} = \tfrac{7}{12} I_3^{-13/6} \lambda I_{3,\boldsymbol{C}} \otimes I_{3,\boldsymbol{C}} - \tfrac{1}{6} I_3^{-7/6} \sum_{k=1}^{3} I_{3,\boldsymbol{C}} \otimes \lambda_{k,\boldsymbol{C}}$$
$$- \tfrac{1}{2} I_3^{-7/6} \lambda I_{3,\boldsymbol{CC}} - \tfrac{1}{6} I_3^{-7/6} \sum_{k=1}^{3} \lambda_{k,\boldsymbol{C}} \otimes I_{3,\boldsymbol{C}} + I_3^{-1/6} \sum_{k=1}^{3} \lambda_{k,\boldsymbol{CC}}. \tag{4.221}$$

In this way, the sums of the derivatives of the reduced stretches are written in terms of the derivatives of the unreduced stretches. Now, Equations (4.198), (4.199), (4.200) and (4.216) show that

$$\sum_{k=1}^{3} \lambda_{k,\boldsymbol{C}} = \tfrac{1}{2\lambda} \sum_{k=1}^{3} \Lambda_{k,\boldsymbol{C}} = \tfrac{1}{2\lambda} \boldsymbol{G}^{\sharp} \tag{4.222}$$

$$\sum_{k=1}^{3} \lambda_{k,\boldsymbol{CC}} = -\tfrac{1}{4\lambda^3} \sum_{k=1}^{3} \Lambda_{k,\boldsymbol{C}} \otimes \Lambda_{k,\boldsymbol{C}} = -\tfrac{1}{4\lambda^3} \sum_{k=1}^{3} \boldsymbol{M}_k \otimes \boldsymbol{M}_k. \tag{4.223}$$

For $\lambda = \lambda_1 = \lambda_2 = \lambda_3$, Equation (4.202) reduces to rank one and, consequently, $\boldsymbol{A} = 0$

$$\sum_{k=1}^{3} \boldsymbol{M}^k \otimes \boldsymbol{M}_k = \boldsymbol{C}_{,\boldsymbol{C}} \tag{4.224}$$

which is equivalent to

$$\sum_{k=1}^{3} \boldsymbol{M}_k \otimes \boldsymbol{M}_k = (\boldsymbol{G}^{\sharp} \otimes \boldsymbol{G}^{\sharp}) : \boldsymbol{C}_{,\boldsymbol{C}}. \tag{4.225}$$

Accordingly,

$$\sum_{k=1}^{3} \lambda_{k,\boldsymbol{CC}} = -\tfrac{1}{4\lambda^3} (\boldsymbol{G}^{\sharp} \otimes \boldsymbol{G}^{\sharp}) : \mathbb{I}_I. \tag{4.226}$$

In a similar way, one arrives at

$$\sum_{k=1}^{3} \lambda_{k,\boldsymbol{C}} \otimes \lambda_{k,\boldsymbol{C}} = \frac{1}{4\lambda^2} \sum_{k=1}^{3} \Lambda_{k,\boldsymbol{C}} \otimes \Lambda_{k,\boldsymbol{C}}$$

$$= \frac{1}{4\lambda^2} (\boldsymbol{G}^\sharp \otimes \boldsymbol{G}^\sharp) : \mathbb{I}_I. \tag{4.227}$$

Hence, Equations (4.219) to (4.221) yield

$$\sum_{k=1}^{3} \bar{\lambda}_{k,\boldsymbol{C}} = -\tfrac{1}{2} I_3^{-1/6} \lambda \boldsymbol{C}^{-1} + I_3^{-1/6} \frac{1}{2\lambda} \boldsymbol{G}^\sharp \tag{4.228}$$

$$\sum_{k=1}^{3} \bar{\lambda}_{k,\boldsymbol{C}} \otimes \bar{\lambda}_{k,\boldsymbol{C}} = \frac{1}{12} I_3^{-1/3} \lambda^2 \boldsymbol{C}^{-1} \otimes \boldsymbol{C}^{-1} + \frac{1}{4\lambda^2} I_3^{-1/3} \mathbb{I}_{G^\sharp}$$

$$- \tfrac{1}{12} I_3^{-1/3} \boldsymbol{C}^{-1} \otimes \boldsymbol{G}^\sharp - \frac{1}{12} I_3^{-1/3} \boldsymbol{G}^\sharp \otimes \boldsymbol{C}^{-1} \tag{4.229}$$

$$\sum_{k=1}^{3} \bar{\lambda}_{k,\boldsymbol{CC}} = \tfrac{7}{12} I_3^{-1/6} \lambda \boldsymbol{C}^{-1} \otimes \boldsymbol{C}^{-1} - \frac{1}{12\lambda} I_3^{-1/6} \boldsymbol{C}^{-1} \otimes \boldsymbol{G}^\sharp$$

$$- \tfrac{1}{2} \lambda I_3^{-1/6} \boldsymbol{C}^{-1} \otimes \boldsymbol{C}^{-1} + \tfrac{1}{2} \lambda I_3^{-1/6} \mathbb{I}_{C^{-1}}$$

$$- \frac{1}{12\lambda} I_3^{-1/6} \boldsymbol{G}^\sharp \otimes \boldsymbol{C}^{-1} - \frac{1}{4\lambda^3} I_3^{-1/6} \mathbb{I}_{G^\sharp}. \tag{4.230}$$

In a similar way, the expressions for the stress and stiffness for an elastomeric foam for $\hat{\lambda} = \hat{\lambda}_1 = \hat{\lambda}_2 = \hat{\lambda}_3$ reduce to

$$\boldsymbol{S} = 2 \sum_{i=1}^{N} \frac{2\mu_i}{\alpha_i} \left[\hat{\lambda} \sum_{k=1}^{3} \hat{\lambda}_{k,\boldsymbol{C}} - (J^{\mathrm{el}})^{-\alpha_i \beta_i - 1} J^{\mathrm{el}}_{,\boldsymbol{C}} \right] \tag{4.231}$$

$$\boldsymbol{\Sigma} = 8 \sum_{i=1}^{N} \frac{\mu_i}{\alpha_i} \left[(\alpha_i - 1) \hat{\lambda}^{\alpha_i - 2} \sum_{k=1}^{3} \hat{\lambda}_{k,\boldsymbol{C}} \otimes \hat{\lambda}_{k,\boldsymbol{C}} + \hat{\lambda}^{\alpha_i - 1} \sum_{k=1}^{3} \hat{\lambda}_{k,\boldsymbol{CC}} \right.$$

$$\left. + (\alpha_i \beta_i + 1)(J^{\mathrm{el}})^{-\alpha_i \beta_i - 2} J^{\mathrm{el}}_{,\boldsymbol{C}} \otimes J^{\mathrm{el}}_{,\boldsymbol{C}} - (J^{\mathrm{el}})^{-\alpha_i \beta_i - 1} J^{\mathrm{el}}_{,\boldsymbol{CC}} \right]. \tag{4.232}$$

Since J^{th} depends on the temperature only, Equations (4.222) and (4.227) also apply to $\hat{\lambda}$.

4.5 Tangent Stiffness Matrix at Zero Deformation

The tangent stiffness matrix at zero deformation can be obtained by substituting $\boldsymbol{F} = \boldsymbol{I}$ in the expression for Σ^{KLMN}. The expressions in the previous section take the form

$$I_1 = 3 = \overline{I}_1 \tag{4.233}$$

$$I_2 = 3 = \overline{I}_2 \tag{4.234}$$

$$I_3 = 1 \tag{4.235}$$

$$\lambda_1 = \lambda_2 = \lambda_3 = \overline{\lambda}_1 = \overline{\lambda}_2 = \overline{\lambda}_3 = 1. \tag{4.236}$$

The derivatives of the invariants take the value

$$\frac{\partial I_1}{\partial C_{KL}} = G^{KL} \tag{4.237}$$

$$\frac{\partial I_2}{\partial C_{KL}} = 2G^{KL} \tag{4.238}$$

$$\frac{\partial I_3}{\partial C_{KL}} = G^{KL} \tag{4.239}$$

$$\frac{\partial^2 I_1}{\partial C_{KL}\partial C_{MN}} = 0 \tag{4.240}$$

$$\frac{\partial^2 I_2}{\partial C_{KL}\partial C_{MN}} = G^{KL}G^{MN} - \tfrac{1}{2}(G^{KM}G^{LN} + G^{KN}G^{LM}) \tag{4.241}$$

$$\frac{\partial^2 I_3}{\partial C_{KL}\partial C_{MN}} = \frac{\partial^2 I_2}{\partial C_{KL}\partial C_{MN}} \tag{4.242}$$

$$\frac{\partial \overline{I}_1}{\partial C_{KL}} = 0 \tag{4.243}$$

$$\frac{\partial \overline{I}_2}{\partial C_{KL}} = 0 \tag{4.244}$$

$$\frac{\partial J^{\text{el}}}{\partial C_{KL}} = \tfrac{1}{2}G^{KL} \tag{4.245}$$

$$\frac{\partial^2 \overline{I}_1}{\partial C_{KL}\partial C_{MN}} = -\tfrac{1}{3}G^{KL}G^{MN} + \tfrac{1}{2}(G^{KM}G^{LN} + G^{KN}G^{LM}) \tag{4.246}$$

$$\frac{\partial^2 \overline{I}_2}{\partial C_{KL}\partial C_{MN}} = \frac{\partial^2 \overline{I}_1}{\partial C_{KL}\partial C_{MN}} \tag{4.247}$$

$$\frac{\partial^2 J^{\text{el}}}{\partial C_{KL}\partial C_{MN}} = \tfrac{1}{4}G^{KL}G^{MN} - \tfrac{1}{4}(G^{KM}G^{LN} + G^{KN}G^{LM}). \tag{4.248}$$

Finally, the derivatives of the principal stretches satisfy for $\boldsymbol{F} = \boldsymbol{I}$:

$$\sum_{j=1}^{3} \lambda_{j,\boldsymbol{C}} = \tfrac{1}{2}\boldsymbol{G}^{\sharp}. \tag{4.249}$$

$$\sum_{j=1}^{3} \lambda_{j,\boldsymbol{CC}} = -\tfrac{1}{4}\mathbb{I}_{G^{\sharp}} \tag{4.250}$$

$$\sum_{j=1}^{3} \lambda_{j,\boldsymbol{C}} \otimes \lambda_{j,\boldsymbol{C}} = \tfrac{1}{4}\mathbb{I}_{G^{\sharp}} \tag{4.251}$$

$$\sum_{j=1}^{3} \bar{\lambda}_{j,\boldsymbol{C}} = 0 \tag{4.252}$$

$$\sum_{j=1}^{3} \bar{\lambda}_{j,\boldsymbol{CC}} = -\tfrac{1}{12}\boldsymbol{G}^{\sharp} \otimes \boldsymbol{G}^{\sharp} + \tfrac{1}{4}\mathbb{I}_{G^{\sharp}} \tag{4.253}$$

$$\sum_{j=1}^{3} \bar{\lambda}_{j,\boldsymbol{C}} \otimes \bar{\lambda}_{j,\boldsymbol{C}} = -\tfrac{1}{12}\boldsymbol{G}^{\sharp} \otimes \boldsymbol{G}^{\sharp} + \tfrac{1}{4}\mathbb{I}_{G^{\sharp}}. \tag{4.254}$$

Comparison with Equation (1.436) reveals that, in the initial configuration, an equivalent λ and μ can be defined as the coefficient of the terms $G^{KL}G^{MN}$ and $G^{KM}G^{LN} + G^{KN}G^{LM}$ respectively.

4.5.1 Polynomial form

Substitution of the above expressions into Equation (4.123) yields

$$\Sigma^{KLMN} = \left[-\tfrac{4}{3}(B_{10} + B_{01}) + \frac{2}{D_1}\right] G^{KL}G^{MN} + 2(B_{10} + B_{01})(G^{KM}G^{LN} + G^{KN}G^{LM}). \tag{4.255}$$

Hence,

$$\lambda_{\text{eq}} = -\tfrac{4}{3}(B_{10} + B_{01}) + \frac{2}{D_1} \tag{4.256}$$

$$\mu_{\text{eq}} = 2(B_{10} + B_{01}). \tag{4.257}$$

Frequently, an equivalent bulk modulus K_{eq} is defined, satisfying (cf Equation (1.454))

$$K_{\text{eq}} = \lambda_{\text{eq}} + \tfrac{2}{3}\mu_{\text{eq}}. \tag{4.258}$$

Hence,

$$K_{\text{eq}} = \frac{2}{D_1}. \tag{4.259}$$

4.5.2 Arruda–Boyce form

Equation (4.129) yields

$$\Sigma^{KLMN} = \left[-\frac{4\mu}{3}\left(\frac{1}{2} + \frac{3}{10\lambda_m^2} + \frac{297}{1050\lambda_m^4} + \frac{2052}{7000\lambda_m^6} + \frac{210\,195}{673\,750\lambda_m^8}\right) - \frac{2}{D}\right] G^{KL}G^{MN}$$
$$+ 2\mu\left(\frac{1}{2} + \frac{3}{10\lambda_m^2} + \frac{297}{1050\lambda_m^4} + \frac{2052}{7000\lambda_m^6} + \frac{210\,195}{673\,750\lambda_m^8}\right)\cdot$$
$$(G^{KM}G^{LN} + G^{KN}G^{LM}) \tag{4.260}$$

Hence,

$$\mu_{\text{eq}} = 2\mu\left(\frac{1}{2} + \frac{3}{10\lambda_m^2} + \frac{297}{1050\lambda_m^4} + \frac{2052}{7000\lambda_m^6} + \frac{210\,195}{673\,750\lambda_m^8}\right) \tag{4.261}$$

$$K_{\text{eq}} = \frac{2}{D}. \tag{4.262}$$

4.5.3 Ogden form

Substitution into Equation (4.218) leads to

$$\boldsymbol{\Sigma} = \sum_{i=1}^{N} \frac{8\mu_i}{\alpha_i}\left[\alpha_i\left(-\frac{1}{12}\boldsymbol{G}^\sharp \otimes \boldsymbol{G}^\sharp + \frac{1}{4}\mathbb{I}_{G^\sharp}\right)\right] + \frac{2}{D_1}\boldsymbol{G}^\sharp \otimes \boldsymbol{G}^\sharp \tag{4.263}$$

from which

$$\mu_{\text{eq}} = \sum_{i=1}^{N} \mu_i \tag{4.264}$$

$$K_{\text{eq}} = \frac{2}{D_1}. \tag{4.265}$$

4.5.4 Elastomeric foam behavior

Equation (4.232) yields

$$\boldsymbol{\Sigma} = 8\sum_{i=1}^{N}\frac{\mu_i}{\alpha_i}\left[(\alpha_i - 1)\frac{1}{4}\mathbb{I}_{G^\sharp} - \frac{1}{4}\mathbb{I}_{G^\sharp} + \frac{(\alpha_i\beta_i + 1)}{4}\boldsymbol{G}^\sharp \otimes \boldsymbol{G}^\sharp - \frac{1}{4}\boldsymbol{G}^\sharp \otimes \boldsymbol{G}^\sharp + \frac{1}{2}\mathbb{I}_{G^\sharp}\right]$$
$$= \sum_{i=1}^{N}\mu_i\left(2\mathbb{I}_{G^\sharp} + 2\beta_i\boldsymbol{G}^\sharp \otimes \boldsymbol{G}^\sharp\right). \tag{4.266}$$

This leads to the following expressions for the equivalent constants:

$$\mu_{\text{eq}} = \sum_{i=1}^{N} \mu_i \tag{4.267}$$

$$\lambda_{eq} = 2\sum_{i=1}^{N} \beta_i \mu_i \tag{4.268}$$

$$K_{eq} = \sum_{i=1}^{N} 2\mu_i \quad \beta_i + \tfrac{1}{3}\Big) . \tag{4.269}$$

4.5.5 Closure

For the polynomial model, the Arruda–Boyce model and the Ogden model, the equivalent bulk modulus is related to the coefficient D_1 (polynomial model, Ogden model) or D (Arruda–Boyce model). Incompressible behavior corresponds to $D_1 = D = 0$. To avoid the resulting singularities in the material law, the CalculiX® code (CalculiX 2003) replaces this behavior by a nearly incompressible behavior corresponding to an equivalent Poisson coefficient $\mu_{eq} = 0.475$ at zero deformation. One finds (Equation (1.455)),

$$K_{eq} = \frac{2\mu_{eq}(1 + \nu_{eq})}{3(1 - 2\nu_{eq})} . \tag{4.270}$$

Accordingly, for a polynomial material,

$$D_1 = \frac{3(1 - 2\nu_{eq})}{\mu_{eq}(1 + \nu_{eq})} = \frac{0.1017}{\mu_{eq}} . \tag{4.271}$$

If $N > 1$ in the polynomial model, the following numerical relationship (disregarding the dimensions) is proposed:

$$D_i = \left[\frac{3(1 - 2\nu_{eq})}{\mu_{eq}(1 + \nu_{eq})}\right]^i = \left(\frac{0.1017}{\mu_{eq}}\right)^i . \tag{4.272}$$

4.6 Inflation of a Balloon

This is a classical example discussed in different places in the literature (see e.g. (Holzapfel 2000), (Beatty 1987) and (Verron *et al.* 2001)). The geometry is depicted in Figure 4.8. The undeformed radius and thickness are 10 m and 0.1 m respectively.

Assume that we select a St Venant–Kirchhoff material, that is, a linear elastic isotropic material, satisfying

$$\boldsymbol{E} = \frac{1}{E} [-\nu(\mathrm{tr}\boldsymbol{S})\boldsymbol{G} + (1 + \nu)\boldsymbol{S}] . \tag{4.273}$$

Consider a material particle of the balloon on the X-axis. Because of symmetry conditions, we have $S_{12} = S_{23} = S_{13} = 0$ and $S_{22} = S_{33} = S$. Furthermore, the balloon is assumed to be in plane stress, $S_{11} = 0$. Accordingly, Equation (4.273) leads to the following strains:

$$E_{22} = E_{33} = \left(\frac{1 - \nu}{E}\right) S \tag{4.274}$$

$$E_{11} = -\frac{2\nu}{E} S \tag{4.275}$$

$$E_{12} = E_{13} = E_{23} = 0. \tag{4.276}$$

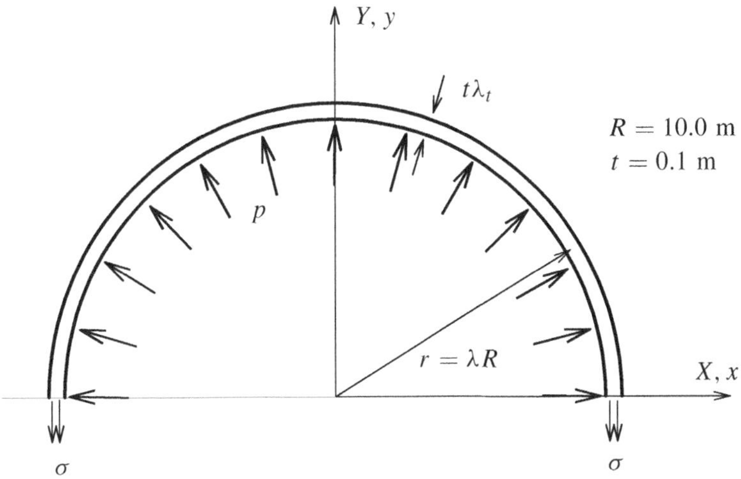

Figure 4.8 Geometry of the balloon

The stretch can be obtained from Equation (1.45):

$$\lambda_{(N)} = \sqrt{(2E_{KL} + 1)N^K N^L} \tag{4.277}$$

yielding for the circumferential stretch (take N in Y-direction)

$$\lambda = \sqrt{2\left(\frac{1-\nu}{E}\right)S + 1} \tag{4.278}$$

and for the thickness stretch

$$\lambda_t = \sqrt{\left(1 - \frac{4\nu}{E}S\right)}. \tag{4.279}$$

Taking $\lambda = r/R$ as the independent variable during the inflation of the balloon, the Piola–Kirchhoff stress of the second kind can be obtained from Equation (4.278):

$$S = \frac{E(\lambda^2 - 1)}{2(1-\nu)}. \tag{4.280}$$

Expressing the equilibrium of a hemisphere, one obtains

$$p\pi r^2 = S(2\pi R)t \tag{4.281}$$

from which the pressure p results:

$$p = S\frac{(2Rt)}{r^2} = \frac{Et}{R(1-\nu)}\left(1 - \frac{1}{\lambda^2}\right). \tag{4.282}$$

This is a monotonic increasing function of λ. Substituting Equation (4.278) into Equation (4.279), one obtains an expression for the thickness stretch of the balloon:

$$\lambda_t = \sqrt{1 - \frac{2\nu(\lambda^2 - 1)}{(1 - \nu)}}. \tag{4.283}$$

Surprisingly enough, λ_t is zero for

$$\lambda = \sqrt{\frac{1 - \nu}{2\nu} + 1}. \tag{4.284}$$

If $\nu = 0.5$, the thickness of the balloon is reduced to zero for $\lambda = \sqrt{3/2}$. This corresponds to an infinite circumferential Cauchy stress. It is well known that the circumferential stretch during inflation can reach values up to 10 and higher. Accordingly, the St Venant–Kirchhoff material is not suited to model balloon behavior.

In Chapter 1, it was emphasized that $\nu = 0.5$ represents isochoric deformation for infinitesimal strains only. The real isochoric condition is $J = 1$. This can be illustrated by noticing that for the balloon

$$J = \lambda_t \lambda^2 = \lambda^2 \sqrt{1 - \frac{2\nu(\lambda^2 - 1)}{(1 - \nu)}}. \tag{4.285}$$

Substituting $\nu = 0.5$, one obtains

$$J_{\nu=0.5} = \lambda^2 \sqrt{3 - 2\lambda^2}. \tag{4.286}$$

The plot of this function in Figure 4.9 shows that $\nu = 0.5$ is indeed a bad approximation for isochoric behavior as soon as the stretch deviates markedly from $\lambda = 1$.

To model true balloon behavior, recourse must be taken to hyperelastic laws such as the Neo-Hooke or Mooney–Rivlin law. The following constants are taken

$$\text{neo-Hooke: } B_{10} = 211\,250.\ \text{Pa}, \quad D_1 = 0.2367 \times 10^{-6}\ \text{Pa}^{-1} \tag{4.287}$$

$$\text{Mooney–Rivlin: } B_{10} = 184\,843.75\ \text{Pa}, \quad B_{01} = 26\,406.25\ \text{Pa},$$
$$D_1 = 0.2367 \times 10^{-6}\ \text{Pa}^{-1}. \tag{4.288}$$

The isochoric constants are taken from (Holzapfel 2000), the volumetric constants are such that the equivalent Poisson coefficient amounts to $\nu = 0.475$. Only one-eighth of the balloon was modeled using seventy-five 20-node brick elements with reduced integration (one layer across the thickness). The pressure as a function of the circumferential stretch is plotted in Figure 4.10 and agrees well overall with the analytical predictions for incompressible material in (Holzapfel 2000). The neo-Hooke curve is about 6% lower than the analytical prediction. This also applies to the Mooney–Rivlin model for small stretches up to the local maximum at a stretch of approximately 1.5. For higher stretch, the present curve does not show the local minimum with the renewed pressure increase obtained in (Holzapfel 2000). This is attributed to the volumetric term. Notice that both models predict a local pressure maximum. This phenomenon, which was not predicted by the St Venant–Kirchhoff

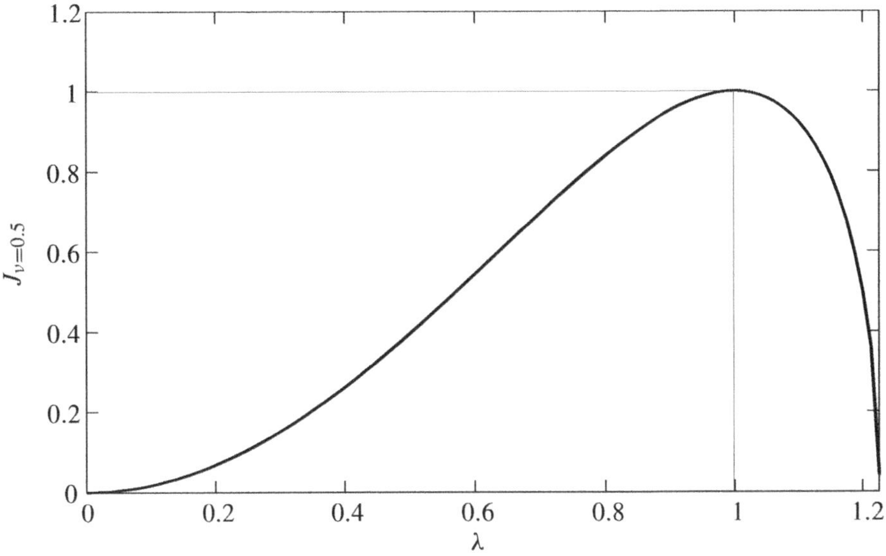

Figure 4.9 Jacobian determinant during deformation for a St Venant–Kirchhoff material

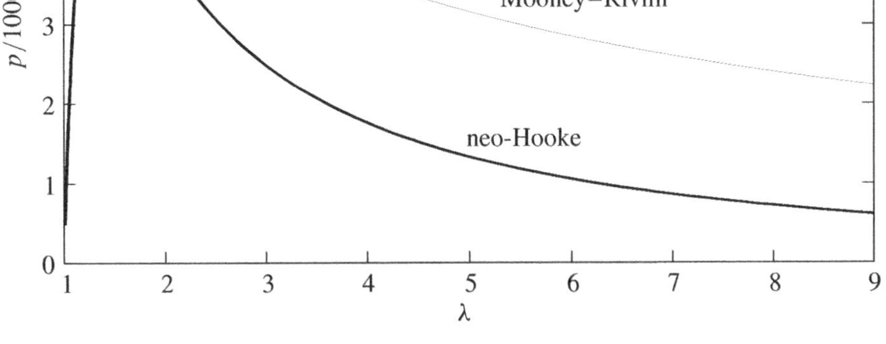

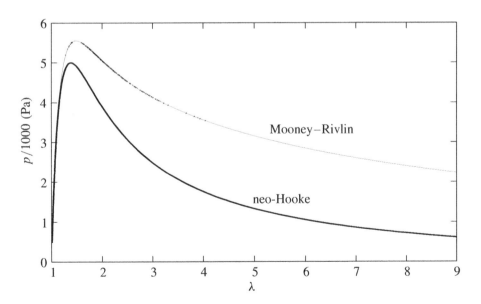

Figure 4.10 Pressure in the balloon

material description, is well known: inflating a party balloon, the initial pressure is quite high, but decreases significantly after some stretching takes place. Because its behavior is not monotonic, the pressure cannot be taken as an independent variable during the finite element calculation. The radial forces in the nodes of the mesh are taken instead since they continuously increase.

4.7 Anisotropic Hyperelasticity

Recently developed materials are frequently anisotropic, such as fabrics embedded in a matrix material (Reese *et al.* 2001). For these applications, the theory of the previous section has to be extended. Anisotropic materials are characterized by the fact that

$$\Sigma(\boldsymbol{F} \cdot \boldsymbol{Q}) - \Sigma(\boldsymbol{F}) \tag{4.289}$$

does not apply for arbitrary rotation tensors $\boldsymbol{Q} \in SO(3)$ (the group of all rotations without reflection in three-dimensional space). The group $\mathcal{G}$ for which Equation (4.289) applies, if any, is called the *material symmetry group* and characterizes the material:

$$\mathcal{G} = \left\{ \boldsymbol{Q} | \Sigma(\boldsymbol{F} \cdot \boldsymbol{Q}) = \Sigma(\boldsymbol{F}) \right\} \subset SO(3). \tag{4.290}$$

For instance, transverse isotropic materials are characterized by a preferred unit direction $\boldsymbol{A}$ about which the material is isotropic, that is,

$$\mathcal{G} = \left\{ \boldsymbol{Q}(\alpha, \boldsymbol{A}) | 0 < \alpha < 2\pi \right\} \tag{4.291}$$

where $\boldsymbol{Q}(\alpha, \boldsymbol{A})$ denotes a rotation about $\boldsymbol{A}$ covering an angle α. Anisotropic materials are frequently characterized by so-called structural tensors $\boldsymbol{M}$, which are invariant under the material symmetry group:

$$\boldsymbol{Q} \cdot \boldsymbol{M} = \boldsymbol{M}, \quad \forall \boldsymbol{Q} \in \mathcal{G}. \tag{4.292}$$

For transversely isotropic materials, there is one structural tensor defined by

$$\boldsymbol{M} := \boldsymbol{A} \otimes \boldsymbol{A}, \quad \|\boldsymbol{A}\| = 1. \tag{4.293}$$

Indeed,

$$\begin{aligned} \boldsymbol{Q} \cdot \boldsymbol{M} &= \boldsymbol{Q} \cdot (\boldsymbol{A} \otimes \boldsymbol{A}) = (\boldsymbol{Q} \cdot \boldsymbol{A}) \otimes \boldsymbol{A} \\ &= \boldsymbol{A} \otimes \boldsymbol{A} = \boldsymbol{M}. \end{aligned} \tag{4.294}$$

The behavior of anisotropic materials is not invariant under the proper orthogonal group of transformations SO(3). However, if the structural tensors are also transformed, that is, the preferred material directions undergo the same rotation, one obtains

$$\Sigma(\boldsymbol{F}, \boldsymbol{M}) = \Sigma(\boldsymbol{F} \cdot \boldsymbol{Q}, \boldsymbol{Q}^{\mathrm{T}} \cdot \boldsymbol{M} \cdot \boldsymbol{Q}) \tag{4.295}$$

and taking objectivity into account,

$$\Sigma(\boldsymbol{C}, \boldsymbol{M}) = \Sigma(\boldsymbol{Q}^{\mathrm{T}} \cdot \boldsymbol{C} \cdot \boldsymbol{Q}, \boldsymbol{Q}^{\mathrm{T}} \cdot \boldsymbol{M} \cdot \boldsymbol{Q}). \tag{4.296}$$

Consequently, Σ is an isotropic function of $\boldsymbol{C}$ and $\boldsymbol{M}$. Here, $\boldsymbol{M}$ stands for all structural matrices appropriate for the material. An application of the above concept to anisotropic viscoplasticity can be found in (Schröder *et al.* 2002). In what follows, we will concentrate on transversely isotropic materials and deal with only one structural tensor. For further details, the reader is referred to (Schröder and Neff 2001).

4.7.1 Transversely isotropic materials

To proceed, we will express Σ as a function of a polynomial basis. For isotropic scalar functions of two symmetric tensors, such a basis consists of the following terms (Spencer 1971):

$$J_1 := \mathrm{tr}\boldsymbol{C} \tag{4.297}$$

$$J_2 := \mathrm{tr}\boldsymbol{C}^2 \tag{4.298}$$

$$J_3 := \mathrm{tr}\boldsymbol{C}^3 \tag{4.299}$$

$$J_4 := \mathrm{tr}(\boldsymbol{C} \cdot \boldsymbol{M}) \tag{4.300}$$

$$J_5 := \mathrm{tr}(\boldsymbol{C}^2 \cdot \boldsymbol{M}) \tag{4.301}$$

$$J_6 := \mathrm{tr}(\boldsymbol{C} \cdot \boldsymbol{M}^2) \tag{4.302}$$

$$J_7 := \mathrm{tr}(\boldsymbol{C}^2 \cdot \boldsymbol{M}^2). \tag{4.303}$$

Since $\boldsymbol{M}^2 = \boldsymbol{M}$, we have $J_6 = J_4$ and $J_7 = J_5$. One recognizes J_1, J_2 and J_3 as invariants of $\boldsymbol{C}$, although more frequently I_1, I_2 and I_3 are used, defined by

$$I_1 = J_1 = \|\boldsymbol{F}\|^2 \tag{4.304}$$

$$I_2 = \mathrm{tr}(\mathrm{Cof}\boldsymbol{C}) = \tfrac{1}{2}(J_1^2 - J_2) = \|\mathrm{Cof}\boldsymbol{F}\|^2 \tag{4.305}$$

$$I_3 = \det \boldsymbol{C} = \tfrac{1}{6}(2J_3 + J_1^3 - 3J_1J_2) = (\det \boldsymbol{F})^2. \tag{4.306}$$

In previous sections, it was shown that I_1, I_2 and I_3 are polyconvex functions. The invariant J_4 can be expressed as

$$\begin{aligned} J_4 &= \mathrm{tr}(\boldsymbol{F}^{\mathrm{T}} \cdot \boldsymbol{F} \cdot (\boldsymbol{A} \otimes \boldsymbol{A})) \\ &= \boldsymbol{A}^{\mathrm{T}} \cdot \boldsymbol{F}^{\mathrm{T}} \cdot \boldsymbol{F} \cdot \boldsymbol{A} \\ &= \|\boldsymbol{F} \cdot \boldsymbol{A}\|^2. \end{aligned} \tag{4.307}$$

This is a convex function of $\boldsymbol{F}$ due to the norm properties and the convex monotonic increasing behavior of x^2 for $x \geq 0$. Accordingly, J_4 is convex.

It can be proved that J_5 is not polyconvex (Schröder and Neff 2001). However, it is clear that

$$K_1 := \|\mathrm{Cof}\boldsymbol{F} \cdot \boldsymbol{A}\|^2 \tag{4.308}$$

is polyconvex. K_1 can also be written as

$$\begin{aligned} K_1 &= \boldsymbol{A}^{\mathrm{T}} \cdot (\mathrm{Cof}\boldsymbol{F})^{\mathrm{T}} \cdot (\mathrm{Cof}\boldsymbol{F}) \cdot \boldsymbol{A} \\ &= \boldsymbol{A}^{\mathrm{T}} \cdot \mathrm{Cof}(\boldsymbol{F}^{\mathrm{T}} \cdot \boldsymbol{F}) \cdot \boldsymbol{A} \\ &= \mathrm{tr}\,[(\mathrm{Cof}\boldsymbol{C}) \cdot (\boldsymbol{A} \otimes \boldsymbol{A})] \\ &= \mathrm{tr}\,[(\mathrm{Cof}\boldsymbol{C}) \cdot \boldsymbol{M}]. \end{aligned} \tag{4.309}$$

The tensor $\boldsymbol{C}$ satisfies its characteristic equation, accordingly,

$$\boldsymbol{C}^3 - I_1\boldsymbol{C}^2 + I_2\boldsymbol{C} - I_3 = 0$$

$$\Updownarrow$$

$$\boldsymbol{C}^2 \cdot \boldsymbol{M} - I_1\boldsymbol{C} \cdot \boldsymbol{M} + I_2\boldsymbol{M} - I_3\boldsymbol{C}^{-1} \cdot \boldsymbol{M} = 0$$

$$\Updownarrow$$

$$\mathrm{tr}(\boldsymbol{C}^2 \cdot \boldsymbol{M}) - I_1\mathrm{tr}(\boldsymbol{C} \cdot \boldsymbol{M}) + I_2 = \mathrm{tr}(\mathrm{Cof}\boldsymbol{C} \cdot \boldsymbol{M})$$

$$\Updownarrow$$

$$J_5 - I_1 J_4 + I_2 = K_1 \tag{4.310}$$

which shows that K_1 can be written as a function of J_5. Notice the nice analogy between J_4 and K_1, and I_1 and I_2 respectively (Equations (4.304), (4.305), (4.307) and (4.308)). The physical significance of J_4 and K_1 is also noteworthy: Equation (1.30) reveals that J_4 is a measure of the change of length of a unit vector along the structural axis, whereas Equation (1.65) shows that K_1 can be interpreted as the area change of a unit area perpendicular to the structural axis. On the basis of these physical observations, sometimes the change in length of vectors perpendicular to the material axis, and the change in area of area elements whose normal is perpendicular to the material axis are considered. They are defined by

$$K_2 := \mathrm{tr}(\boldsymbol{C} \cdot \boldsymbol{D}) = I_1 - J_4 = \|\boldsymbol{F}\|^2 - \|\boldsymbol{F} \cdot \boldsymbol{A}\|^2 \tag{4.311}$$

$$K_3 := \mathrm{tr}[(\mathrm{Cof}\boldsymbol{C}) \cdot \boldsymbol{D}] = I_1 J_4 - J_5 = \|\mathrm{Cof}\boldsymbol{F}\|^2 - \|(\mathrm{Cof}\boldsymbol{F}) \cdot \boldsymbol{A}\|^2 \tag{4.312}$$

where

$$\boldsymbol{D} := \boldsymbol{G}^\sharp - \boldsymbol{M}. \tag{4.313}$$

It can be proved that K_2 and K_3 are polyconvex (Schröder and Neff 2001). Indeed,

$$\begin{aligned} K_2 &= \mathrm{tr}[\boldsymbol{F}^{\mathrm{T}} \cdot \boldsymbol{F} \cdot (\boldsymbol{G}^\sharp - \boldsymbol{M})] \\ &= \|\boldsymbol{F} \cdot (\boldsymbol{G}^\sharp - \boldsymbol{M})\|^2 \end{aligned} \tag{4.314}$$

since

$$(\boldsymbol{G}^\sharp - \boldsymbol{M}) \cdot (\boldsymbol{G}^\sharp - \boldsymbol{M}) = \boldsymbol{G}^\sharp - \boldsymbol{M} \tag{4.315}$$

and

$$(\boldsymbol{G}^\sharp - \boldsymbol{M})^{\mathrm{T}} = (\boldsymbol{G}^\sharp - \boldsymbol{M}). \tag{4.316}$$

Similarly, one finds

$$K_3 = \|\mathrm{Cof}\boldsymbol{F} \cdot (\boldsymbol{G}^\sharp - \boldsymbol{M})\|^2. \tag{4.317}$$

Analogous to the proof leading to Equation (4.64), one can prove that

$$\frac{J_4}{(\det \boldsymbol{F})^{2/3}} \tag{4.318}$$

and

$$\frac{J_4^{3/2}}{(\det \boldsymbol{F})^{2}} \tag{4.319}$$

are polyconvex as well as the same terms with J_4 replaced by K_1, K_2 or K_3. In particular,

$$\overline{J}_4 := \frac{J_4}{(\det \boldsymbol{F})^{2/3}} \tag{4.320}$$

$$\overline{K}_1^{3/2} := \left[\frac{K_1}{(\det \boldsymbol{F})^{4/3}}\right]^{3/2} \tag{4.321}$$

$$\overline{K}_2 := \frac{K_2}{(\det \boldsymbol{F})^{2/3}} \tag{4.322}$$

and

$$\overline{K}_3^{3/2} := \left[\frac{K_3}{(\det \boldsymbol{F})^{4/3}}\right]^{3/2} \tag{4.323}$$

are polyconvex and consequently, also $\overline{J}_4^n$, $\overline{K}_1^{3n/2}$, $\overline{K}_2^n$ and $\overline{K}_3^{3n/2}$ (Theorem 4.1.8) (notice that J_4, K_1, K_2, $K_3 \geq 0$). For $\boldsymbol{C} = \boldsymbol{G}$, one obtains

$$J_4 = \overline{J}_4 = 1 \tag{4.324}$$

$$K_1 = \overline{K}_1 = 1 \tag{4.325}$$

$$K_2 = \overline{K}_2 = 2 \tag{4.326}$$

$$K_3 = \overline{K}_3 = 2 \tag{4.327}$$

but $\overline{J}_4 \geq 1$ is not guaranteed, hence we cannot argue that terms of the form $(\overline{J}_4 - 1)^k$ are polyconvex. On the other hand, terms such as

$$\mathrm{e}^{(\overline{J}_4-1)}, \mathrm{e}^{(\overline{K}_1^{3/2}-1)}, \mathrm{e}^{(\overline{K}_2-2)}, \mathrm{e}^{(\overline{K}_3^{3/2}-2\sqrt{2})} \tag{4.328}$$

are polyconvex, since e^x is a convex, monotonic increasing function in $\mathbb{R}$.

4.7.2 Fiber-reinforced material

In this section, an anisotropic hyperelastic model for fiber-reinforced materials will be discussed. It is a model that was developed for arteries by Holzapfel (Holzapfel *et al.* 2000) but which seems promising for other applications as well. It consists of an isotropic neo-Hooke part superimposed by strengthening terms in the fiber direction (the volumetric

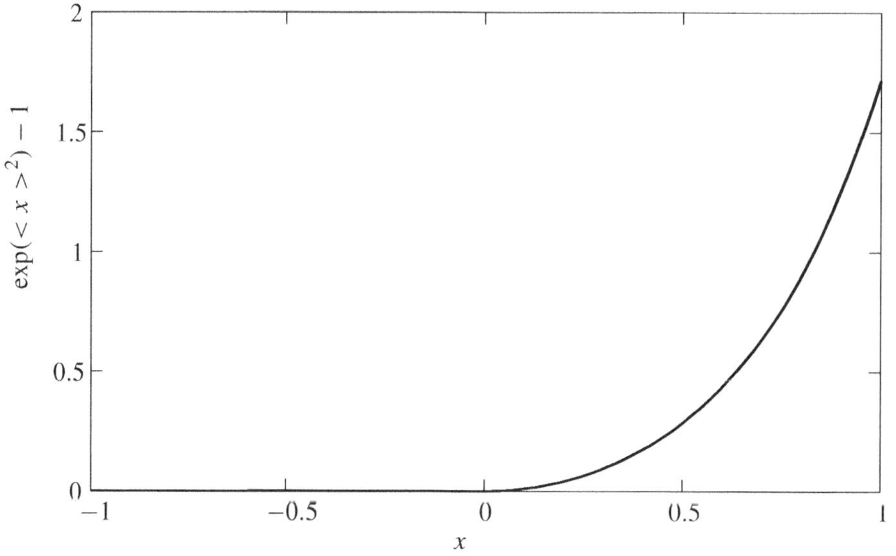

Figure 4.11 Generic form of the anisotropic term

term is not a part of the original Holzapfel model):

$$\Sigma = B_{10}(\bar{I}_1 - 3) + \frac{1}{D_1}(J-1)^2 + \sum_{i=1}^{N} \frac{k_{1i}}{2k_{2i}} \left[e^{k_{2i} <\bar{J}_{4i}-1>^2} - 1 \right] \tag{4.329}$$

where

$$\begin{aligned} < x > &= x \quad \text{for } x > 0 \\ &= 0 \quad \text{for } x \leq 0. \end{aligned} \tag{4.330}$$

and

$$k_{2i} > 0. \tag{4.331}$$

Notice that the anisotropic term applies only if the fibers are extended. Under compression the fibers do not contribute any strength. Under tension, however, the strengthening is exponential. There are as many terms as there are fiber directions, each with its own constants k_{1i} and k_{2i}. Notice that $e^{a<x>^2} - 1$ $(a > 0)$ is a C^1 monotonically increasing convex function (Figure 4.11). Accordingly, the anisotropic terms in Equation (4.329) are polyconvex.

Differentiation of Equation (4.329) leads to S^{KL} and Σ^{KLMN},

$$S^{KL} = B_{10} \frac{\partial \bar{I}_1}{\partial C_{KL}} + \frac{1}{D_1}(1 - I_3^{-1/2}) \frac{\partial I_3}{\partial C_{KL}} + \sum_{i=1}^{N} k_{1i}(\bar{J}_{4i} - 1) \left[e^{k_{2i}(\bar{J}_{4i}-1)^2} \right] \frac{\partial \bar{J}_{4i}}{\partial C_{KL}} \tag{4.332}$$

$$
\begin{aligned}
\Sigma^{KLMN} &= \frac{\partial^2 \Sigma}{\partial C_{KL} C_{MN}} = B_{10} \frac{\partial^2 \overline{I}_1}{\partial C_{KL} \partial C_{MN}} + \frac{1}{2D_1} I_3^{-3/2} \frac{\partial I_3}{\partial C_{KL}} \frac{\partial I_3}{\partial C_{MN}} \\
&\quad + \frac{1}{D_1}(1 - I_3^{-1/2}) \frac{\partial^2 I_3}{\partial C_{KL} C_{MN}} + \sum_{i=1}^{N} k_{1i} \left[\mathrm{e}^{k_{2i}(\overline{J}_{4i}-1)^2} \right] \cdot \\
&\quad \cdot \left[\frac{\partial \overline{J}_{4i}}{\partial C_{KL}} \frac{\partial \overline{J}_{4i}}{\partial C_{MN}} \quad 1 + 2k_{2i}(\overline{J}_{4i} - 1)^2 \Big) + (\overline{J}_{4i} - 1) \frac{\partial^2 \overline{J}_{4i}}{\partial C_{KL} C_{MN}} \right]
\end{aligned}
\tag{4.333}
$$

and the derivatives of J_{4i} and $\overline{J}_{4i}$ yield

$$J_{4i} = M_i^{IJ} C_{IJ} \tag{4.334}$$

$$\frac{\partial J_{4i}}{\partial C_{KL}} = M_i^{KL} \tag{4.335}$$

$$\frac{\partial^2 J_{4i}}{\partial C_{KL} \partial C_{MN}} = 0 \tag{4.336}$$

$$\overline{J}_{4i} = I_3^{-1/3} J_{4i} \tag{4.337}$$

$$\frac{\partial \overline{J}_{4i}}{\partial C_{KL}} = -\tfrac{1}{3} I_3^{-4/3} J_{4i} \frac{\partial I_3}{\partial C_{KL}} + I_3^{-1/3} \frac{\partial J_{4i}}{\partial C_{KL}} \tag{4.338}$$

$$
\begin{aligned}
\frac{\partial^2 \overline{J}_{4i}}{\partial C_{KL} \partial C_{MN}} &= \tfrac{4}{9} I_3^{-7/3} J_{4i} \frac{\partial I_3}{\partial C_{KL}} \frac{\partial I_3}{\partial C_{MN}} \\
&\quad - \tfrac{1}{3} I_3^{-4/3} \left(\frac{\partial J_{4i}}{\partial C_{MN}} \frac{\partial I_3}{\partial C_{KL}} + \frac{\partial J_{4i}}{\partial C_{KL}} \frac{\partial I_3}{\partial C_{MN}} \right) \\
&\quad - \tfrac{1}{3} I_3^{-4/3} J_{4i} \frac{\partial^2 I_3}{\partial C_{KL} \partial C_{MN}} + I_3^{-1/3} \frac{\partial^2 J_{4i}}{\partial C_{KL} \partial C_{MN}}.
\end{aligned}
\tag{4.339}
$$

To investigate the effect of the anisotropic terms on the initial stiffness, the limit $\boldsymbol{C} \to \boldsymbol{G}$ is taken

$$J_{4i}\big|_{\boldsymbol{C}=\boldsymbol{G}} = 1 \tag{4.340}$$

$$\left.\frac{\partial J_{4i}}{\partial C_{KL}}\right|_{\boldsymbol{C}=\boldsymbol{G}} = M_i^{KL} \tag{4.341}$$

$$\left.\frac{\partial^2 J_{4i}}{\partial C_{KL} \partial C_{MN}}\right|_{\boldsymbol{C}=\boldsymbol{G}} = 0 \tag{4.342}$$

$$\overline{J}_{4i}\big|_{\boldsymbol{C}=\boldsymbol{G}} = 1 \tag{4.343}$$

$$\left.\frac{\partial \overline{J}_{4i}}{\partial C_{KL}}\right|_{\boldsymbol{C}=\boldsymbol{G}} = M_i^{KL} - \tfrac{1}{3} G^{KL} \tag{4.344}$$

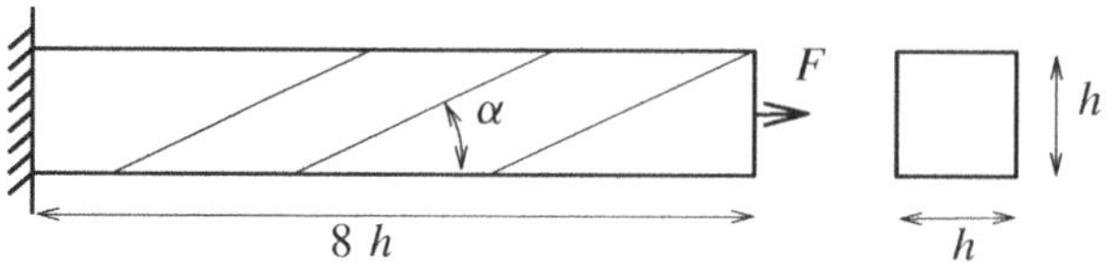

Figure 4.12 Geometry of the cantilever beam

$$\left.\frac{\partial^2 \overline{J}_{4i}}{\partial C_{KL}\partial C_{MN}}\right|_{C=G} = \tfrac{1}{9}G^{KL}G^{MN} + \tfrac{1}{6}(G^{KM}G^{LN} + G^{KN}G^{LM}) - \tfrac{1}{3}(M_i^{MN}G^{KL} + M_i^{KL}G^{MN}). \quad (4.345)$$

Substitution in Equation (4.333) yields for the anisotropic terms

$$\left.\Sigma^{KLMN}\right|_{C=G,\text{anisotropic}} = \sum_{i=1}^{N} k_{1i}(M_i^{KL} - \tfrac{1}{3}G^{KL})(M_i^{MN} - \tfrac{1}{3}G^{MN}). \quad (4.346)$$

For a fiber aligned with the 1-direction ($M^{11} = 1$, all other $M^{KL} = 0$) one obtains

$$\left.\Sigma^{KLMN}\right|_{C=G,\text{anisotropic}} = \begin{bmatrix} \begin{bmatrix} 4/9 & -2/9 & -2/9 \\ -2/9 & 1/9 & 1/9 \\ -2/9 & 1/9 & 1/9 \end{bmatrix} & [0]_{3\times 3} \\ [0]_{3\times 3} & [0]_{3\times 3} \end{bmatrix} k_{1i}. \quad (4.347)$$

Accordingly, the initial stiffness in fiber direction is increased by $\frac{4}{9}k_{1i}$. The parameter k_{1i} has the unit of stress, k_{2i} is dimensionless and governs the strength increase at increasing deformation.

Consider the cantilever beam in Figure 4.12 subject to a force F evenly distributed at its end. The force keeps its magnitude and direction during deformation. The material consists of an isotropic neo-Hooke substrate strengthened by fibers. It is assumed to satisfy Equation (4.329) with constants:

$$B_{10}h^2/F = 0.192505 \quad (4.348)$$

$$D_1 F/h^2 = 0.26 \quad (4.349)$$

$$N = 1 \quad (4.350)$$

$$k_{11}h^2/F = 0.23632 \quad (4.351)$$

$$k_{21} = 0.8393. \quad (4.352)$$

There is only one layer of fibers making an angle α with the axis of the beam. The relative axial displacement u_a/h and transversal displacement $100u_t/h$ at the end of the

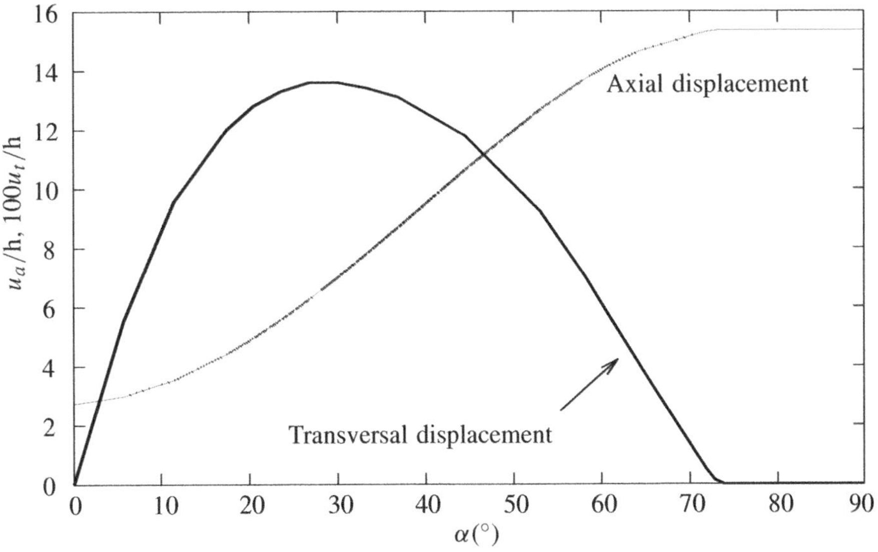

Figure 4.13 Longitudinal and transversal displacement of the end of the beam

beam are shown in Figure 4.13. If the fibers are parallel to the axis of the beam, the transversal displacement is zero and the axial displacement exhibits a minimum because of the strengthening effect of the fibers. As the angle with the axis increases, the strengthening effect decreases steadily. The transversal displacement exhibits a maximum at about 27° because of the asymmetry induced by the fibers.

5

Infinitesimal Strain Plasticity

5.1 Introduction

The materials treated so far had no memory. The instantaneous deformation for such materials is a function of the instantaneous loading only. The previous loading history is of no importance. However, for a lot of practical materials this assumption does not hold. A piece of metal that has been forged into a car component will react differently on loading because of the forging process: the component remembers it has been forged. Furthermore, the activation of memory allows for the simulation of another new phenomenon: irreversibility of deformation. Without this feature, it would be impossible to deform a body into a new form without continuously applying loads: on releasing pressure, a car would return into ore!

Memory and irreversibility of deformation are two important characteristics of plasticity. Although the term is most often applied to metals, it is also used to describe irreversible behavior in soils, biological tissue, and so on. Here, we treat metals only. Furthermore, attention is focused on the infinitesimal theory, that is, strains and rotations are assumed to be so small that material and spatial quantities coincide. First, the general framework is derived using the one-dimensional example as a guide. Then, the isotropic viscoplastic theory is deduced. Finally, a detailed analysis is presented of single-crystal viscoplasticity and von Mises plasticity of elastically anisotropic materials. The treatment of other viscoplastic formulations, such as Drucker–Prager or Gurson, runs along the same lines. For fundamental reference works on plasticity, see ((Kachanov 1971); (Lemaitre and Chaboche 1990); and (Save and Massonnet 1972)).

5.2 The General Framework of Plasticity

5.2.1 Theoretical derivation

Throughout the present chapter, it will be assumed that the strains and rotations are small, such that the infinitesimal quantities $\tilde{e}_{KL}$ and $\tilde{E}_{KL}$ can be used. One recalls

The Finite Element Method for Three-dimensional Thermomechanical Applications Guido Dhondt
 ISBN: 0-470-85752-8

(Section (1.14.4)),

$$\tilde{E}_{KL} \approx \tilde{e}_{KL}\delta^k_K\delta^l_L \tag{5.1}$$

$$S^{KL} \approx \sigma^{kl}\delta_k{}^K\delta_l{}^L. \tag{5.2}$$

Furthermore, it will be assumed that a rectangular spatial coordinate system and a rectangular material coordinate system are used and that both coincide. To emphasize this, the infinitesimal strain tensor will be represented by the new symbol ϵ_{kl}.

How can we characterize a viscoplastic material? A simple stress–strain test on steel yields Figure 5.1. One notices that the material is elastic for small stresses. For growing stress ($\sigma > \sigma_A$), the curve deviates from the elastic straight line by an amount that will be called the *plastic strain* ϵ^{p}. It is this amount that is not recovered after unloading (point C). When loading again, the material remains elastic up to $\sigma = \sigma_B$ before accumulating further plastic strain. Accordingly, the elastic range depends on the previous plastic flow.

From these considerations, the following assumptions (some of which are valid for the infinitesimal range only) seem plausible:

1. The total strain can be decomposed in an additive way into an elastic part and a plastic part:

$$\epsilon = \epsilon^{\mathrm{e}} + \epsilon^{\mathrm{p}}. \tag{5.3}$$

2. Plastic flow starts as soon as the stress (axial stress in a tensile specimen, an appropriate stress invariant in the two- and three-dimensional case) reaches a specific value, which can change as a function of the previous amount of plastic deformation. Accordingly, this value is a function and will be denoted $-q$. The amount of plastic flow is represented by α. Hence,

$$q = q(\alpha). \tag{5.4}$$

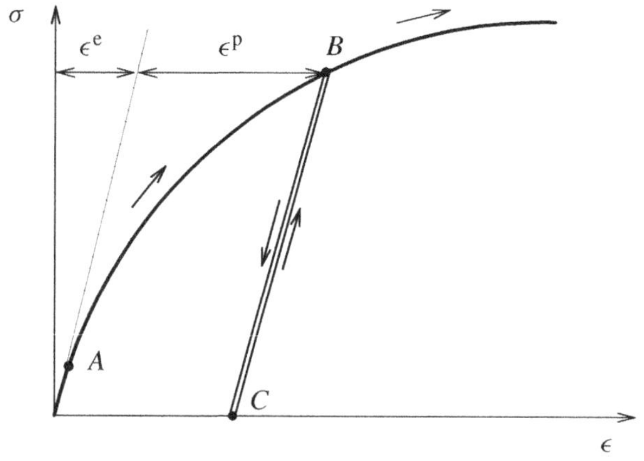

Figure 5.1 Elastoplastic stress–strain curve

q is an internal dynamic variable (comparable to a stress), α is the kinematic counterpart (comparable to a strain). The material remains elastic if

$$\sigma < -q \Leftrightarrow \sigma + q < 0. \tag{5.5}$$

The function

$$f(\sigma, q) = \sigma + q = 0 \tag{5.6}$$

represents the boundary of the elastic range and is called the *yield surface*.

In theory, plastic deformation takes place if the yield surface is exceeded. However, because of the plastic flow, the yield surface is expanded such that the stress stays on the yield surface (in the absence of viscosity). This can be made obvious by simply performing an unloading–loading experiment, as shown in Figure 5.1. Consequently, the increasing stress drags the yield surface along, and Equation (5.6) keeps its validity during the plastic flow. The reverse statement is not true: Equation (5.6) does not necessarily imply plastic deformation. Imagine we load until $\sigma = -q$ and freeze the loading at that point: this is a purely elastic process. Summarizing, during plastic deformation, one has

$$\sigma = E(\epsilon - \epsilon^{\mathrm{p}}) \tag{5.7}$$

$$q = q(\alpha) \tag{5.8}$$

$$\sigma + q = 0. \tag{5.9}$$

Assume that we know the total strain and want to know the stress. Equations (5.7) to (5.9) yield three equations in the four unknowns σ, ϵ^{p}, q and α: one equation is lacking. Physically, we do not know at what rate the plastic flow is accumulated: an evolution equation for ϵ^{p} (and actually, also for α) is lacking. This equation will be obtained by maximizing the plastic dissipation.

Recall from Chapter 1 that the entropy rate was obtained by substituting an expression for the free energy, Equation (1.390), into the entropy inequality, yielding Equation (1.392) and subsequently Equation (1.396). For our infinitesimal (one-dimensional) considerations, Equation (1.411) reduces to

$$\psi = \psi(\epsilon, \theta, \nabla\theta, X). \tag{5.10}$$

In the previous analysis, it was shown that the total strain of the elastic theory should be replaced by ϵ^{e}. Furthermore, a dependence of ψ on the internal kinematic variable is assumed. Hence,

$$\psi = \psi(\epsilon - \epsilon^{\mathrm{p}}, \alpha, \theta, \nabla\theta, X). \tag{5.11}$$

Note that the time derivative of the internal kinematic variable α is not necessarily continuous: at unloading, the time rate of α discontinuously drops to zero. This also applies to ϵ^{p}. Taking the time derivative yields

$$\dot{\psi} = \frac{\partial\psi}{\partial(\epsilon - \epsilon^{\mathrm{p}})}\dot{\epsilon} - \frac{\partial\psi}{\partial(\epsilon - \epsilon^{\mathrm{p}})}\dot{\epsilon}^{\mathrm{p}} + \frac{\partial\psi}{\partial\alpha}\dot{\alpha} + \frac{\partial\psi}{\partial\theta}\dot{\theta} + \frac{\partial\psi}{\partial\nabla\theta}\dot{\overline{\nabla\theta}}. \tag{5.12}$$

Substituting this expression into Equation (1.389) yields ($\rho_0 \approx \rho$)

$$\frac{1}{\theta}\left(-\rho\frac{\partial\psi}{\partial\epsilon^{\mathrm{e}}}+\sigma\right)\dot{\epsilon}+\frac{\rho}{\theta}\frac{\partial\psi}{\partial\epsilon^{\mathrm{e}}}\dot{\epsilon}^{\mathrm{p}}-\frac{\rho}{\theta}\frac{\partial\psi}{\partial\alpha}\dot{\alpha}-\frac{\rho}{\theta}\left(\frac{\partial\psi}{\partial\theta}+\eta\right)\dot{\theta}-\frac{\rho}{\theta}\frac{\partial\psi}{\partial\nabla\theta}\dot{\overline{\nabla\theta}}-\frac{1}{\theta^2}q^{\theta}\nabla\theta\geq 0 \tag{5.13}$$

where q^{θ} stands for the heat conduction (the superscript θ is introduced in this chapter to avoid confusion between the heat conduction q^{θ} and the internal dynamic variable q). This inequality is satisfied if

$$\sigma=\rho\frac{\partial\psi}{\partial\epsilon^{\mathrm{e}}} \tag{5.14}$$

$$\eta=-\frac{\partial\psi}{\partial\theta} \tag{5.15}$$

$$\frac{\partial\psi}{\partial\nabla\theta}=0 \tag{5.16}$$

and

$$\sigma\dot{\epsilon}^{\mathrm{p}}-\rho\frac{\partial\psi}{\partial\alpha}\dot{\alpha}-\frac{1}{\theta}q^{\theta}\nabla\theta\geq 0. \tag{5.17}$$

The first two terms do not vanish because of the time history of plastic deformation. As in Equation (5.14), q is now defined by

$$q=-\rho\frac{\partial\psi}{\partial\alpha} \tag{5.18}$$

(or, the other way around, Equation (5.18) can be viewed as the definition of $\psi(\alpha)$ through Equation (5.4)). Inequality (5.17) is satisfied if

$$\sigma\dot{\epsilon}^{\mathrm{p}}+q\dot{\alpha}\geq 0 \tag{5.19}$$

and

$$-\frac{1}{\theta}q^{\theta}\nabla\theta\geq 0. \tag{5.20}$$

Now, the evolution equations for $\dot{\epsilon}^{\mathrm{p}}$ and $\dot{\alpha}$ are obtained by postulating that for a given $\dot{\epsilon}^{\mathrm{p}}$ and $\dot{\alpha}$, the state (σ, q) will prevail, which maximizes (Simo and Hughes 1997) (Halphen and Nguyen Quoc Son 1975)

$$d^{\mathrm{p}}=\sigma\dot{\epsilon}^{\mathrm{p}}+q\dot{\alpha} \tag{5.21}$$

subject to

$$f(\sigma, q)\leq 0. \tag{5.22}$$

This postulate is accepted to hold for metals, but does not necessarily hold for all kinds of material. This also amounts to minimizing

$$l^{\mathrm{p}}:=-\sigma\dot{\epsilon}^{\mathrm{p}}-q\dot{\alpha}+\dot{\gamma}f(\sigma, q) \tag{5.23}$$

with respect to σ and q where

$$\dot{\gamma} \geq 0 \tag{5.24}$$

and subject to

$$\dot{\gamma} f(\sigma, q) = 0 \tag{5.25}$$

(Luenberger 1989). By taking the derivative of l^{p} with respect to σ and q, one obtains the evolution equations

$$\dot{\epsilon}^{\mathrm{p}} = \dot{\gamma} \frac{\partial f(\sigma, q)}{\partial \sigma} \tag{5.26}$$

$$\dot{\alpha} = \dot{\gamma} \frac{\partial f(\sigma, q)}{\partial q}. \tag{5.27}$$

Equations (5.22), (5.24) and (5.25) are called the *Kuhn–Tucker conditions.* Although there are two new equations, Equations (5.26) and (5.27), there is also a new unknown $\dot{\gamma}$, called the *consistency parameter* (sometimes called the *plastic rate parameter*). By Equation (5.24), $\dot{\gamma}$ cannot be negative, and by Equation (5.25), it can be strictly positive only if yielding takes place. The notation $\dot{\gamma}$ was chosen because it is easiest to consider it as a rate of accumulated plastic flow. To emphasize that, for plastic deformation to occur, the stress state has to persist on the yield surface the following equation is added

$$\dot{\gamma} \dot{f}(\sigma, q) = 0. \tag{5.28}$$

This is also called the *consistency condition.* Summarizing, the following equations apply:

1. Elastic stress–strain relations

$$\sigma = E(\epsilon - \epsilon^{\mathrm{p}}) \tag{5.29}$$

2. Internal variable relationship

$$q = -h(\alpha) \tag{5.30}$$

3. Yield surface

$$f(\sigma, q) = 0 \tag{5.31}$$

4. Evolution equations

$$\dot{\epsilon}^{\mathrm{p}} = \dot{\gamma} \frac{\partial f(\sigma, q)}{\partial \sigma} \tag{5.32}$$

$$\dot{\alpha} = \dot{\gamma} \frac{\partial f(\sigma, q)}{\partial q} \tag{5.33}$$

5. Kuhn–Tucker equations

$$\dot{\gamma} \geq 0, \quad f(\sigma, q) \leq 0, \quad \dot{\gamma} f(\sigma, q) = 0 \tag{5.34}$$

6. Consistency condition

$$\dot{\gamma}\dot{f}(\sigma, q) = 0. \tag{5.35}$$

The evolution equations are also called the *flow rule* (Equation (5.32)) and the *hardening law* (Equation (5.33)). Equations (5.29) to (5.35) are quite general and are easy to extend to higher dimensions. Notice particularly that the maximum-dissipation principle implies that the evolution equations can be derived from the yield surface. This kind of model is called *associative*.

Returning to our basic yield surface, Equation (5.9), extending it to negative stresses in the form

$$|\sigma| + q \leq 0 \tag{5.36}$$

and substituting this equation into (5.32) and (5.33) yields

$$\dot{\epsilon}^{\mathrm{p}} = \dot{\gamma}\,\mathrm{sgn}(\sigma) \tag{5.37}$$

$$\dot{\alpha} = \dot{\gamma}. \tag{5.38}$$

Equations (5.37) and (5.38) reveal that, for our simple example, $\dot{\gamma}$ and $\dot{\alpha}$ are the magnitude of the plastic strain rate:

$$\dot{\gamma} = \left|\dot{\epsilon}^{\mathrm{p}}\right|. \tag{5.39}$$

Hence, since $\sigma = h(\alpha)$ and $\alpha = \epsilon^{\mathrm{p}}$ for monotonic loading ($\dot{\alpha} = \dot{\epsilon}^{\mathrm{p}}$ and let $\alpha = 0$ for $\epsilon^{\mathrm{p}} = 0$), one can derive $h(\alpha)$ from Figure 5.1 by subtracting the elastic strain at a given stress level (Figure 5.2).

The type of hardening considered so far is called *isotropic hardening*, since it equally applies to positive and negative stresses (Figure 5.3). In practice, the Bauschinger effect is frequently observed: after plastic deformation in the tensile range, plastic deformation in the compressive range takes place at higher stress levels than expected. Pure kinematic hardening implies that the size of the elastic range has not changed, just its origin: $BD = 2\,OA$ in Figure 5.3. Introducing an internal variable $-q_2$ for the center of the elastic range, this amounts to a yield surface of the form

$$|\sigma + q_2| + c \leq 0 \tag{5.40}$$

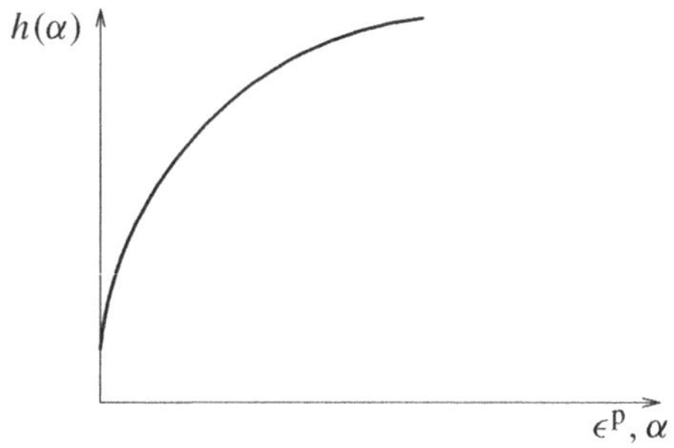

Figure 5.2 Isotropic hardening curve

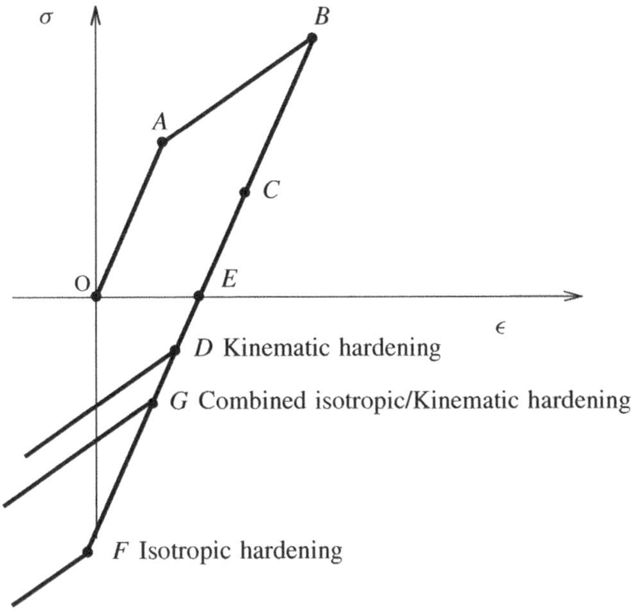

Figure 5.3 Types of hardening

where c is a constant. If $BF = 2\ BE$, pure isotropic hardening applies. Point G symbolizes a state in between, that is, combined hardening. For this general case, the yield surface looks like

$$|\sigma + q_2| + q_1 \leq 0 \tag{5.41}$$

leading to the flow rule

$$\dot{\epsilon}^{\mathrm{p}} = \dot{\gamma}\,\mathrm{sgn}(\sigma + q_2) \tag{5.42}$$

and the evolution equations

$$\dot{\alpha}_1 = \dot{\gamma} \tag{5.43}$$

$$\dot{\alpha}_2 = \dot{\gamma}\,\mathrm{sgn}(\sigma + q_2). \tag{5.44}$$

From Equation (5.42), one again notices that $\dot{\gamma}$ is the rate of the accumulated plastic strain (in absolute value). The only functions left to be determined are $q_1(\alpha_1)$ and $q_2(\alpha_2)$, both of which are material characteristics to be obtained by experiments. Notice that the inclusion of kinematic hardening does not substantially change the governing relations (Equations (5.29)–(5.35)): instead of q, one now deals with q_1 and q_2 or, equivalently, a two component vector $\boldsymbol{q}$.

The theory derived so far can also be extended to include viscous effects. For many materials, one observes that during loading the stress state exceeds the yield surface and slowly creeps back with time until the yield surface is reached from above. In Figure 5.4,

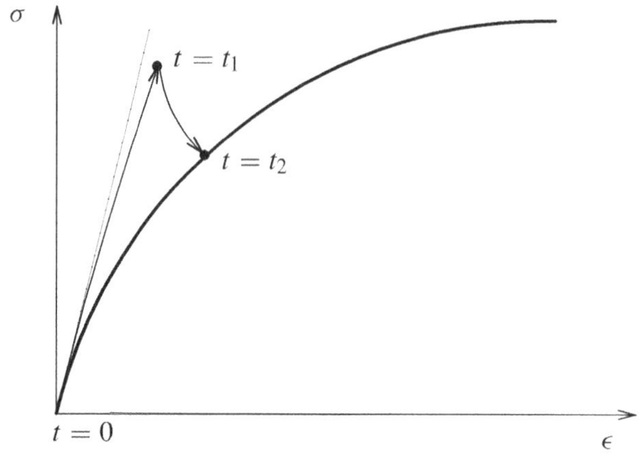

Figure 5.4 Viscoplasticity concept

the external loading (forces, temperature etc.) is frozen at $t = t_1$. Because of the rate at which the loading was applied, the material did not have time to accumulate enough plastic deformation to attain equilibrium conditions. Physically, this means that since plastic flow is largely equivalent to dislocation motion, the dislocations did not have enough time to redistribute in accordance with the applied load. As time goes by, an equilibrium configuration is reached. This phenomenon is also called *creep* and is usually modeled by a creep law of the form

$$\dot{\epsilon}^{\mathrm{p}} = g^{-1}(\sigma^0) \tag{5.45}$$

where σ^0 is the overload. This is the stress amount by which the yield surface is exceeded and which constitutes the driving force for the creep process. Consequently, if plastic deformation occurs, Equation (5.31) has to be replaced by

$$f(\sigma, q) = g(\dot{\epsilon}^{\mathrm{p}}). \tag{5.46}$$

An example of a creep law is the Norton law, which is of the form

$$\dot{\epsilon}^{\mathrm{p}} = A(\sigma^0)^n. \tag{5.47}$$

The parameter A is usually a small number (depending on the unit of σ^0), and $n > 1$, for example, $n = 5$. Both A and n are material parameters.

5.2.2 Numerical implementation

Ultimately, Equations (5.29) to (5.35) have to be solved; for viscous problems, Equation (5.31) is to be replaced by Equation (5.46). Notice that Equations (5.34) and (5.35) have a kind of regulating character: they characterize the discrete split into elastic deformation and plastic deformation.

The usual finite-element procedure consists of the creation of a tangent-stiffness matrix at an instantaneous deformation field and solving the resulting linear equations to obtain a new deformation field, after which the procedure can be repeated until convergence. Thus, two major tasks ensue: the calculation of a consistent tangent-stiffness matrix and the calculation of the stress corresponding to a given deformation field in order to check convergence (satisfaction of the equilibrium conditions). In both cases, the deformation field is given. Consequently, the total strain is known. This is an extremely important fact and the starting point of most numerical algorithms in present finite-element implementations. Accordingly, ϵ_{n+1} is known, where $t = t_{n+1}$ is the new time step and all quantities at $t = t_n$ are known. Further strategy consists of trial and error. One first assumes that no plasticity occurs in the present step, that is,

$$\epsilon_{n+1}^{\mathrm{p,trial}} = \epsilon_n^{\mathrm{p}} \tag{5.48}$$

$$q_{n+1}^{\mathrm{trial}} = q_n \tag{5.49}$$

$$\gamma_{n+1}^{\mathrm{trial}} = \gamma_n. \tag{5.50}$$

Hence (Equation (5.29)),

$$\sigma_{n+1}^{\mathrm{trial}} = E(\epsilon_{n+1} - \epsilon_n^{\mathrm{p}}). \tag{5.51}$$

If the new stress state lies within the elastic range, that is, if

$$f(\sigma_{n+1}^{\mathrm{trial}}, q_{n+1}^{\mathrm{trial}}) \leq 0 \tag{5.52}$$

then the solution is found

$$\epsilon_{n+1}^{\mathrm{p}} = \epsilon_{n+1}^{\mathrm{p,trial}} \tag{5.53}$$

$$q_{n+1} = q_{n+1}^{\mathrm{trial}} \tag{5.54}$$

$$\gamma_{n+1} = \gamma_{n+1}^{\mathrm{trial}} \tag{5.55}$$

$$\sigma_{n+1} = \sigma_{n+1}^{\mathrm{trial}} \tag{5.56}$$

and the tangent modulus is the elastic one. If this is not the case, plasticity takes place and the following equations must be satisfied at $t = t_{n+1}$:

$$f(\sigma, q) = g(\dot{\epsilon} - E^{-1}\dot{\sigma}) \tag{5.57}$$

$$\dot{\epsilon} - E^{-1}\dot{\sigma} = \dot{\gamma}\frac{\partial f(\sigma, q)}{\partial \sigma} \tag{5.58}$$

$$\overline{h^{-1}(-q)}^{\,\cdot} = \dot{\gamma}\frac{\partial f(\sigma, q)}{\partial q}. \tag{5.59}$$

These are (for the one-dimensional case) three equations in the three unknowns σ, q and $\dot{\gamma}$. Now, a backward Euler scheme is applied to turn the differential equations (5.57) to (5.59) into difference equations. To this end, the equations are evaluated at $t = t_{n+1}$ and the first derivative $\dot{\epsilon}$ is replaced by

$$\dot{\epsilon} = \frac{\epsilon_{n+1} - \epsilon_n}{\Delta t} + O(\Delta t) \tag{5.60}$$

and similarly for the other first derivatives. One can prove that the backward Euler scheme is first-order accurate and unconditionally stable. Applying this to Equations (5.57) to (5.59) leads to

$$f(\sigma_{n+1}, q_{n+1}) = g[\Delta\epsilon_{n+1} - E^{-1}(\sigma_{n+1} - \sigma_n)] \tag{5.61}$$

$$\Delta\epsilon_{n+1} - E^{-1}(\sigma_{n+1} - \sigma_n) = \Delta\gamma_{n+1} \frac{\partial f(\sigma_{n+1}, q_{n+1})}{\partial \sigma} \tag{5.62}$$

$$h^{-1}(-q_{n+1}) - h^{-1}(-q_n) = \Delta\gamma_{n+1} \frac{\partial f(\sigma_{n+1}, q_{n+1})}{\partial q} \tag{5.63}$$

where

$$\Delta\epsilon_{n+1} := \epsilon_{n+1} - \epsilon_n \tag{5.64}$$

$$\Delta\gamma_{n+1} := \gamma_{n+1} - \gamma_n. \tag{5.65}$$

Equations (5.61) to (5.63) are three nonlinear equations in three unknowns σ_{n+1}, q_{n+1} and $\Delta\gamma_{n+1}$. They can be solved using the customary mathematical techniques to solve sets of nonlinear equations. To concretize the further derivation, the yield surface of Equation (5.36) is taken, no creep is assumed and a linear isotropic hardening law is chosen of the form

$$q = -\sigma_0 - K\alpha \tag{5.66}$$

where σ_0 and K are constants. Consequently,

$$|\sigma_{n+1}| + q_{n+1} = 0 \tag{5.67}$$

$$\Delta\epsilon_{n+1} - E^{-1}(\sigma_{n+1} - \sigma_n) = \Delta\gamma_{n+1}\text{sgn}(\sigma_{n+1}) \tag{5.68}$$

$$(-\sigma_0 - q_{n+1}) - (-\sigma_0 - q_n) = K\Delta\gamma_{n+1}. \tag{5.69}$$

The sign of σ_{n+1} is the same as that of $\sigma_{n+1}^{\text{trial}}$ (Figure 5.5).

$$\text{sgn}(\sigma_{n+1}) = \text{sgn}(\sigma_{n+1}^{\text{trial}}). \tag{5.70}$$

Accordingly (Equations (5.68) and (5.69)),

$$\sigma_{n+1} = E\Delta\epsilon_{n+1} - E\Delta\gamma_{n+1}\text{sgn}(\sigma_{n+1}^{\text{trial}}) + \sigma_n \tag{5.71}$$

$$q_{n+1} = q_n - K\Delta\gamma_{n+1} \tag{5.72}$$

yielding (Equation (5.67))

$$\Delta\gamma_{n+1} = \frac{1}{E+K}\left[E\Delta\epsilon_{n+1}\text{sgn}(\sigma_{n+1}^{\text{trial}}) + \sigma_n\text{sgn}(\sigma_{n+1}^{\text{trial}}) + q_n\right]. \tag{5.73}$$

Substitution of $\Delta\gamma_{n+1}$ into Equations (5.71) and (5.72) yields σ_{n+1} and q_{n+1}. In particular (Equation (5.71)),

$$\sigma_{n+1} = E(\epsilon_{n+1} - \epsilon_n) - \frac{E^2(\epsilon_{n+1} - \epsilon_n)}{E+K} - \frac{E\sigma_n}{E+K} - \frac{Eq_n\text{sgn}(\sigma_{n+1}^{\text{trial}})}{E+K} + \sigma_n \tag{5.74}$$

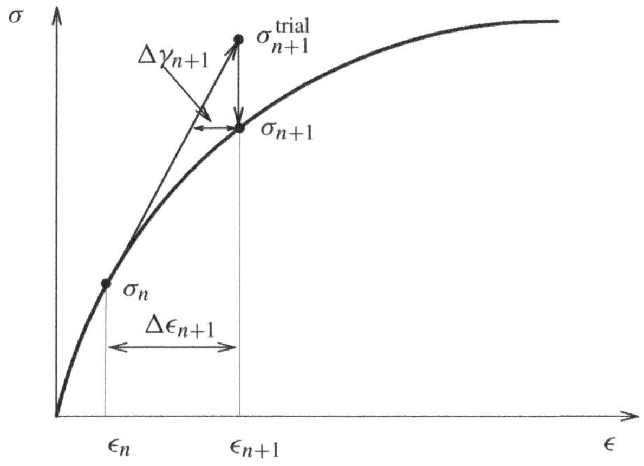

Figure 5.5 Trial-and-error method

from which the plastic tangent is obtained:

$$\frac{\mathrm{d}\sigma_{n+1}}{\mathrm{d}\epsilon_{n+1}} = \frac{EK}{E+K}. \tag{5.75}$$

Because of the particular choice of q and f, the resulting equation, Equation (5.73), was linear and easy to solve.

5.3 Three-dimensional Single Surface Viscoplasticity

5.3.1 Theoretical derivation

The governing equations for three-dimensional applications are the same as for one-dimensional applications and their derivation is similar (cf Equations (5.29)–(5.35)):

1. Elastic stress–strain relations

$$\boldsymbol{\sigma} = \frac{\partial \Sigma}{\partial \boldsymbol{\epsilon}^{\mathrm{e}}} \tag{5.76}$$

2. Internal variable relationships

$$\boldsymbol{q} = -h(\boldsymbol{\alpha}) \tag{5.77}$$

3. Yield surface

$$f(\boldsymbol{\sigma}, \boldsymbol{q}) = 0 \tag{5.78}$$

4. Evolution equations

$$\dot{\boldsymbol{\epsilon}}^{\mathrm{p}} = \dot{\gamma}\frac{\partial f(\boldsymbol{\sigma}, \boldsymbol{q})}{\partial \boldsymbol{\sigma}} \tag{5.79}$$

$$\dot{\boldsymbol{\alpha}} = \dot{\gamma}\frac{\partial f(\boldsymbol{\sigma}, \boldsymbol{q})}{\partial \boldsymbol{q}} \tag{5.80}$$

5. Kuhn–Tucker equations

$$\dot{\gamma} \geq 0, \quad f(\boldsymbol{\sigma}, \boldsymbol{q}) \leq 0, \quad \dot{\gamma} f(\boldsymbol{\sigma}, \boldsymbol{q}) = 0 \tag{5.81}$$

6. Consistency condition

$$\dot{\gamma} \dot{f}(\boldsymbol{\sigma}, \boldsymbol{q}) = 0. \tag{5.82}$$

If viscous effects are to be taken into account, Equation (5.78) is replaced by

$$f(\boldsymbol{\sigma}, \boldsymbol{q}) = g(\dot{\boldsymbol{\epsilon}}^{\mathrm{p}}) \tag{5.83}$$

and $f > 0$ is feasible. Although the form of the equations is similar, due to the tensorial character of the quantities involved, one arrives at a much larger set of nonlinear equations than in the one-dimensional case. Indeed, Equations (5.78) to (5.80) lead to a set of nonlinear equations the size of which is the sum of the number of independent stress components and the number of internal variables plus one. Here, an example covering the usual isotropic metal plasticity will be given.

To fully determine the plasticity model defined by Equations (5.76) to (5.83), one has to define the yield surface, choose the internal variables and possibly define a creep law.

In three dimensions, the yield surface is not so trivial as in the one-dimensional case. Indeed, one has to find one scalar equation connecting the tensorial quantities $\boldsymbol{\sigma}$ and $\boldsymbol{q}$. Furthermore, for isotropic materials, the yield surface should contain invariants only, that is, $I_{1\sigma}$, $I_{2\sigma}$ and $I_{3\sigma}$. Practical observations have shown that the hydrostatic pressure p does not significantly lead to plasticity in metals (does not apply to soils!). Since

$$p := -\tfrac{1}{3} I_{1\sigma} \tag{5.84}$$

the first invariant does not enter the yield condition explicitly. Therefore, a new stress tensor, the deviatoric stress $\boldsymbol{s}$ is defined by

$$\boldsymbol{s} := \operatorname{dev} \boldsymbol{\sigma} := \boldsymbol{\sigma} + p\boldsymbol{I}. \tag{5.85}$$

On the basis of the deviatoric stress, a new invariant is defined called the *von Mises stress* σ_{vm} by

$$\sigma_{\mathrm{vm}} := \sqrt{\tfrac{3}{2} \|\boldsymbol{s}\|^2} = \sqrt{\tfrac{3}{2} \|\operatorname{dev} \boldsymbol{\sigma}\|^2}. \tag{5.86}$$

Since

$$\begin{aligned} \|\boldsymbol{s}\|^2 &= \|\boldsymbol{\sigma}\|^2 - \tfrac{1}{3}(I_{1\sigma})^2 \\ &= \tfrac{2}{3}(I_{1\sigma})^2 - 2I_{2\sigma} \end{aligned} \tag{5.87}$$

one finds

$$\sigma_{\mathrm{vm}} = \sqrt{(I_{1\sigma})^2 - 3I_{2\sigma}}. \tag{5.88}$$

The von Mises stress is a measure of the shear energy. The factor $\sqrt{2/3}$ is introduced such that the von Mises stress in a tensile test coincides with the applied tensile stress. Including isotropic and kinematic hardening (internal variables q_1 and $\boldsymbol{q}_2$ respectively), the following yield surface is proposed (Huber–von Mises yield surface):

$$\|\text{dev}\ (\boldsymbol{\sigma}) + \boldsymbol{q}_2\| + \sqrt{\tfrac{2}{3}}q_1 = 0. \tag{5.89}$$

In deviatoric space (axes s_1, s_2 and s_3), Equation (5.89) is a sphere with radius $\sqrt{2/3}\ q_1$ and center $-\boldsymbol{q}_2$ (Figure 5.6). The inside of the sphere is the elastic range. During plasticity, the sphere can both expand (isotropic hardening) and move (kinematic hardening). The internal variable q_1 is a scalar, whereas $\boldsymbol{q}_2$ is a tensor. Since only the deviatoric part of the stress is relevant in Equation (5.89), the hydrostatic part of $\boldsymbol{q}_2$ remains arbitrary throughout the analysis.

Defining

$$\boldsymbol{\xi} := \text{dev}\ (\boldsymbol{\sigma}) + \boldsymbol{q}_2 \tag{5.90}$$

and since

$$\boldsymbol{q}_2 = \text{dev}\ (\boldsymbol{q}_2) \tag{5.91}$$

$$\frac{\partial}{\partial \boldsymbol{\xi}}\|\boldsymbol{\xi}\| = \frac{\boldsymbol{\xi}}{\|\boldsymbol{\xi}\|} \tag{5.92}$$

$$\frac{\partial}{\partial \boldsymbol{\sigma}}(\text{dev}\ (\boldsymbol{\sigma})) = \mathbb{I} - \tfrac{1}{3}\boldsymbol{I} \otimes \boldsymbol{I} \tag{5.93}$$

where

$$(\mathbb{I})^{ij}{}_{kl} := \tfrac{1}{2}(\delta^i{}_k\delta^j{}_l + \delta^i{}_l\delta^j{}_k) \tag{5.94}$$

is the fourth-order identity tensor and $\boldsymbol{I}$ is the second-order identity tensor, straightforward application of Equations (5.79) and (5.80) yields

$$\dot{\boldsymbol{\epsilon}}^{\text{p}} = \dot{\gamma}\frac{\boldsymbol{\xi}}{\|\boldsymbol{\xi}\|} \tag{5.95}$$

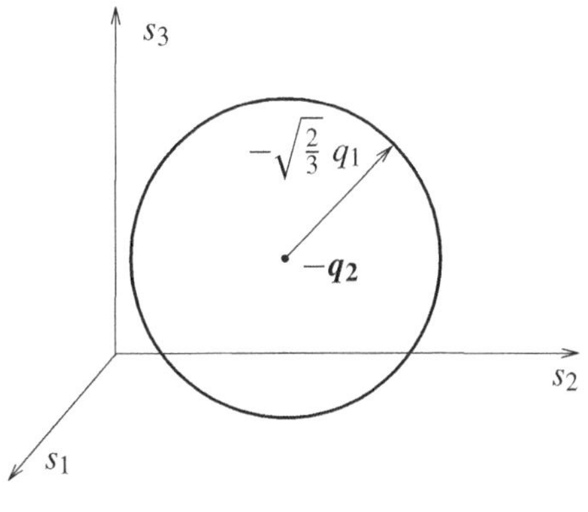

Figure 5.6 Yield surface in deviatoric stress space

$$\dot{\alpha}_1 = \dot{\gamma}\sqrt{\tfrac{2}{3}} \tag{5.96}$$

$$\dot{\boldsymbol{\alpha}}_2 = \dot{\gamma}\frac{\boldsymbol{\xi}}{\|\boldsymbol{\xi}\|}. \tag{5.97}$$

Similar to the von Mises stress, an equivalent plastic strain is defined by

$$\epsilon^{\mathrm{peq}} := \sqrt{\tfrac{2}{3}}\|\boldsymbol{\epsilon}^{\mathrm{p}}\|. \tag{5.98}$$

The factor $\sqrt{2/3}$ is introduced such that in a tensile test, the equivalent plastic strain is equal to the plastic strain in the tensile direction. Indeed, since a hydrostatic pressure does not lead to plasticity, the plastic deformation is volume preserving, and hence,

$$\epsilon_{11}^{\mathrm{p}} + \epsilon_{22}^{\mathrm{p}} + \epsilon_{33}^{\mathrm{p}} = 0. \tag{5.99}$$

Consequently, for uniaxial plastic strain, one has

$$\epsilon_{22}^{\mathrm{p}} = \epsilon_{33}^{\mathrm{p}} = -\tfrac{1}{2}\epsilon_{11}^{\mathrm{p}} \tag{5.100}$$

and

$$\epsilon^{\mathrm{peq}} = \sqrt{\tfrac{2}{3}}\sqrt{(\epsilon_{11}^{\mathrm{p}})^2 + (\epsilon_{22}^{\mathrm{p}})^2 + (\epsilon_{33}^{\mathrm{p}})^2} = \epsilon_{11}^{\mathrm{p}}. \tag{5.101}$$

From Equations (5.95) and (5.98), it is apparent that the physical meaning of $\sqrt{2/3}\,\dot{\gamma}$ is the equivalent plastic strain rate.

Finally, the relationships $q_1(\alpha_1)$ and $\boldsymbol{q}_2(\boldsymbol{\alpha}_2)$ are left to be defined. The variable $-q_1$ means the von Mises stress with respect to the reference stress $\boldsymbol{q}_2$ at yield (cf Equation (5.89)), α_1 is the accumulated plastic strain. This relationship must be obtained from experiments and will be written as

$$q_1 = -h_1(\alpha_1). \tag{5.102}$$

The second set of internal variables $\boldsymbol{q}_2$ is tensorial and the relationship

$$\boldsymbol{q}_2 = -\boldsymbol{h}_2(\boldsymbol{\alpha}_2) \tag{5.103}$$

is more difficult to obtain. Time differentiation of Equation (5.103) yields

$$\dot{\boldsymbol{q}}_2 = -\frac{\partial \boldsymbol{h}_2}{\partial \boldsymbol{\alpha}_2} : \dot{\boldsymbol{\alpha}}_2 \tag{5.104}$$

which implies that the tensor $\dot{\boldsymbol{q}}_2$ is not necessarily parallel to $\dot{\boldsymbol{\alpha}}_2$. This complicates the subsequent analysis. Since the material is isotropic, it seems convenient to assume that the kinematic hardening is also isotropic, that is, we write the relationships for Equations (5.103) and (5.104) for the equivalent properties:

$$q_2^{\mathrm{eq}} = h_2^{\mathrm{eq}}(\alpha_2^{\mathrm{eq}}) \tag{5.105}$$

leading to

$$\dot{q}_2^{\mathrm{eq}} = \frac{\partial h_2^{\mathrm{eq}}}{\partial \alpha_2^{\mathrm{eq}}}\dot{\alpha}_2^{\mathrm{eq}} \tag{5.106}$$

or

$$\sqrt{\tfrac{3}{2}}\|\dot{\boldsymbol{q}}_2\| = \frac{\partial h_2^{\mathrm{eq}}}{\partial \alpha_2^{\mathrm{eq}}}\sqrt{\tfrac{2}{3}}\|\dot{\boldsymbol{\alpha}}_2\|. \tag{5.107}$$

Equation (5.107) suggests the following isotropic tensorial relationship:

$$\dot{\boldsymbol{q}}_2 = -\tfrac{2}{3}\frac{\partial h_2^{\mathrm{eq}}}{\partial \alpha_2^{\mathrm{eq}}}\dot{\boldsymbol{\alpha}}_2. \tag{5.108}$$

Comparison with Equation (5.104) leads to

$$\frac{\partial \boldsymbol{h}_2}{\partial \boldsymbol{\alpha}_2} = \tfrac{2}{3}\frac{\partial h_2^{\mathrm{eq}}}{\partial \alpha_2^{\mathrm{eq}}}\mathbb{I}_I. \tag{5.109}$$

One finally obtains

$$\dot{q}_1 = -\dot{h}_1 \tag{5.110}$$

$$\dot{\boldsymbol{q}}_2 = -\tfrac{2}{3}\frac{\partial h_2^{\mathrm{eq}}}{\partial \alpha_2^{\mathrm{eq}}}\dot{\gamma}\frac{\boldsymbol{\xi}}{\|\boldsymbol{\xi}\|} = -\sqrt{\tfrac{2}{3}}\dot{h}_2^{\mathrm{eq}}\frac{\boldsymbol{\xi}}{\|\boldsymbol{\xi}\|} \tag{5.111}$$

since

$$\dot{\alpha}_2^{\mathrm{eq}} = \sqrt{\tfrac{2}{3}}\dot{\gamma} \tag{5.112}$$

and Equation (5.96). From Equations (5.95) to (5.97), it is obvious that

$$\epsilon^{\mathrm{peq}} = \alpha_1 = \alpha_2^{\mathrm{eq}}. \tag{5.113}$$

5.3.2 Numerical procedure

Just as in the one-dimensional case, the total strain is assumed to be given, all quantities are known at $t = t_n$, they are to be determined at $t = t_{n+1}$. Again the trial-and-error procedure is used. At first, it is assumed that no plasticity occurs:

$$\boldsymbol{\epsilon}_{n+1}^{\mathrm{p}} = \boldsymbol{\epsilon}_n^{\mathrm{p}} \tag{5.114}$$

$$q_{1,n+1} = q_{1,n} \tag{5.115}$$

$$\boldsymbol{q}_{2,n+1} = \boldsymbol{q}_{2,n} \tag{5.116}$$

$$\gamma_{n+1} = \gamma_n \tag{5.117}$$

$$\boldsymbol{\sigma}_{n+1} = \left.\frac{\partial \Sigma}{\partial \boldsymbol{\epsilon}^{\mathrm{e}}}\right|_{n+1}. \tag{5.118}$$

If

$$\|\mathrm{dev}\,(\boldsymbol{\sigma}_{n+1}) + \boldsymbol{q}_{2,n+1}\| + \sqrt{\tfrac{2}{3}}q_{1,n+1} \leq 0 \tag{5.119}$$

the solution is found. Else, the following equations have to be solved at $t = t_{n+1}$:

$$\boldsymbol{\sigma} = \frac{\partial \Sigma}{\partial \boldsymbol{\epsilon}^{\mathrm{e}}} \quad \text{(6 equations)} \tag{5.120}$$

$$\|\mathrm{dev} + \boldsymbol{q}_2\| + \sqrt{\tfrac{2}{3}} q_1 = \tfrac{2}{3} g(\dot{\epsilon}^{\mathrm{peq}}) \quad \text{(1 equation)} \tag{5.121}$$

$$\dot{\boldsymbol{\epsilon}}^{\mathrm{p}} = \dot{\gamma} \frac{\boldsymbol{\xi}}{\|\boldsymbol{\xi}\|} \quad \text{(5 equations)} \tag{5.122}$$

$$\dot{q}_1 = -\dot{h}_1 \quad \text{(1 equation)} \tag{5.123}$$

$$\dot{\boldsymbol{q}}_2 = -\sqrt{\tfrac{2}{3}} \dot{h}_2^{\mathrm{eq}} \frac{\boldsymbol{\xi}}{\|\boldsymbol{\xi}\|} \quad \text{(5 equations).} \tag{5.124}$$

These are 18 equations in 18 unknowns: $\boldsymbol{\sigma}$ (6), $\boldsymbol{\epsilon}^{\mathrm{p}}$ (5), q_1 (1), $\boldsymbol{q}_2$ (5), $\dot{\gamma}$ (1) (recall that $\boldsymbol{\epsilon}^{\mathrm{p}}$ and $\boldsymbol{q}_2$ are deviatoric in nature). If we assume that the material is isotropic in the elastic regime, the equations can be further simplified. Indeed, the elastic stress–strain relationship for a linear elastic isotropic material satisfies

$$\boldsymbol{\sigma} = \lambda \mathrm{tr}(\boldsymbol{\epsilon}^{\mathrm{e}})\boldsymbol{I} + 2\mu\boldsymbol{\epsilon}^{\mathrm{e}}. \tag{5.125}$$

Hence,

$$\boldsymbol{s} = \mathrm{dev}\ \boldsymbol{\sigma} = 2\mu \mathrm{dev}\ \boldsymbol{\epsilon}^{\mathrm{e}}. \tag{5.126}$$

Equation (5.122) shows that $\dot{\boldsymbol{\epsilon}}^{\mathrm{p}}$ is deviatoric, consequently,

$$\dot{\boldsymbol{\epsilon}}^{\mathrm{p}} = \mathrm{dev}\ (\dot{\boldsymbol{\epsilon}}^{\mathrm{p}}) = \mathrm{dev}\ (\dot{\boldsymbol{\epsilon}} - \dot{\boldsymbol{\epsilon}}^{\mathrm{e}}) = \mathrm{dev}\ (\dot{\boldsymbol{\epsilon}}) - \mathrm{dev}\ (\dot{\boldsymbol{\epsilon}}^{\mathrm{e}}) \tag{5.127}$$

and

$$2\mu \mathrm{dev}\ (\dot{\boldsymbol{\epsilon}}) - \dot{\boldsymbol{s}} = 2\mu\dot{\gamma} \frac{\boldsymbol{\xi}}{\|\boldsymbol{\xi}\|}. \tag{5.128}$$

Replacing the time derivatives by backward Euler differences and defining

$$\Delta\boldsymbol{\epsilon}_{n+1} := \boldsymbol{\epsilon}_{n+1} - \boldsymbol{\epsilon}_n \tag{5.129}$$

and similarly for the other expressions, one obtains

$$2\mu \mathrm{dev}\ (\Delta\boldsymbol{\epsilon}_{n+1}) - \boldsymbol{s}_{n+1} + \boldsymbol{s}_n = 2\mu\Delta\gamma_{n+1} \frac{\boldsymbol{\xi}_{n+1}}{\|\boldsymbol{\xi}_{n+1}\|}. \tag{5.130}$$

Now, $\boldsymbol{s}_{n+1}^{\mathrm{trial}}$ is obtained from $\boldsymbol{s}_n$ by assuming that $\Delta\epsilon$ is purely elastic, that is,

$$\boldsymbol{s}_{n+1}^{\mathrm{trial}} = \boldsymbol{s}_n + 2\mu \mathrm{dev}\ (\Delta\boldsymbol{\epsilon}_{n+1}) \tag{5.131}$$

which leads to

$$\boldsymbol{s}_{n+1}^{\mathrm{trial}} - \boldsymbol{s}_{n+1} = 2\mu\Delta\gamma_{n+1} \frac{\boldsymbol{\xi}_{n+1}}{\|\boldsymbol{\xi}_{n+1}\|} \tag{5.132}$$

for Equation (5.130). Backward Euler for Equation (5.124) yields

$$\boldsymbol{q}_{2,n+1} - \boldsymbol{q}_{2,n} = -\sqrt{\tfrac{2}{3}}\left[h_2^{\text{eq}}(\epsilon_{n+1}^{\text{peq}}) - h_2^{\text{eq}}(\epsilon_n^{\text{peq}})\right]\frac{\boldsymbol{\xi}_{n+1}}{\|\boldsymbol{\xi}_{n+1}\|}. \tag{5.133}$$

Subtracting Equation (5.132) from Equation (5.133), one gets

$$\boldsymbol{\xi}_{n+1} - \boldsymbol{\xi}_{n+1}^{\text{trial}} = \left[-2\mu\Delta\gamma_{n+1} - \sqrt{\tfrac{2}{3}}\Delta h_{2,n+1}^{\text{eq}}\right]\frac{\boldsymbol{\xi}_{n+1}}{\|\boldsymbol{\xi}_{n+1}\|} \tag{5.134}$$

where

$$\boldsymbol{\xi}_{n+1}^{\text{trial}} := \boldsymbol{s}_{n+1}^{\text{trial}} + \boldsymbol{q}_{2,n}. \tag{5.135}$$

Equation (5.134) shows that the vectors $\boldsymbol{\xi}_{n+1}$ and $\boldsymbol{\xi}_{n+1}^{\text{trial}}$ are parallel (therefore, the algorithm is sometimes called the *radial return method*). This result is crucial in the present derivation. If the kinematic hardening had not been isotropic, this simplification would not apply! Since all terms in Equation (5.134) are parallel, the equation applies to their size equally well:

$$\|\boldsymbol{\xi}_{n+1}\| - \|\boldsymbol{\xi}_{n+1}^{\text{trial}}\| = -2\mu\Delta\gamma_{n+1} - \sqrt{\tfrac{2}{3}}\Delta h_{2,n+1}^{\text{eq}}. \tag{5.136}$$

There is only one equation left to be satisfied: the yield condition, which reads

$$\|\boldsymbol{\xi}_{n+1}\| + \sqrt{\tfrac{2}{3}}q_{1,n+1} = \sqrt{\tfrac{2}{3}}g(\Delta\epsilon_{n+1}^{\text{peq}}). \tag{5.137}$$

One finds

$$h_{1,n+1} = h_1(\epsilon_n^{\text{peq}} + \Delta\epsilon_{n+1}^{\text{peq}}) = h_1\left(\epsilon_n^{\text{peq}} + \sqrt{\tfrac{2}{3}}\Delta\gamma_{n+1}\right). \tag{5.138}$$

Finally, one gets for the yield condition

$$\|\boldsymbol{\xi}_{n+1}^{\text{trial}}\| - \left\{2\mu\Delta\gamma_{n+1} + \sqrt{\tfrac{2}{3}}\left[h_2^{\text{eq}}\left(\epsilon_n^{\text{peq}} + \sqrt{\tfrac{2}{3}}\Delta\gamma_{n+1}\right) - h_2^{\text{eq}}(\epsilon_n^{\text{peq}})\right]\right\}$$
$$- \sqrt{\tfrac{2}{3}}h_1\left(\epsilon_n^{\text{peq}} + \sqrt{\tfrac{2}{3}}\Delta\gamma_{n+1}\right) = \sqrt{\tfrac{2}{3}}g\left(\sqrt{\tfrac{2}{3}}\Delta\gamma_{n+1}\right). \tag{5.139}$$

Consequently, we finally arrive at one nonlinear equation in $\Delta\gamma_{n+1}$. This equation can be solved using a Newton–Raphson technique. Denoting the initial value for the unknown $\Delta\gamma_{n+1}$ by $\Delta\gamma_{n+1}^{(0)}$ and writing

$$\Delta\gamma_{n+1}^{(k+1)} = \Delta\gamma_{n+1}^{(k)} + \Delta\Delta\gamma_{n+1}^{(k)} \tag{5.140}$$

linearization of Equation (5.139) about $\Delta\gamma_{n+1}^{(k)}$ yields

$$\|\boldsymbol{\xi}_{n+1}^{\text{trial}}\| - \left\{2\mu\Delta\gamma_{n+1}^{(k)} + \sqrt{\tfrac{2}{3}}\left[h_2^{\text{eq}}\left(\epsilon_n^{\text{peq}} + \sqrt{\tfrac{2}{3}}\Delta\gamma_{n+1}^{(k)}\right) - h_2^{\text{eq}}(\epsilon_n^{\text{peq}})\right]\right\}$$
$$- \sqrt{\tfrac{2}{3}}h_1\left(\epsilon_n^{\text{peq}} + \sqrt{\tfrac{2}{3}}\Delta\gamma_{n+1}^{(k)}\right) - \sqrt{\tfrac{2}{3}}g\left(\sqrt{\tfrac{2}{3}}\Delta\gamma_{n+1}^{(k)}\right)$$
$$- \left[2\mu + \tfrac{2}{3}\left.\frac{\partial(h_1 + h_2^{\text{eq}})}{\partial\epsilon^{\text{peq}}}\right|_{\epsilon_n^{\text{peq}} + \sqrt{\frac{2}{3}}\Delta\gamma_{n+1}^{(k)}} + \tfrac{2}{3}\left.\frac{\partial g}{\partial\Delta\epsilon^{\text{peq}}}\right|_{\sqrt{\frac{2}{3}}\Delta\gamma_{n+1}^{(k)}}\right]\Delta\Delta\gamma_{n+1}^{(k)} = 0. \tag{5.141}$$

Once $\Delta\gamma_{n+1}$ is known, one finds for the other variables

$$\epsilon_{n+1}^{\text{peq}} = \epsilon_n^{\text{peq}} + \sqrt{\tfrac{2}{3}}\Delta\gamma_{n+1} \tag{5.142}$$

$$q_{1,n+1} = -h_1(\epsilon_{n+1}^{\text{peq}}) \tag{5.143}$$

$$\boldsymbol{q}_{2,n+1} = \boldsymbol{q}_{2,n} - \sqrt{\tfrac{2}{3}}\left[h_2^{\text{eq}}(\epsilon_{n+1}^{\text{peq}}) - h_2^{\text{eq}}(\epsilon_n^{\text{peq}})\right]\frac{\boldsymbol{\xi}_{n+1}^{\text{trial}}}{\|\boldsymbol{\xi}_{n+1}^{\text{trial}}\|} \tag{5.144}$$

$$\boldsymbol{\epsilon}_{n+1}^{\text{p}} = \boldsymbol{\epsilon}_n^{\text{p}} + \Delta\gamma_{n+1}\frac{\boldsymbol{\xi}_{n+1}^{\text{trial}}}{\|\boldsymbol{\xi}_{n+1}^{\text{trial}}\|} \tag{5.145}$$

$$\boldsymbol{\sigma}_{n+1} = \left.\frac{\partial\Sigma}{\partial\boldsymbol{\epsilon}^{\text{e}}}\right|_{n+1}. \tag{5.146}$$

5.3.3 Determination of the consistent elastoplastic tangent matrix

The consistent elastoplastic tangent matrix is the instantaneous slope of the stress–total strain relationship. The term "consistent" points to the fact that the slope has to be determined for the actual numerical scheme used, that is, it depends on the numerical procedure (using another scheme, e.g. the midpoint rule instead of backward Euler, will lead to another slope). It is well known (Simo and Hughes 1997) that the slope derived for the present numerical scheme deviates from the continuum tangent. Consistency of the slope is a prerequisite for the quadratic convergence of the Newton–Raphson scheme.

For materials that are linear in the elastic range, Equation (5.146) reduces to

$$\boldsymbol{\sigma}_{n+1} = \boldsymbol{C} : \boldsymbol{\epsilon}_{n+1}^{\text{e}} = \boldsymbol{C} : (\boldsymbol{\epsilon}_{n+1} - \boldsymbol{\epsilon}_{n+1}^{\text{p}}) \tag{5.147}$$

where

$$\boldsymbol{C} := \frac{\partial^2\Sigma}{\partial\boldsymbol{\epsilon}^{\text{e}}\partial\boldsymbol{\epsilon}^{\text{e}}}. \tag{5.148}$$

Substituting Equation (5.145), one now arrives at the following stress–strain relationship:

$$\boldsymbol{\sigma}_{n+1} = \boldsymbol{C} : (\boldsymbol{\epsilon}_{n+1} - \boldsymbol{\epsilon}_n^{\text{p}} - \Delta\gamma_{n+1}\boldsymbol{n}_{n+1}) \tag{5.149}$$

where

$$\boldsymbol{n}_{n+1} := \frac{\boldsymbol{\xi}_{n+1}}{\|\boldsymbol{\xi}_{n+1}\|} = \frac{\boldsymbol{\xi}_{n+1}^{\text{trial}}}{\|\boldsymbol{\xi}_{n+1}^{\text{trial}}\|}. \tag{5.150}$$

Thus, the tangent relation at $t = t_{n+1}$ takes the form

$$\text{d}\boldsymbol{\sigma}_{n+1} = \boldsymbol{C} : (\text{d}\boldsymbol{\epsilon}_{n+1} - \text{d}\Delta\gamma_{n+1}\boldsymbol{n}_{n+1} - \Delta\gamma_{n+1}\,\text{d}\boldsymbol{n}_{n+1}). \tag{5.151}$$

For an isotropic elastic material, $\boldsymbol{C}$ amounts to

$$\boldsymbol{C} = 2\mu\mathbb{I} + \lambda\boldsymbol{I}\otimes\boldsymbol{I} \tag{5.152}$$

where λ, μ are Lamé's constants. Since $\boldsymbol{n}$ is deviatoric $\boldsymbol{I} : \boldsymbol{n} = 0$, one can write

$$\boldsymbol{C} : \boldsymbol{n} = 2\mu\boldsymbol{n} \tag{5.153}$$

and Equation (5.151) reduces to

$$\mathrm{d}\boldsymbol{\sigma}_{n+1} = \boldsymbol{C} : \mathrm{d}\boldsymbol{\epsilon}_{n+1} - 2\mu(\mathrm{d}\Delta\gamma_{n+1}\boldsymbol{n}_{n+1} + \Delta\gamma_{n+1}\,\mathrm{d}\boldsymbol{n}_{n+1}) \tag{5.154}$$

or

$$\mathrm{d}\boldsymbol{\sigma}_{n+1} = \left[\boldsymbol{C} - 2\mu\left(\boldsymbol{n}_{n+1} \otimes \frac{\partial\Delta\gamma_{n+1}}{\partial\boldsymbol{\epsilon}_{n+1}} + \Delta\gamma_{n+1}\frac{\partial\boldsymbol{n}_{n+1}}{\partial\boldsymbol{\epsilon}_{n+1}}\right)\right] : \mathrm{d}\boldsymbol{\epsilon}_{n+1}. \tag{5.155}$$

The consistent elastoplastic tangent is the expression in square braces. To determine $\partial\Delta\gamma_{n+1}/\partial\boldsymbol{\epsilon}_{n+1}$, Equation (5.139) is differentiated with respect to $\boldsymbol{\epsilon}_{n+1}$:

$$\frac{\partial\|\boldsymbol{\xi}^{\text{trial}}_{n+1}\|}{\partial\boldsymbol{\epsilon}_{n+1}} - \left. 2\mu + \tfrac{2}{3}\partial_{\epsilon^{\text{peq}}_{n+1}} h^{\text{eq}}_2 + \tfrac{2}{3}\partial_{\epsilon^{\text{peq}}_{n+1}} h_1 + \tfrac{2}{3}\partial_{\Delta\epsilon^{\text{peq}}_{n+1}} g\right)\frac{\partial\Delta\gamma_{n+1}}{\partial\boldsymbol{\epsilon}_{n+1}} = 0. \tag{5.156}$$

Since

$$\frac{\partial\|\boldsymbol{\xi}^{\text{trial}}_{n+1}\|}{\partial\boldsymbol{\epsilon}_{n+1}} = \frac{\partial\|\boldsymbol{\xi}^{\text{trial}}_{n+1}\|}{\partial\boldsymbol{\xi}^{\text{trial}}_{n+1}} : \frac{\partial\boldsymbol{\xi}^{\text{trial}}_{n+1}}{\partial\boldsymbol{\epsilon}_{n+1}} \tag{5.157}$$

and (combining Equation (5.135) with Equation (5.131))

$$\boldsymbol{\xi}^{\text{trial}}_{n+1} = \boldsymbol{s}_n + 2\mu\text{dev}\ (\boldsymbol{\epsilon}_{n+1} - \boldsymbol{\epsilon}_n) + \boldsymbol{q}_{2,n} \tag{5.158}$$

one obtains (Equations (5.92) and (5.93))

$$\frac{\partial\|\boldsymbol{\xi}^{\text{trial}}_{n+1}\|}{\partial\boldsymbol{\epsilon}_{n+1}} = \boldsymbol{n}_{n+1} : 2\mu\ \left.\mathbb{I} - \tfrac{1}{3}\boldsymbol{I} \otimes \boldsymbol{I}\right) = 2\mu\boldsymbol{n}_{n+1}. \tag{5.159}$$

Consequently,

$$\frac{\partial\Delta\gamma_{n+1}}{\partial\boldsymbol{\epsilon}_{n+1}} = \left. 2\mu + \tfrac{2}{3}\partial_{\epsilon^{\text{peq}}_{n+1}} h^{\text{eq}}_2 + \tfrac{2}{3}\partial_{\epsilon^{\text{peq}}_{n+1}} h_1 + \tfrac{2}{3}\partial_{\Delta\epsilon^{\text{peq}}_{n+1}} g\right)^{-1} 2\mu\boldsymbol{n}_{n+1}. \tag{5.160}$$

The derivative in the last term of Equation (5.155) yields

$$\begin{aligned}
\frac{\partial\boldsymbol{n}_{n+1}}{\partial\boldsymbol{\epsilon}_{n+1}} &= \frac{\partial\boldsymbol{n}_{n+1}}{\partial\boldsymbol{\xi}^{\text{trial}}_{n+1}} : \frac{\partial\boldsymbol{\xi}^{\text{trial}}_{n+1}}{\partial\boldsymbol{\epsilon}_{n+1}} \\
&= \left(\frac{1}{\|\boldsymbol{\xi}^{\text{trial}}_{n+1}\|}\mathbb{I} - \frac{1}{\|\boldsymbol{\xi}^{\text{trial}}_{n+1}\|^2}\boldsymbol{\xi}^{\text{trial}}_{n+1} \otimes \frac{\boldsymbol{\xi}^{\text{trial}}_{n+1}}{\|\boldsymbol{\xi}^{\text{trial}}_{n+1}\|}\right) : 2\mu\ \left.\mathbb{I} - \tfrac{1}{3}\boldsymbol{I} \otimes \boldsymbol{I}\right) \\
&= \frac{2\mu}{\|\boldsymbol{\xi}^{\text{trial}}_{n+1}\|}(\mathbb{I} - \boldsymbol{n}_{n+1} \otimes \boldsymbol{n}_{n+1}) :\ \left.\mathbb{I} - \tfrac{1}{3}\boldsymbol{I} \otimes \boldsymbol{I}\right) \\
&= \frac{2\mu}{\|\boldsymbol{\xi}^{\text{trial}}_{n+1}\|}\ \left.\mathbb{I} - \tfrac{1}{3}\boldsymbol{I} \otimes \boldsymbol{I} - \boldsymbol{n}_{n+1} \otimes \boldsymbol{n}_{n+1}\right).
\end{aligned} \tag{5.161}$$

Substituting Equations (5.160) and (5.161) into Equation (5.155) finally yields

$$\boldsymbol{C}^{\text{ep}} = \boldsymbol{C} - \frac{(2\mu)^2 \Delta\gamma_{n+1}}{\|\boldsymbol{\xi}_{n+1}^{\text{trial}}\|} \quad \mathbb{I} - \tfrac{1}{3}\boldsymbol{I} \otimes \boldsymbol{I} - \boldsymbol{n}_{n+1} \otimes \boldsymbol{n}_{n+1}\Big)$$

$$- (2\mu)^2 \boldsymbol{n}_{n+1} \otimes \boldsymbol{n}_{n+1} \quad 2\mu + \tfrac{2}{3}\partial_{\epsilon_{n+1}^{\text{peq}}} h_2^{\text{eq}} + \tfrac{2}{3}\partial_{\epsilon_{n+1}^{\text{peq}}} h_1 + \tfrac{2}{3}\partial_{\Delta\epsilon_{n+1}^{\text{peq}}} g\Big)^{-1}. \tag{5.162}$$

It can be shown that for $\Delta\gamma_{n+1} = 0$, the continuum elastoplastic tangent is obtained (Simo and Hughes 1997). Accordingly, because of the finite size of the increments, the consistent numerical tangent deviates from the continuum tangent.

5.4 Three-dimensional Multisurface Viscoplasticity: the Cailletaud Single Crystal Model

Single crystals are advanced metallic materials consisting of just one crystal. Substantial progress made in the last two decades in casting technology enables the manufacturers to control crystal growth in the liquid metal by carefully monitoring the cooling conditions. Thus, a highly anisotropic material ensues, in contrast with the usual metallic materials (polycrystals), in which the different orientations of the many crystals assure the isotropic properties. In this section, the focus will be on nickel-base alloys, exhibiting a face cube centered (FCC) crystal structure. A good reference on crystalline plasticity is (Havner 1992).

5.4.1 Theoretical considerations

In single crystals, viscoplasticity is mainly due to a crystallographic dislocation slip. Other mechanisms will not be considered here. The crystallographic slip planes and directions for nickel-base superalloys at high temperature are known (Méric *et al.* 1991) and can be divided into octahedral slip systems (12 systems consisting of 4 $\{111\}$ planes with 3 $< 011 >$ directions per plane, Figure 5.7) and cubic slip systems (6 systems consisting of 3 $\{001\}$ planes with 2 $< 011 >$ directions per plane, Figure 5.8).

Accordingly, the slip planes and directions are explicitly known and are generally denoted by their normal $\boldsymbol{n}^\beta$ and unit vector $\boldsymbol{l}^\beta$ respectively. Here, β stands for any of the 18 slip systems. For each slip system, an orientation tensor $\boldsymbol{m}^\beta$ is defined by

$$\boldsymbol{m}^\beta := \tfrac{1}{2}(\boldsymbol{n}^\beta \otimes \boldsymbol{l}^\beta + \boldsymbol{l}^\beta \otimes \boldsymbol{n}^\beta). \tag{5.163}$$

Whether dislocations move along a slip system basically depends on the shear stress component τ^β in the slip direction:

$$\tau^\beta = \boldsymbol{m}^\beta : \boldsymbol{\sigma} = (\boldsymbol{n}^\beta \otimes \boldsymbol{l}^\beta) : \boldsymbol{\sigma} = \boldsymbol{n}^{\beta T} \cdot \boldsymbol{\sigma} \cdot \boldsymbol{l}^\beta. \tag{5.164}$$

This is really a one-dimensional system and we can fall back on yield-surface formulations such as in Equation (5.41):

$$f^\beta(\boldsymbol{\sigma}, \boldsymbol{q}) = |\tau^\beta + q_2^\beta| + q_1^\beta \leq 0. \tag{5.165}$$

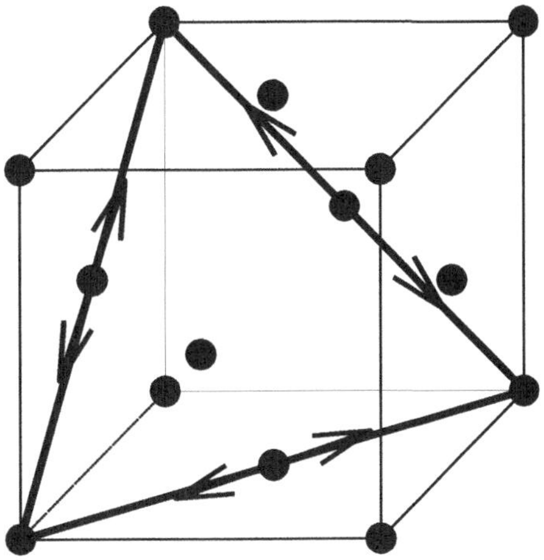

Figure 5.7 Octahedral slip systems

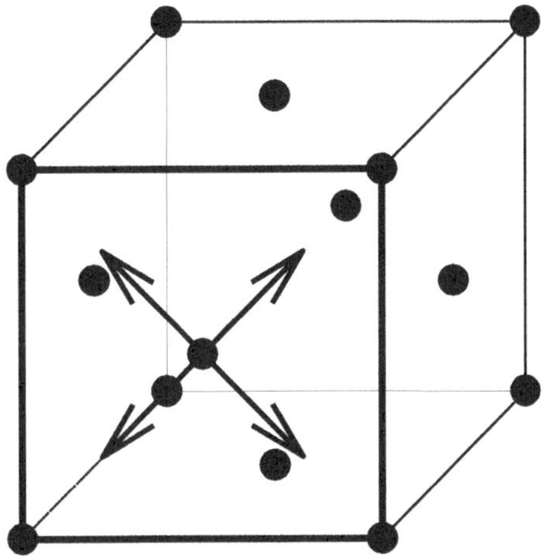

Figure 5.8 Cubic slip systems

However, now we are dealing with 18 yield surfaces at the same time, which may intersect each other at so-called corner points. This is an example of multisurface viscoplasticity. The underlying theory has been developed in the 1950s and 1960s (Koiter 1960), and is a straightforward extension of the one-dimensional derivation in Section 5.2. The governing equations for m slip systems are

1. Elastic stress–strain relations

$$\boldsymbol{\sigma} = \frac{\partial \Sigma}{\partial \boldsymbol{\epsilon}^{\mathrm{e}}} \tag{5.166}$$

2. Internal variable relationships

$$\boldsymbol{q} = -\boldsymbol{h}(\boldsymbol{\alpha}) \tag{5.167}$$

3. Yield surfaces

$$f^{\beta}(\boldsymbol{\sigma}, \boldsymbol{q}) = 0 \tag{5.168}$$

4. Evolution equations

$$\dot{\boldsymbol{\epsilon}}^{\mathrm{p}} = \sum_{\beta=1}^{m} \dot{\gamma}^{\beta} \frac{\partial f^{\beta}(\boldsymbol{\sigma}, \boldsymbol{q})}{\partial \boldsymbol{\sigma}} \tag{5.169}$$

$$\dot{\boldsymbol{\alpha}} = \sum_{\beta=1}^{m} \dot{\gamma}^{\beta} \frac{\partial f^{\beta}(\boldsymbol{\sigma}, \boldsymbol{q})}{\partial \boldsymbol{q}} \tag{5.170}$$

5. Kuhn–Tucker conditions

$$\dot{\gamma}^{\beta} \geq 0, \quad f^{\beta}(\boldsymbol{\sigma}, \boldsymbol{q}) \leq 0, \quad \dot{\gamma}^{\beta} f^{\beta}(\boldsymbol{\sigma}, \boldsymbol{q}) = 0 \tag{5.171}$$

6. Consistency conditions

$$\dot{\gamma}^{\beta} \dot{f}^{\beta}(\boldsymbol{\sigma}, \boldsymbol{q}) = 0. \tag{5.172}$$

$\boldsymbol{q}$ is a vector of internal variables with a size that is usually a multiple of the number of slip systems. In Equations (5.169) and (5.170), $\dot{\gamma}^{\beta}$ is only nonzero for the active slip systems. In what follows, we will concentrate on a particular single crystal viscoplastic model developed by Georges Cailletaud and coworkers, (Méric *et al.* 1991), (Méric and Cailletaud 1991). For other single crystal models, see (Fedelich 2002) and (Meissonnier *et al.* 2001). The Cailletaud model does not completely fit into the theory described by Equations (5.166) to (5.171) because of the following two aspects:

1. The model is not associative, which means that the evolution equations are derived from a function that differs from the yield surface.
2. The evolution equation, Equation (5.170), is modified by a term depending on the total accumulated plasticity.

Denoting the yield surfaces by h^β (not to confuse with h in Equation (5.167)), the extra term for $\dot{\boldsymbol{\alpha}}$ by $\boldsymbol{w}$ and the creep term (viscous term) by $g^\beta(\dot{\gamma}^\beta)$, one obtains

1. Elastic stress–strain relations

$$\boldsymbol{\sigma} = \frac{\partial \Sigma}{\partial \boldsymbol{\epsilon}^{\mathrm{e}}} \tag{5.173}$$

2. Internal variable relationships

$$\boldsymbol{q} = -\boldsymbol{h}(\boldsymbol{\alpha}) \tag{5.174}$$

3. Yield surfaces

$$h^\beta(\boldsymbol{\sigma}, \boldsymbol{q}) = g^\beta(\dot{\gamma}^\beta) \tag{5.175}$$

4. Evolution equations

$$\dot{\boldsymbol{\epsilon}}^{\mathrm{p}} = \sum_{\beta=1}^{m} \dot{\gamma}^\beta \frac{\partial f^\beta(\boldsymbol{\sigma}, \boldsymbol{q})}{\partial \boldsymbol{\sigma}} \tag{5.176}$$

$$\dot{\boldsymbol{\alpha}} = \sum_{\beta=1}^{m} \dot{\gamma}^\beta \left[\frac{\partial f^\beta(\boldsymbol{\sigma}, \boldsymbol{q})}{\partial \boldsymbol{q}} + \boldsymbol{w}^\beta \left(\int_0^t \dot{\gamma}^\beta \, \mathrm{d}t \right) \right] \tag{5.177}$$

5. Kuhn–Tucker conditions

$$\dot{\gamma}^\beta \geq 0, \quad h^\beta(\boldsymbol{\sigma}, \boldsymbol{q}) \leq 0, \quad \gamma^\beta h^\beta(\boldsymbol{\sigma}, \boldsymbol{q}) = 0 \tag{5.178}$$

6. Consistency conditions

$$\dot{\gamma}^\beta \dot{h}^\beta(\boldsymbol{\sigma}, \boldsymbol{q}) = 0. \tag{5.179}$$

Specifically, in the Cailletaud model, there are $2m$ internal dynamic variables, which will be denoted by $q_1^\beta, q_2^\beta, \beta = 1, \ldots, m$. The internal variable relationships take the form

$$q_1^\beta = -b^\beta Q^\beta \alpha_1^\beta \tag{5.180}$$

$$q_2^\beta = -c^\beta \alpha_2^\beta \tag{5.181}$$

where b^β, Q^β and $c^\beta, \beta = 1, \ldots, m$ are constants. Consequently, Equations (5.180) and (5.181) are linear. The yield surfaces are defined by

$$h^\beta(\boldsymbol{\sigma}, \boldsymbol{q}) := \left| \tau^\beta + q_2^\beta \right| - r_0^\beta + \sum_{\alpha=1}^{m} H_{\beta\alpha} q_1^\alpha. \tag{5.182}$$

The parameters r_0^β are the initial yield values and $H_{\beta\alpha}$ are the interaction coefficients between the slip systems. Equation (5.182) is a slightly more complicated form than Equation (5.165). The potential function for the evolution equations reads

$$f^\beta(\boldsymbol{\sigma}, \boldsymbol{q}) := \left| \tau^\beta + q_2^\beta \right| + q_1^\beta + \frac{d^\beta}{2c^\beta}(q_2^\beta)^2 + \frac{1}{2Q^\beta}(q_1^\beta)^2. \tag{5.183}$$

This is the yield function in Equation (5.165), augmented by quadratic terms in q_1^β and q_2^β. The parameters d^β, $\beta = 1, \dots, m$ are constants, c^β and Q^β already appeared in the internal variable relationships. The only functions left are g^β and $\boldsymbol{w}^\beta$. They will be specified in the next section.

5.4.2 Numerical aspects

The numerical procedure to solve Equations (5.173) to (5.179) is totally similar to the methods treated in the previous sections. The two basic considerations in the analysis are

1. The procedure is strain-driven, that is, we start from a given increment $\Delta\epsilon_{n+1} = \epsilon_{n+1} - \epsilon_n$ and look for the corresponding stress and tangent-stiffness matrix.
2. A trial-and-error procedure is used, starting from the assumption that the step is purely elastic. A verification of the yield condition tells us whether this assumption is right.

Consequently, we assume in step $n+1$

$$\boldsymbol{\alpha}_{n+1} = \boldsymbol{\alpha}_n \tag{5.184}$$

$$\boldsymbol{\epsilon}^{\mathrm{p}}_{n+1} = \boldsymbol{\epsilon}^{\mathrm{p}} \tag{5.185}$$

$$\gamma^\beta_{n+1} = \gamma^\beta_n, \quad \beta = 1, \dots, m. \tag{5.186}$$

The stress is obtained from Equation (5.173). Now, the yield surfaces are verified. If

$$h^\beta(\boldsymbol{\sigma}, \boldsymbol{q}) \le 0, \quad \forall\, \beta \in \{1, \dots, m\} \tag{5.187}$$

then the step is elastic and the solution is found. Equation (5.173) yields the stress, the tangent-stiffness matrix is the elasticity tensor. If

$$B^{(0)}_{\mathrm{act}} := \{\beta | h^\beta(\boldsymbol{\sigma}, \boldsymbol{q}) > 0\} \ne \emptyset \tag{5.188}$$

plastic flow takes place. $B^{(0)}_{\mathrm{act}}$ is the initial set of active slip planes. Equations (5.173) and (5.174) lead to

$$\dot{\boldsymbol{\sigma}} = \boldsymbol{C} : (\dot{\boldsymbol{\epsilon}} - \dot{\boldsymbol{\epsilon}}^{\mathrm{p}}) \tag{5.189}$$

$$\dot{\boldsymbol{q}} = -\frac{\partial \boldsymbol{h}}{\partial \boldsymbol{\alpha}} : \dot{\boldsymbol{\alpha}} =: -\boldsymbol{D} : \dot{\boldsymbol{\alpha}}. \tag{5.190}$$

For the Cailletaud model, $\boldsymbol{D}$ is a constant matrix. Using a backward Euler scheme, Equations (5.189) and (5.190) and Equations (5.175) to (5.177) can be rewritten as

$$\Delta\boldsymbol{\sigma}_{n+1} = \boldsymbol{C}_{n+1} : (\Delta\boldsymbol{\epsilon}_{n+1} - \Delta\boldsymbol{\epsilon}^{\mathrm{p}}_{n+1}) \tag{5.191}$$

$$\Delta\boldsymbol{q}_{n+1} = -\boldsymbol{D}_{n+1} : \Delta\boldsymbol{\alpha}_{n+1} \tag{5.192}$$

$$h^\beta(\boldsymbol{\sigma}_{n+1}, \boldsymbol{q}_{n+1}) = g^\beta(\Delta\gamma^\beta_{n+1}), \quad \beta \in B^{(0)}_{\mathrm{act}} \tag{5.193}$$

$$\Delta\boldsymbol{\epsilon}^{\mathrm{p}}_{n+1} = \sum_{\beta \in B^{(0)}_{\mathrm{act}}} \Delta\gamma^\beta_{n+1} \partial_\sigma f^\beta(\boldsymbol{\sigma}_{n+1}, \boldsymbol{q}_{n+1}) \tag{5.194}$$

$$\Delta \boldsymbol{\alpha}_{n+1} = \sum_{\beta \in B_{\text{act}}^{(0)}} \Delta \gamma_{n+1}^{\beta} \partial_{\boldsymbol{q}} f^{\beta}(\boldsymbol{\sigma}_{n+1}, \boldsymbol{q}_{n+1}). \tag{5.195}$$

The abbreviations

$$\partial_{\sigma} f^{\beta} := \frac{\partial f^{\beta}}{\partial \boldsymbol{\sigma}} \tag{5.196}$$

and likewise for the derivative with respect to $\boldsymbol{q}$ were used, and the functions $\boldsymbol{w}^{\beta}$ were dropped for now. Notice that only the active slip planes are considered in Equations (5.193) to (5.195)! Substituting Equations (5.191) and (5.192) into Equations (5.193) to (5.195) one finally obtains

$$h^{\beta}(\boldsymbol{\sigma}_{n+1}, \boldsymbol{q}_{n+1}) = g^{\beta}(\Delta \gamma_{n+1}^{\beta}), \quad \beta \in B_{\text{act}}^{(0)} \tag{5.197}$$

$$\Delta \boldsymbol{\epsilon}_{n+1} - \boldsymbol{C}_{n+1}^{-1} : \Delta \boldsymbol{\sigma}_{n+1} = \sum_{\beta \in B_{\text{act}}^{(0)}} \Delta \gamma_{n+1}^{\beta} \partial_{\sigma} f^{\beta}(\boldsymbol{\sigma}_{n+1}, \boldsymbol{q}_{n+1}) \tag{5.198}$$

$$-\boldsymbol{D}_{n+1}^{-1} : \Delta \boldsymbol{q}_{n+1} = \sum_{\beta \in B_{\text{act}}^{(0)}} \Delta \gamma_{n+1}^{\beta} \partial_{\boldsymbol{q}} f^{\beta}(\boldsymbol{\sigma}_{n+1}, \boldsymbol{q}_{n+1}). \tag{5.199}$$

If m_{act} is the number of active slip planes, Equation (5.197) represents m_{act} equations, Equation (5.198) represents 6 equations and Equation (5.199) represents $2 \times m_{\text{act}}$ equations in the unknowns $\boldsymbol{\sigma}_{n+1}$ (6), $\boldsymbol{q}_{n+1}$ ($2 \times m_{\text{act}}$) and $\Delta \gamma_{n+1}^{\beta}$ (m_{act}). Hence, we obtain $3 \times m_{\text{act}} + 6$ equations in $3 \times m_{\text{act}} + 6$ unknowns. For the inactive slip planes, Equations (5.184) to (5.186) apply. Equations (5.197) to (5.199) are the basis for our further consideration.

5.4.3 Stress update algorithm

The stress can be determined by solving Equations (5.197) to (5.199) for $\Delta \boldsymbol{\sigma}_{n+1}$. Since these equations are nonlinear, a Newton–Raphson iterative technique is used for their solution (cf Section 3.1). Assume that we have an intermediate solution denoted by a superscript (k). To obtain a better approximation, the Equations (5.197) to (5.199) are linearized at the solution (k) and solved. Denoting

$$h_{n+1}^{\beta} := h^{\beta}(\boldsymbol{\sigma}_{n+1}, \boldsymbol{q}_{n+1}) \tag{5.200}$$

and similarly for f and g, linearization yields

$$\left[h_{n+1}^{\beta(k)} - g_{n+1}^{(k)}\right] + \partial_{\sigma} h_{n+1}^{\beta(k)} : \Delta \boldsymbol{\sigma}_{n+1}^{(k)} + \partial_{\boldsymbol{q}} h_{n+1}^{\beta(k)} : \Delta \boldsymbol{q}_{n+1}^{(k)} - \partial_{\Delta\gamma} g_{n+1}^{\beta(k)} \Delta\Delta\gamma_{n+1}^{\beta(k)} = 0 \tag{5.201}$$

$$\begin{aligned}
&\left[-\boldsymbol{\epsilon}_{n+1}^{\text{p}} + \boldsymbol{\epsilon}_{n}^{\text{p}} + \sum_{\beta \in B_{\text{act}}^{(k)}} \Delta \gamma_{n+1}^{\beta} \partial_{\sigma} f_{n+1}^{\beta}\right]^{(k)} + \boldsymbol{C}_{n+1}^{-1(k)} : \Delta \boldsymbol{\sigma}_{n+1}^{(k)} \\
&\quad + \sum_{\beta \in B_{\text{act}}^{(k)}} \Delta\Delta \gamma_{n+1}^{\beta(k)} \partial_{\sigma} f_{n+1}^{\beta(k)} \\
&+ \sum_{\beta \in B_{\text{act}}^{(k)}} \Delta \gamma_{n+1}^{\beta(k)} \left[\partial_{\sigma\sigma}^{2} f_{n+1}^{\beta} : \Delta \boldsymbol{\sigma}_{n+1} + \partial_{\sigma \boldsymbol{q}}^{2} f_{n+1}^{\beta} : \Delta \boldsymbol{q}_{n+1}\right]^{(k)} = 0
\end{aligned} \tag{5.202}$$

$$\left[-\boldsymbol{\alpha}_{n+1}+\boldsymbol{\alpha}_n+\sum_{\beta\in B_{\text{act}}^{(k)}}\Delta\gamma_{n+1}^{\beta}\partial_{\boldsymbol{q}}f_{n+1}^{\beta}\right]^{(k)}+\boldsymbol{D}_{n+1}^{-1(k)}:\Delta\boldsymbol{q}_{n+1}^{(k)}$$
$$+\sum_{\beta\in B_{\text{act}}^{(k)}}\Delta\Delta\gamma_{n+1}^{\beta(k)}\partial_{\boldsymbol{q}}f_{n+1}^{\beta(k)}$$
$$+\sum_{\beta\in B_{\text{act}}^{(k)}}\Delta\gamma_{n+1}^{\beta(k)}\left[\partial_{\boldsymbol{q\sigma}}^2 f_{n+1}^{\beta}:\Delta\boldsymbol{\sigma}_{n+1}+\partial_{\boldsymbol{qq}}^2 f_{n+1}^{\beta}:\Delta\boldsymbol{q}_{n+1}\right]^{(k)}=0 \tag{5.203}$$

where

$$\Delta\boldsymbol{\sigma}_{n+1}^{(k)}:=\boldsymbol{\sigma}_{n+1}^{(k+1)}-\boldsymbol{\sigma}_{n+1}^{(k)} \tag{5.204}$$
$$\Delta\boldsymbol{q}_{n+1}^{(k)}:=\boldsymbol{q}_{n+1}^{(k+1)}-\boldsymbol{q}_{n+1}^{(k)} \tag{5.205}$$
$$\Delta\Delta\gamma_{n+1}^{\beta(k)}:=\Delta\gamma_{n+1}^{\beta(k+1)}-\Delta\gamma_{n+1}^{\beta(k)}. \tag{5.206}$$

Notice that in the above derivation, $\Delta\boldsymbol{\epsilon}_{n+1}$ as well as all quantities with the superscript (k) are assumed to be given (the process is strain-driven). The first terms in square brackets in each equation are the function values of Equations (5.193) to (5.195), equivalent to $f(x_0)-F$ in Equation (3.2). If the equations are satisfied, these function values should be zero. Therefore, they are also called the *residual*. The other terms in the equations are the gradients. Equations (5.201) to (5.203) are linear in $\Delta\Delta\gamma_{n+1}^{\beta(k)}$, $\Delta\boldsymbol{\sigma}_{n+1}^{(k)}$ and $\Delta\boldsymbol{q}_{n+1}^{(k)}$. Defining

$$\left\{R_{n+1}^{(k)}\right\}:=\left\{\begin{matrix}-\boldsymbol{\epsilon}_{n+1}^{\mathrm{p}}+\boldsymbol{\epsilon}_n^{\mathrm{p}}\\-\boldsymbol{\alpha}_{n+1}+\boldsymbol{\alpha}_n\end{matrix}\right\}^{(k)}+\sum_{\beta\in B_{\text{act}}^{(k)}}\Delta\gamma_{n+1}^{\beta(k)}\left\{\begin{matrix}\partial_{\boldsymbol{\sigma}}f_{n+1}^{\beta}\\\partial_{\boldsymbol{q}}f_{n+1}^{\beta}\end{matrix}\right\}^{(k)} \tag{5.207}$$

which is the residual of Equations (5.202) to (5.203),

$$\left[A_{n+1}^{(k)}\right]^{-1}$$
$$:=\begin{bmatrix}\boldsymbol{C}_{n+1}^{-1}+\sum_{\beta\in B_{\text{act}}^{(k)}}\Delta\gamma_{n+1}^{\beta}\partial_{\boldsymbol{\sigma\sigma}}^2 f_{n+1}^{\beta} & \sum_{\beta\in B_{\text{act}}^{(k)}}\Delta\gamma_{n+1}^{\beta}\partial_{\boldsymbol{\sigma q}}^2 f_{n+1}^{\beta}\\ \sum_{\beta\in B_{\text{act}}^{(k)}}\Delta\gamma_{n+1}^{\beta}\partial_{\boldsymbol{q\sigma}}^2 f_{n+1}^{\beta} & \boldsymbol{D}_{n+1}^{-1}+\sum_{\beta\in B_{\text{act}}^{(k)}}\Delta\gamma_{n+1}^{\beta}\partial_{\boldsymbol{qq}}^2 f_{n+1}^{\beta}\end{bmatrix}^{(k)} \tag{5.208}$$

and finally

$$\left\{F_{n+1}^{\beta(k)}\right\}:=\left\{\begin{matrix}\partial_{\boldsymbol{\sigma}}f_{n+1}^{\beta}\\\partial_{\boldsymbol{q}}f_{n+1}^{\beta}\end{matrix}\right\}^{(k)} \tag{5.209}$$

$$\left\{H_{n+1}^{\beta(k)}\right\}:=\left\{\begin{matrix}\partial_{\boldsymbol{\sigma}}h_{n+1}^{\beta}\\\partial_{\boldsymbol{q}}h_{n+1}^{\beta}\end{matrix}\right\}^{(k)} \tag{5.210}$$

then Equations (5.202) and (5.203) can be written as

$$\left\{R_{n+1}^{(k)}\right\} + \left[A_{n+1}^{(k)}\right]^{-1} : \left\{\begin{array}{c}\Delta\boldsymbol{\sigma}_{n+1}^{(k)} \\ \Delta\boldsymbol{q}_{n+1}^{(k)}\end{array}\right\} + \sum_{\beta\in B_{\text{act}}^{(k)}} \Delta\Delta\gamma_{n+1}^{\beta(k)} \left\{F_{n+1}^{\beta(k)}\right\} = 0 \tag{5.211}$$

which is equivalent to

$$\left[A_{n+1}^{(k)}\right] : \left\{R_{n+1}^{(k)}\right\} + \left\{\begin{array}{c}\Delta\boldsymbol{\sigma}_{n+1}^{(k)} \\ \Delta\boldsymbol{q}_{n+1}^{(k)}\end{array}\right\} + \sum_{\beta\in B_{\text{act}}^{(k)}} \Delta\Delta\gamma_{n+1}^{\beta(k)} \left[A_{n+1}^{(k)}\right] : \left\{F_{n+1}^{\beta(k)}\right\} = 0. \tag{5.212}$$

From Equation (5.201), one gets

$$h_{n+1}^{\beta(k)} - g_{n+1}^{\beta(k)}\Big) + \left\{H_{n+1}^{\beta(k)}\right\}^{\mathrm{T}} : \left\{\begin{array}{c}\Delta\boldsymbol{\sigma}_{n+1}^{(k)} \\ \Delta\boldsymbol{q}_{n+1}^{(k)}\end{array}\right\} - \partial_{\Delta\gamma} g_{n+1}^{\beta(k)} \Delta\Delta\gamma_{n+1}^{\beta(k)} = 0. \tag{5.213}$$

Premultiplying Equation (5.212) by $\left\{H_{n+1}^{\alpha(k)}\right\}^{\mathrm{T}}$ and inserting Equation (5.213) leads to

$$\begin{aligned}\left\{H_{n+1}^{\alpha(k)}\right\}^{\mathrm{T}} : \left[A_{n+1}^{(k)}\right] : \left\{R_{n+1}^{(k)}\right\} + \partial_{\Delta\gamma} g_{n+1}^{\alpha(k)} \Delta\Delta\gamma_{n+1}^{\alpha(k)} - \ h_{n+1}^{\alpha(k)} - g_{n+1}^{\alpha(k)}\Big) \\ + \sum_{\beta\in B_{\text{act}}^{(k)}} \Delta\Delta\gamma_{n+1}^{\beta(k)} \left\{H_{n+1}^{\alpha(k)}\right\}^{\mathrm{T}} : \left[A_{n+1}^{(k)}\right] : \left\{F_{n+1}^{\beta(k)}\right\} = 0, \quad \alpha \in B_{\text{act}}^{(k)}.\end{aligned} \tag{5.214}$$

Defining

$$(G_{\alpha\beta})_{n+1}^{(k)} := \left\{H_{n+1}^{\alpha(k)}\right\}^{\mathrm{T}} : \left[A_{n+1}^{(k)}\right] : \left\{F_{n+1}^{\beta(k)}\right\} + \partial_{\Delta\gamma} g_{n+1}^{\beta(k)} \delta_{\alpha\beta} \tag{5.215}$$

Equation (5.214) can be rewritten as

$$\sum_{\beta\in B_{\text{act}}^{(k)}} (G_{\alpha\beta})_{n+1}^{(k)} \Delta\Delta\gamma_{n+1}^{\beta(k)} = \ h_{n+1}^{\alpha(k)} - g_{n+1}^{\alpha(k)}\Big) - \left\{H_{n+1}^{\alpha(k)}\right\}^{\mathrm{T}} : \left[A_{n+1}^{(k)}\right] : \left\{R_{n+1}^{(k)}\right\}, \alpha \in B_{\text{act}}^{(k)}. \tag{5.216}$$

These are $m_{\text{act}}^{(k)}$ linear equations in $m_{\text{act}}^{(k)}$ unknowns. Their solution yields $\Delta\Delta\gamma_{n+1}^{\beta(k)}$, $\beta \in B_{\text{act}}^{(k)}$. Substituting into Equation (5.212) yields an expression for $\Delta\boldsymbol{\sigma}_{n+1}^{(k)}$ and $\Delta\boldsymbol{q}_{n+1}^{(k)}$ and iteration (k) seems to be finished. However, there is one further consideration to be taken into account. Equation (5.188), which defines the active slip systems, is not completely correct in the sense that it constitutes a necessary condition to be an active system but not a sufficient one. Indeed, because of the presence of corner points in the yield surface, the consistency parameter after iteration $k+1$ (Equation (5.206)),

$$\Delta\overline{\gamma}_{n+1}^{(k+1)} := \Delta\gamma_{n+1}^{(k)} + \Delta\Delta\gamma_{n+1}^{(k)} \tag{5.217}$$

is not necessarily positive. For details the reader is referred to (Simo and Hughes 1997). All active planes for which $\Delta\overline{\gamma}_{n+1}^{(k+1)} \leq 0$ have to be removed from $B_{\text{act}}^{(k)}$ and Equation (5.216) has to be solved again until for all active slip systems $\Delta\overline{\gamma}_{n+1}^{(k+1)} > 0$. Accordingly, the number of active slip systems can decrease from iteration to iteration, which is symbolized by the superscript (k) on $B_{\text{act}}^{(k)}$.

What form do the above equations take in the Cailletaud model? Recall that the potential for the evolution equations is defined by

$$f^\beta(\boldsymbol{\sigma}, \boldsymbol{q}) := |\boldsymbol{\sigma} : \boldsymbol{m}^\beta + q_2^\beta| + q_1^\beta + \frac{d^\beta}{2c^\beta}(q_2^\beta)^2 + \frac{1}{2Q^\beta}(q_1^\beta)^2. \tag{5.218}$$

Consequently,

$$\partial_{\boldsymbol{\sigma}} f_{n+1}^\beta = \boldsymbol{m}^\beta \operatorname{sgn}(\tau_{n+1}^\beta + q_{2,n+1}^\beta) \tag{5.219}$$

$$\partial_{\boldsymbol{q}} f_{n+1}^\beta = \begin{Bmatrix} 0 \\ \vdots \\ 0 \\ 1 + \frac{q_1^\beta}{Q^\beta} \\ \operatorname{sgn}(\tau^\beta + q_2^\beta) + \frac{d^\beta}{c^\beta} q_2^\beta \\ 0 \\ \vdots \\ 0 \end{Bmatrix} \begin{matrix} \\ \\ \\ \leftarrow \text{row}(2\beta - 1) \\ \leftarrow \text{row}(2\beta) \\ \\ \\ \leftarrow \text{row}(2m) \end{matrix} . \tag{5.220}$$

In Equation (5.177), $\partial_{\boldsymbol{q}} f_{n+1}^\beta$ was modified by a function $\boldsymbol{w}^\beta$ taking the total plasticity into account. In the Cailletaud model, $\boldsymbol{w}^\beta$ is defined by

$$\boldsymbol{w}^\beta := \begin{Bmatrix} 0 \\ \vdots \\ 0 \\ (\varphi^\beta - 1)\operatorname{sgn}(\tau^\beta + q_2^\beta) \\ 0 \\ \vdots \\ 0 \end{Bmatrix} \begin{matrix} \\ \\ \\ \leftarrow \text{row}(2\beta) \\ \\ \\ \\ \end{matrix} \tag{5.221}$$

where

$$\varphi^\beta := \phi^\beta + (1 - \phi^\beta)\mathrm{e}^{-\delta^\beta \int_0^t \dot{\gamma}^\beta \, \mathrm{d}t}. \tag{5.222}$$

The parameters ϕ^β and δ^β are material constants. In the numerical procedure, the accumulated plasticity in step $n + 1$ can be approximated by

$$\int_0^t \dot{\gamma}^\beta \, \mathrm{d}t \approx \sum_{i=1}^{n} \Delta\gamma_i^\beta + \Delta\gamma_{n+1}^{\beta(k)}. \tag{5.223}$$

In our derivation, the effect of $\boldsymbol{w}^\beta$ will be incorporated into a modified $\partial_{\boldsymbol{q}} f_{n+1}^\beta$:

$$\partial_{\boldsymbol{q}} f_{n+1}^{\beta *} = \left\{ \begin{array}{c} 0 \\ \vdots \\ 0 \\ 1 + \frac{q_1^\beta}{Q^\beta} \\ \varphi^\beta \operatorname{sgn}(\tau^\beta + q_2^\beta) + \frac{d^\beta}{c^\beta} q_2^\beta \\ 0 \\ \vdots \\ 0 \end{array} \right\} \begin{array}{l} \\ \\ \\ \leftarrow \text{row}(2\beta - 1) \\ \leftarrow \text{row}(2\beta) \\ \\ \\ \leftarrow \text{row}(2m) \end{array} . \tag{5.224}$$

This modified value will be used instead of the original value in all previously derived formulas. Notice that theoretically we now have

$$\partial_{\boldsymbol{q}} f_{n+1}^\beta(\boldsymbol{\sigma}_{n+1}, \boldsymbol{q}_{n+1}, \Delta\gamma_{n+1}^\beta) \tag{5.225}$$

that is, $\partial_{\boldsymbol{q}} f_{n+1}^\beta$ is not only a function of $\boldsymbol{\sigma}_{n+1}$ and $\boldsymbol{q}_{n+1}$, but also of $\Delta\gamma_{n+1}^\beta$. Accordingly, the linearization of Equation (5.199) is not completely correct any more. This effect will be neglected. It will at most slow down convergence. The second derivatives yield

$$\partial_{\boldsymbol{\sigma}\boldsymbol{\sigma}}^2 f_{n+1}^\beta = 0 \quad (6 \times 6 \text{ matrix}) \tag{5.226}$$

$$\partial_{\boldsymbol{\sigma}\boldsymbol{q}}^2 f_{n+1}^\beta = 0 \quad (6 \times 2m \text{ matrix}) \tag{5.227}$$

$$\partial_{\boldsymbol{q}\boldsymbol{\sigma}}^2 f_{n+1}^\beta = 0 \quad (2m \times 6 \text{ matrix}) \tag{5.228}$$

$$\partial_{\boldsymbol{q}\boldsymbol{q}}^2 f_{n+1}^\beta = \left[\begin{array}{c|c|c} 0 & 0 & 0 \\ \hline 0 & \begin{array}{cc} \frac{1}{Q^\beta} & 0 \\ 0 & \frac{d^\beta}{c^\beta} \end{array} & 0 \\ \hline 0 & 0 & 0 \end{array} \right] \quad (2m \times 2m \text{ matrix}) \tag{5.229}$$

where the submatrix in Equation (5.229) occupies rows and columns $(2\beta - 1)$ and 2β. From Equations (5.180), (5.181) and (5.190) one finds

$$\boldsymbol{D} = \operatorname{Diag}(b^1 Q^1, c^1, b^2 Q^2, \ldots, b^m Q^m, c^m). \tag{5.230}$$

Consequently (Equation (5.173)),

$$\left\{F_{n+1}^{\beta(k)}\right\} = \left\{\begin{array}{c} \boldsymbol{m}^{\beta}\operatorname{sgn}(\tau_{n+1}^{\beta} + q_{2,n+1}^{\beta}) \\ \cdots\cdots\cdots\cdots\cdots\cdots \\ 0 \\ \vdots \\ 0 \\ 1 + \frac{q_{1,n+1}^{\beta}}{Q^{\beta}} \\ \varphi^{\beta}\operatorname{sgn}(\tau_{n+1}^{\beta} + q_{2,n+1}^{\beta}) + \frac{d^{\beta}}{c^{\beta}} q_{2,n+1}^{\beta} \\ 0 \\ \vdots \\ 0 \end{array}\right\}^{(k)} \begin{array}{l} \\ \\ \\ \\ \\ \leftarrow \text{row}(2\beta + 5) \\ \leftarrow \text{row}(2\beta + 6) \\ \\ \\ \\ \end{array} \tag{5.231}$$

and (Equation (5.209))

$$\left[A_{n+1}^{(k)}\right]^{-1} = \operatorname{Diag}\left(\boldsymbol{C}_{n+1}^{-1}:, \frac{1}{b^1 Q^1} + \Delta\gamma_{n+1}^{1}\frac{1}{Q^1}, \frac{1}{c^1} + \Delta\gamma_{n+1}^{1}\frac{d^1}{c^1}, \frac{1}{b^2 Q^2} + \Delta\gamma_{n+1}^{2}\frac{1}{Q^2}, \cdots\right)^{(k)}. \tag{5.232}$$

The sum over $B_{\text{act}}^{(k)}$ in Equation (5.208) was replaced by a sum over all slip systems since $\Delta\gamma_{n+1} = 0$ for an inactive slip system. Combining Equations (5.231) and (5.232), one obtains

$$\left[A_{n+1}^{(k)}\right] : \left\{F_{n+1}^{\beta(k)}\right\} = \left\{\begin{array}{c} \boldsymbol{C}_{n+1} : \boldsymbol{m}^{\beta}\operatorname{sgn}(\tau_{n+1}^{\beta} + q_{2,n+1}^{\beta}) \\ \cdots\cdots\cdots\cdots\cdots\cdots \\ 0 \\ \vdots \\ 0 \\ \frac{Q^{\beta} + q_1^{\beta}}{\frac{1}{b^{\beta}} + \Delta\gamma_{n+1}^{\beta}} \\ \frac{\varphi^{\beta} c^{\beta}\operatorname{sgn}(\tau_{n+1}^{\beta} + q_{2,n+1}^{\beta}) + d^{\beta} q_{2,n+1}^{\beta}}{1 + \Delta\gamma_{n+1}^{\beta} d^{\beta}} \\ 0 \\ \vdots \\ 0 \end{array}\right\}^{(k)} \begin{array}{l} \\ \\ \\ \\ \\ \leftarrow \text{row}(2\beta + 5)\ . \\ \leftarrow \text{row}(2\beta + 6) \\ \\ \\ \\ \end{array} \tag{5.233}$$

Now recall the expression for the yield surface in the Cailletaud model:

$$h^{\beta}(\boldsymbol{\sigma}, \boldsymbol{q}) := |\boldsymbol{\sigma} : \boldsymbol{m}^{\beta} + q_2^{\beta}| - r_0^{\beta} + \sum_{\alpha=1}^{m} H_{\beta\alpha} q_1^{\alpha}. \tag{5.234}$$

Differentiation yields

$$\partial_{\sigma} h^{\beta}_{n+1} = \boldsymbol{m}^{\beta} \operatorname{sgn}(\tau^{\beta}_{n+1} + q^{\beta}_{2,n+1}) \tag{5.235}$$

$$\partial_{q_1^{\delta}} h^{\beta}_{n+1} = \sum_{\alpha=1}^{m} H_{\beta\alpha} \delta_{\alpha\delta} = H_{\beta\delta} \tag{5.236}$$

$$\partial_{q_2^{\delta}} h^{\beta}_{n+1} = \operatorname{sgn}(\tau^{\beta}_{n+1} + q^{\beta}_{2,n+1}) \cdot \delta_{\beta\delta}. \tag{5.237}$$

In matrix form,

$$\left\{ H^{\beta(k)}_{n+1} \right\} = \left\{ \begin{array}{c} \boldsymbol{m}^{\beta} \operatorname{sgn}(\tau^{\beta}_{n+1} + q^{\beta}_{2,n+1}) \\ \cdots\cdots\cdots\cdots\cdots\cdots \\ H_{\beta 1} \\ 0 \\ H_{\beta 2} \\ 0 \\ \vdots \\ H_{\beta\beta} \\ \operatorname{sgn}(\tau^{\beta}_{n+1} + q^{\beta}_{2,n+1}) \\ H_{\beta(\beta+1)} \\ \vdots \\ H_{\beta m} \\ 0 \end{array} \right\}^{(k)}. \tag{5.238}$$

Hence (Equation (5.215)),

$$(G_{\beta\beta})^{(k)}_{n+1} = \boldsymbol{m}^{\beta} : \boldsymbol{C}_{n+1} : \boldsymbol{m}^{\beta} + H_{\beta\beta} \frac{\left[Q^{\beta} + q^{\beta(k)}_{1,n+1} \right]}{\left[\frac{1}{b^{\beta}} + \Delta\gamma^{\beta(k)}_{n+1} \right]} + \frac{\left[\varphi^{\beta} c^{\beta} + d^{\beta} q^{\beta(k)}_{2,n+1} \operatorname{sgn}(\tau^{\beta(k)}_{n+1} + q^{\beta(k)}_{2,n+1}) \right]}{\left[1 + \Delta\gamma^{\beta(k)}_{n+1} d^{\beta} \right]} + \partial_{\Delta\gamma} g^{\beta(k)}_{n+1} \tag{5.239}$$

$$(G_{\alpha\beta})^{(k)}_{n+1} = \boldsymbol{m}^{\alpha} : \boldsymbol{C}_{n+1} : \boldsymbol{m}^{\beta} \operatorname{sgn}(\tau^{\beta(k)}_{n+1} + q^{\beta(k)}_{2,n+1}) \operatorname{sgn}(\tau^{\alpha(k)}_{n+1} + q^{\alpha(k)}_{2,n+1}) + H_{\alpha\beta} \frac{\left[Q^{\beta} + q^{\beta(k)}_{1,n+1} \right]}{\left[\frac{1}{b^{\beta}} + \Delta\gamma^{\beta(k)}_{n+1} \right]}, \quad \alpha \neq \beta. \tag{5.240}$$

For the right-hand side of Equation (5.216), the following term is needed:

$$\left\{T^{\alpha}_{n+1}\right\} := \left(\left\{H^{\alpha(k)}_{n+1}\right\}^{\mathrm{T}} : \left[A^{(k)}_{n+1}\right]\right)^{\mathrm{T}} = \begin{Bmatrix} \boldsymbol{C}_{n+1} : \boldsymbol{m}^{\alpha}\operatorname{sgn}(\tau^{\alpha}_{n+1} + q^{\alpha}_{2,n+1}) \\ \cdots\cdots\cdots\cdots\cdots\cdots \\ \frac{H_{\alpha 1} Q^1}{(1/b^1)+\Delta\gamma^1_{n+1}} \\ 0 \\ \frac{H_{\alpha 2} Q^2}{(1/b^2)+\Delta\gamma^2_{n+1}} \\ 0 \\ \vdots \\ \frac{H_{\alpha\alpha} Q^{\alpha}}{(1/b^{\alpha})+\Delta\gamma^{\alpha}_{n+1}} \\ \frac{c^{\alpha}\operatorname{sgn}(\tau^{\alpha}_{n+1}+q^{\alpha}_{2,n+1})}{1+\Delta\gamma^{\alpha}_{n+1} d^{\alpha}} \\ \vdots \\ \frac{H_{\alpha m} Q^m}{(1/b^m)+\Delta\gamma^m_{n+1}} \\ 0 \end{Bmatrix}^{(k)} . \tag{5.241}$$

The only function left to be specified is the creep function g. In the Cailletaud model, the relationship between the viscous shear strain rate and the shear stress in a slip plane is defined by

$$\left|\dot{\epsilon}^{\mathrm{p}\beta}_{\tau}\right| = \left\langle \frac{\tau^{\beta}}{K^{\beta}} \right\rangle^{n^{\beta}} \tag{5.242}$$

where $< x >= x$ for $x \geq 0$ and $< x >= 0$ for $x < 0$. K^{β} and n^{β} are material constants. The total viscous strain rate is related to the slip plane shear strain rates by (Koiter 1960)

$$\dot{\boldsymbol{\epsilon}}^{\mathrm{p}} = \sum_{\beta=1}^{m} \dot{\epsilon}^{\mathrm{p}\beta}_{\tau} \boldsymbol{m}^{\beta}. \tag{5.243}$$

Now, combining Equations (5.169) and (5.219) yields

$$\dot{\boldsymbol{\epsilon}}^{\mathrm{p}} = \sum_{\beta=1}^{m} \dot{\gamma}^{\beta} \boldsymbol{m}^{\beta} \operatorname{sgn}(\tau^{\beta}_{n+1} + q^{\beta}_{2,n+1}). \tag{5.244}$$

Comparison of Equations (5.243) and (5.244) yields

$$\dot{\gamma}^{\beta} = \left|\dot{\epsilon}^{\mathrm{p}\beta}_{\tau}\right| \tag{5.245}$$

since the applied shear stress and resulting shear strain rate have the same sign. Hence, Equation (5.242) can be written as

$$\dot{\gamma}^{\beta} = \left\langle \frac{\tau^{\beta}}{K^{\beta}} \right\rangle^{n^{\beta}} . \tag{5.246}$$

In the viscoplastic theory, the stress g^β by which the yield surface is exceeded is to be relaxed by creep, that is,

$$< \tau^\beta >= g^\beta. \tag{5.247}$$

Consequently,

$$g^\beta(\dot{\gamma}^\beta) = K^\beta(\dot{\gamma}^\beta)^{(1/n^\beta)} \tag{5.248}$$

$$= K^\beta\left(\frac{\Delta\gamma^\beta}{\Delta t}\right)^{(1/n^\beta)} \tag{5.249}$$

and

$$\partial_{\Delta\gamma} g_{n+1}^{\beta(k)} = \frac{K^\beta}{n^\beta \Delta t}\left(\frac{\Delta\gamma_{n+1}^{\beta(k)}}{\Delta t}\right)^{\frac{1}{n^\beta}-1}. \tag{5.250}$$

The last expression is used in Equation (5.239).

Summarizing, one arrives at the following algorithm to obtain $\boldsymbol{\sigma}_{n+1}$ from $\boldsymbol{\sigma}_n$:

1. Compute the elastic predictor and the value of the yield surfaces

$$\boldsymbol{\epsilon}_{n+1}^{\mathrm{p,trial}} = \boldsymbol{\epsilon}_n^{\mathrm{p}} \tag{5.251}$$

$$\boldsymbol{q}_{n+1}^{\mathrm{trial}} = \boldsymbol{q}_n \tag{5.252}$$

$$\Delta\gamma_{n+1}^{\mathrm{trial}} = 0 \tag{5.253}$$

$$\boldsymbol{\sigma}_{n+1}^{\mathrm{trial}} = \boldsymbol{C}_n : (\boldsymbol{\epsilon}_{n+1} - \boldsymbol{\epsilon}_{n+1}^{\mathrm{p,trial}}) \tag{5.254}$$

$$h_{n+1}^{\beta,\mathrm{trial}} = \left|\boldsymbol{m}^\beta : \boldsymbol{\sigma}_{n+1}^{\mathrm{trial}} + q_{2,n+1}^{\mathrm{trial}}\right| - r_0^\beta + \sum_{\alpha=1}^{m} H_{\beta\alpha} q_{1,n+1}^{\alpha,\mathrm{trial}}. \tag{5.255}$$

   Notice that $g^\beta(\Delta\gamma_{n+1}^{\mathrm{trial}}) = 0$.

2. Check for plasticity.

   If $h_{n+1}^{\beta,\mathrm{trial}} \leq 0, \quad \forall\beta$: step $n+1$ is elastic, that is,

$$\boldsymbol{\sigma}_{n+1} = \boldsymbol{\sigma}_{n+1}^{\mathrm{trial}} \tag{5.256}$$

   the values at $t = t_{n+1}$ are the trial values: the solution is found.

   else

$$B_{\mathrm{act}}^{(0)} = \left\{\beta \in \{1, \ldots, m\} | h_{n+1}^{\beta,\mathrm{trial}} > 0\right\} \tag{5.257}$$

$$\boldsymbol{\epsilon}_{n+1}^{\mathrm{p}(0)} = \boldsymbol{\epsilon}_n^{\mathrm{p}} \tag{5.258}$$

$$\boldsymbol{\alpha}_{n+1}^{(0)} = \boldsymbol{\alpha}_n \tag{5.259}$$

$$\Delta\gamma_{n+1}^{\beta(0)} = 0, \quad \beta = 1, 2, \ldots, m. \tag{5.260}$$

   endif

3. Start of the outer loop

   calculate $\boldsymbol{\sigma}_{n+1}^{(k)}$ and $\boldsymbol{q}_{n+1}^{(k)}$ from $\boldsymbol{\epsilon}_{n+1}^{\mathrm{p}(k)}$ and $\boldsymbol{\alpha}_{n+1}^{(k)}$ and check if the flow rule and the evolution equations are satisfied.

$$\boldsymbol{\sigma}_{n+1}^{(k)} = \boldsymbol{C}_{n+1} : (\boldsymbol{\epsilon}_{n+1} - \boldsymbol{\epsilon}_{n+1}^{\mathrm{p}(k)}) \tag{5.261}$$

$$\boldsymbol{q}_{n+1}^{(k)} = -\boldsymbol{D}_{n+1} : \boldsymbol{\alpha}_{n+1}^{(k)} \tag{5.262}$$

$$\left\{R_{n+1}^{(k)}\right\} = \left\{\begin{matrix} -\boldsymbol{\epsilon}_{n+1}^{\mathrm{p}} + \boldsymbol{\epsilon}_{n}^{\mathrm{p}} \\ -\boldsymbol{\alpha}_{n+1} + \boldsymbol{\alpha}_{n} \end{matrix}\right\}^{(k)} + \sum_{\beta \in B_{\mathrm{act}}^{(k)}} \left\{F_{n+1}^{\beta(k)}\right\} \tag{5.263}$$

$$\begin{aligned} h_{n+1}^{\beta(k)} - g_{n+1}^{\beta(k)} = {} & \left|\boldsymbol{m}^{\beta} : \boldsymbol{\sigma}_{n+1}^{(k)} + q_{2,n+1}^{\beta(k)}\right| - r_0^{\beta} + \sum_{\alpha=1}^{m} H_{\beta\alpha} q_{1,n+1}^{\alpha(k)} \\ & - K^{\beta}\left(\frac{\Delta\gamma_{n+1}^{\beta(k)}}{\Delta t}\right)^{\frac{1}{n^{\beta}}} \end{aligned} \tag{5.264}$$

   if $\left|h_{n+1}^{\beta(k)} - g_{n+1}^{\beta(k)}\right| < \mathrm{TOL}$ and $\left\|R_{n+1}^{(k)}\right\| < \mathrm{TOL}$, leave the outer loop.

4. Start of the inner loop

   determine $\Delta\Delta\gamma_{n+1}^{\beta(k)}$ by Equation (5.216) without creep effects.

   If $\Delta\overline{\gamma}_{n+1}^{\beta(k+1)} := \Delta\gamma_{n+1}^{\beta(k)} + \Delta\Delta\gamma_{n+1}^{\beta(k)} > 0, \quad \forall\beta \in B_{\mathrm{act}}^{(k)}$ then

      recalculate $\Delta\Delta\gamma_{n+1}^{\beta(k)}$ by Equation (5.216) with creep effects and exit inner loop.

   else

      remove the inactive slip planes from $B_{\mathrm{act}}^{(k)}$ and reiterate the inner loop.

   endif

   End of the inner loop

5. Update the internal variables

   Determine $\left\{\begin{matrix} \Delta\boldsymbol{\sigma}_{n+1}^{(k)} \\ \Delta\boldsymbol{q}_{n+1}^{(k)} \end{matrix}\right\}$ from Equation (5.212).

$$\left\{\begin{matrix} \Delta\boldsymbol{\epsilon}_{n+1}^{\mathrm{p}} \\ \Delta\boldsymbol{\alpha}_{n+1} \end{matrix}\right\}^{(k)} = -\begin{bmatrix} \boldsymbol{C}_{n+1}^{-1} & 0 \\ 0 & \boldsymbol{D}_{n+1}^{-1} \end{bmatrix}^{(k)} : \left\{\begin{matrix} \Delta\boldsymbol{\sigma}_{n+1}^{(k)} \\ \Delta\boldsymbol{q}_{n+1}^{(k)} \end{matrix}\right\} \tag{5.265}$$

$$\boldsymbol{\epsilon}_{n+1}^{\mathrm{p}(k+1)} = \boldsymbol{\epsilon}_{n+1}^{\mathrm{p}(k)} + \Delta\boldsymbol{\epsilon}_{n+1}^{\mathrm{p}(k)} \tag{5.266}$$

$$\boldsymbol{\alpha}_{n+1}^{(k+1)} = \boldsymbol{\alpha}_{n+1}^{(k)} + \Delta\boldsymbol{\alpha}_{n+1}^{(k)} \tag{5.267}$$

$$\Delta\gamma_{n+1}^{\beta(k+1)} = \Delta\gamma_{n+1}^{\beta(k)} + \Delta\Delta\gamma_{n+1}^{\beta(k)} \tag{5.268}$$

   set $k \leftarrow k + 1$ and reiterate the outer loop.

6. End of outer loop. Now, the determination of the plastic tangent modulus can start.

Finally, two more remarks:

(a) It is advantageous to substitute Equation (5.265) directly into Equation (5.212) yielding

$$\left\{\begin{matrix}\Delta\boldsymbol{\epsilon}^{\mathrm{p}}_{n+1}\\ \Delta\boldsymbol{\alpha}_{n+1}\end{matrix}\right\}^{(k)} = \begin{bmatrix}\boldsymbol{C}^{-1}_{n+1} & 0\\ 0 & \boldsymbol{D}^{-1}_{n+1}\end{bmatrix}^{(k)} : \left[A^{(k)}_{n+1}\right] :$$

$$: \left\{\left\{R^{(k)}_{n+1}\right\} + \sum\nolimits_{\beta\in B^{(k)}_{\mathrm{act}}}\left\{F^{\beta(k)}_{n+1}\right\}\Delta\Delta\gamma^{\beta(k)}_{n+1}\right\} \quad (5.269)$$

where (Equation (5.232)),

$$\begin{bmatrix}\boldsymbol{C}^{-1} & 0\\ 0 & \boldsymbol{D}^{-1}\end{bmatrix}^{(k)}_{n+1} : \left[A^{(k)}_{n+1}\right]$$

$$= \mathrm{Diag}\left(\boldsymbol{I}: \frac{1}{1+b^1\Delta\gamma^1_{n+1}}, \frac{1}{1+d^1\Delta\gamma^1_{n+1}}, \frac{1}{1+b^2\Delta\gamma^2_{n+1}}, \ldots\right)^{(k)}. \quad (5.270)$$

(b) The inner loop is necessary to determine the active slip planes (cf (Simo and Hughes 1997) for more details). In the determination process, the viscous terms are dropped to make sure that the viscous procedure converges in the limit to the same point on the yield surface as the inviscid formulation.

5.4.4 Determination of the consistent elastoplastic tangent matrix

The determination of the consistent elastoplastic moduli also starts from Equations (5.197) to (5.199). We have attained equilibrium for $t = t_{n+1}$, that is, Equations (5.197) to (5.199) are identically satisfied and we would like to know how $\boldsymbol{\sigma}$ changes if $\boldsymbol{\epsilon}$ is perturbed. Therefore, we differentiate these equations:

$$\partial_{\sigma} h^{\beta}_{n+1} : \mathrm{d}\boldsymbol{\sigma}_{n+1} + \partial_{\boldsymbol{q}} h^{\beta}_{n+1} : \mathrm{d}q_{n+1} - \partial_{\Delta\gamma} g^{\beta}_{n+1} \cdot \mathrm{d}\Delta\gamma^{\beta}_{n+1} = 0 \quad (5.271)$$

$$-\,\mathrm{d}\boldsymbol{\epsilon}_{n+1} + \boldsymbol{C}^{-1}_{n+1} : \mathrm{d}\boldsymbol{\sigma}_{n+1} + \sum_{\beta\in B^{(k)}_{\mathrm{act}}} \mathrm{d}\Delta\gamma^{\beta}_{n+1}\partial_{\sigma} f^{\beta}_{n+1}$$

$$+ \sum_{\beta\in B^{(k)}_{\mathrm{act}}} \Delta\gamma^{\beta}_{n+1} \;\; \partial^2_{\sigma\sigma} f^{\beta}_{n+1} : \mathrm{d}\boldsymbol{\sigma}_{n+1} + \partial^2_{\sigma\boldsymbol{q}} f^{\beta}_{n+1} : \mathrm{d}q_{n+1}\Big) = 0 \quad (5.272)$$

$$\boldsymbol{D}^{-1}_{n+1} : \mathrm{d}\boldsymbol{q}_{n+1} + \sum_{\beta\in B^{(k)}_{\mathrm{act}}} \mathrm{d}\Delta\gamma^{\beta}_{n+1}\partial_{\boldsymbol{q}} f^{\beta}_{n+1}$$

$$+ \sum_{\beta\in B^{(k)}_{\mathrm{act}}} \Delta\gamma^{\beta}_{n+1} \;\; \partial^2_{\boldsymbol{q}\sigma} f^{\beta}_{n+1} : \mathrm{d}\boldsymbol{\sigma}_{n+1} + \partial^2_{\boldsymbol{q}\boldsymbol{q}} f^{\beta}_{n+1} : \mathrm{d}\boldsymbol{q}_{n+1}\Big) = 0. \quad (5.273)$$

These equations are very similar to Equations (5.201) to (5.203). In fact, by replacing $\left\{R_{n+1}^{(k)}\right\}$ by $-\left\{\begin{matrix} \mathrm{d}\boldsymbol{\epsilon}_{n+1} \\ 0 \end{matrix}\right\}$, $h_{n+1}^{\alpha(k)} - g_{n+1}^{\alpha(k)}$ by 0, Δ by d and dropping the superindex (k), they are identical. Hence, by comparing with Equation (5.216), one arrives at the following set of equations:

$$\sum_{\beta \in B_{\text{act}}} (G_{\alpha\beta})_{n+1}\, \mathrm{d}\Delta\gamma_{n+1}^{\beta} = \left\{T_{n+1}^{\alpha}\right\}^{\mathrm{T}} : \left\{\begin{matrix} \mathrm{d}\boldsymbol{\epsilon}_{n+1} \\ 0 \end{matrix}\right\}, \quad \alpha \in B_{\text{act}} \tag{5.274}$$

yielding

$$\mathrm{d}\Delta\gamma_{n+1}^{\beta} - \left(\sum_{\alpha \in B_{\text{act}}} (G^{-1\beta\alpha})_{n+1} \left\{T_{n+1}^{\alpha}\right\}^{\mathrm{T}}\right) : \left\{\begin{matrix} \mathrm{d}\boldsymbol{\epsilon}_{n+1} \\ 0 \end{matrix}\right\}, \quad \beta \in B_{\text{act}}. \tag{5.275}$$

The equivalent equation of Equation (5.212) reads

$$\begin{aligned} \left\{\begin{matrix} \mathrm{d}\boldsymbol{\sigma}_{n+1} \\ \mathrm{d}\boldsymbol{q}_{n+1} \end{matrix}\right\} &= \left[A_{n+1}\right] : \left\{\left\{\begin{matrix} \mathrm{d}\boldsymbol{\epsilon}_{n+1} \\ 0 \end{matrix}\right\} - \textstyle\sum_{\beta \in B_{\text{act}}} \left\{F_{n+1}^{\beta}\right\} \mathrm{d}\gamma_{n+1}^{\beta}\right\} \\ &= \left[A_{n+1}\right] : \left[[I] - \sum_{\beta \in B_{\text{act}}} \sum_{\alpha \in B_{\text{act}}} (G^{-1\beta\alpha})_{n+1} \left\{F_{n+1}^{\beta}\right\} \otimes \left\{T_{n+1}^{\alpha}\right\}^{\mathrm{T}}\right] : \left\{\begin{matrix} \mathrm{d}\boldsymbol{\epsilon}_{n+1} \\ 0 \end{matrix}\right\}. \end{aligned} \tag{5.276}$$

Only the relationship between $\mathrm{d}\boldsymbol{\sigma}_{n+1}$ and $\mathrm{d}\boldsymbol{\epsilon}_{n+1}$ is needed, hence (Equations (5.231), (5.232) and (5.241)),

$$\begin{aligned} \left\{\mathrm{d}\boldsymbol{\sigma}_{n+1}\right\} = [\boldsymbol{C}_{n+1} - \sum_{\beta \in B_{\text{act}}} \sum_{\alpha \in B_{\text{act}}} \boldsymbol{C}_{n+1} : \boldsymbol{m}^{\beta} \operatorname{sgn}(\tau_{n+1}^{\beta} + q_{2,n+1}^{\beta})(G^{-1\beta\alpha})_{n+1} \otimes \\ \otimes\, \boldsymbol{m}^{\alpha} : \boldsymbol{C}_{n+1} \operatorname{sgn}(\tau_{n+1}^{\alpha} + q_{2,n+1}^{\alpha})] : \left\{\mathrm{d}\boldsymbol{\epsilon}_{n+1}\right\}. \end{aligned} \tag{5.277}$$

Hence, the consistent elastoplastic stiffness matrix $\boldsymbol{C}_{n+1}^{\text{ep}}$ satisfies

$$\boldsymbol{C}_{n+1}^{\text{ep}} = \boldsymbol{C}_{n+1} - \sum_{\beta \in B_{\text{act}}} \sum_{\alpha \in B_{\text{act}}} (G^{-1\beta\alpha})_{n+1} \boldsymbol{M}^{\beta} \otimes \boldsymbol{M}^{\alpha\mathrm{T}} \tag{5.278}$$

where

$$\boldsymbol{M}^{\alpha} := \boldsymbol{C}_{n+1} : \boldsymbol{m}^{\alpha} \operatorname{sgn}(\tau_{n+1}^{\alpha} + q_{2,n+1}^{\alpha}). \tag{5.279}$$

Equation (5.278) shows that each active slip plane modifies the stiffness matrix (without plastic flow, the consistent elastoplastic stiffness matrix reduces to the elasticity matrix). Since $\left[G\right]$ is not necessarily symmetric (cf Equation (5.240)), $\boldsymbol{C}^{\text{ep}}$ is not necessarily symmetric either. In practice, the matrix is often symmetrized by adding the transpose and dividing by two.

5.4.5 Tensile test on an anisotropic material

Consider the tensile specimen in Figure 5.9. The axis of the specimen coincides with the z-axis. The orientation of the anisotropic material is defined by the x'-y'-z' axis system.

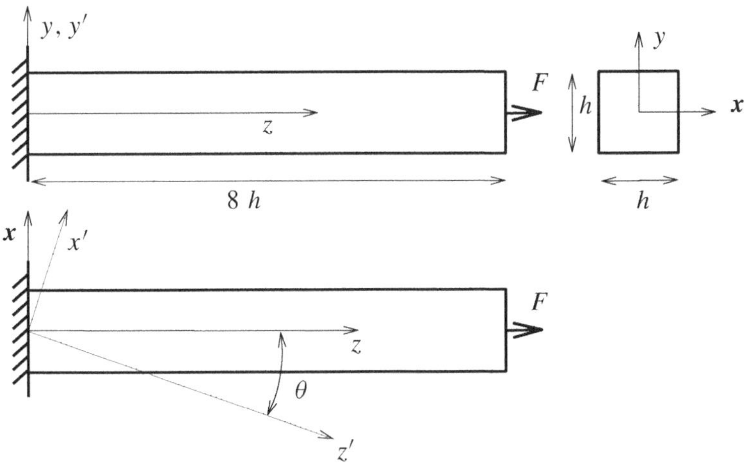

Figure 5.9 Geometry of the tensile specimen

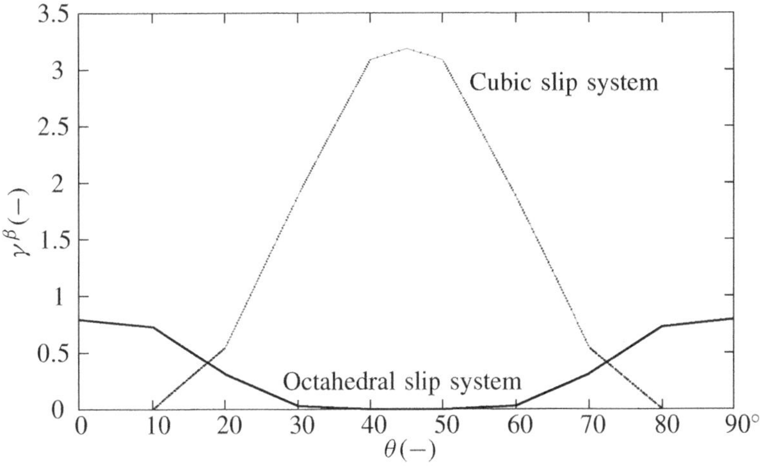

Figure 5.10 Accumulated plastic slip

The y- and y'-axes coincide, whereas the z- and z'-axis include an angle θ. A constant force F is applied at the end of the specimen in the z-direction. Now we look at what happens if we vary the angle θ from 0° to 90°. In particular, we investigate the accumulated plastic slip in two different slip systems: the first slip system is octahedral and is characterized by $\boldsymbol{n} = (1, 1, 1)$ and $\boldsymbol{l} = (1, 0, -1)$, the second is a cubic slip system and is defined by $\boldsymbol{n} = (0, 0, 1)$ and $\boldsymbol{l} = (1, -1, 0)$. The slip systems are defined in the local x'-y'-z' system.

Figure 5.10 shows that the octahedral slip system is activated if the global axes and the material axes are aligned. Then, the slip direction that is considered includes an angle of 45° with the loading axis, leading to a large slip. The cubic slip system is activated especially for angles close to $\theta = 45°$. Here again, the angle between the slip direction and the loading direction is maximized.

5.5 Anisotropic Elasticity with a von Mises–type Yield Surface

In the previous section, we introduced the Cailletaud model for single crystals. In order to use the model, 21 parameters must be determined (3 elastic constants and 9 viscoplastic constants for each slip system) for the relevant temperature range. This is a huge and expensive task. Therefore, one frequently resorts to the following approximation: the elastic range is properly described by the anisotropic elasticity tensor. The yield surface, however, is assumed to be isotropic of the von Mises form. In this respect, the equations are similar to the ones in Section 5.3.1. However, because of the anisotropic elastic behavior, the solution method is more complex and closely linked to the solution procedure in the Cailletaud model.

5.5.1 Basic equations

The governing equations are merely a concretization of Equations (5.76) to (5.83):

1. Elastic stress–strain relation

$$\boldsymbol{\sigma} = \boldsymbol{C} : (\boldsymbol{\epsilon} - \boldsymbol{\epsilon}^{\mathrm{p}}). \tag{5.280}$$

2. Internal variable relationships

   Two internal variables are used: an isotropic scalar variable q_1 denoting the radius of the elastic domain in deviatoric stress space and a kinematic tensor variable $\boldsymbol{q}_2$ denoting its center. The relationship between the internal variables in stress space $\{q_1, \boldsymbol{q}_2\}$ and the corresponding ones in strain space $\{\alpha_1, \boldsymbol{\alpha}_2\}$ is assumed to be linear. A generalization to other functional relationships is no problem.

$$q_1 = -d_1\alpha_1 \tag{5.281}$$

$$\boldsymbol{q}_2 = -\tfrac{2}{3}d_2\boldsymbol{\alpha}_2. \tag{5.282}$$

   The factor $\frac{2}{3}$ is introduced such that the equivalent quantities satisfy (cf Equation (5.10))

$$q_2^{\mathrm{eq}} = d_2\alpha_2^{\mathrm{eq}}. \tag{5.283}$$

3. Yield surface (Equation (5.89))

$$\|\mathrm{dev}\ (\boldsymbol{\sigma}) + \boldsymbol{q}_2\| + \sqrt{\tfrac{2}{3}}(q_1 - r_0) = 0. \tag{5.284}$$

   The parameter r_0 is the yield stress at zero-equivalent plastic strain.

4. Evolution equations

   In the associative theory, they are derived from the yield surface in the form of Equations (5.79) and (5.80) and, for a von Mises type surface, Equations (5.95) to (5.97):

$$\dot{\boldsymbol{\epsilon}}^{\mathrm{p}} = \dot{\gamma}\boldsymbol{n} \tag{5.285}$$

$$\dot{\alpha}_1 = \dot{\gamma}\sqrt{\tfrac{2}{3}} \tag{5.286}$$

$$\dot{\boldsymbol{\alpha}}_2 = \dot{\gamma}\boldsymbol{n} \tag{5.287}$$

   where

$$\boldsymbol{n} := \frac{\boldsymbol{\xi}}{\|\boldsymbol{\xi}\|} \tag{5.288}$$

   and

$$\boldsymbol{\xi} := \operatorname{dev}(\boldsymbol{\sigma}) + \boldsymbol{q}_2. \tag{5.289}$$

5. Kuhn–Tucker equations

$$\dot{\gamma} \geq 0, \quad f(\boldsymbol{\sigma}, q_1, \boldsymbol{q}_2) \leq 0, \quad \dot{\gamma} f(\boldsymbol{\sigma}, q_1, \boldsymbol{q}_2) = 0. \tag{5.290}$$

6. Consistency condition

$$\dot{\gamma}\dot{f}(\boldsymbol{\sigma}, q_1, \boldsymbol{q}_2) = 0. \tag{5.291}$$

Viscous effects are taken into account by a Norton-type law

$$\dot{\epsilon}^{\mathrm{peq}} = A(\sigma_{\mathrm{vm}})^n \tag{5.292}$$

or

$$\sigma_{\mathrm{vm}} = g(\dot{\epsilon}^{\mathrm{peq}}) = \left(\frac{\dot{\epsilon}^{\mathrm{peq}}}{A}\right)^{(1/n)} \tag{5.293}$$

and Equation (5.284) is replaced by

$$\|\operatorname{dev}(\boldsymbol{\sigma}) + \boldsymbol{q}_2\| + \sqrt{\tfrac{2}{3}}\left[q_1 - r_0 - g(\dot{\epsilon}^{\mathrm{peq}})\right] = 0. \tag{5.294}$$

Finally, recall that (Equations (5.112) and (5.113))

$$\alpha_1 = \alpha_2^{\mathrm{eq}} = \epsilon^{\mathrm{peq}} = \sqrt{\tfrac{2}{3}}\gamma. \tag{5.295}$$

5.5.2 Numerical procedure

Starting from known quantities at time $t = t_n$, the solution at $t = t_{n+1}$ is what is being looked for. Using the trial-and-error procedure explained in previous sections, we first assume that no plasticity takes place in $[t_n, t_{n+1}]$. Consequently (Equations (5.114)–(5.118)),

$$\boldsymbol{\epsilon}^{\mathrm{p}}_{n+1} = \boldsymbol{\epsilon}^{\mathrm{p}}_{n} \tag{5.296}$$

$$q_{1,n+1} = q_{1,n} \tag{5.297}$$

$$\boldsymbol{q}_{2,n+1} = \boldsymbol{q}_{2,n} \tag{5.298}$$

$$\gamma_{n+1} = \gamma_n \tag{5.299}$$

$$\boldsymbol{\sigma}_{n+1} = \boldsymbol{C} : (\boldsymbol{\epsilon}_{n+1} - \boldsymbol{\epsilon}^{\mathrm{p}}_{n+1}). \tag{5.300}$$

If

$$\|\mathrm{dev}\ (\boldsymbol{\sigma}_{n+1}) + \boldsymbol{q}_{2,n+1}\| + \sqrt{\tfrac{2}{3}}(q_{1,n+1} - r_0) \leq 0 \tag{5.301}$$

the assumption was right and the solution at $t = t_{n+1}$ is found. If Equation (5.301) is not satisfied, the following set of 24 equations in 24 unknowns, obtained by backward Euler discretization of Equations (5.279) to (5.282), (5.285) to (5.287) and (5.294) has to be solved:

$$\Delta\boldsymbol{\sigma}_{n+1} = \boldsymbol{C}_{n+1} : (\Delta\boldsymbol{\epsilon}_{n+1} - \Delta\boldsymbol{\epsilon}^{\mathrm{p}}_{n+1}) \quad \text{(6 equations)} \tag{5.302}$$

$$\Delta q_{1,n+1} = -d_1 \Delta\alpha_{1,n+1} \quad \text{(1 equation)} \tag{5.303}$$

$$\Delta\boldsymbol{q}_{2,n+1} = -\tfrac{2}{3} d_2 \Delta\boldsymbol{\alpha}_{2,n+1} \quad \text{(5 equations)} \tag{5.304}$$

$$\|\mathrm{dev}\ (\boldsymbol{\sigma}_{n+1}) + \boldsymbol{q}_{2,n+1}\| + \sqrt{\tfrac{2}{3}}\left[q_{1,n+1} - r_0 - g\left(\sqrt{\tfrac{2}{3}}\Delta\gamma_{n+1}\right)\right] = 0 \quad \text{(1 equation)} \tag{5.305}$$

$$\Delta\boldsymbol{\epsilon}^{\mathrm{p}}_{n+1} = \Delta\gamma_{n+1}\boldsymbol{n}_{n+1} \quad \text{(5 equations)} \tag{5.306}$$

$$\Delta\alpha_{1,n+1} = \Delta\gamma_{n+1}\sqrt{\tfrac{2}{3}} \quad \text{(1 equation)} \tag{5.307}$$

$$\Delta\boldsymbol{\alpha}_{2,n+1} = \Delta\gamma_{n+1}\boldsymbol{n}_{n+1} \quad \text{(5 equations)} \tag{5.308}$$

in the unknowns $\Delta\boldsymbol{\sigma}_{n+1}$ (6), $\Delta q_{1,n+1}$ (1), $\Delta\boldsymbol{q}_{2,n+1}$ (5), $\Delta\gamma_{n+1}$ (1), $\Delta\boldsymbol{\epsilon}^{\mathrm{p}}_{n+1}$ (5), $\Delta\alpha_{1,n+1}$ (1) and $\Delta\boldsymbol{\alpha}_{2,n+1}$ (5). Because of the anisotropic character of Equation (5.302), the solution method of Section 5.3 cannot be used. However, notice the similarity of Equations (5.302) to (5.308) to Equations (5.191) to (5.195). Indeed, since

$$\boldsymbol{n}_{n+1} = \partial_{\sigma} f(\boldsymbol{\sigma}_{n+1}, q_{1,n+1}, \boldsymbol{q}_{2,n+1}) \tag{5.309}$$

the present set of equations can be considered as a special case of Equations (5.191) to (5.195) for $h = f$, $\boldsymbol{q} := \{q_1, \boldsymbol{q}_2\}$, g replaced by $\sqrt{2/3}\, g$ and just 1 slip system. Focusing on the solution method starting at Equation (5.207), one obtains the following residual:

$$\left\{R^{(k)}_{n+1}\right\} = \begin{Bmatrix} -\boldsymbol{\epsilon}^{\mathrm{p}}_{n+1} + \boldsymbol{\epsilon}^{\mathrm{p}}_{n} \\ -\alpha_{1,n+1} + \alpha_{1,n} \\ -\boldsymbol{\alpha}_{2,n+1} + \boldsymbol{\alpha}_{2,n} \end{Bmatrix}^{(k)} + \Delta\gamma^{(k)}_{n+1} \begin{Bmatrix} \boldsymbol{n}_{n+1} \\ \sqrt{\tfrac{2}{3}} \\ \boldsymbol{n}_{n+1} \end{Bmatrix}^{(k)}. \tag{5.310}$$

For the determination of $\left[A_{n+1}^{(k)}\right]^{-1}$, the second derivatives of f with respect to $\boldsymbol{\sigma}$ and $\boldsymbol{q}$ are needed. Using Equations (5.90) to (5.93) one arrives at

$$\partial^2_{\sigma\sigma} f = \frac{\partial}{\partial \boldsymbol{\sigma}} \frac{\boldsymbol{\xi}}{\|\boldsymbol{\xi}\|} = \frac{1}{\|\boldsymbol{\xi}\|} \frac{\partial}{\partial \boldsymbol{\sigma}}(\boldsymbol{\xi}) - \frac{1}{\|\boldsymbol{\xi}\|^2} \boldsymbol{\xi} \otimes \frac{\partial}{\partial \boldsymbol{\sigma}} \|\boldsymbol{\xi}\|$$

$$= \frac{1}{\|\boldsymbol{\xi}\|} \quad \mathbb{I} - \tfrac{1}{3} \boldsymbol{I} \otimes \boldsymbol{I} \Big) - \frac{1}{\|\boldsymbol{\xi}\|^2} \boldsymbol{\xi} \otimes \frac{\boldsymbol{\xi}}{\|\boldsymbol{\xi}\|} \tag{5.311}$$

$$= \frac{1}{\|\boldsymbol{\xi}\|} \quad \mathbb{I} - \tfrac{1}{3} \boldsymbol{I} \otimes \boldsymbol{I} - \boldsymbol{n} \otimes \boldsymbol{n} \Big) \tag{5.312}$$

and in a similar way,

$$\partial^2_{q_2 q_2} f = \partial^2_{\sigma q_2} f = \partial^2_{q_2 \sigma} f = \partial^2_{\sigma\sigma} f = \frac{\boldsymbol{\chi}}{\|\boldsymbol{\xi}\|} \tag{5.313}$$

where

$$\boldsymbol{\chi} := \mathbb{I} - \tfrac{1}{3} \boldsymbol{I} \otimes \boldsymbol{I} - \boldsymbol{n} \otimes \boldsymbol{n}. \tag{5.314}$$

Let $\boldsymbol{B}$ be a fourth-order tensor of the form

$$\boldsymbol{B} := a\mathbb{I} + b\boldsymbol{I} \otimes \boldsymbol{I} + c\boldsymbol{n} \otimes \boldsymbol{n} \tag{5.315}$$

with $a, b, c \in \mathbb{R}$,

$$\|\boldsymbol{n}\| = 1 \tag{5.316}$$

and $\boldsymbol{n}$ is deviatoric:

$$\boldsymbol{n} : \boldsymbol{I} = 0. \tag{5.317}$$

Since fourth-order tensors in three-dimensional space can be viewed as 9×9 matrices, we know that tensor contraction $(\boldsymbol{A} : \boldsymbol{B})$ is associative and that there is a neutral element $\mathbb{I}$. However, for tensors of the form in Equation (5.315), contraction is also commutative. Indeed,

$$\begin{aligned} &(a_1\mathbb{I} + b_1\boldsymbol{I} \otimes \boldsymbol{I} + c_1\boldsymbol{n} \otimes \boldsymbol{n}) : (a_2\mathbb{I} + b_2\boldsymbol{I} \otimes \boldsymbol{I} + c_2\boldsymbol{n} \otimes \boldsymbol{n}) \\ &\quad = a_1 a_2 \mathbb{I} + (a_1 b_2 + b_1 a_2 + 3 b_1 b_2)\boldsymbol{I} \otimes \boldsymbol{I} + (a_1 c_2 + c_1 a_2 + c_1 c_2)\boldsymbol{n} \otimes \boldsymbol{n} \end{aligned} \tag{5.318}$$

which is symmetric in the indices 1 and 2. Straightforward calculation shows that

$$\boldsymbol{B}^{-1} = \frac{1}{a}\left(\mathbb{I} - \frac{b}{a+3b}\boldsymbol{I} \otimes \boldsymbol{I} - \frac{c}{a+c}\boldsymbol{n} \otimes \boldsymbol{n}\right) \tag{5.319}$$

and

$$\boldsymbol{A} : \boldsymbol{\chi} = \boldsymbol{\chi} : \boldsymbol{A} = a\boldsymbol{\chi}. \tag{5.320}$$

Now, Equation (5.208) reduces to the following form:

$$\left[A_{n+1}^{(k)}\right]^{-1} = \begin{bmatrix} \boldsymbol{C}_{n+1}^{-1} + \dfrac{\Delta\gamma_{n+1}}{\|\boldsymbol{\xi}_{n+1}\|}\boldsymbol{\chi}_{n+1} & 0 & \dfrac{\Delta\gamma_{n+1}}{\|\boldsymbol{\xi}_{n+1}\|}\boldsymbol{\chi}_{n+1} \\ 0 & \dfrac{1}{d_1} & 0 \\ \dfrac{\Delta\gamma_{n+1}}{\|\boldsymbol{\xi}_{n+1}\|}\boldsymbol{\chi}_{n+1} & 0 & \dfrac{3}{2d_2}\mathbb{I} + \dfrac{\Delta\gamma_{n+1}}{\|\boldsymbol{\xi}_{n+1}\|}\boldsymbol{\chi}_{n+1} \end{bmatrix}^{(k)}. \tag{5.321}$$

Defining

$$a := \frac{3}{2d_2} \tag{5.322}$$

$$b := \frac{\Delta\gamma_{n+1}^{(k)}}{\|\boldsymbol{\xi}_{n+1}\|} \tag{5.323}$$

(notice that b is a function of n and k, although not explicitly indicated!) and dropping the indices $n+1$ and (k) for simplicity, one arrives at

$$\left[A\right]^{-1} = \begin{bmatrix} \boldsymbol{C}^{-1} + b\boldsymbol{\chi} & 0 & b\boldsymbol{\chi} \\ 0 & d_1^{-1} & 0 \\ b\boldsymbol{\chi} & 0 & a\mathbb{I} + b\boldsymbol{\chi} \end{bmatrix}. \tag{5.324}$$

In the further derivation, $\left[A\right]$ will be needed and it is clearly numerically advantageous if this inversion can be performed in a largely analytical way. Denoting

$$\left[A\right] := \begin{bmatrix} \boldsymbol{P} & 0 & \boldsymbol{R} \\ 0 & d_1 & 0 \\ \boldsymbol{Q} & 0 & \boldsymbol{S} \end{bmatrix}, \tag{5.325}$$

the submatrices $\boldsymbol{P}$, $\boldsymbol{Q}$, $\boldsymbol{R}$ and $\boldsymbol{S}$ satisfy

$$\begin{bmatrix} \boldsymbol{C}^{-1} + b\boldsymbol{\chi} & b\boldsymbol{\chi} \\ b\boldsymbol{\chi} & a\mathbb{I} + b\boldsymbol{\chi} \end{bmatrix} : \begin{bmatrix} \boldsymbol{P} & \boldsymbol{R} \\ \boldsymbol{Q} & \boldsymbol{S} \end{bmatrix} = \begin{bmatrix} \mathbb{I} & 0 \\ 0 & \mathbb{I} \end{bmatrix}. \tag{5.326}$$

To solve this system, the block in the first row and second column of the left matrix will be reduced by premultiplying the first block equation by $(a\mathbb{I} + b\boldsymbol{\chi})$, the second by $b\boldsymbol{\chi}$ and subtracting the second from the first. This results in

$$\begin{bmatrix} (a\mathbb{I} + b\boldsymbol{\chi}) : (\boldsymbol{C}^{-1} + b\boldsymbol{\chi}) - (b\boldsymbol{\chi}) : (b\boldsymbol{\chi}) & 0 \\ b\boldsymbol{\chi} & a\mathbb{I} + b\boldsymbol{\chi} \end{bmatrix} : \begin{bmatrix} \boldsymbol{P} & \boldsymbol{R} \\ \boldsymbol{Q} & \boldsymbol{S} \end{bmatrix} = \begin{bmatrix} a\mathbb{I} + b\boldsymbol{\chi} & -b\boldsymbol{\chi} \\ 0 & \mathbb{I} \end{bmatrix}. \tag{5.327}$$

Notice that the block in the first row and second column actually reads

$$(a\mathbb{I} + b\boldsymbol{\chi}) : [(b\boldsymbol{\chi}) : \boldsymbol{Q}] - (b\boldsymbol{\chi}) : [(a\mathbb{I} + b\boldsymbol{\chi}) : \boldsymbol{Q}] = 0 \tag{5.328}$$

which is only true by virtue of the associativity and above all the commutativity of the tensor-contraction operation for this kind of tensors.

$\boldsymbol{P}$ can be obtained from Equation (5.327) by solving the following equation:

$$\left[(a\mathbb{I} + b\boldsymbol{\chi}) : \boldsymbol{C}^{-1} + ab\boldsymbol{\chi}\right] : \boldsymbol{P} = a\mathbb{I} + b\boldsymbol{\chi}. \tag{5.329}$$

In order to find $\Delta\Delta\gamma_{n+1}^{(k)}$, the equivalent of Equation (5.216) has to be solved. Since

$$\left\{F_{n+1}^{(k)}\right\} = \left\{H_{n+1}^{(k)}\right\} = \left\{\begin{matrix} \boldsymbol{n}_{n+1} \\ \sqrt{\frac{2}{3}} \\ \boldsymbol{n}_{n+1} \end{matrix}\right\}^{(k)} \tag{5.330}$$

it is clear that $\boldsymbol{P} : \boldsymbol{n}$, $\boldsymbol{Q} : \boldsymbol{n}$, $\boldsymbol{R} : \boldsymbol{n}$ and $\boldsymbol{S} : \boldsymbol{n}$ are needed, and not $\boldsymbol{P}$, $\boldsymbol{Q}$, $\boldsymbol{R}$ and $\boldsymbol{S}$. This is an easier task to accomplish. Since (Equation (5.327))

$$[a\mathbb{I} + b\boldsymbol{\chi}] : \boldsymbol{Q} = -[b\boldsymbol{\chi} : \boldsymbol{P}] \tag{5.331}$$

$$[(a\mathbb{I} + b\boldsymbol{\chi}) : \boldsymbol{C}^{-1} + ab\boldsymbol{\chi}] : \boldsymbol{R} = -b\boldsymbol{\chi} \tag{5.332}$$

$$[a\mathbb{I} + b\boldsymbol{\chi}] : \boldsymbol{S} = [\mathbb{I} - b\boldsymbol{\chi} : \boldsymbol{R}] \tag{5.333}$$

and

$$\boldsymbol{\chi} : \boldsymbol{n} = \boldsymbol{\chi} : \boldsymbol{I} = 0 \tag{5.334}$$

$$[a\mathbb{I} + b\boldsymbol{\chi}]^{-1} = \frac{1}{a+b}\left[\mathbb{I} + \frac{b}{3a}\boldsymbol{I} \otimes \boldsymbol{I} + \frac{b}{a}\boldsymbol{n} \otimes \boldsymbol{n}\right] \tag{5.335}$$

one arrives at

$$[(a\mathbb{I} + b\boldsymbol{\chi}) : \boldsymbol{C}^{-1} + ab\boldsymbol{\chi}] : (\boldsymbol{P} : \boldsymbol{n}) = a\boldsymbol{n} \tag{5.336}$$

$$\boldsymbol{Q} : \boldsymbol{n} = -\frac{b}{a+b}\boldsymbol{\chi} : (\boldsymbol{P} : \boldsymbol{n}) \tag{5.337}$$

$$\boldsymbol{R} : \boldsymbol{n} = 0 \tag{5.338}$$

$$\boldsymbol{S} : \boldsymbol{n} = \frac{1}{a}\boldsymbol{n}. \tag{5.339}$$

Consequently, only one 6×6 set of equations must be solved (Equation (5.336), because of symmetry conditions, the nine equations reduce to six), the other equations are explicit. The equivalent of Equation (5.216) now reads

$$\left[\left\{\begin{matrix} \boldsymbol{n}_{n+1}^{(k)} \\ \sqrt{\frac{2}{3}} \\ \boldsymbol{n}_{n+1}^{(k)} \end{matrix}\right\}^{\mathrm{T}} : \left[A_{n+1}^{(k)}\right] : \left\{\begin{matrix} \boldsymbol{n}_{n+1}^{(k)} \\ \sqrt{\frac{2}{3}} \\ \boldsymbol{n}_{n+1}^{(k)} \end{matrix}\right\} + \sqrt{\frac{2}{3}}\partial_{\Delta\gamma} g_{n+1}^{(k)}\right] \Delta\Delta\gamma_{n+1}^{(k)}$$

$$= \left(f_{n+1}^{(k)} - \sqrt{\frac{2}{3}} g_{n+1}^{(k)}\right) - \left\{R_{n+1}^{(k)}\right\}^{\mathrm{T}} : \left[A_{n+1}^{(k)}\right]^{\mathrm{T}} : \left\{\begin{matrix} \boldsymbol{n}_{n+1}^{(k)} \\ \sqrt{\frac{2}{3}} \\ \boldsymbol{n}_{n+1}^{(k)} \end{matrix}\right\}. \tag{5.340}$$

Since

$$\left[A_{n+1}^{(k)}\right] : \begin{Bmatrix} \boldsymbol{n}_{n+1}^{(k)} \\ \sqrt{\frac{2}{3}} \\ \boldsymbol{n}_{n+1}^{(k)} \end{Bmatrix} := \begin{Bmatrix} \boldsymbol{P} : \boldsymbol{n} \\ \sqrt{\frac{2}{3}} d_1 \\ \boldsymbol{Q} : \boldsymbol{n} + \boldsymbol{S} : \boldsymbol{n} \end{Bmatrix}_{n+1}^{(k)} \tag{5.341}$$

and

$$\left[A_{n+1}^{(k)}\right]^{\mathrm{T}} = \left[A_{n+1}^{(k)}\right] \tag{5.342}$$

($\left[A_{n+1}^{(k)}\right]^{-1}$ is symmetric, Equation (5.324), and the inverse of a symmetric matrix is also symmetric), one obtains

$$\left[(\boldsymbol{n} : \boldsymbol{P} : \boldsymbol{n})_{n+1}^{(k)} + \tfrac{2}{3} d_1 + \tfrac{2}{3} d_2 + \sqrt{\tfrac{2}{3}} \partial_{\Delta\gamma} g_{n+1}^{(k)}\right] \Delta\Delta\gamma_{n+1}^{(k)}$$
$$= \left(f_{n+1}^{(k)} - \sqrt{\tfrac{2}{3}} g_{n+1}^{(k)}\right) - \left\{R_{n+1}^{(k)}\right\}^{\mathrm{T}} : \begin{Bmatrix} \boldsymbol{P} : \boldsymbol{n} \\ \sqrt{\frac{2}{3}} d_1 \\ \boldsymbol{Q} : \boldsymbol{n} + \boldsymbol{S} : \boldsymbol{n} \end{Bmatrix}_{n+1}^{(k)} \tag{5.343}$$

where

$$g_{n+1}^{(k)} = \left(\sqrt{\frac{2}{3}} \frac{\Delta\gamma_{n+1}^{(k)}}{A \Delta t}\right)^{\frac{1}{n}} \tag{5.344}$$

and

$$\partial_{\Delta\gamma} g_{n+1}^{(k)} = \sqrt{\frac{2}{3}} \frac{1}{A n \Delta t} \left(\sqrt{\frac{2}{3}} \frac{\Delta\gamma_{n+1}^{(k)}}{A \Delta t}\right)^{\frac{1}{n}-1} \tag{5.345}$$

represent the viscous effects. Equation (5.343) is a linear equation in $\Delta\Delta\gamma_{n+1}^{(k)}$. Once the correction to the consistency parameter is known, the corrections to the internal variables can be calculated using the following equivalent of Equation (5.269) (again dropping the indices $n+1$ and (k) for simplicity):

$$\begin{Bmatrix} \Delta\boldsymbol{\epsilon}^{\mathrm{p}} \\ \Delta\alpha_1 \\ \Delta\boldsymbol{\alpha}_2 \end{Bmatrix} = \begin{bmatrix} \boldsymbol{C}^{-1} & 0 & 0 \\ 0 & d_1^{-1} & 0 \\ 0 & 0 & a\mathbb{I} \end{bmatrix} : \begin{bmatrix} \boldsymbol{P} & 0 & \boldsymbol{R} \\ 0 & d_1 & 0 \\ \boldsymbol{Q} & 0 & \boldsymbol{S} \end{bmatrix} : \begin{Bmatrix} \boldsymbol{R}_{\epsilon} \\ R_{\alpha_1} \\ \boldsymbol{R}_{\alpha_2} \end{Bmatrix} \tag{5.346}$$

$$= \begin{Bmatrix} \boldsymbol{C}^{-1} : (\boldsymbol{P} : \boldsymbol{R}_{\epsilon} + \boldsymbol{R} : \boldsymbol{R}_{\alpha_2}) \\ R_{\alpha_1} \\ a(\boldsymbol{Q} : \boldsymbol{R}_{\epsilon} + \boldsymbol{S} : \boldsymbol{R}_{\alpha_2}) \end{Bmatrix} \tag{5.347}$$

where

$$\begin{Bmatrix} \boldsymbol{R}_\epsilon \\ R_{\alpha_1} \\ \boldsymbol{R}_{\alpha_2} \end{Bmatrix} = \{R\} + \Delta\Delta\gamma \begin{Bmatrix} \boldsymbol{n} \\ \sqrt{\frac{2}{3}} \\ \boldsymbol{n} \end{Bmatrix} \tag{5.348}$$

is the update of the residual. Substitution of Equations (5.329) and (5.332) into the first block equation of Equation (5.347) yields

$$[(a\mathbb{I} + b\boldsymbol{\chi}) : \boldsymbol{C}^{-1} + ab\boldsymbol{\chi}] : [\boldsymbol{C} : \{\Delta\boldsymbol{\epsilon}^{\mathrm{p}}\}] = [(a\mathbb{I} + b\boldsymbol{\chi}) : \boldsymbol{R}_\epsilon - b\boldsymbol{\chi} : \boldsymbol{R}_{\alpha_2}]. \tag{5.349}$$

Notice that the left-hand matrix of Equation (5.349) is the same as in Equation (5.336) and consequently the LU decomposition (in an upper and lower matrix (Zienkiewicz and Taylor 1989)) can be reused. Furthermore, $\boldsymbol{C}^{-1}$, needed to obtain $\{\Delta\boldsymbol{\epsilon}^{\mathrm{p}}\}$ from $\boldsymbol{C} : \{\Delta\boldsymbol{\epsilon}^{\mathrm{p}}\}$, was already computed to obtain the left-hand side in Equation (5.336). Substituting Equations (5.331) and (5.333) into the lower block equation leads to

$$\{\Delta\boldsymbol{\alpha}_2\} = a[a\mathbb{I} + b\boldsymbol{\chi}]^{-1} : [\{\boldsymbol{R}_{\alpha_2}\} - b\boldsymbol{\chi} : (\boldsymbol{P} : \{\boldsymbol{R}_\epsilon\} + \boldsymbol{R} : \{\boldsymbol{R}_{\alpha_2}\})]. \tag{5.350}$$

Using the first block equation of Equation (5.347), this is equivalent to

$$\{\Delta\boldsymbol{\alpha}_2\} = a[a\mathbb{I} + b\boldsymbol{\chi}]^{-1} : [\{\boldsymbol{R}_{\alpha_2}\} - b\boldsymbol{\chi} : \{\boldsymbol{C} : \Delta\boldsymbol{\epsilon}^{\mathrm{p}}\}] \tag{5.351}$$

or

$$\{\Delta\boldsymbol{\alpha}_2\} = \frac{a}{a+b}\left[\mathbb{I} + \frac{b}{3a}\boldsymbol{I} \otimes \boldsymbol{I} + \frac{b}{a}\boldsymbol{n} \otimes \boldsymbol{n}\right] : \{\boldsymbol{R}_{\alpha_2}\} - \frac{ab}{a+b}\boldsymbol{\chi} : \{\boldsymbol{C} : \Delta\boldsymbol{\epsilon}^{\mathrm{p}}\}. \tag{5.352}$$

Accordingly, reintroducing the indices ($\boldsymbol{C}$ is no function of n), one finds for the corrections of the internal variables

$$\left[(a\mathbb{I} + b\boldsymbol{\chi}_{n+1}^{(k)}) : \boldsymbol{C}^{-1} + ab\boldsymbol{\chi}_{n+1}^{(k)}\right] : \left[\boldsymbol{C} : \{\Delta\boldsymbol{\epsilon}^{\mathrm{p}}\}_{n+1}^{(k)}\right] \\ = \left[(a\mathbb{I} + b\boldsymbol{\chi}) : \{\boldsymbol{R}_\epsilon\}_{n+1}^{(k)} - b\boldsymbol{\chi}_{n+1}^{(k)} : \{\boldsymbol{R}_{\alpha_2}\}_{n+1}^{(k)}\right] \tag{5.353}$$

and

$$\{\Delta\alpha_2\}_{n+1}^{(k)} = \frac{a}{a+b}\left[\mathbb{I} + \frac{b}{3a}\boldsymbol{I} \otimes \boldsymbol{I} + \frac{b}{a}\boldsymbol{n}_{n+1}^{(k)} \otimes \boldsymbol{n}_{n+1}^{(k)}\right] : \{\boldsymbol{R}_{\alpha_2}\}_{n+1}^{(k)} \\ - \frac{ab}{a+b}\boldsymbol{\chi}_{n+1}^{(k)} : \left[\boldsymbol{C} : \{\Delta\boldsymbol{\epsilon}^{\mathrm{p}}\}_{n+1}^{(k)}\right]. \tag{5.354}$$

Recall that b is also a function of k and n. As soon as all the corrections are determined, the satisfaction of the flow rule

$$\left| f_{n+1}^{(k+1)} - \sqrt{\tfrac{2}{3}} g_{n+1}^{(k+1)} \right| < \mathrm{TOL} \tag{5.355}$$

can be checked. If satisfied, convergence is reached and the loop can be left. If not, a new correction must be determined. Notice that this loop corresponds to the outer loop in Section 5.4. There is no inner loop since we deal with single surface plasticity. Once convergence is reached, the consistent elastoplastic tangent matrix can be determined. This is obtained from the equations equivalent to Equations (5.276) and (5.278):

$$\left\{\mathrm{d}\boldsymbol{\sigma}_{n+1}\right\} = \boldsymbol{P}_{n+1} : \ \mathbb{I} - G_{n+1}^{-1}[\boldsymbol{n}_{n+1} \otimes \boldsymbol{P}_{n+1} : \boldsymbol{n}_{n+1}]\Big) : \left\{\mathrm{d}\boldsymbol{\epsilon}_{n+1}\right\} \tag{5.356}$$

and

$$\boldsymbol{C}_{n+1}^{\mathrm{ep}} = \boldsymbol{P}_{n+1} - G_{n+1}^{-1}(\boldsymbol{P}_{n+1} : \boldsymbol{n}_{n+1}) \otimes (\boldsymbol{P}_{n+1} : \boldsymbol{n}_{n+1}). \tag{5.357}$$

Here, $\boldsymbol{P} : \boldsymbol{n}$ and $\boldsymbol{P}$ are both needed. For the calculation of $\boldsymbol{P}$, Equation (5.329) can be used. Notice that $\boldsymbol{P}$ is needed only after convergence is reached. The quantity G_{n+1} is defined by (Equation (5.343)):

$$G_{n+1} = (\boldsymbol{n} : \boldsymbol{P} : \boldsymbol{n})_{n+1} + \tfrac{2}{3}d_1 + \tfrac{2}{3}d_2 + \sqrt{\tfrac{2}{3}}\partial_{\Delta\gamma} g_{n+1}. \tag{5.358}$$

Notice that in the absence of kinematic hardening, d_2 can be zero. In that case, a is undetermined (Equation (5.322)). Hence, care must be taken in the implementation to express the equations in terms of a^{-1}. For instance, Equation (5.336) then reads

$$\left[(\mathbb{I} + ba^{-1}\boldsymbol{\chi}) : \boldsymbol{C}^{-1} + b\boldsymbol{\chi}\right] : (\boldsymbol{P} : \boldsymbol{n}) = \boldsymbol{n}. \tag{5.359}$$

Summarizing, the algorithm runs as follows:

1. Compute the elastic predictor and the value of the yield surface (Equations (5.296)–(5.300))
2. Check for plasticity (Equation (5.301)). If satisfied, the solution is found. Else, go to (3).
3. Loop construct
   (a) Calculate the residuals of the flow rule, Equation (5.305), and evolution laws, Equation (5.310). If small enough, exit.
   (b) Calculate a correction to the consistency parameter, Equation (5.343).
   (c) Calculate a correction to the internal variables, Equations (5.353) and (5.354); go to 3a.
4. Determine the consistent elastoplastic tangent matrix, Equation (5.357).

5.5.3 Special case: isotropic elasticity

For isotropic materials, the above equations can be substantially simplified. Indeed,

$$\boldsymbol{C} = \lambda \boldsymbol{I} \otimes \boldsymbol{I} + 2\mu\mathbb{I} \tag{5.360}$$

where μ and λ are Lamé's constants. Using Equation (5.319), one obtains

$$\boldsymbol{C}^{-1} = \left(\frac{1}{9K} - \frac{1}{6\mu}\right)\boldsymbol{I} \otimes \boldsymbol{I} + \frac{1}{2\mu}\mathbb{I} \tag{5.361}$$

where

$$K := \lambda + \tfrac{2}{3}\mu. \tag{5.362}$$

Defining

$$\alpha := \frac{1}{2\mu} \tag{5.363}$$

$$\beta := \frac{1}{9K} - \frac{1}{6\mu} \tag{5.364}$$

one obtains

$$\boldsymbol{C}^{-1} = \alpha\mathbb{I} + \beta\boldsymbol{I} \otimes \boldsymbol{I}. \tag{5.365}$$

From Equation (5.329), the following expression for $\boldsymbol{P}$ results:

$$\boldsymbol{P} = \left[(a\mathbb{I} + b\boldsymbol{\chi}) : \boldsymbol{C}^{-1} + ab\boldsymbol{\chi}\right]^{-1} : [a\mathbb{I} + b\boldsymbol{\chi}]. \tag{5.366}$$

Substitution of Equation (5.365) into (5.366) and taking into account the laws applicable to tensors of the type at stake (such as Equation (5.319)), one arrives after some algebra at

$$\boldsymbol{P} = \frac{1}{[a\alpha + (a+\alpha)b]}\left[(a+b)\mathbb{I} + \frac{ab - 3\beta(a+b)}{3(\alpha + 3\beta)}\boldsymbol{I} \otimes \boldsymbol{I} + \frac{ab}{\alpha}\boldsymbol{n} \otimes \boldsymbol{n}\right] \tag{5.367}$$

and

$$\boldsymbol{P} : \boldsymbol{n} = \tfrac{1}{\alpha}\boldsymbol{n} = 2\mu\boldsymbol{n}. \tag{5.368}$$

Accordingly, the coefficient of $\Delta\Delta\gamma_{n+1}^{(k)}$ in Equation (5.343) reduces to

$$G_{n+1} = 2\mu + \tfrac{2}{3}d_1 + \tfrac{2}{3}d_2 + \sqrt{\tfrac{2}{3}}\partial_{\Delta\gamma} g_{n+1}^{(k)} \tag{5.369}$$

which is identical to the corresponding coefficient in Equation (5.141) since

$$\partial_{\Delta\gamma} g_{n+1}^{(k)} = \sqrt{\tfrac{2}{3}}\partial_{\Delta\epsilon^{\mathrm{peq}}} g_{n+1}^{(k)}. \tag{5.370}$$

The equivalent consistent elastoplastic tangent matrix takes the form (Equation (5.357)

$$\boldsymbol{C}_{n+1}^{\mathrm{ep}} = \boldsymbol{P}_{n+1} - \frac{4\mu^2\boldsymbol{n}_{n+1} \otimes \boldsymbol{n}_{n+1}}{2\mu + \tfrac{2}{3}d_1 + \tfrac{2}{3}d_2 + \tfrac{2}{3}\partial_{\Delta\epsilon^{\mathrm{p}}} g_{n+1}}. \tag{5.371}$$

To check whether Equation (5.371) coincides with Equation (5.162), α and β are substituted into P_{n+1} (the index $n+1$ is dropped for simplicity):

$$\boldsymbol{P} = \frac{1}{1+(2\mu + a^{-1})b}[2\mu\mathbb{I} + \lambda \boldsymbol{I} \otimes \boldsymbol{I}] \\ + \frac{2\mu b}{1+(2\mu + a^{-1})b}\left[a^{-1}\mathbb{I} + K(1 - 3\beta a^{-1})\boldsymbol{I} \otimes \boldsymbol{I} + 2\mu \boldsymbol{n} \otimes \boldsymbol{n}\right] \quad (5.372)$$

which is equivalent to

$$\boldsymbol{P} = (2\mu\mathbb{I} + \lambda \boldsymbol{I} \otimes \boldsymbol{I}) - \frac{(2\mu)^2 b}{1+(2\mu + a^{-1})b}[\mathbb{I} - \tfrac{1}{3}\boldsymbol{I} \otimes \boldsymbol{I} - \boldsymbol{n} \otimes \boldsymbol{n}]. \quad (5.373)$$

Recall that $a = \frac{2}{3}d_2$ and $b = \Delta\gamma/\|\boldsymbol{\xi}\|$. Notice that b contains $\|\boldsymbol{\xi}\|$, whereas Equation (5.162) contains $\|\boldsymbol{\xi}^{\text{trial}}\|$. The connection between both is given by Equation (5.134), in which the term Δh_2^{eq} takes the form

$$\Delta h_2^{\text{eq}} = d_2 \Delta\epsilon^{\text{peq}} = d_2\sqrt{\tfrac{2}{3}}\Delta\gamma \quad (5.374)$$

in the present context of linear hardening laws. Consequently, Equation (5.134) leads to

$$\|\boldsymbol{\xi}^{\text{trial}}\| = \|\boldsymbol{\xi}\|[1 + b(2\mu + a^{-1})] \quad (5.375)$$

and Equation (5.373) yields

$$\boldsymbol{P} = (2\mu\mathbb{I} + \lambda \boldsymbol{I} \otimes \boldsymbol{I}) - \frac{(2\mu)^2 \Delta\gamma}{\|\boldsymbol{\xi}^{\text{trial}}\|} \ \mathbb{I} - \tfrac{1}{3}\boldsymbol{I} \otimes \boldsymbol{I} - \boldsymbol{n} \otimes \boldsymbol{n}\Big). \quad (5.376)$$

Equations (5.371) and (5.376) reproduce Equation (5.162). The isotropic case is recovered as a special case of the anisotropic formulation.

6

Finite Strain Elastoplasticity

Finite strain plasticity implies the existence of large strains or rotations. Therefore, the concept of objectivity (Section 1.6) plays a major role in the development of a finite strain elastoplasticity theory. There are two major classes of models. The first class extends the additive strain concept of the infinitesimal theory to the deformation rate tensor $\boldsymbol{d}$, that is, $\boldsymbol{d} = \boldsymbol{d}^{\mathrm{e}} + \boldsymbol{d}^{\mathrm{p}}$. This, however, leads to a hypoelastic formulation, which means that the elastic stress–strain relations cannot be derived from a stored energy function. For a discussion of this type of models the reader is referred to (Simo and Hughes 1997). The second class of models involves a multiplicative decomposition of the deformation gradient into a plastic and an elastic part and goes back to the work by Lee and Liu (Lee and Liu 1967), (Lee 1969), see also (Simo 1988a), (Simo 1988b) and (Simo and Miehe 1992). Thereby, the elastoplastic motion is viewed as a composition of stress-free plastic flow and stress-inducing elastic deformation. Because of its physical relevance and hyperelastic description of the elastic deformation, this type of model has grown very popular. The theory has been extended to anisotropic viscoplasticity (Miehe 1996a), (Miehe 1996b), (Reese and Svendsen 2003) and nonlocal gradient-enhanced elastoplasticity (Geers *et al.* 2003). The multiplicative concept is also applicable to the inelastic deformation of non-metallic materials such as rubber (Lubliner 1985), (Reese 2003b) and is micromechanically motivated (Reese 2001).

6.1 Multiplicative Decomposition of the Deformation Gradient

The multiplicative decomposition states that the deformation in an elastoplastic material consists of a purely plastic part due to dislocation motion, leading to an intermediate stress-free configuration, followed by a purely elastic deformation rotating and distorting the crystal lattice (Figures 6.1 and 6.2).

$$\boldsymbol{F} = \boldsymbol{F}^{\mathrm{e}} \cdot \boldsymbol{F}^{\mathrm{p}}. \tag{6.1}$$

The Finite Element Method for Three-dimensional Thermomechanical Applications Guido Dhondt

ISBN: 0-470-85752-8

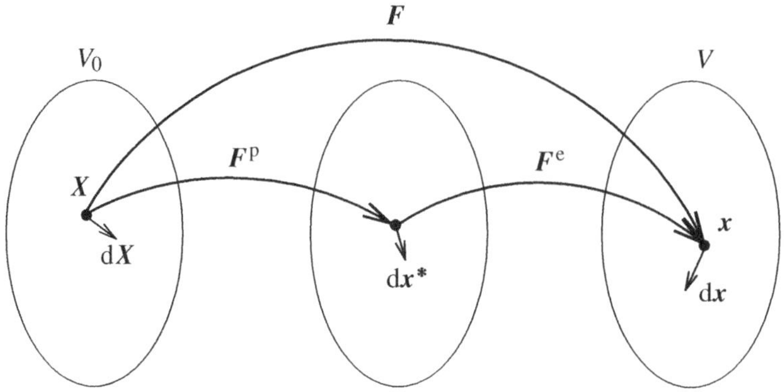

Figure 6.1 Multiplicative decomposition of the deformation gradient

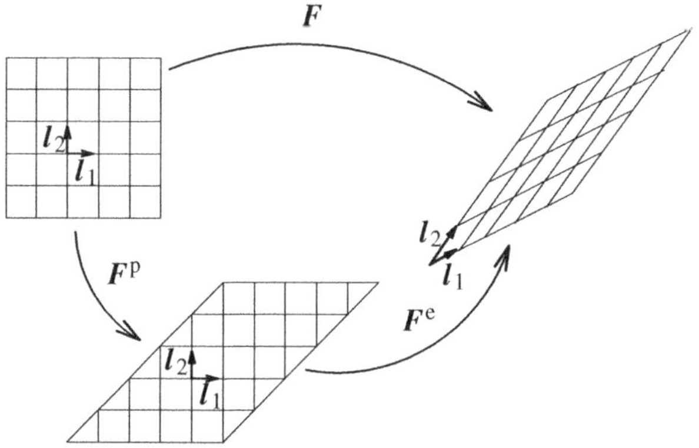

Figure 6.2 Deformation of the crystal lattice

The total left and right Cauchy–Green tensors satisfy (Chapter 1)

$$\boldsymbol{b} = \boldsymbol{F} \cdot \boldsymbol{F}^{\mathrm{T}} \tag{6.2}$$

$$\boldsymbol{C} = \boldsymbol{F}^{\mathrm{T}} \cdot \boldsymbol{F} \tag{6.3}$$

whereas the elastic left Cauchy–Green tensor $\boldsymbol{b}^{\mathrm{e}}$ and the plastic right Cauchy–Green tensor $\boldsymbol{C}^{\mathrm{p}}$ are defined as follows:

$$\boldsymbol{b}^{\mathrm{e}} = \boldsymbol{F}^{\mathrm{e}} \cdot \boldsymbol{F}^{\mathrm{eT}} \tag{6.4}$$

$$\boldsymbol{C}^{\mathrm{p}} = \boldsymbol{F}^{\mathrm{pT}} \cdot \boldsymbol{F}^{\mathrm{p}}. \tag{6.5}$$

The inverse left elastic Cauchy–Green tensor or elastic Finger tensor $\boldsymbol{b}^{\mathrm{e}-1}$ and the right plastic Cauchy–Green tensor $\boldsymbol{C}^{\mathrm{p}}$ are a push-forward/pull-back pair:

$$\boldsymbol{C}^{\mathrm{p}} = \boldsymbol{F}^{\mathrm{pT}} \cdot \boldsymbol{F}^{\mathrm{p}} = \boldsymbol{F}^{\mathrm{T}} \cdot \boldsymbol{F}^{\mathrm{e-T}} \cdot \boldsymbol{F}^{\mathrm{e}-1} \cdot \boldsymbol{F} = \boldsymbol{F}^{\mathrm{T}} \cdot \boldsymbol{b}^{\mathrm{e}-1} \cdot \boldsymbol{F}. \tag{6.6}$$

Accordingly, $\boldsymbol{C}^{\mathrm{p}}$ is the pull-back of $\boldsymbol{b}^{\mathrm{e}-1}$ and $\boldsymbol{b}^{\mathrm{e}-1}$ is the push-forward of $\boldsymbol{C}^{\mathrm{p}}$. Recall that $\boldsymbol{C}$ and the spatial metric tensor $\boldsymbol{g}$ are push-forward/pull-back pairs, as well as $\boldsymbol{C}^{-1}$ and $\boldsymbol{g}^{-1}$. From Figure 6.1, the following relationships prevail:

$$\mathrm{d}\boldsymbol{x} = \boldsymbol{F} \cdot \mathrm{d}\boldsymbol{X} \tag{6.7}$$

$$\mathrm{d}\boldsymbol{x}^* = \boldsymbol{F}^{\mathrm{p}} \cdot \mathrm{d}\boldsymbol{X} = \boldsymbol{F}^{\mathrm{e}-1} \cdot \mathrm{d}\boldsymbol{x} \tag{6.8}$$

and consequently,

$$\mathrm{d}s^2 = C_{KL}\, \mathrm{d}X^K\, \mathrm{d}X^L \tag{6.9}$$

$$\mathrm{d}s^{*2} = C^{\mathrm{p}}_{KL}\, \mathrm{d}X^K\, \mathrm{d}X^L = b^{\mathrm{e}-1}_{kl}\, \mathrm{d}x^k\, \mathrm{d}x^l. \tag{6.10}$$

Hence, the plastic right Cauchy–Green tensor plays the role of a metric tensor in the intermediate configuration with respect to the material frame of reference.

6.2 Deriving the Flow Rule

6.2.1 Arguments of the free-energy function and yield condition

Concentrating on mechanical applications, we start from a general energy function of mechanical grade 1(cf Equation (1.377))

$$\Sigma = \Sigma(\boldsymbol{F}, \boldsymbol{F}^{\mathrm{p}}, \boldsymbol{X}). \tag{6.11}$$

Objectivity in the spatial configuration requires (cf Chapter 1)

$$\Sigma = \Sigma(\boldsymbol{C}, \boldsymbol{F}^{\mathrm{p}}, \boldsymbol{X}). \tag{6.12}$$

Now, in addition, invariance under arbitrary rigid motions in the intermediate configuration is postulated. This is only satisfied if Σ is a function of the inner product of any two vectors in the intermediate configuration:

$$\Sigma = \Sigma(\boldsymbol{C}, \boldsymbol{F}^{\mathrm{pT}} \cdot \boldsymbol{F}^{\mathrm{p}}, \boldsymbol{X}) \tag{6.13}$$

$$= \Sigma(\boldsymbol{C}, \boldsymbol{C}^{\mathrm{p}}, \boldsymbol{X}) \tag{6.14}$$

or, dropping $\boldsymbol{X}$ for convenience,

$$\Sigma = \Sigma(\boldsymbol{C}, \boldsymbol{C}^{\mathrm{p}}). \tag{6.15}$$

In the theory of plasticity, additional internal variables are frequently defined, which we will denote by $\boldsymbol{A}$ in their kinematic form. Accordingly,

$$\Sigma = \Sigma(\boldsymbol{C}, \boldsymbol{C}^{\mathrm{p}}, \boldsymbol{A}) \tag{6.16}$$

and

$$\dot{\Sigma} = \frac{\partial \Sigma}{\partial \boldsymbol{C}} : \dot{\boldsymbol{C}} + \frac{\partial \Sigma}{\partial \boldsymbol{C}^{\mathrm{p}}} : \dot{\boldsymbol{C}}^{\mathrm{p}} + \frac{\partial \Sigma}{\partial \boldsymbol{A}} : \dot{\boldsymbol{A}}. \tag{6.17}$$

In Equation (6.16), $\boldsymbol{C}^{\mathrm{p}}$ and $\boldsymbol{A}$ represent the time–history dependence of plastic deformation. Substitution of Equation (6.17) into the Clausius–Duhem inequality yields

$$\frac{1}{\theta}\left(-\frac{\partial \Sigma}{\partial \boldsymbol{C}}+\frac{1}{2}\boldsymbol{S}\right):\dot{\boldsymbol{C}}-\frac{1}{\theta}\left(\frac{\partial \Sigma}{\partial \boldsymbol{C}^{\mathrm{p}}}:\dot{\boldsymbol{C}}^{\mathrm{p}}+\frac{\partial \Sigma}{\partial \boldsymbol{A}}:\dot{\boldsymbol{A}}\right)-\frac{\rho_0}{\theta}\dot{\theta}\eta-\frac{1}{\theta^2}\boldsymbol{Q}^{\theta}\cdot\nabla_0\theta\geq 0 \quad (6.18)$$

where $\boldsymbol{Q}^{\theta}$ is the heat flux. Assuming similar relationships as in Equation (6.16) for η and $\boldsymbol{Q}^{\theta}$, Equation (6.18) is satisfied if

$$\boldsymbol{S}=2\frac{\partial \Sigma}{\partial \boldsymbol{C}}(\boldsymbol{C},\boldsymbol{C}^{\mathrm{p}},\boldsymbol{A}) \quad (6.19)$$

$$\eta=0 \quad (6.20)$$

$$\boldsymbol{Q}^{\theta}=0 \quad (6.21)$$

$$-\frac{\partial \Sigma}{\partial \boldsymbol{C}^{\mathrm{p}}}:\dot{\boldsymbol{C}}^{\mathrm{p}}-\frac{\partial \Sigma}{\partial \boldsymbol{A}}:\dot{\boldsymbol{A}}\geq 0. \quad (6.22)$$

Equation (6.19) is the classical expression for the second Piola–Kirchhoff stress $\boldsymbol{S}$. Equations (6.20) and (6.21) result from the fact that no temperature dependence is assumed. The crucial equation left to be satisfied is the dissipation inequality (Equation (6.22)). It suggests the definition of the dynamic form $\boldsymbol{Q}$ of the internal variables by

$$\boldsymbol{Q}:=-\frac{\partial \Sigma}{\partial \boldsymbol{A}}:=-\boldsymbol{h}(\boldsymbol{A}) \quad (6.23)$$

reducing Equation (6.22) to

$$-\frac{\partial \Sigma}{\partial \boldsymbol{C}^{\mathrm{p}}}:\dot{\boldsymbol{C}}^{\mathrm{p}}+\boldsymbol{Q}:\dot{\boldsymbol{A}}\geq 0. \quad (6.24)$$

The main goal is to derive expressions for the evolution of $\boldsymbol{C}^{\mathrm{p}}$ and $\boldsymbol{A}$, that is, expressions for $\dot{\boldsymbol{C}}^{\mathrm{p}}$ and $\dot{\boldsymbol{A}}$.

From the previous chapter, we know that an additional equation in the form of a yield condition is required to describe plasticity. The yield condition is usually written in terms of the stress $\boldsymbol{S}$ and the dynamic internal variables $\boldsymbol{Q}$:

$$\Phi(\boldsymbol{S},\boldsymbol{Q})\leq 0. \quad (6.25)$$

Because of Equations (6.19) and (6.23), this is equivalent to

$$\Phi(\boldsymbol{C},\boldsymbol{C}^{\mathrm{p}},\boldsymbol{Q})\leq 0. \quad (6.26)$$

6.2.2 Principle of maximum plastic dissipation

The previous section has shown that the thermodynamic state is characterized by the variables $\{\boldsymbol{C},\boldsymbol{C}^{\mathrm{p}},\boldsymbol{A}\}$. Now, an uncoupled free energy in the internal state variables $\boldsymbol{A}$ is assumed of the form

$$\Sigma=\Sigma(\boldsymbol{C},\boldsymbol{C}^{\mathrm{p}})+\Xi(\boldsymbol{A}). \quad (6.27)$$

The plastic dissipation $\mathbb{D}^{\mathrm{p}}$ amounts to Equation (6.24):

$$\mathbb{D}^{\mathrm{p}}(\boldsymbol{C}, \boldsymbol{C}^{\mathrm{p}}, \boldsymbol{A}; \dot{\boldsymbol{C}}^{\mathrm{p}}; \dot{\boldsymbol{A}}) := -\frac{\partial \Sigma}{\partial \boldsymbol{C}^{\mathrm{p}}} : \dot{\boldsymbol{C}}^{\mathrm{p}} - \frac{\partial \Xi}{\partial \boldsymbol{A}} : \dot{\boldsymbol{A}}. \tag{6.28}$$

To derive the flow rule, the principle of maximum dissipation is invoked. It states that, for fixed $\{\boldsymbol{C}^{\mathrm{p}}, \boldsymbol{A}\}$, the field $\boldsymbol{C}$ will take such a value that for all other fields $\boldsymbol{C}$ satisfying the yield condition, the plastic dissipation is smaller. Hence, defining the cone

$$\mathbb{K}_{\phi} := \{\tilde{\boldsymbol{C}} \in \mathbb{R}^6 | \Phi(\tilde{\boldsymbol{C}}, \boldsymbol{C}^{\mathrm{p}}, \boldsymbol{Q}) \leq 0\} \tag{6.29}$$

of all the states satisfying the yield condition, we have

$$\mathbb{D}^{\mathrm{p}}(\boldsymbol{C}, \boldsymbol{C}^{\mathrm{p}}, \boldsymbol{A}; \dot{\boldsymbol{C}}^{\mathrm{p}}, \dot{\boldsymbol{A}}) \geq \mathbb{D}^{\mathrm{p}}(\tilde{\boldsymbol{C}}, \boldsymbol{C}^{\mathrm{p}}, \boldsymbol{A}; \dot{\boldsymbol{C}}^{\mathrm{p}}, \dot{\boldsymbol{A}}) \quad \forall \tilde{\boldsymbol{C}} \in \mathbb{K}_{\phi} \tag{6.30}$$

or, by use of Equation (6.28),

$$-\frac{\partial \Sigma(\boldsymbol{C}, \boldsymbol{C}^{\mathrm{p}})}{\partial \boldsymbol{C}^{\mathrm{p}}} : \dot{\boldsymbol{C}}^{\mathrm{p}} \geq -\frac{\partial \Sigma(\tilde{\boldsymbol{C}}, \boldsymbol{C}^{\mathrm{p}})}{\partial \boldsymbol{C}^{\mathrm{p}}} : \dot{\boldsymbol{C}}^{\mathrm{p}}, \quad \forall \tilde{\boldsymbol{C}} \in \mathbb{K}_{\phi}. \tag{6.31}$$

Accordingly, $\boldsymbol{C}$ satisfies

$$\boldsymbol{C} = \arg\left\{\max_{\tilde{\boldsymbol{C}} \in \mathbb{K}_{\phi}} \left[-\frac{\partial \Sigma(\tilde{\boldsymbol{C}}, \boldsymbol{C}^{\mathrm{p}})}{\partial \boldsymbol{C}^{\mathrm{p}}} : \dot{\boldsymbol{C}}^{\mathrm{p}}\right]\right\} \tag{6.32}$$

or

$$\boldsymbol{C} = \arg\left\{\min_{\tilde{\boldsymbol{C}} \in \mathbb{K}_{\phi}} \left[\frac{\partial \Sigma(\tilde{\boldsymbol{C}}, \boldsymbol{C}^{\mathrm{p}})}{\partial \boldsymbol{C}^{\mathrm{p}}} : \dot{\boldsymbol{C}}^{\mathrm{p}}\right]\right\} \tag{6.33}$$

where "arg" denotes the argument of the function. This is a constrained minimization problem amenable to mathematical analysis. Indeed, one can prove (Luenberger 1989) that the solution of Equation (6.33) is equivalent to the minimization of the functional

$$\mathbb{L}^{\mathrm{p}} := \frac{\partial \Sigma(\boldsymbol{C}, \boldsymbol{C}^{\mathrm{p}})}{\partial \boldsymbol{C}^{\mathrm{p}}} : \dot{\boldsymbol{C}}^{\mathrm{p}} + \dot{\gamma}\Phi(\boldsymbol{C}, \boldsymbol{C}^{\mathrm{p}}, \boldsymbol{Q}) \tag{6.34}$$

subject to

$$\dot{\gamma} \geq 0 \tag{6.35}$$

$$\dot{\gamma}\Phi(\boldsymbol{C}, \boldsymbol{C}^{\mathrm{p}}, \boldsymbol{Q}) = 0. \tag{6.36}$$

The minimization of $\mathbb{L}^{\mathrm{p}}$ is equivalent to

$$\frac{\partial \mathbb{L}^{\mathrm{p}}}{\partial \boldsymbol{C}} = 0 \tag{6.37}$$

or

$$\frac{\partial^2 \Sigma(\boldsymbol{C}, \boldsymbol{C}^{\mathrm{p}})}{\partial \boldsymbol{C} \partial \boldsymbol{C}^{\mathrm{p}}} : \dot{\boldsymbol{C}}^{\mathrm{p}} = -\dot{\gamma}\frac{\partial \Phi(\boldsymbol{C}, \boldsymbol{C}^{\mathrm{p}}, \boldsymbol{Q})}{\partial \boldsymbol{C}}. \tag{6.38}$$

Equation (6.38) is the flow rule! The principle of maximum plastic dissipation leads to a flow rule, which is a function of the hyperelastic free-energy potential and the yield surface only. Accordingly, as soon as the hyperelastic free energy and yield surface are known, the flow rule is uniquely defined.

6.2.3 Uncoupled volumetric/deviatoric response

The elastoplastic theory can be further simplified if one assumes a completely uncoupled volumetric/deviatoric response throughout the entire range of deformation. It is obtained through a multiplicative decomposition of the deformation gradient:

$$\boldsymbol{F} = J^{1/3}\overline{\boldsymbol{F}} \tag{6.39}$$

and accordingly

$$\det(\overline{\boldsymbol{F}}) = 1. \tag{6.40}$$

The associated right Cauchy–Green tensor takes the form

$$\overline{\boldsymbol{C}} = \overline{\boldsymbol{F}}^{\mathrm{T}} \cdot \overline{\boldsymbol{F}} = J^{-2/3}\boldsymbol{C}. \tag{6.41}$$

By using the chain rule and taking into account that

$$\frac{\partial J}{\partial \boldsymbol{C}} = \frac{J}{2}\boldsymbol{C}^{-1} \tag{6.42}$$

one obtains

$$\frac{\partial \overline{\boldsymbol{C}}}{\partial \boldsymbol{C}} = J^{-2/3} \quad \mathbb{I} - \tfrac{1}{3}\boldsymbol{C} \otimes \boldsymbol{C}^{-1}) \tag{6.43}$$

and in general

$$\frac{\partial(\cdot)}{\partial \boldsymbol{C}} = J^{-2/3}\mathrm{DEV}\left[\frac{\partial(\cdot)}{\partial \overline{\boldsymbol{C}}}\right] \tag{6.44}$$

where

$$\mathrm{DEV}[\cdot] := (\cdot) - \tfrac{1}{3}[\boldsymbol{C} : (\cdot)]\boldsymbol{C}^{-1} \tag{6.45}$$

is the pull-back of the deviator in spatial coordinates. For example, we know that $\boldsymbol{C}^{\mathrm{p}^{-1}}$ is the pull-back of $\boldsymbol{b}^{\mathrm{e}}$:

$$\boldsymbol{C}^{\mathrm{p}^{-1}} = \boldsymbol{F}^{-1} \cdot \boldsymbol{b}^{\mathrm{e}} \cdot \boldsymbol{F}^{-\mathrm{T}}. \tag{6.46}$$

Accordingly,

$$\mathrm{DEV}\boldsymbol{C}^{\mathrm{p}^{-1}} = \boldsymbol{F}^{-1} \cdot \mathrm{dev}\boldsymbol{b}^{\mathrm{e}} \cdot \boldsymbol{F}^{-\mathrm{T}}, \tag{6.47}$$

which leads to

$$\begin{aligned}
\mathrm{DEV}\boldsymbol{C}^{\mathrm{p}^{-1}} &= \boldsymbol{F}^{-1} \cdot \left[\boldsymbol{b}^{\mathrm{e}} - \tfrac{1}{3}(\boldsymbol{b}^{\mathrm{e}} : \boldsymbol{g})\right] \cdot \boldsymbol{F}^{-\mathrm{T}} \\
&= \boldsymbol{C}^{\mathrm{p}^{-1}} - \tfrac{1}{3}(\boldsymbol{b}^{\mathrm{e}} : \boldsymbol{g})\boldsymbol{C}^{-1} \\
&= \boldsymbol{C}^{\mathrm{p}^{-1}} - \tfrac{1}{3}(\boldsymbol{C}^{\mathrm{p}^{-1}} : \boldsymbol{C})\boldsymbol{C}^{-1}.
\end{aligned} \tag{6.48}$$

For metals, the plastic deformation is considered to be isochoric, and consequently the volumetric response is purely elastic. Hence,

$$J = J^{\mathrm{e}}, \quad J^{\mathrm{p}} = 1. \tag{6.49}$$

6.3 Isotropic Hyperelasticity with a von Mises–type Yield Surface

6.3.1 Uncoupled isotropic hyperelastic model

Next, the attention is focused on an isotropic hyperelastic model of the form (Simo 1988a)

$$\Sigma(\boldsymbol{g}, \boldsymbol{b}^{\mathrm{e}-1}, \boldsymbol{F}) = \tfrac{1}{2}\mu(J^{-2/3} I_{1b^{\mathrm{e}}} - 3) + U(J) \tag{6.50}$$

where $I_{b^{\mathrm{e}}}$ is the first invariant of the elastic left Cauchy–Green tensor. The first term on the right-hand side of Equation (6.50) is isochoric, the second is volumetric. The choice of $U(J)$ is not unique. Here, we will take

$$U(J) = \tfrac{1}{2}K\left[\tfrac{1}{2}(J^2 - 1) - \ln J\right] \tag{6.51}$$

which satisfies the asymptotic requirements

$$\lim_{J\to+\infty} U(J) = +\infty \tag{6.52}$$

$$\lim_{J\to 0} U(J) = +\infty \tag{6.53}$$

discussed in Chapter 4. The parameter K is the bulk modulus.

For $\boldsymbol{C} = \boldsymbol{G}$, Equation (6.50) leads to the classical isotropic Hooke law. To prove this, the derivative with respect to $\boldsymbol{C}$ is taken. Using relationships derived in Section 4.4 and noting that $I_{1b} = I_{1C} = I_1$, one obtains

$$\frac{\partial\Sigma}{\partial C_{KL}} = \tfrac{1}{2}\mu J^{-2/3}\left(-\tfrac{1}{3}I_1 C^{-1^{KL}} + G^{KL}\right) + \tfrac{1}{4}K\left(J^2 - 1\right)C^{-1^{KL}} \tag{6.54}$$

and

$$\begin{aligned}\frac{\partial^2\Sigma}{\partial C_{KL}\partial C_{MN}} &= -\tfrac{1}{6}\mu J^{-2/3}C^{-1^{MN}}\left(G^{KL} - \tfrac{1}{3}I_1 C^{-1^{KL}}\right)\\ &+ \tfrac{1}{2}\mu J^{-2/3}\left[-\tfrac{1}{3}G^{MN}C^{-1^{KL}} + \tfrac{1}{6}I_1(C^{-1^{KM}}C^{-1^{LN}} + C^{-1^{KN}}C^{-1^{LM}})\right]\\ &+ \tfrac{1}{4}KJ^2C^{-1^{MN}}C^{-1^{KL}} - \tfrac{1}{8}K(J^2-1)(C^{-1^{KM}}C^{-1^{LN}} + C^{-1^{KN}}C^{-1^{LM}}).\end{aligned} \tag{6.55}$$

For $\boldsymbol{C} = \boldsymbol{G}$, the first and last term drop out and one obtains

$$4\left.\frac{\partial^2\Sigma}{\partial C_{KL}\partial C_{MN}}\right|_{\boldsymbol{C}=\boldsymbol{G}} = \left(K - \tfrac{2}{3}\mu\right)G^{KL}G^{MN} + \mu(G^{KM}G^{LN} + G^{KN}G^{LM}) \tag{6.56}$$

where $K - \frac{2}{3}\mu = \lambda$. This is the classical Hooke law for linear isotropic materials.

Using the appropriate push-forward/pull-back pairs, Equation (6.50) can be reformulated as

$$\Sigma(\boldsymbol{C}, \boldsymbol{C}^{\mathrm{p}}) = \tfrac{1}{2}\mu(J^{-2/3}\mathrm{TR}\boldsymbol{C}^{\mathrm{p}-1} - 3) + U(J) \tag{6.57}$$

where

$$\mathrm{TR}[\cdot] = [\cdot] : \boldsymbol{C} \tag{6.58}$$

is the pull-back of the trace operator in spatial coordinates:

$$\mathrm{tr}(\boldsymbol{b}^{\mathrm{e}}) = \boldsymbol{b}^{\mathrm{e}} : \boldsymbol{g} = \boldsymbol{C}^{\mathrm{p}} : \boldsymbol{C} =: \mathrm{TR}(\boldsymbol{C}^{\mathrm{p}^{-1}}). \tag{6.59}$$

Equation (6.57) can also be written as

$$\Sigma(\boldsymbol{C}, \boldsymbol{C}^{\mathrm{p}}) = \tfrac{1}{2}\mu(\overline{\boldsymbol{C}} : \boldsymbol{C}^{\mathrm{p}^{-1}} - 3) + U(J). \tag{6.60}$$

Applying Equations (6.19),(6.42) and (6.44), one obtains for the second Piola–Kirchhoff stress

$$\boldsymbol{S} = pJ\boldsymbol{C}^{-1} + \mu\mathrm{DEV}(\overline{\boldsymbol{C}}^{\mathrm{p}^{-1}}) \tag{6.61}$$

where

$$p = \frac{\mathrm{d}U}{\mathrm{d}J} = \tfrac{1}{2}K\left[\frac{J^2 - 1}{J}\right] \tag{6.62}$$

and

$$\overline{\boldsymbol{C}}^{\mathrm{p}^{-1}} := J^{-2/3}\boldsymbol{C}^{\mathrm{p}^{-1}}. \tag{6.63}$$

Accordingly (Equation (6.38)),

$$\frac{\partial^2\Sigma(\boldsymbol{C}, \boldsymbol{C}^{\mathrm{p}})}{\partial\boldsymbol{C}\partial\boldsymbol{C}^{\mathrm{p}}} = \frac{1}{2}\frac{\partial\boldsymbol{S}}{\partial\boldsymbol{C}^{\mathrm{p}}} = \frac{\mu}{2}\mathrm{DEV}\left(\frac{\partial\overline{\boldsymbol{C}}^{\mathrm{p}^{-1}}}{\partial\boldsymbol{C}^{\mathrm{p}}}\right) \tag{6.64}$$

and

$$\frac{\partial^2\Sigma(\boldsymbol{C}, \boldsymbol{C}^{\mathrm{p}})}{\partial\boldsymbol{C}\partial\boldsymbol{C}^{\mathrm{p}}} : \dot{\boldsymbol{C}}^{\mathrm{p}} = -\mu J^{-2/3}\mathrm{DEV}\left(\frac{\partial\boldsymbol{C}^{\mathrm{p}^{-1}}}{\partial t}\right). \tag{6.65}$$

6.3.2 Yield surface and derivation of the flow rule

One of the frequently used forms of the yield surface is due to von Mises:

$$\Phi(\boldsymbol{S}, \boldsymbol{C}, \boldsymbol{Q}) = \|\mathrm{DEV}\,\boldsymbol{S}\| + \sqrt{\tfrac{2}{3}}q_1 \tag{6.66}$$

where q_1 is a scalar internal plastic variable satisfying $q_1 = -h_1(\alpha_1)$. The variables q_1 and α_1 are spatial quantities. Since $\boldsymbol{C}^{-1}$ is a symmetric matrix, one finds

$$\mathrm{DEV}(\boldsymbol{C}^{-1}) = \boldsymbol{C}^{-1} - \tfrac{1}{3}(\boldsymbol{C} : \boldsymbol{C}^{-1})\boldsymbol{C}^{-1} = \boldsymbol{0} \tag{6.67}$$

leading to (Equation (6.61))

$$\mathrm{DEV}\,\boldsymbol{S} = \mu\mathrm{DEV}(\overline{\boldsymbol{C}}^{\mathrm{p}^{-1}}). \tag{6.68}$$

Hence, Equation (6.66) can be rewritten as

$$\Phi(\boldsymbol{C}, \boldsymbol{C}^{\mathrm{p}}, \boldsymbol{Q}) = \mu \|\mathrm{DEV}(\overline{\boldsymbol{C}}^{\mathrm{p}-1})\| - \sqrt{\tfrac{2}{3}} h_1(\alpha_1). \tag{6.69}$$

The unit of Φ is stress, and in the present chapter we will assume that this is the Kirchhoff stress. Since $(\boldsymbol{S}, \boldsymbol{\tau})$ and $(\boldsymbol{C}, \boldsymbol{g})$ are push-forward/pull-back pairs, one can write

$$\|\mathrm{DEV}\,\boldsymbol{S}\| = \sqrt{(\mathrm{DEV}\,S)^{IJ}(\mathrm{DEV}\,S)^{KL} C_{IK} C_{JL}} \tag{6.70}$$

$$= \sqrt{(\mathrm{dev}\,\tau)^{ij}(\mathrm{dev}\,\tau)^{kl} g_{ik} g_{jl}}. \tag{6.71}$$

For the flow rule (Equation (6.38)), we need the derivative of Φ with respect to $\boldsymbol{C}$. One can write

$$\frac{\partial \Phi}{\partial \boldsymbol{C}} = \frac{\partial \|\mathrm{DEV}\,\boldsymbol{S}\|}{\partial \boldsymbol{C}} \tag{6.72}$$

$$= \frac{1}{2\|\mathrm{DEV}\,\boldsymbol{S}\|} \frac{\partial}{\partial \boldsymbol{C}} \|\mathrm{DEV}\,\boldsymbol{S}\|^2. \tag{6.73}$$

Furthermore (Equation (6.44)),

$$\frac{\partial}{\partial \boldsymbol{C}} \|\mathrm{DEV}\,\boldsymbol{S}\|^2 = J^{-2/3} \mathrm{DEV}\left(\frac{\partial}{\partial \overline{\boldsymbol{C}}} \|\mathrm{DEV}\,\boldsymbol{S}\|^2\right). \tag{6.74}$$

Now (Equation (6.68)),

$$\frac{\partial}{\partial \overline{\boldsymbol{C}}} \|\mathrm{DEV}\,\boldsymbol{S}\|^2 = \mu^2 \frac{\partial}{\partial \overline{\boldsymbol{C}}} \|\mathrm{DEV}\,\overline{\boldsymbol{C}}^{\mathrm{p}-1}\|^2 \tag{6.75}$$

$$= \mu^2 \frac{\partial}{\partial \overline{\boldsymbol{C}}} \left[(\mathrm{DEV}\,\boldsymbol{C}^{\mathrm{p}-1})^{IJ} (\mathrm{DEV}\,\boldsymbol{C}^{\mathrm{p}-1})^{KL} \overline{\boldsymbol{C}}_{IK} \overline{\boldsymbol{C}}_{JL}\right] \tag{6.76}$$

$$= 2\mu^2 \left[\frac{\partial}{\partial \overline{\boldsymbol{C}}} (\mathrm{DEV}\,\boldsymbol{C}^{\mathrm{p}-1})^{IJ}\right] (\mathrm{DEV}\,\boldsymbol{C}^{\mathrm{p}-1})^{KL} \overline{\boldsymbol{C}}_{IK} \overline{\boldsymbol{C}}_{JL}$$
$$+ 2\mu^2 (\mathrm{DEV}\,\boldsymbol{C}^{\mathrm{p}-1})^{IJ} (\mathrm{DEV}\,\boldsymbol{C}^{\mathrm{p}-1})^{KL} \overline{\boldsymbol{C}}_{JL}. \tag{6.77}$$

Since

$$\frac{\partial}{\partial \overline{\boldsymbol{C}}} (\mathrm{DEV}\,\boldsymbol{C}^{\mathrm{p}-1}) = \frac{\partial}{\partial \overline{\boldsymbol{C}}} \left[\boldsymbol{C}^{\mathrm{p}-1} - \tfrac{1}{3}(\boldsymbol{C} : \boldsymbol{C}^{\mathrm{p}-1}) \boldsymbol{C}^{-1}\right] \tag{6.78}$$

$$= \frac{\partial}{\partial \overline{\boldsymbol{C}}} \left[\boldsymbol{C}^{\mathrm{p}-1} - \tfrac{1}{3}(\overline{\boldsymbol{C}} : \boldsymbol{C}^{\mathrm{p}-1}) \overline{\boldsymbol{C}}^{-1}\right] \tag{6.79}$$

$$= -\tfrac{1}{3}(\mathbb{I} : \boldsymbol{C}^{\mathrm{p}-1}) \overline{\boldsymbol{C}}^{-1} + \tfrac{1}{3}(\overline{\boldsymbol{C}} : \boldsymbol{C}^{\mathrm{p}-1}) \mathbb{I}_{\overline{C}^{-1}} \tag{6.80}$$

$$= -\tfrac{1}{3} \overline{\boldsymbol{C}}^{-1} \otimes \boldsymbol{C}^{\mathrm{p}-1} + \tfrac{1}{3}(\boldsymbol{C}^{\mathrm{p}-1} : \overline{\boldsymbol{C}}) \mathbb{I}_{\overline{C}^{-1}} \tag{6.81}$$

where

$$\mathbb{I}_{\overline{C}^{-1}}\Big)^{IJKL} := \tfrac{1}{2}\left[(\overline{\boldsymbol{C}}^{-1})^{IK} (\overline{\boldsymbol{C}}^{-1})^{JL} + (\overline{\boldsymbol{C}}^{-1})^{IL} (\overline{\boldsymbol{C}}^{-1})^{JK}\right] \tag{6.82}$$

and

$$\mathbb{I} := \mathbb{I}_I \tag{6.83}$$

one obtains

$$\left[\frac{\partial}{\partial \overline{\boldsymbol{C}}}\|\mathrm{DEV}\boldsymbol{S}\|^2\right]^{AB}$$

$$= \tfrac{2}{3}\mu^2[\overbrace{(\boldsymbol{C}^{\mathrm{p}-1} : \boldsymbol{C})\mathbb{I}_{\overline{C}^{-1}}}^{(i)} - \overbrace{\overline{\boldsymbol{C}}^{-1} \otimes \boldsymbol{C}^{\mathrm{p}-1}}^{(ii)}]^{IJAB}(\mathrm{DEV}\boldsymbol{C}^{\mathrm{p}-1})^{KL}\overline{\boldsymbol{C}}_{IK}\overline{\boldsymbol{C}}_{JL}$$
$$+ \underbrace{2\mu^2(\mathrm{DEV}\boldsymbol{C}^{\mathrm{p}-1})^{AJ}(\mathrm{DEV}\boldsymbol{C}^{\mathrm{p}-1})^{BL}\overline{\boldsymbol{C}}_{JL}}_{(iii)}. \tag{6.84}$$

Substituting Equation (6.84) into

$$\frac{\partial\Phi}{\partial \boldsymbol{C}} = \frac{J^{-2/3}}{2\|\mathrm{DEV}\boldsymbol{S}\|}\mathrm{DEV}\left(\frac{\partial}{\partial \overline{\boldsymbol{C}}}\|\mathrm{DEV}\boldsymbol{S}\|^2\right) \tag{6.85}$$

one notices that $\partial\Phi/\partial\boldsymbol{C}$ consists of three additive terms, each of which will be treated separately:

1.

$$\mathrm{DEV}\left[\frac{J^{-2/3}\mu^2}{3\|\mathrm{DEV}\boldsymbol{S}\|}(\boldsymbol{C}^{\mathrm{p}-1} : \overline{\boldsymbol{C}})\mathbb{I}_{\overline{C}^{-1}}^{IJAB}\overline{\boldsymbol{C}}_{IK}\overline{\boldsymbol{C}}_{JL}(\mathrm{DEV}\boldsymbol{C}^{\mathrm{p}-1})^{KL}\right]$$

$$= \mathrm{DEV}\left[\frac{J^{-4/3}\mu^2}{3\|\mathrm{DEV}\boldsymbol{S}\|}\tfrac{1}{2}(\boldsymbol{C}^{\mathrm{p}-1} : \overline{\boldsymbol{C}})(\delta^A{}_K\delta^B{}_L + \delta^B{}_K\delta^A{}_L)(\mathrm{DEV}\boldsymbol{C}^{\mathrm{p}-1})^{KL}\right] \tag{6.86}$$

$$= \mathrm{DEV}\left[\frac{J^{-2/3}\mu}{3\|\mathrm{DEV}\boldsymbol{S}\|}\mathrm{TR}\boldsymbol{C}^{\mathrm{p}-1}(\mathrm{DEV}\boldsymbol{S})^{AB}\right] \tag{6.87}$$

$$= \overline{\mu}\boldsymbol{N} \tag{6.88}$$

where

$$\overline{\mu} := \tfrac{1}{3}\mu J^{-2/3}\mathrm{TR}\boldsymbol{C}^{\mathrm{p}-1} \tag{6.89}$$

$$\boldsymbol{N} := \frac{\mathrm{DEV}\boldsymbol{S}}{\|\mathrm{DEV}\boldsymbol{S}\|}. \tag{6.90}$$

2.

$$-\mathrm{DEV}\left[\frac{J^{-2/3}}{2\|\mathrm{DEV}\boldsymbol{S}\|}\tfrac{2}{3}\mu^2\ \overline{\boldsymbol{C}}^{-1} \otimes \boldsymbol{C}^{\mathrm{p}-1}\Big)^{IJAB}(\mathrm{DEV}\boldsymbol{C}^{\mathrm{p}-1})^{KL}\overline{\boldsymbol{C}}_{IK}\overline{\boldsymbol{C}}_{JL}\right]$$

$$= -\mathrm{DEV}\left[\frac{\mu^2 J^{-4/3}}{\|\mathrm{DEV}\boldsymbol{S}\|}(\boldsymbol{C}^{-1})^{IJ}(\boldsymbol{C}^{\mathrm{p}-1})^{AB}\boldsymbol{C}_{IK}\boldsymbol{C}_{JL}(\mathrm{DEV}\boldsymbol{C}^{\mathrm{p}-1})^{KL}\right] \tag{6.91}$$

$$= -\mathrm{DEV}\left[\frac{\mu^2 J^{-4/3}}{\|\mathrm{DEV}\boldsymbol{S}\|}(\boldsymbol{C}^{\mathrm{p}-1})^{AB}\boldsymbol{C}_{KL}(\mathrm{DEV}\boldsymbol{C}^{\mathrm{p}-1})^{KL}\right] = 0 \tag{6.92}$$

since

$$\boldsymbol{C} : (\mathrm{DEV}\boldsymbol{C}^{\mathrm{p}-1}) = 0. \tag{6.93}$$

To obtain Equation (6.93), recall that $\boldsymbol{C}^{-1} : \boldsymbol{C} = 3$ since $\boldsymbol{C}$ is symmetric.

3.

$$\mathrm{DEV}\left[\frac{J^{-2/3}}{\|\mathrm{DEV}\boldsymbol{S}\|}\mu^2(\mathrm{DEV}\boldsymbol{C}^{\mathrm{p}-1})^{AJ}(\mathrm{DEV}\boldsymbol{C}^{\mathrm{p}-1})^{BL}\overline{\boldsymbol{C}}_{JL}\right]$$

$$= \mathrm{DEV}\left[\frac{\mu^2}{\|\mathrm{DEV}\boldsymbol{S}\|}(\mathrm{DEV}\overline{\boldsymbol{C}}^{\mathrm{p}-1})^{AJ}(\mathrm{DEV}\overline{\boldsymbol{C}}^{\mathrm{p}-1})^{BL}\boldsymbol{C}_{JL}\right] \tag{6.94}$$

$$= \mathrm{DEV}\left[\frac{(\mathrm{DEV}\boldsymbol{S})^{AJ}(\mathrm{DEV}\boldsymbol{S})^{BL}\boldsymbol{C}_{JL}}{\|\mathrm{DEV}\boldsymbol{S}\|}\right] \tag{6.95}$$

$$= \|\mathrm{DEV}\boldsymbol{S}\|\mathrm{DEV}(\boldsymbol{N}^2). \tag{6.96}$$

Accordingly, Equation (6.38) yields

$$-\dot{\gamma}\frac{\partial\overline{\Phi}}{\partial\boldsymbol{C}} = -\dot{\gamma}\overline{\mu}\left[\boldsymbol{N} + \frac{\|\mathrm{DEV}\boldsymbol{S}\|}{\overline{\mu}}\mathrm{DEV}(\boldsymbol{N}^2)\right]. \tag{6.97}$$

Equating Equations (6.65) and (6.97) yields the flow rule:

$$-J^{-2/3}\mu\mathrm{DEV}\left(\frac{\partial\boldsymbol{C}^{\mathrm{p}-1}}{\partial t}\right) = 2\dot{\gamma}\overline{\mu}\left[\boldsymbol{N} + \frac{\|\mathrm{DEV}\boldsymbol{S}\|}{\overline{\mu}}\mathrm{DEV}(\boldsymbol{N}^2)\right]. \tag{6.98}$$

The second term on the right-hand side is much smaller than the first one and is usually dropped.

For infinitesimal strains and rotations, Equation (6.98) reduces to Equation (5.95). Indeed, substituting $\overline{\mu}$ yields (neglecting the term with $\boldsymbol{N}^2$)

$$\mathrm{DEV}(\dot{\boldsymbol{C}}^{\mathrm{p}-1}) = -\frac{2\dot{\gamma}}{3}\mathrm{TR}(\boldsymbol{C}^{\mathrm{p}-1})\boldsymbol{N}. \tag{6.99}$$

In the infinitesimal theory, one can write

$$\boldsymbol{C}^{\mathrm{p}} \approx \boldsymbol{I} + 2\boldsymbol{\epsilon}^{\mathrm{p}} \tag{6.100}$$

$$\boldsymbol{C}^{\mathrm{p}-1} \approx \boldsymbol{I} - 2\boldsymbol{\epsilon}^{\mathrm{p}} \tag{6.101}$$

$$\dot{\boldsymbol{C}}^{\mathrm{p}-1} \approx -2\dot{\boldsymbol{\epsilon}}^{\mathrm{p}}. \tag{6.102}$$

The field $\boldsymbol{\epsilon}^{\mathrm{p}}$ is deviatoric, hence,

$$\mathrm{DEV}(\dot{\boldsymbol{C}}^{\mathrm{p}-1}) \approx -2\dot{\boldsymbol{\epsilon}}^{\mathrm{p}} \tag{6.103}$$

$$\mathrm{TR}(\boldsymbol{C}^{\mathrm{p}-1}) \approx 3 \tag{6.104}$$

and Equation (6.99) reduces to

$$\dot{\boldsymbol{\epsilon}}^{\mathrm{p}} = \dot{\gamma}\boldsymbol{n} \tag{6.105}$$

which coincides with Equation (5.95).

6.4 Extensions

6.4.1 Kinematic hardening

Frequently, a more generalized form of the yield surface is used, in which the center of the yield surface can move. This is accomplished by replacing Equation (6.69) by

$$\Phi(\boldsymbol{C}, \boldsymbol{C}^{\mathrm{p}}, \boldsymbol{Q}) = \|\mathrm{DEV}(\mu\overline{\boldsymbol{C}}^{\mathrm{p}^{-1}} + \overline{\boldsymbol{Q}}_2)\| - \sqrt{\tfrac{2}{3}}h_1(\alpha_1). \tag{6.106}$$

Here, $-\overline{\boldsymbol{Q}}_2$ represents the moving yield-surface center. Equation (6.68) still applies. Replacing $\mathrm{DEV}(\boldsymbol{S})$ by $\mathrm{DEV}(\boldsymbol{S} + \overline{\boldsymbol{Q}}_2)$ and $\overline{\boldsymbol{C}}^{\mathrm{p}^{-1}}$ by $\overline{\boldsymbol{C}}^{\mathrm{p}^{-1}} + \overline{\boldsymbol{Q}}_2/\mu$ in the previous section, one obtains for the flow rule

$$-J^{-2/3}\mu\mathrm{DEV}\left(\frac{\partial \boldsymbol{C}^{\mathrm{p}^{-1}}}{\partial t}\right) = 2\dot{\gamma}\overline{\overline{\mu}}\left(\boldsymbol{N} + \frac{\|\mathrm{DEV}(\boldsymbol{S} + \overline{\boldsymbol{Q}}_2)\|}{\overline{\overline{\mu}}}\mathrm{DEV}(\boldsymbol{N}^2)\right) \tag{6.107}$$

where

$$\overline{\overline{\mu}} := \tfrac{1}{3}J^{-2/3}\mathrm{TR}(\mu\boldsymbol{C}^{\mathrm{p}^{-1}} + \boldsymbol{Q}_2) \tag{6.108}$$

$$= \overline{\mu} + \tfrac{1}{3}J^{-2/3}\mathrm{TR}\,\boldsymbol{Q}_2 \tag{6.109}$$

$$\boldsymbol{N} = \frac{\mathrm{DEV}(\mu\overline{\boldsymbol{C}}^{\mathrm{p}^{-1}} + \overline{\boldsymbol{Q}}_2)}{\|\mathrm{DEV}(\mu\overline{\boldsymbol{C}}^{\mathrm{p}^{-1}} + \overline{\boldsymbol{Q}}_2)\|} = \frac{\mathrm{DEV}(\boldsymbol{S} + \overline{\boldsymbol{Q}}_2)}{\|\mathrm{DEV}(\boldsymbol{S} + \overline{\boldsymbol{Q}}_2)\|} \tag{6.110}$$

$$\overline{\boldsymbol{Q}}_2 = J^{-2/3}\boldsymbol{Q}_2. \tag{6.111}$$

The left-hand side of Equation (6.98) is not changed since it derives from the potential function. Equation (6.107) is an evolution equation for $\boldsymbol{C}^{\mathrm{p}^{-1}}$. The field $\boldsymbol{Q}_2$ is called the *back stress* and represents an internal plastic variable for which an evolution equation is needed as well. Since the fields $\boldsymbol{Q}_2$ and $\mu\boldsymbol{C}^{\mathrm{p}^{-1}}$ are related, an equation similar to Equation (6.107) seems plausible:

$$-J^{-2/3}\mathrm{DEV}\left(\frac{\partial \boldsymbol{Q}_2}{\partial t}\right) = \left(\frac{h_2^{\mathrm{eq}'}}{3\mu}\right)2\dot{\gamma}\overline{\overline{\mu}}\left[\boldsymbol{N} + \frac{\|\mathrm{DEV}(\boldsymbol{S} + \overline{\boldsymbol{Q}}_2)\|}{\overline{\overline{\mu}}}\mathrm{DEV}(\boldsymbol{N}^2)\right]. \tag{6.112}$$

The factor

$$\frac{1}{3\mu}h_2^{\mathrm{eq}'} := \frac{1}{3\mu}\frac{\partial h_2^{\mathrm{eq}}}{\partial \alpha_2^{\mathrm{eq}}} \tag{6.113}$$

was introduced to assure that Equation (6.112) reduces to its infinitesimal equivalent, Equation (5.111). Indeed ($\boldsymbol{q}_2$ is deviatoric,)

$$\mathrm{DEV}(\dot{\boldsymbol{Q}}_2) \approx \mathrm{dev}(\dot{\boldsymbol{q}}_2) \approx \dot{\boldsymbol{q}}_2 \tag{6.114}$$

$$J \approx 1 \tag{6.115}$$

$$\overline{\overline{\mu}} \approx \mu + \tfrac{1}{3}\mathrm{tr}\boldsymbol{q}_2 \approx \mu \tag{6.116}$$

yielding

$$-\dot{\boldsymbol{q}}_2 = \frac{2}{3}\frac{\partial h_2^{\mathrm{eq}}}{\partial \alpha_2^{\mathrm{eq}}}\dot{\gamma}\boldsymbol{n}. \tag{6.117}$$

Here too, one defines

$$h_2^{\mathrm{eq}} := \sqrt{\tfrac{3}{2}}\|\boldsymbol{h}_2\| \tag{6.118}$$

$$\alpha_2^{\mathrm{eq}} := \sqrt{\tfrac{2}{3}}\|\boldsymbol{\alpha}_2\|. \tag{6.119}$$

In the previous chapter, we derived for the infinitesimal theory

$$\alpha_1 = \alpha_2^{\mathrm{eq}} = \epsilon^{\mathrm{peq}} \tag{6.120}$$

and

$$\dot{\epsilon}^{\mathrm{peq}} = \sqrt{\tfrac{2}{3}}\dot{\gamma}. \tag{6.121}$$

These equations will also be used for the finite theory. Accordingly,

$$h_2^{\mathrm{eq}'} := \frac{\partial h_2^{\mathrm{eq}}}{\partial \epsilon^{\mathrm{peq}}}. \tag{6.122}$$

The only curves to be provided by the user are $h_1(\epsilon^{\mathrm{peq}})$ for isotropic hardening and $h_2^{\mathrm{eq}}(\epsilon^{\mathrm{peq}})$ for kinematic hardening.

Equation (6.121) can be interpreted as the definition of ϵ^{peq} for finite strains. Combining Equation (6.121) with the flow rule, Equation (6.99), yields the following kinematic relationship for $\dot{\epsilon}^{\mathrm{peq}}$:

$$\dot{\epsilon}^{\mathrm{peq}} = \sqrt{\tfrac{3}{2}}\frac{\|\mathrm{DEV}(\dot{\boldsymbol{C}}^{\mathrm{p}-1})\|}{\mathrm{TR}(\boldsymbol{C}^{\mathrm{p}-1})}. \tag{6.123}$$

Notice that Equations (6.107) and (6.112) determine only the deviatoric part of $\dot{\boldsymbol{C}}^{\mathrm{p}-1}$ and $\dot{\boldsymbol{Q}}_2$. To guarantee a unique definition of $\dot{\boldsymbol{C}}^{\mathrm{p}-1}$ and $\dot{\boldsymbol{Q}}_2$, the following additional constraints can be defined:

$$\mathrm{TR}(\dot{\boldsymbol{C}}^{\mathrm{p}-1}) = 0. \tag{6.124}$$

$$\mathrm{TR}(\dot{\boldsymbol{Q}}_2) = 0. \tag{6.125}$$

6.4.2 Viscoplastic behavior

For plastic behavior, it is assumed that the stress tensor cannot exceed the yield surface. To illustrate this, the yield surface, as given by Equation (6.106), is shown in Figure 6.3 in deviatoric principal stress space (assuming that $\overline{\boldsymbol{Q}}_2$ and $\boldsymbol{S}$ have the same eigenvectors).

The yield surface can also be written as (cf Equation (6.106))

$$\|\mathrm{DEV}(\boldsymbol{S} + \overline{\boldsymbol{Q}}_2)\| = \sqrt{\tfrac{2}{3}}h_1(\alpha_1) \tag{6.126}$$

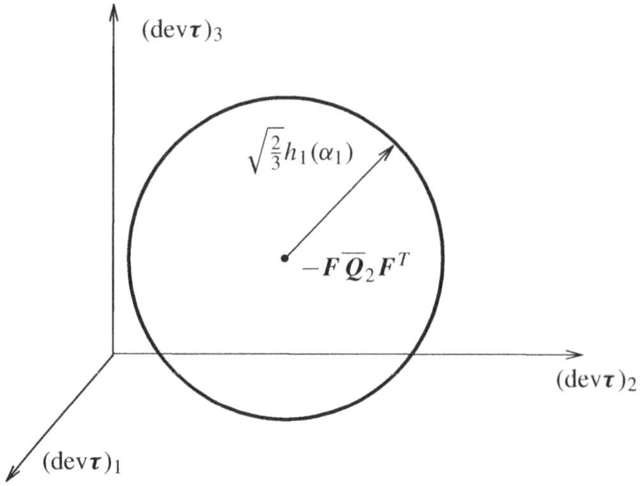

Figure 6.3 von Mises yield surface in deviatoric principal Kirchhoff stress space

which shows that the von Mises surface is indeed a sphere with its center at $-\overline{\boldsymbol{Q}}_2$ and radius $\sqrt{\frac{2}{3}}q_1$. In the classical plasticity theory, any physical stress state must lie on or within the yield surface. However, both q_1 and $\overline{\boldsymbol{Q}}_2$ can change because of isotropic and kinematic hardening respectively. In the viscoplastic theory, the stress state can momentarily lie outside the yield surface; however, it tends asymptotically to the yield surface as time goes by. The way in which the yield surface is approached is generally given by a creep law of the form

$$\tau_{\text{vm}} = f(\dot{\epsilon}^{\text{peq}}). \tag{6.127}$$

The quantity τ_{vm} is the von Mises equivalent stress of the Kirchhoff tensor satisfying

$$\tau_{\text{vm}} := \sqrt{\tfrac{3}{2}}\|\text{dev}\,\boldsymbol{\tau}\| = \sqrt{\tfrac{3}{2}}\sqrt{\text{dev}\,\boldsymbol{\tau} : \text{dev}\,\boldsymbol{\tau}}. \tag{6.128}$$

Accordingly, the plastic equality in Equation (6.126) is replaced in the viscoplastic case by

$$\|\text{DEV}(\boldsymbol{S} + \overline{\boldsymbol{Q}}_2)\| - \sqrt{\tfrac{2}{3}}h_1(\epsilon^{\text{peq}}) = \sqrt{\tfrac{2}{3}}f(\dot{\epsilon}^{\text{peq}}) \tag{6.129}$$

if $\|\text{DEV}(\boldsymbol{S} + \overline{\boldsymbol{Q}}_2)\| - \sqrt{\frac{2}{3}}h_1(\epsilon^{\text{peq}}) > 0$, else the material remains elastic. A typical example of a creep law in the infinitesimal theory is the Norton law

$$\dot{\epsilon}^{\text{peq}} = A\tau_{\text{vm}}^n \tag{6.130}$$

which can also be written as

$$\tau_{\text{vm}} = \left(\frac{\dot{\epsilon}^{\text{peq}}}{A}\right)^{(1/n)}. \tag{6.131}$$

6.5 Summary of the Equations

The following equations result:

1. Hyperelasticity equation

$$S = pJC^{-1} + \mu \mathrm{DEV}(\overline{C}^{\mathrm{p}^{-1}}). \tag{6.132}$$

2. von Mises yield condition

   (a) For plastic materials:

$$\|\Xi\| - \sqrt{\tfrac{2}{3}} h_1(\epsilon^{\mathrm{peq}}) \le 0. \tag{6.133}$$

   (b) For viscoplastic materials:

$$\|\Xi\| - \sqrt{\tfrac{2}{3}} h_1(\epsilon^{\mathrm{peq}}) = \sqrt{\tfrac{2}{3}} f(\dot{\epsilon}^{\mathrm{peq}}) \quad \text{for } \|\Xi\| \ge \sqrt{\tfrac{2}{3}} h_1(\epsilon^{\mathrm{peq}}) \tag{6.134}$$

   where

$$\Xi := \mathrm{DEV}(S + \overline{Q}_2). \tag{6.135}$$

3. Flow rule

$$-J^{-2/3} \mu \mathrm{DEV}(\dot{C}^{\mathrm{p}^{-1}}) = 2\dot{\gamma} \overline{\overline{\mu}} N \tag{6.136}$$

$$\overline{\overline{\mu}} := \overline{\mu} + \tfrac{1}{3} J^{-2/3} \mathrm{TR}\, Q_2 \tag{6.137}$$

$$\overline{\mu} := \tfrac{1}{3} \mu J^{-2/3} \mathrm{TR} C^{\mathrm{p}^{-1}} \tag{6.138}$$

$$N := \frac{\mathrm{DEV}\,\Xi}{\|\mathrm{DEV}\,\Xi\|} \tag{6.139}$$

$$\mathrm{TR}(\dot{C}^{\mathrm{p}^{-1}}) = 0. \tag{6.140}$$

4. Kinematic hardening law

$$-J^{-2/3} \mathrm{DEV}(\dot{Q}_2) = \left(\frac{h_2^{\mathrm{eq}'}(\epsilon^{\mathrm{peq}})}{3\mu} \right) 2\dot{\gamma} \overline{\overline{\mu}} N \tag{6.141}$$

$$\mathrm{TR}(\dot{Q}_2) = 0. \tag{6.142}$$

6.6 Stress Update Algorithm

6.6.1 Derivation

The equations describing viscoplasticity were summarized in the previous section. Ultimately, we would like to transform these equations into a numerical algorithm yielding the solution at time-step $n+1$ if the solution at $t = t_n$ is known. To this end, the backward

Euler rule will be applied to the time derivatives. It is an implicit unconditionally stable scheme expressing the time derivative at $t = t_{n+1}$ in terms of the function values at $t = t_n$ and $t = t_{n+1}$:

$$(\dot{f})_{n+1} \approx \frac{f_{n+1} - f_n}{\Delta t}. \tag{6.143}$$

Applying this to the flow rule, Equation (6.136), yields

$$\mu J_{n+1}^{-2/3} \mathrm{DEV}_{n+1}(\boldsymbol{C}^{\mathrm{p}\,-1}_{n+1} - \boldsymbol{C}^{\mathrm{p}\,-1}_{n}) = -2(\gamma_{n+1} - \gamma_n)\overline{\overline{\mu}}_{n+1}\boldsymbol{N}_{n+1} \tag{6.144}$$

where

$$\mathrm{DEV}_{n+1}\boldsymbol{A} := \boldsymbol{A} - \tfrac{1}{3}(\boldsymbol{A} : \boldsymbol{C}_{n+1})\boldsymbol{C}_{n+1}^{-1}. \tag{6.145}$$

Henceforth, the following definition will be used:

$$\Delta\gamma_{n+1} := \gamma_{n+1} - \gamma_n. \tag{6.146}$$

The hyperelasticity equation, Equation (6.132), yields

$$\mathrm{DEV}_{n+1}\boldsymbol{S}_{n+1} = \mu J_{n+1}^{-2/3} \mathrm{DEV}_{n+1}(\boldsymbol{C}^{\mathrm{p}\,-1}_{n+1}). \tag{6.147}$$

Consequently, the flow rule, Equation (6.144), can be written as

$$\mathrm{DEV}_{n+1}\boldsymbol{S}_{n+1} = \mu J_{n+1}^{-2/3} \mathrm{DEV}_{n+1}\boldsymbol{C}^{\mathrm{p}\,-1}_{n} - 2\Delta\gamma_{n+1}\overline{\overline{\mu}}_{n+1}\boldsymbol{N}_{n+1}. \tag{6.148}$$

Equations (6.138) and (6.137) yield

$$\overline{\overline{\mu}}_{n+1} = \tfrac{1}{3}\mu J_{n+1}^{-2/3} \mathrm{TR}_{n+1}\boldsymbol{C}^{\mathrm{p}\,-1}_{n+1} + \tfrac{1}{3} J_{n+1}^{-2/3} \mathrm{TR}_{n+1}\boldsymbol{Q}_{2,n+1}. \tag{6.149}$$

The auxiliary equations, Equations (6.140) and (6.142), lead to

$$\boldsymbol{C}^{\mathrm{p}\,-1}_{n+1} : \boldsymbol{C}_{n+1} = \boldsymbol{C}^{\mathrm{p}\,-1}_{n} : \boldsymbol{C}_{n+1} \tag{6.150}$$

$$\boldsymbol{Q}_{2,n+1} : \boldsymbol{C}_{n+1} = \boldsymbol{Q}_{2,n} : \boldsymbol{C}_{n+1}. \tag{6.151}$$

Hence, Equation (6.149) can be transformed into

$$\overline{\overline{\mu}}_{n+1} = \tfrac{1}{3}\mu J_{n+1}^{-2/3} \mathrm{TR}_{n+1}\boldsymbol{C}^{\mathrm{p}\,-1}_{n} + \tfrac{1}{3} J_{n+1}^{-2/3} \mathrm{TR}_{n+1}\boldsymbol{Q}_{2,n}. \tag{6.152}$$

In a similar way, Equation (6.141) is transformed into

$$J_{n+1}^{-2/3} \mathrm{DEV}_{n+1}(\boldsymbol{Q}_{2,n+1}) = J_{n+1}^{-2/3} \mathrm{DEV}_{n+1}(\boldsymbol{Q}_{2,n}) - \frac{2h_2^{\mathrm{eq}'}}{3\mu}\Delta\gamma_{n+1}\overline{\overline{\mu}}_{n+1}\boldsymbol{N}_{n+1}. \tag{6.153}$$

Defining

$$\boldsymbol{T}^{\mathrm{trial}}_{n+1} := \mu J_{n+1}^{-2/3} \mathrm{DEV}_{n+1}\boldsymbol{C}^{\mathrm{p}\,-1}_{n} \tag{6.154}$$

$$\boldsymbol{A}^{\mathrm{trial}}_{n+1} := -J_{n+1}^{-2/3} \mathrm{DEV}_{n+1}(\boldsymbol{Q}_{2,n}) \tag{6.155}$$

$$\boldsymbol{\Xi}^{\mathrm{trial}}_{n+1} := \boldsymbol{T}^{\mathrm{trial}}_{n+1} - \boldsymbol{A}^{\mathrm{trial}}_{n+1} \tag{6.156}$$

$$\boldsymbol{T}_{n+1} := \mathrm{DEV}_{n+1}\boldsymbol{S}_{n+1} \tag{6.157}$$

$$\boldsymbol{A}_{n+1} := -J_{n+1}^{-2/3} \mathrm{DEV}_{n+1}(\boldsymbol{Q}_{2,n+1}) \tag{6.158}$$

$$\boldsymbol{\Xi}_{n+1} := \boldsymbol{T}_{n+1} - \boldsymbol{A}_{n+1} \tag{6.159}$$

the evolution equations, Equations (6.148) and (6.153), can be rewritten as

$$\boldsymbol{T}_{n+1} = \boldsymbol{T}^{\text{trial}}_{n+1} - 2\Delta\gamma_{n+1}\overline{\overline{\mu}}_{n+1}\boldsymbol{N}_{n+1} \tag{6.160}$$

$$\boldsymbol{A}_{n+1} = \boldsymbol{A}^{\text{trial}}_{n+1} + \frac{2h_2^{\text{eq'}}}{3\mu}\Delta\gamma_{n+1}\overline{\overline{\mu}}_{n+1}\boldsymbol{N}_{n+1}. \tag{6.161}$$

Notice that as soon as the displacement field at $t = t_{n+1}$ is known, the trial functions can be calculated. They represent the stress state at $t = t_{n+1}$ in the assumption that step $n + 1$ is purely elastic. Next, the yield-surface condition is checked (Equation (6.133)). Using Equations (6.157), (6.158) and (6.159), the yield condition at $t = t_{n+1}$ can be expressed as

$$\|\boldsymbol{\Xi}_{n+1}\| \leq \sqrt{\tfrac{2}{3}}h_1(\epsilon^{\text{peq}}_{n+1}). \tag{6.162}$$

Assuming at first that there is no plastic flow, or

$$\boldsymbol{\Xi}_{n+1} = \boldsymbol{\Xi}^{\text{trial}}_{n+1} \tag{6.163}$$

$$\epsilon^{\text{peq}}_{n+1} = \epsilon^{\text{peq}}_{n} \tag{6.164}$$

Equation (6.162) reduces to

$$\|\boldsymbol{\Xi}^{\text{trial}}_{n+1}\| \leq \sqrt{\tfrac{2}{3}}h_1(\epsilon^{\text{peq}}_{n}). \tag{6.165}$$

If this equation is satisfied, the state is purely elastic and the trial functions are the solution. Furthermore, the plastic internal variables do not change. If, on the other hand, Equation (6.165) is not satisfied, plastic deformation occurs and Equation (6.134) applies, where $f = 0$ for nonviscous deformation. At $t = t_{n+1}$, this equation reads

$$\|\boldsymbol{\Xi}_{n+1}\| = \sqrt{\tfrac{2}{3}}h_1(\epsilon^{\text{peq}}_{n+1}) + \sqrt{\tfrac{2}{3}}f(\epsilon^{\text{peq}}_{n+1}). \tag{6.166}$$

Now $\boldsymbol{\Xi}_{n+1}$ satisfies

$$\boldsymbol{\Xi}_{n+1} = \boldsymbol{T}_{n+1} - \boldsymbol{A}_{n+1} \tag{6.167}$$

$$= \boldsymbol{T}^{\text{trial}}_{n+1} - \boldsymbol{A}^{\text{trial}}_{n+1} - 2\overline{\overline{\mu}}_{n+1}\left(1 + \frac{h_2^{\text{eq'}}}{3\mu}\right)\Delta\gamma_{n+1}\boldsymbol{N}_{n+1} \tag{6.168}$$

$$= \boldsymbol{\Xi}^{\text{trial}}_{n+1} - 2\overline{\overline{\mu}}_{n+1}\left(1 + \frac{h_2^{\text{eq'}}}{3\mu}\right)\Delta\gamma_{n+1}\frac{\boldsymbol{\Xi}_{n+1}}{\|\boldsymbol{\Xi}_{n+1}\|}. \tag{6.169}$$

Equation (6.169) reveals that $\boldsymbol{\Xi}_{n+1}$ and $\boldsymbol{\Xi}^{\text{trial}}_{n+1}$ are parallel, and accordingly,

$$\boldsymbol{N}_{n+1} = \frac{\boldsymbol{\Xi}^{\text{trial}}_{n+1}}{\|\boldsymbol{\Xi}^{\text{trial}}_{n+1}\|} \tag{6.170}$$

which means that $\boldsymbol{N}_{n+1}$ can be calculated using the trial state. Substituting Equation (6.170) into Equation (6.169) and taking the norm, one finds that

$$\|\boldsymbol{\Xi}_{n+1}\| = \|\boldsymbol{\Xi}^{\text{trial}}_{n+1}\| - 2\overline{\overline{\mu}}_{n+1}\left(1 + \frac{h_2^{\text{eq'}}}{3\mu}\right)\Delta\gamma_{n+1} \tag{6.171}$$

and the yield condition, Equation (6.166) leads to

$$\|\Xi^{\rm trial}_{n+1}\| - 2\overline{\overline{\mu}}_{n+1}\left(1+\frac{h_2^{\rm eq'}(\epsilon^{\rm peq}_{n+1})}{3\mu}\right)\Delta\gamma_{n+1} = \sqrt{\tfrac{2}{3}}h_1(\epsilon^{\rm peq}_{n+1}) + \sqrt{\tfrac{2}{3}}f(\dot{\epsilon}^{\rm peq}_{n+1}). \tag{6.172}$$

In this equation, $\|\Xi^{\rm trial}_{n+1}\|$ and $\overline{\overline{\mu}}_{n+1}$ are known, $\epsilon^{\rm peq}_{n+1}$ and $\Delta\gamma_{n+1}$ are unknowns, related by Equation (6.121) , or, equivalently,

$$\epsilon^{\rm peq}_{n+1} = \epsilon^{\rm peq}_{n} + \sqrt{\tfrac{2}{3}}\Delta\gamma_{n+1}. \tag{6.173}$$

Accordingly, Equation (6.172) yields

$$\|\Xi^{\rm trial}_{n+1}\| - 2\overline{\overline{\mu}}_{n+1}\left[1+\frac{h_2^{\rm eq'}\left(\epsilon^{\rm peq}_{n} + \sqrt{\tfrac{2}{3}}\Delta\gamma_{n+1}\right)}{3\mu}\right]\Delta\gamma_{n+1}$$
$$= \sqrt{\tfrac{2}{3}}h_1\left(\epsilon^{\rm peq}_{n} + \sqrt{\tfrac{2}{3}}\Delta\gamma_{n+1}\right) + \sqrt{\tfrac{2}{3}}f\left(\sqrt{\tfrac{2}{3}}\Delta\gamma_{n+1}\right). \tag{6.174}$$

This is a nonlinear equation in $\Delta\gamma_{n+1}$, which can be solved by the Newton–Raphson technique. Once $\Delta\gamma_{n+1}$ is known, all other quantities can be calculated using the equations in this section. Indeed, the definition of DEV leads to

$$\mathrm{DEV}_{n+1}\boldsymbol{C}^{\rm p}{}^{-1}_{n+1} = \boldsymbol{C}^{\rm p}{}^{-1}_{n+1} - \tfrac{1}{3}(\boldsymbol{C}^{\rm p}{}^{-1}_{n+1} : \boldsymbol{C}_{n+1})\boldsymbol{C}^{-1}_{n+1} \tag{6.175}$$

$$\mathrm{DEV}_{n+1}\boldsymbol{C}^{\rm p}{}^{-1}_{n} = \boldsymbol{C}^{\rm p}{}^{-1}_{n} - \tfrac{1}{3}(\boldsymbol{C}^{\rm p}{}^{-1}_{n} : \boldsymbol{C}_{n+1})\boldsymbol{C}^{-1}_{n+1} \tag{6.176}$$

and consequently, since

$$\mathrm{TR}_{n+1}\boldsymbol{C}^{\rm p}{}^{-1}_{n+1} = \mathrm{TR}_{n+1}\boldsymbol{C}^{\rm p}{}^{-1}_{n}, \tag{6.177}$$

we find

$$\mathrm{DEV}_{n+1}\boldsymbol{C}^{\rm p}{}^{-1}_{n+1} - \mathrm{DEV}_{n+1}\boldsymbol{C}^{\rm p}{}^{-1}_{n} = \boldsymbol{C}^{\rm p}{}^{-1}_{n+1} - \boldsymbol{C}^{\rm p}{}^{-1}_{n}. \tag{6.178}$$

Equation (6.144) can be transformed into

$$\mathrm{DEV}_{n+1}\boldsymbol{C}^{\rm p}{}^{-1}_{n+1} - \mathrm{DEV}_{n+1}\boldsymbol{C}^{\rm p}{}^{-1}_{n} = -2\Delta\gamma_{n+1}\frac{\overline{\overline{\mu}}_{n+1}}{\mu}J^{-2/3}_{n+1}\boldsymbol{N}_{n+1}. \tag{6.179}$$

Hence,

$$\boldsymbol{C}^{\rm p}{}^{-1}_{n+1} = \boldsymbol{C}^{\rm p}{}^{-1}_{n} - 2\Delta\gamma_{n+1}\frac{\overline{\overline{\mu}}_{n+1}}{\mu}J^{-2/3}_{n+1}\boldsymbol{N}_{n+1}. \tag{6.180}$$

Similarly, Equation (6.153) yields

$$\boldsymbol{Q}_{2,n+1} = \boldsymbol{Q}_{2,n} - \frac{2h^{\rm eq'}}{3\mu}\Delta\gamma_{n+1}\overline{\overline{\mu}}_{n+1}J^{-2/3}_{n+1}\boldsymbol{N}_{n+1}. \tag{6.181}$$

Finally, the second Piola–Kirchhoff stress follows from Equations (6.148) and (6.132):

$$\boldsymbol{S}_{n+1} = \boldsymbol{T}^{\rm trial}_{n+1} - 2\Delta\gamma_{n+1}\overline{\overline{\mu}}_{n+1}\boldsymbol{N}_{n+1} + \frac{K}{2}(J^2_{n+1} - 1)\boldsymbol{C}^{-1}_{n+1}. \tag{6.182}$$

6.6.2 Summary

Given: $\boldsymbol{C}^{\mathrm{p}-1}_{n}$, $\boldsymbol{Q}_{2,n}$, γ_n, $\epsilon_n^{\mathrm{peq}}$, U_{n+1}.

1. Step 1: Geometric update
   $\boldsymbol{C}_{n+1}$, J_{n+1}, $\boldsymbol{C}^{-1}_{n+1}$.

2. Step 2: Elastic prediction

$$\mathrm{TR}_{n+1}\boldsymbol{C}^{\mathrm{p}-1}_{n} = \boldsymbol{C}^{\mathrm{p}-1}_{n} : \boldsymbol{C}_{n+1} \tag{6.183}$$

$$\mathrm{DEV}_{n+1}\boldsymbol{C}^{\mathrm{p}-1}_{n} = \boldsymbol{C}^{\mathrm{p}-1}_{n} - \tfrac{1}{3}(\mathrm{TR}_{n+1}\boldsymbol{C}^{\mathrm{p}-1}_{n})\boldsymbol{C}^{-1}_{n+1} \tag{6.184}$$

$$\mathrm{TR}_{n+1}(\boldsymbol{Q}_{2,n}) = \boldsymbol{Q}_{2,n} : \boldsymbol{C}_{n+1} \tag{6.185}$$

$$\mathrm{DEV}_{n+1}(\boldsymbol{Q}_{2,n}) = \boldsymbol{Q}_{2,n} - \tfrac{1}{3}\mathrm{TR}_{n+1}(\boldsymbol{Q}_{2,n})\boldsymbol{C}^{-1}_{n+1} \tag{6.186}$$

$$\boldsymbol{T}^{\mathrm{trial}}_{n+1} = \mu J_{n+1}^{-2/3}\mathrm{DEV}_{n+1}\boldsymbol{C}^{\mathrm{p}-1}_{n} \tag{6.187}$$

$$\boldsymbol{A}^{\mathrm{trial}}_{n+1} = -J_{n+1}^{-2/3}\mathrm{DEV}_{n+1}(\boldsymbol{Q}_{2,n}) \tag{6.188}$$

$$\boldsymbol{\Xi}^{\mathrm{trial}}_{n+1} = \boldsymbol{T}^{\mathrm{trial}}_{n+1} - \boldsymbol{A}^{\mathrm{trial}}_{n+1}. \tag{6.189}$$

3. Step 3: Check for yielding
   If

$$\|\boldsymbol{\Xi}^{\mathrm{trial}}_{n+1}\| - \sqrt{\tfrac{2}{3}}h_1(\epsilon_n^{\mathrm{peq}}) \le 0 \tag{6.190}$$

$(\cdot)_{n+1} = (\cdot)^{\mathrm{trial}}_{n+1}$ and EXIT.

4. Step 4: Radial return scheme

$$\overline{\overline{\mu}}_{n+1} = \tfrac{1}{3}J_{n+1}^{-2/3}\left[\mu\mathrm{TR}_{n+1}(\boldsymbol{C}^{\mathrm{p}-1}_{n}) + \mathrm{TR}_{n+1}(\boldsymbol{Q}_{2,n})\right] \tag{6.191}$$

$$\boldsymbol{N}_{n+1} = \frac{\boldsymbol{\Xi}^{\mathrm{trial}}_{n+1}}{\|\boldsymbol{\Xi}^{\mathrm{trial}}_{n+1}\|} \tag{6.192}$$

$$\|\boldsymbol{\Xi}^{\mathrm{trial}}_{n+1}\| - 2\overline{\overline{\mu}}_{n+1}\left[1 + \frac{h_2^{\mathrm{eq}'}\left(\epsilon_n^{\mathrm{peq}} + \sqrt{\tfrac{2}{3}}\Delta\gamma_{n+1}\right)}{3\mu}\right]\Delta\gamma_{n+1} = \sqrt{\tfrac{2}{3}}h_1\left(\epsilon_n^{\mathrm{peq}} + \sqrt{\tfrac{2}{3}}\Delta\gamma_{n+1}\right) + \sqrt{\tfrac{2}{3}}f\left(\sqrt{\tfrac{2}{3}}\Delta\gamma_{n+1}\right) \tag{6.193}$$

from which $\Delta\gamma_{n+1}$ can be determined.

5. Update of the plastic state variables

$$\epsilon_{n+1}^{\mathrm{peq}} = \epsilon_n^{\mathrm{peq}} + \sqrt{\tfrac{2}{3}}\Delta\gamma_{n+1} \tag{6.194}$$

$$\boldsymbol{C}^{\mathrm{p}-1}_{n+1} = \boldsymbol{C}^{\mathrm{p}-1}_{n} - 2\Delta\gamma_{n+1}\frac{\overline{\overline{\mu}}_{n+1}}{\mu}J_{n+1}^{2/3}\boldsymbol{N}_{n+1} \tag{6.195}$$

$$\boldsymbol{Q}_{2,n+1} = \boldsymbol{Q}_n - \frac{2h_2^{\mathrm{eq}'}}{3\mu}\Delta\gamma_{n+1}\overline{\overline{\mu}}_{n+1}J_{n+1}^{2/3}\boldsymbol{N}_{n+1}. \tag{6.196}$$

6. Update of the stress

$$\boldsymbol{S}_{n+1} = \boldsymbol{T}_{n+1}^{\mathrm{trial}} - 2\Delta\gamma_{n+1}\overline{\overline{\mu}}_{n+1}\boldsymbol{N}_{n+1} + \frac{K}{2}(J_{n+1}^2 - 1)\boldsymbol{C}_{n+1}^{-1}. \tag{6.197}$$

Sometimes, Equation (6.193) is written in a different way. Defining

$$f_{n+1}^{\mathrm{trial}} := \|\boldsymbol{\Xi}_{n+1}^{\mathrm{trial}}\| - \sqrt{\tfrac{2}{3}}h_1(\epsilon_n^{\mathrm{peq}}) \tag{6.198}$$

one gets

$$\begin{aligned} f_{n+1}^{\mathrm{trial}} &= \sqrt{\tfrac{2}{3}}\left[h_1\left(\epsilon_n^{\mathrm{peq}} + \sqrt{\tfrac{2}{3}}\Delta\gamma_{n+1}\right) - h_1(\epsilon_n^{\mathrm{peq}})\right] \\ &\quad + \sqrt{\tfrac{2}{3}}f\left(\sqrt{\tfrac{2}{3}}\Delta\gamma_{n+1}\right) + 2\overline{\overline{\mu}}_{n+1}\left[1 + \frac{h_2^{\mathrm{eq}'}\left(\epsilon_n^{\mathrm{peq}} + \sqrt{\tfrac{2}{3}}\Delta\gamma_{n+1}\right)}{3\mu}\right]\Delta\gamma_{n+1}. \end{aligned} \tag{6.199}$$

For linear hardening and creep laws of the form

$$h_1\left(\epsilon_n^{\mathrm{peq}} + \sqrt{\tfrac{2}{3}}\Delta\gamma_{n+1}\right) = h_1(\epsilon_n^{\mathrm{peq}}) + h_1'\Delta\gamma_{n+1}\sqrt{\tfrac{2}{3}} \tag{6.200}$$

$$h_2^{\mathrm{eq}}\left(\epsilon_n^{\mathrm{peq}} + \sqrt{\tfrac{2}{3}}\Delta\gamma_{n+1}\right) = h_2^{\mathrm{eq}}(\epsilon_n^{\mathrm{peq}}) + h_2^{\mathrm{eq}'}\Delta\gamma_{n+1}\sqrt{\tfrac{2}{3}} \tag{6.201}$$

$$f\left(\sqrt{\tfrac{2}{3}}\Delta\gamma_{n+1}\right) = \eta\sqrt{\tfrac{2}{3}}\Delta\gamma_{n+1} \tag{6.202}$$

where h_1', $h_2^{\mathrm{eq}'}$ and η are constants. Equation (6.199) further reduces to

$$2\overline{\overline{\mu}}\Delta\gamma_{n+1} = \frac{f_{n+1}^{\mathrm{trial}}}{1 + \frac{h_2^{\mathrm{eq}'}}{3\mu} + \frac{h_1'}{3\overline{\overline{\mu}}} + \frac{\eta}{3\overline{\overline{\mu}}}}. \tag{6.203}$$

For nonlinear laws, Equation (6.199) is first written as

$$\begin{aligned} g(\Delta\gamma_{n+1}) &:= f_{n+1}^{\mathrm{trial}} - \sqrt{\tfrac{2}{3}}\left[h_1\left(\epsilon_n^{\mathrm{peq}} + \sqrt{\tfrac{2}{3}}\Delta\gamma_{n+1}\right) - h_1(\epsilon_n^{\mathrm{peq}})\right] - \sqrt{\tfrac{2}{3}}f\left(\sqrt{\tfrac{2}{3}}\Delta\gamma_{n+1}\right) \\ &\quad - \overline{\overline{\mu}}_{n+1}\left\{2\Delta\gamma_{n+1} + \sqrt{\tfrac{2}{3}}\left[h_2^{\mathrm{eq}}\left(\epsilon_n^{\mathrm{peq}} + \sqrt{\tfrac{2}{3}}\Delta\gamma_{n+1}\right) - h_2^{\mathrm{eq}}(\epsilon_n^{\mathrm{peq}})\right]\frac{1}{\mu}\right\} \end{aligned} \tag{6.204}$$

since (backward Euler)

$$h_2^{\mathrm{eq}'}\left(\epsilon_n^{\mathrm{peq}}+\sqrt{\tfrac{2}{3}}\Delta\gamma_{n+1}\right)\approx\frac{h_2^{\mathrm{eq}}\left(\epsilon_n^{\mathrm{peq}}+\sqrt{\tfrac{2}{3}}\Delta\gamma_{n+1}\right)-h_2^{\mathrm{eq}}\left(\epsilon_n^{\mathrm{peq}}\right)}{\sqrt{\tfrac{2}{3}}\Delta\gamma_{n+1}}. \tag{6.205}$$

For a Newton–Raphson type solution of Equation (6.204), the derivative of g is also needed:

$$\begin{aligned}\frac{\mathrm{d}g}{\mathrm{d}(\Delta\gamma_{n+1})}&=-\tfrac{2}{3}h_1'\left(\epsilon_n^{\mathrm{peq}}+\sqrt{\tfrac{2}{3}}\Delta\gamma_{n+1}\right)-\tfrac{2}{3}f'\left(\sqrt{\tfrac{2}{3}}\Delta\gamma_{n+1}\right)\\&\quad-2\overline{\overline{\mu}}_{n+1}\left[1+\left(\frac{1}{3\mu}\right)h_2^{\mathrm{eq}'}\left(\epsilon_n^{\mathrm{peq}}+\sqrt{\tfrac{2}{3}}\Delta\gamma_{n+1}\right)\right]\end{aligned} \tag{6.206}$$

where h_1', $h_2^{\mathrm{eq}'}$ and f' denote derivatives with respect to their arguments. Since h_1, h_2^{eq} and f are user-defined functions, the derivatives can be determined too (analytically or numerically). The Newton–Raphson scheme can be started with $\Delta\gamma_{n+1}^{(0)}=0$. Subsequent iterations yield

$$\Delta\gamma_{n+1}^{(k+1)}=\Delta\gamma_{n+1}^{(k)}-\frac{g(\Delta\gamma_{n+1}^{(k)})}{g'(\Delta\gamma_{n+1}^{(k)})} \tag{6.207}$$

until $\Delta\gamma_{n+1}^{(\mathrm{final})}$ is small enough. Occasionally, depending on the form of the creep and hardening functions, the Newton–Raphson procedure does not converge (cf Chapter 3). Then, other techniques such as bisection (Press *et al.* 1990) (Lührs *et al.* 1997) can be used.

6.6.3 Expansion of a thick-walled cylinder

Consider the expansion of a long, thick-walled cylinder with inner radius r_i of 10 mm and an outer radius r_o of 20 mm, subject to internal pressure p. The material constants are E=11050 MPa, $\nu=0.454$ and $\sigma_{\mathrm{vm}}=0.5$ MPa at zero equivalent plastic strain. In the plastic range, the material does not harden (perfect plastic behavior). Consequently, the von Mises stress at the zero equivalent plastic strain applies to the complete plastic range.

A quarter of the cylinder is modeled with three 20-node brick elements in the radial direction and 5 in the circumferential direction. In the axial direction, only one element layer is modeled, with its upper and lower layers of nodes fixed in the axial direction (plane strain assumption). Reduced integration is used throughout. Instead of applying an internal pressure, the nodes at the inner radius are moved in the radial direction in a uniform way. The reason for this is shown in Figure 6.4: as soon as the cylinder is fully in the plastic regime, the internal pressure steadily decreases. Therefore, it cannot be used as the loading parameter. Also shown is the thickness of the cylinder.

Figure 6.5 shows the change in volume. Notice that during plastic deformation, the volume decreases slightly. Accordingly, the plastic flow is not completely isochoric. This is discussed in more detail in Section 6.8.

Comparison with the results published by Simo (Simo 1988b) shows good agreement. Accordingly, 20-node brick elements with reduced integration can be used for large strain plasticity. The use of fully integrated 20-node brick elements leads to divergence.

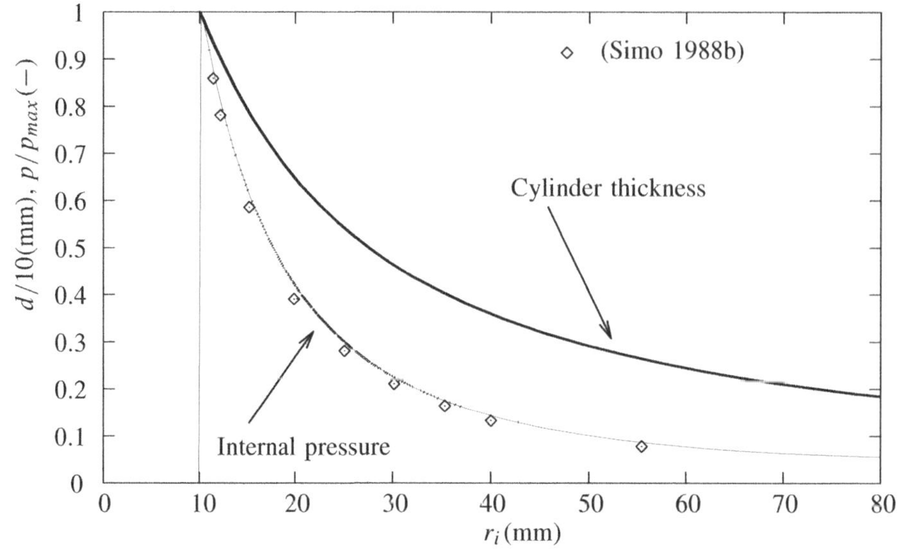

Figure 6.4 Variation of the internal pressure and thickness

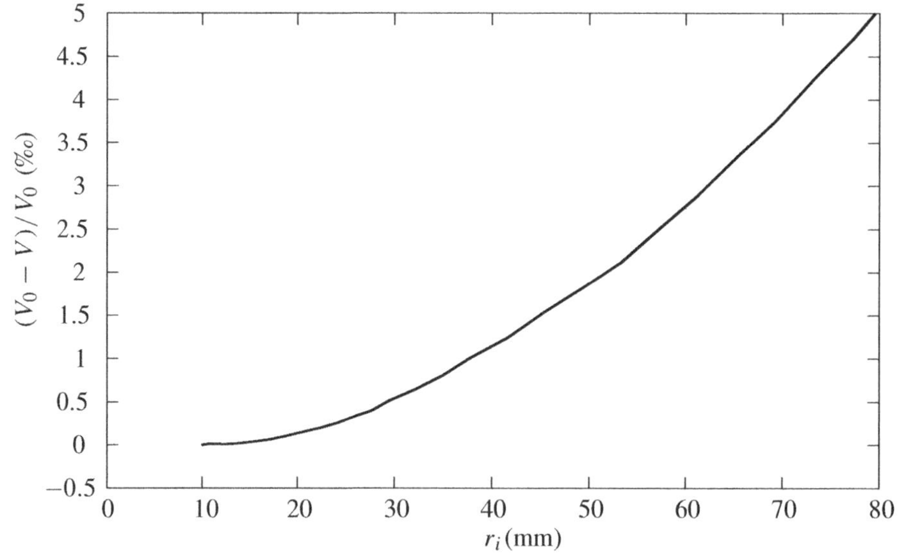

Figure 6.5 Variation of the volume

6.7 Derivation of Consistent Elastoplastic Moduli

For finite element calculations, we also need to determine the consistent elastoplastic moduli at $t = t_{n+1}$. These moduli are the derivatives of the second Piola–Kirchhoff stress with respect to the Lagrange strain:

$$\boldsymbol{B}_{n+1} := \frac{\partial \boldsymbol{S}_{n+1}}{\partial \boldsymbol{E}_{n+1}} = 2\frac{\partial \boldsymbol{S}_{n+1}}{\partial \boldsymbol{C}_{n+1}}. \tag{6.208}$$

Recall that $\boldsymbol{S}_{n+1}$ takes the form (Equation (6.197)

$$\boldsymbol{S}_{n+1} = J_{n+1}U'(J_{n+1})\boldsymbol{C}_{n+1}^{-1} + \mu J_{n+1}^{-2/3}\mathrm{DEV}_{n+1}\boldsymbol{C}^{\mathrm{p}-1}_{n} - 2\Delta\gamma_{n+1}\overline{\overline{\mu}}_{n+1}\boldsymbol{N}_{n+1}. \tag{6.209}$$

The first term on the right-hand side of Equation (6.209) is the volumetric part, the second is the deviatoric trial stress and the third is the plastic correction.

6.7.1 The volumetric stress

Taking into account that

$$\frac{\partial J}{\partial \boldsymbol{C}} = \frac{J}{2}\boldsymbol{C}^{-1} \tag{6.210}$$

$$\frac{\partial \boldsymbol{C}^{-1}}{\partial \boldsymbol{C}} = -\mathbb{I}_{C^{-1}} \tag{6.211}$$

with $\mathbb{I}_{C^{-1}}$ defined in Equation (6.82), one gets

$$2\frac{\partial}{\partial \boldsymbol{C}}[JU'(J)\boldsymbol{C}^{-1}] = JU'(J)\boldsymbol{C}^{-1}\otimes\boldsymbol{C}^{-1} + J^2U''(J)\boldsymbol{C}^{-1}\otimes\boldsymbol{C}^{-1} - 2JU'(J)\mathbb{I}_{C^{-1}} \tag{6.212}$$

$$= J^2U''(J)\boldsymbol{C}^{-1}\otimes\boldsymbol{C}^{-1} + Jp(\boldsymbol{C}^{-1}\otimes\boldsymbol{C}^{-1} - 2\mathbb{I}_{C^{-1}}). \tag{6.213}$$

The index $n+1$ was dropped for convenience. For $U(J)$ defined in Equation (6.51), Equation (6.213) takes the form

$$2\frac{\partial}{\partial \boldsymbol{C}}[JU'(J)\boldsymbol{C}^{-1}] = KJ^2\boldsymbol{C}^{-1}\otimes\boldsymbol{C}^{-1} - K(J^2-1)\mathbb{I}_{C^{-1}}. \tag{6.214}$$

6.7.2 Trial stress

Since

$$\frac{\partial J^{-2/3}}{\partial \boldsymbol{C}} = -\tfrac{1}{3}J^{-2/3}\boldsymbol{C}^{-1} \tag{6.215}$$

and

$$\mathrm{DEV}(\cdot) = (\cdot) - \tfrac{1}{3}[(\cdot):\boldsymbol{C}]\boldsymbol{C}^{-1} \tag{6.216}$$

one gets

$$2\frac{\partial \boldsymbol{T}^{\mathrm{trial}}}{\partial \boldsymbol{C}} = -\tfrac{2}{3}\mu J^{-2/3}\mathrm{DEV}(\boldsymbol{C}^{\mathrm{p}-1})\otimes\boldsymbol{C}^{-1} - \tfrac{2}{3}\mu J^{-2/3}\boldsymbol{C}^{-1}\otimes(\boldsymbol{C}^{\mathrm{p}-1}:\mathbb{I}_I) + \tfrac{2}{3}\mu J^{-2/3}(\boldsymbol{C}:\boldsymbol{C}^{\mathrm{p}-1})I_{C^{-1}} \tag{6.217}$$

$$= -\tfrac{2}{3}\mu J^{-2/3}\left[C^{\mathrm{p}-1}\otimes C^{-1} - \tfrac{1}{3}(C:C^{\mathrm{p}-1})C^{-1}\otimes C^{-1}\right]$$

$$-\tfrac{2}{3}\mu J^{-2/3}C^{-1}\otimes C^{\mathrm{p}-1} + \tfrac{2}{3}\mu J^{-2/3}(C:C^{\mathrm{p}-1})\mathbb{I}_{C^{-1}} \qquad (6.218)$$

$$= \tfrac{2}{3}\mu J^{-2/3}(C:C^{\mathrm{p}-1})\left[\mathbb{I}_{C^{-1}} + \tfrac{1}{3}C^{-1}\otimes C^{-1}\right]$$

$$-\tfrac{2}{3}\mu J^{-2/3}\left[C^{\mathrm{p}-1}\otimes C^{-1} + C^{-1}\otimes C^{\mathrm{p}-1}\right] \qquad (6.219)$$

$$= \tfrac{2}{3}\mu J^{-2/3}(C:C^{\mathrm{p}-1})\left[\mathbb{I}_{C^{-1}} - \tfrac{1}{3}C^{-1}\otimes C^{-1}\right]$$

$$-\tfrac{2}{3}\mu J^{-2/3}\left\{\left[C^{\mathrm{p}-1} - \tfrac{1}{3}(C:C^{\mathrm{p}-1})C^{-1}\right]\otimes C\right.$$

$$\left.+C\otimes\left[C^{\mathrm{p}-1} - \tfrac{1}{3}(C:C^{\mathrm{p}-1})C^{-1}\right]\right\} \qquad (6.220)$$

$$-\tfrac{2}{3}\mu J^{-2/3}(C:C^{\mathrm{p}-1})\left[\mathbb{I}_{C^{-1}} \quad \tfrac{1}{3}C^{-1}\otimes C^{-1}\right]$$

$$-\tfrac{2}{3}\mu J^{-2/3}\left[\mathrm{DEV}(C^{\mathrm{p}-1})\otimes C^{-1} + C^{-1}\otimes\mathrm{DEV}(C^{\mathrm{p}-1})\right]. \qquad (6.221)$$

Accordingly,

$$2\frac{\partial T^{\mathrm{trial}}_{n+1}}{\partial C_{n+1}} = \tfrac{2}{3}\mu J^{-2/3}_{n+1}(C_{n+1}:C^{\mathrm{p}-1}_n)\left[\mathbb{I}_{C^{-1}_{n+1}} - \tfrac{1}{3}C^{-1}_{n+1}\otimes C^{-1}_{n+1}\right]$$

$$-\tfrac{2}{3}\mu J^{-2/3}_{n+1}\left[\mathrm{DEV}C^{\mathrm{p}-1}_n\otimes C^{-1}_{n+1} + C^{-1}_{n+1}\otimes\mathrm{DEV}C^{\mathrm{p}-1}_n\right] := B^{\mathrm{trial}}_{n+1}. \quad (6.222)$$

6.7.3 Plastic correction

This is the most difficult part. One obtains

$$2\frac{\partial}{\partial C_{n+1}}\left(-2\Delta\gamma_{n+1}\overline{\overline{\mu}}_{n+1}N_{n+1}\right) = -4\overline{\overline{\mu}}_{n+1}N_{n+1}\otimes\frac{\partial\Delta\gamma_{n+1}}{\partial C_{n+1}}$$

$$-4\Delta\gamma_{n+1}N_{n+1}\otimes\frac{\partial\overline{\overline{\mu}}_{n+1}}{\partial C} - 4\Delta\gamma_{n+1}\overline{\overline{\mu}}_{n+1}\frac{\partial N_{n+1}}{\partial C_{n+1}}. \quad (6.223)$$

Concentrating on the last term,

$$\frac{\partial N_{n+1}}{\partial C_{n+1}} = \frac{\partial}{\partial C_{n+1}}\frac{\Xi^{\mathrm{trial}}_{n+1}}{\|\Xi^{\mathrm{trial}}_{n+1}\|}$$

$$= \frac{1}{\|\Xi^{\mathrm{trial}}_{n+1}\|}\left(\frac{\partial\Xi^{\mathrm{trial}}_{n+1}}{\partial C_{n+1}} - N_{n+1}\otimes\frac{\partial\|\Xi^{\mathrm{trial}}_{n+1}\|}{\partial C_{n+1}}\right). \qquad (6.224)$$

In complete analogy to Equation (6.222), one finds

$$2\frac{\partial \boldsymbol{A}^{\text{trial}}_{n+1}}{\partial \boldsymbol{C}_{n+1}} = -\tfrac{2}{3} J^{-2/3}_{n+1} (\boldsymbol{C}_{n+1} : \boldsymbol{Q}_{2,n}) \ \ \mathbb{I}_{C^{-1}_{n+1}} - \tfrac{1}{3} \boldsymbol{C}^{-1}_{n+1} \otimes \boldsymbol{C}^{-1}_{n+1}\Big)$$

$$+ \tfrac{2}{3} J^{-2/3}_{n+1} \Big[\mathrm{DEV}(\boldsymbol{Q}_{2,n}) \otimes \boldsymbol{C}^{-1}_{n+1} + \boldsymbol{C}^{-1}_{n+1} \otimes \mathrm{DEV}(\boldsymbol{Q}_{2,n})\Big] \tag{6.225}$$

and accordingly,

$$2\frac{\partial \boldsymbol{\Xi}^{\text{trial}}_{n+1}}{\partial \boldsymbol{C}_{n+1}} = 2\overline{\overline{\mu}}_{n+1} \ \ \mathbb{I}_{C^{-1}_{n+1}} - \tfrac{1}{3} \boldsymbol{C}^{-1}_{n+1} \otimes \boldsymbol{C}^{-1}_{n+1}\Big)$$

$$- \tfrac{2}{3} \ \ \boldsymbol{\Xi}^{\text{trial}}_{n+1} \otimes \boldsymbol{C}^{-1}_{n+1} + \boldsymbol{C}^{-1}_{n+1} \otimes \boldsymbol{\Xi}^{-1}_{n+1}\Big) := \boldsymbol{H}^{\text{trial}}_{n+1}. \tag{6.226}$$

Furthermore,

$$\frac{\partial \|\boldsymbol{\Xi}^{\text{trial}}_{n+1}\|}{\partial \boldsymbol{C}_{n+1}} = \frac{\partial}{\partial \boldsymbol{C}_{n+1}} \sqrt{\Xi^{IJ,\text{trial}}_{n+1} \Xi^{KL,\text{trial}}_{n+1} C_{IJ,n+1} C_{KL,n+1}} \tag{6.227}$$

$$= \tfrac{1}{2} \boldsymbol{H}^{\text{trial}}_{n+1} : \boldsymbol{N}_{n+1} + \|\boldsymbol{\Xi}^{\text{trial}}_{n+1}\| \boldsymbol{N}^2_{n+1} \tag{6.228}$$

and

$$\boldsymbol{H}^{\text{trial}}_{n+1} : \boldsymbol{N}_{n+1} = 2\overline{\overline{\mu}}_{n+1} \boldsymbol{N}_{n+1} - \tfrac{2}{3} \|\boldsymbol{\Xi}^{\text{trial}}_{n+1}\| \boldsymbol{C}^{-1}_{n+1}. \tag{6.229}$$

Equations (6.224), (6.226), (6.227) and (6.229) yield

$$\frac{\partial \boldsymbol{N}_{n+1}}{\partial \boldsymbol{C}_{n+1}} = \frac{1}{\|\boldsymbol{\Xi}^{\text{trial}}_{n+1}\|} \left\{ \boldsymbol{H}^{\text{trial}}_{n+1} - \boldsymbol{N}_{n+1} \otimes \left[\overline{\overline{\mu}}_{n+1} \boldsymbol{N}_{n+1} + \|\boldsymbol{\Xi}^{\text{trial}}_{n+1}\| \mathrm{DEV}_{n+1}(\boldsymbol{N}^2_{n+1})\right]\right\} \tag{6.230}$$

since

$$\mathrm{DEV}(\boldsymbol{N}^2) = \boldsymbol{N}^2 - \tfrac{1}{3}(\boldsymbol{N} \cdot \boldsymbol{N} : \boldsymbol{C})\boldsymbol{C}^{-1} = \boldsymbol{N}^2 - \tfrac{1}{3}\boldsymbol{C}^{-1}. \tag{6.231}$$

For the second term, one starts from the expression for $\overline{\mu}_{n+1}$:

$$\overline{\mu}_{n+1} = \tfrac{1}{3} \mu J^{-2/3}_{n+1} \boldsymbol{C}^{\mathrm{p}-1} : \boldsymbol{C}_{n+1} \tag{6.232}$$

$$\Downarrow$$

$$\frac{\partial \overline{\mu}_{n+1}}{\partial \boldsymbol{C}_{n+1}} = -\tfrac{1}{3} \overline{\mu}_{n+1} \boldsymbol{C}^{-1}_{n+1} + \tfrac{1}{3} \mu J^{-2/3}_{n+1} \boldsymbol{C}^{\mathrm{p}-1}_{n} \tag{6.233}$$

and

$$\overline{\overline{\mu}}_{n+1} = \overline{\mu}_{n+1} + \tfrac{1}{3} J^{-2/3}_{n+1} (\boldsymbol{Q}_{2,n} : \boldsymbol{C}_{n+1}) \tag{6.234}$$

$$\Downarrow$$

$$\frac{\partial \overline{\overline{\mu}}_{n+1}}{\partial \boldsymbol{C}_{n+1}} = \frac{\partial \overline{\mu}_{n+1}}{\partial \boldsymbol{C}_{n+1}} + \tfrac{1}{3} J^{-2/3}_{n+1} \mathrm{DEV}_{n+1}(\boldsymbol{Q}_{2,n}). \tag{6.235}$$

Note the following interesting expression:

$$\frac{\partial}{\partial \boldsymbol{C}_{n+1}}\left[J_{n+1}^{-2/3}\mathrm{TR}_{n+1}(\cdot)\right] = J_{n+1}^{-2/3}\mathrm{DEV}_{n+1}(\cdot). \tag{6.236}$$

The first term is obtained by taking the derivative of Equation (6.193):

$$\frac{\partial \|\boldsymbol{\Xi}_{n+1}^{\mathrm{trial}}\|}{\partial \boldsymbol{C}_{n+1}} - 2\frac{\partial \overline{\overline{\mu}}_{n+1}}{\partial \boldsymbol{C}_{n+1}}\left(1+\frac{h_2^{\mathrm{eq}'}}{3\mu}\right)\Delta\gamma_{n+1} - 2\overline{\overline{\mu}}_{n+1}\left(1+\frac{h_2^{\mathrm{eq}'}}{3\mu}\right)\frac{\partial \Delta\gamma_{n+1}}{\partial \boldsymbol{C}_{n+1}}$$
$$- \tfrac{2}{3}h_1'\frac{\partial \Delta\gamma_{n+1}}{\partial \boldsymbol{C}_{n+1}} - \tfrac{2}{3}f'\frac{\partial \Delta\gamma_{n+1}}{\partial \boldsymbol{C}_{n+1}} = 0 \tag{6.237}$$

($h_2^{\mathrm{eq}'}$ is assumed to be constant) from which

$$2\overline{\overline{\mu}}_{n+1}\frac{\partial \Delta\gamma_{n+1}}{\partial \boldsymbol{C}_{n+1}} = \frac{1}{\left(1+\frac{h_2^{\mathrm{eq}'}}{3\mu}+\frac{h_1'}{3\overline{\overline{\mu}}_{n+1}}+\frac{f'}{3\overline{\overline{\mu}}_{n+1}}\right)}\cdot$$
$$\cdot\left[\frac{\partial \|\boldsymbol{\Xi}_{n+1}^{\mathrm{trial}}\|}{\partial \boldsymbol{C}_{n+1}} - 2\Delta\gamma_{n+1}\left(1+\frac{h_2^{\mathrm{eq}'}}{3\mu}\right)\frac{\partial \overline{\overline{\mu}}_{n+1}}{\partial \boldsymbol{C}_{n+1}}\right]. \tag{6.238}$$

Collecting terms, the derivative of the plastic correction yields

$$2\frac{\partial}{\partial \boldsymbol{C}_{n+1}}\left(-2\Delta\gamma_{n+1}\overline{\overline{\mu}}_{n+1}\boldsymbol{N}_{n+1}\right) = \frac{-1}{\left(1+\frac{h_2^{\mathrm{eq}'}}{3\mu}+\frac{h_1'}{3\overline{\overline{\mu}}_{n+1}}+\frac{f'}{3\overline{\overline{\mu}}_{n+1}}\right)}\cdot$$
$$\cdot\left[\boldsymbol{N}_{n+1}\otimes 2\frac{\partial \|\boldsymbol{\Xi}_{n+1}^{\mathrm{trial}}\|}{\partial \boldsymbol{C}_{n+1}} - 2\Delta\gamma_{n+1}\left(1+\frac{h_2^{\mathrm{eq}'}}{3\mu}\right)\boldsymbol{N}_{n+1}\otimes 2\frac{\partial \overline{\overline{\mu}}_{n+1}}{\partial \boldsymbol{C}_{n+1}}\right]$$
$$- 2\Delta\gamma_{n+1}\boldsymbol{N}_{n+1}\otimes 2\frac{\partial \overline{\overline{\mu}}_{n+1}}{\partial \boldsymbol{C}_{n+1}} - \frac{2\Delta\gamma_{n+1}\overline{\overline{\mu}}_{n+1}}{\|\boldsymbol{\Xi}_{n+1}^{\mathrm{trial}}\|}\left[\boldsymbol{H}_{n+1}^{\mathrm{trial}} - \boldsymbol{N}_{n+1}\otimes 2\frac{\partial \|\boldsymbol{\Xi}_{n+1}^{\mathrm{trial}}\|}{\partial \boldsymbol{C}_{n+1}}\right] \tag{6.239}$$

$$= -\ \tfrac{1}{\delta_0} - f_0\Big)\,\boldsymbol{N}_{n+1}\otimes 2\frac{\partial \|\boldsymbol{\Xi}_{n+1}^{\mathrm{trial}}\|}{\partial \boldsymbol{C}_{n+1}}$$
$$+ 2\gamma_{n+1}\left[\tfrac{1}{\delta_0}\left(1+\frac{h_2^{\mathrm{eq}'}}{3\mu}\right) - 1\right]\boldsymbol{N}_{n+1}\otimes 2\frac{\partial \overline{\overline{\mu}}_{n+1}}{\partial \boldsymbol{C}_{n+1}} - f_0\boldsymbol{H}_{n+1}^{\mathrm{trial}} \tag{6.240}$$

where

$$f_0 := \frac{2\overline{\overline{\mu}}_{n+1}\Delta\gamma_{n+1}}{\|\boldsymbol{\Xi}_{n+1}^{\mathrm{trial}}\|} \tag{6.241}$$

$$\delta_0 := 1 + \frac{h_2^{\mathrm{eq}'}}{3\mu} + \frac{h_1'}{3\overline{\overline{\mu}}_{n+1}} + \frac{f'}{3\overline{\overline{\mu}}_{n+1}}. \tag{6.242}$$

Substituting Equations (6.228), (6.229), (6.231) and (6.235) into Equation (6.240) yields

$$
\begin{aligned}
2\frac{\partial}{\partial \boldsymbol{C}_{n+1}}\left(-2\Delta\gamma_{n+1}\overline{\overline{\mu}}_{n+1}\boldsymbol{N}_{n+1}\right)
&= -f_1\left[2\overline{\overline{\mu}}_{n+1}\boldsymbol{N}_{n+1}\otimes\boldsymbol{N}_{n+1} + 2\|\boldsymbol{\Xi}^{\text{trial}}_{n+1}\|\boldsymbol{N}_{n+1}\otimes\text{DEV}_{n+1}(\boldsymbol{N}^2_{n+1})\right]\\
&\quad + 2\gamma_{n+1}\left[\tfrac{1}{\delta_0}\left(1+\frac{h_2^{\text{eq}'}}{3\mu}\right)-1\right]\tfrac{2}{3}\|\boldsymbol{\Xi}^{\text{trial}}_{n+1}\|\boldsymbol{N}_{n+1}\otimes\boldsymbol{N}_{n+1} - f_0\boldsymbol{H}^{\text{trial}}_{n+1}
\end{aligned}
\tag{6.243}
$$

where

$$
f_1 := \tfrac{1}{\delta_0} - f_0 \tag{6.244}
$$

or

$$
\begin{aligned}
2\frac{\partial}{\partial \boldsymbol{C}_{n+1}}\left(-2\Delta\gamma_{n+1}\overline{\overline{\mu}}_{n+1}\boldsymbol{N}_{n+1}\right)
&= -\delta_1\boldsymbol{N}_{n+1}\otimes\boldsymbol{N}_{n+1} - \delta_2\boldsymbol{N}_{n+1}\otimes\text{DEV}_{n+1}(\boldsymbol{N}^2_{n+1}) - f_0\boldsymbol{H}^{\text{trial}}_{n+1}
\end{aligned}
\tag{6.245}
$$

where

$$
\delta_1 := f_1 2\overline{\overline{\mu}}_{n+1} - \left[\tfrac{1}{\delta_0}\left(1+\frac{h_2^{\text{eq}'}}{3\mu}\right)-1\right]\tfrac{4}{3}\gamma_{n+1}\|\boldsymbol{\Xi}^{\text{trial}}_{n+1}\| \tag{6.246}
$$

and

$$
\delta_2 := 2\|\boldsymbol{\Xi}^{\text{trial}}_{n+1}\| f_1. \tag{6.247}
$$

Summarizing,

$$
\begin{aligned}
\boldsymbol{B}_{n+1} &= K J^2_{n+1}\boldsymbol{C}^{-1}_{n+1}\otimes\boldsymbol{C}^{-1}_{n+1} - K(J^2_{n+1}-1)\mathbb{I}_{C^{-1}_{n+1}} + \boldsymbol{B}^{\text{trial}}_{n+1}\\
&\quad - \delta_1\boldsymbol{N}_{n+1}\otimes\boldsymbol{N}_{n+1} - \delta_2\boldsymbol{N}_{n+1}\otimes\text{DEV}_{n+1}(\boldsymbol{N}^2_{n+1}) - f_0\boldsymbol{H}^{\text{trial}}_{n+1}
\end{aligned}
\tag{6.248}
$$

where

$$
\boldsymbol{B}^{\text{trial}}_{n+1} = 2\overline{\mu}_{n+1}\ \ \mathbb{I}_{C^{-1}_{n+1}} - \tfrac{1}{3}\boldsymbol{C}^{-1}_{n+1}\otimes\boldsymbol{C}^{-1}_{n+1}\big) - \tfrac{2}{3}\ \ \boldsymbol{T}^{\text{trial}}_{n+1}\otimes\boldsymbol{C}^{-1}_{n+1} + \boldsymbol{C}^{-1}_{n+1}\otimes\boldsymbol{T}^{\text{trial}}_{n+1}\big) \tag{6.249}
$$

$$
\boldsymbol{H}^{\text{trial}}_{n+1} = 2\overline{\overline{\mu}}_{n+1}\ \ \mathbb{I}_{C^{-1}_{n+1}} - \tfrac{1}{3}\boldsymbol{C}^{-1}_{n+1}\otimes\boldsymbol{C}^{-1}_{n+1}\big) - \tfrac{2}{3}\ \ \boldsymbol{\Xi}^{\text{trial}}_{n+1}\otimes\boldsymbol{C}^{-1}_{n+1} + \boldsymbol{C}^{-1}_{n+1}\otimes\boldsymbol{\Xi}^{\text{trial}}_{n+1}\big) \tag{6.250}
$$

$$
f_0 = \frac{2\overline{\overline{\mu}}_{n+1}\Delta\gamma_{n+1}}{\|\boldsymbol{\Xi}^{\text{trial}}_{n+1}\|} \tag{6.251}
$$

$$
f_1 = \tfrac{1}{\delta_0} - f_0 \tag{6.252}
$$

$$
\delta_0 = 1 + \frac{h_2^{\text{eq}'}}{3\mu} + \frac{h_1'}{3\overline{\overline{\mu}}_{n+1}} + \frac{f'}{3\overline{\overline{\mu}}_{n+1}}. \tag{6.253}
$$

$$\delta_1 = f_1 2\overline{\overline{\mu}}_{n+1} - \left[\frac{1}{\delta_0}\left(1 + \frac{h_2^{\text{eq}'}}{3\mu}\right) - 1\right]\frac{4}{3}\Delta\gamma_{n+1}\|\mathbf{\Xi}_{n+1}^{\text{trial}}\| \tag{6.254}$$

$$\delta_2 = 2\|\mathbf{\Xi}_{n+1}^{\text{trial}}\| f_1. \tag{6.255}$$

This concludes a long and tedious calculation. Notice that the tangent modulus is usually not isotropic, although the material is isotropic in the elastic range. Plasticity induces anisotropy.

The expression for $\boldsymbol{B}_{n+1}$ is not symmetric because of the $\boldsymbol{N}_{n+1} \otimes \text{DEV}_{n+1}(\boldsymbol{N}_{n+1}^2)$ term. In practice, this term is often symmetrized:

$$\left[\boldsymbol{N}_{n+1} \otimes \text{DEV}_{n+1}(\boldsymbol{N}_{n+1}^2)\right]^S$$
$$:= \frac{1}{2}\left[\boldsymbol{N}_{n+1} \otimes \text{DEV}_{n+1}(\boldsymbol{N}_{n+1}^2) + \text{DEV}_{n+1}(\boldsymbol{N}_{n+1}^2) \otimes \boldsymbol{N}_{n+1}\right]. \tag{6.256}$$

This does not lead to wrong solutions, but may decrease the rate of convergence of the scheme. However, the effect is deemed to be small.

6.8 Isochoric Plastic Deformation

In the previous derivation, the volume-preserving aspect of plastic deformation (Equation (6.49)) has not been taken into account (Simo and Miehe 1992). Indeed, $J^{\text{p}} = 1$ implies

$$\det \boldsymbol{C}^{\text{p}^{-1}} = 1 \tag{6.257}$$

and accordingly,

$$\overline{\det \boldsymbol{C}^{\text{p}^{-1}}}^{\,\cdot} = 0 \tag{6.258}$$

or

$$\frac{\partial \det \boldsymbol{C}^{\text{p}^{-1}}}{\partial \boldsymbol{C}^{\text{p}^{-1}}} : \dot{\boldsymbol{C}}^{\text{p}^{-1}} = 0. \tag{6.259}$$

Using Equation (1.509) for the derivative of the third invariant of a matrix, this yields

$$\dot{\boldsymbol{C}}^{\text{p}^{-1}} : \boldsymbol{C}^{\text{p}} = 0 \tag{6.260}$$

which does not agree with the assumption in Equation (6.124):

$$\text{TR}(\dot{\boldsymbol{C}}^{\text{p}^{-1}}) = \dot{\boldsymbol{C}}^{\text{p}^{-1}} : \boldsymbol{C} = 0. \tag{6.261}$$

Accordingly, it looks as if Equation (6.177) does not hold and Equation (6.179) yields $\text{DEV}_{n+1}\boldsymbol{C}^{\text{p}^{-1}}_{n+1}$ and not $\boldsymbol{C}^{\text{p}^{-1}}_{n+1}$. However, we know that (Equation (6.48))

$$\boldsymbol{C}^{\text{p}^{-1}} = \text{DEV}\boldsymbol{C}^{\text{p}^{-1}} + \frac{1}{3}\text{TR}(\boldsymbol{C}^{\text{p}^{-1}})\boldsymbol{C}^{-1} \tag{6.262}$$

which implies that the knowledge of $\mathrm{TR}_{n+1}(\boldsymbol{C}^{\mathrm{p}-1}_{n+1})$ suffices to determine $\boldsymbol{C}^{\mathrm{p}-1}_{n+1}$. Defining the invariants of $\boldsymbol{C}^{\mathrm{p}-1}_{n+1}$ by

$$J_{1C^{\mathrm{p}-1}} := \boldsymbol{C}^{\mathrm{p}-1} : \boldsymbol{C} = \mathrm{tr}\boldsymbol{b}^{\mathrm{e}} \tag{6.263}$$

$$J_{2C^{\mathrm{p}-1}} := (\boldsymbol{C}^{\mathrm{p}-1} \cdot \boldsymbol{C} \cdot \boldsymbol{C}^{\mathrm{p}-1}) : \boldsymbol{C} = \mathrm{tr}\boldsymbol{b}^{\mathrm{e}2} \tag{6.264}$$

$$J_{3C^{\mathrm{p}-1}} := (\boldsymbol{C}^{\mathrm{p}-1} \cdot \boldsymbol{C} \cdot \boldsymbol{C}^{\mathrm{p}-1} \cdot \boldsymbol{C} \cdot \boldsymbol{C}^{\mathrm{p}-1}) : \boldsymbol{C} = \mathrm{tr}\boldsymbol{b}^{\mathrm{e}3} \tag{6.265}$$

one arrives at, Equations (4.304) to (4.306)

$$I_{1C^{\mathrm{p}-1}} = J_{1C^{\mathrm{p}-1}} = I_{1b^{\mathrm{e}}} \tag{6.266}$$

$$I_{2C^{\mathrm{p}-1}} = \tfrac{1}{2}(J^2_{1C^{\mathrm{p}-1}} - J_{2C^{\mathrm{p}-1}}) = I_{2b^{\mathrm{e}}} \tag{6.267}$$

$$I_{3C^{\mathrm{p}-1}} = \mathrm{DET}\boldsymbol{C}^{\mathrm{p}-1} = \tfrac{1}{6}(2J_{3C^{\mathrm{p}-1}} + J^3_{1C^{\mathrm{p}-1}} - 3J_{1C^{\mathrm{p}-1}}J_{2C^{\mathrm{p}-1}}) = I_{3b^{\mathrm{e}}}. \tag{6.268}$$

Since

$$\boldsymbol{C}^{\mathrm{p}-1} = \boldsymbol{F}^{-1} \cdot \boldsymbol{b}^{\mathrm{e}} \cdot \boldsymbol{F}^{-\mathrm{T}} \tag{6.269}$$

one finds

$$\det \boldsymbol{C}^{\mathrm{p}-1} = 1 \quad \Leftrightarrow \quad \det \boldsymbol{b}^{\mathrm{e}} = \mathrm{DET}\boldsymbol{C}^{\mathrm{p}-1} = J^2. \tag{6.270}$$

Let us, for the simplicity of notation, denote $\boldsymbol{C}^{\mathrm{p}-1}$ by $\boldsymbol{A}$ in what follows. The eigenvalues satisfy the characteristic equation:

$$\Lambda_A^3 - I_{1A}\Lambda_A^2 + I_{2A}\Lambda_A - I_{3A} = 0. \tag{6.271}$$

The same applies to the eigenvalues and invariants of $\mathrm{DEV}\boldsymbol{A}$:

$$\Lambda^3_{\mathrm{DEV}A} + I_{2\mathrm{DEV}A}\Lambda_{\mathrm{DEV}A} - I_{3\mathrm{DEV}A} = 0 \tag{6.272}$$

since

$$I_{1A} = \mathrm{TR}(\mathrm{DEV}\boldsymbol{A}) = 0. \tag{6.273}$$

The eigenvalues of $\boldsymbol{A}$ and $\mathrm{DEV}\boldsymbol{A}$ are related by

$$\Lambda_{\mathrm{DEV}A} = \Lambda_A - \tfrac{1}{3}I_{1A}. \tag{6.274}$$

Accordingly, Equation (6.272) reduces to

$$(\Lambda_A - \tfrac{1}{3}I_{1A})^3 + I_{2\mathrm{DEV}A}(\Lambda_A - \tfrac{1}{3}I_{1A}) - I_{3A} = 0. \tag{6.275}$$

Expanding Equation (6.275) and substituting Equation (6.272) yields

$$I_{3A} - I_{2A}\Lambda_A + \tfrac{1}{3}\Lambda_A I^2_{1A} - (\tfrac{1}{3}I_{1A})^3 + I_{2\mathrm{DEV}A}\Lambda_A - (\tfrac{1}{3}I_{1A})I_{2\mathrm{DEV}A} - I_{3\mathrm{DEV}A} = 0. \tag{6.276}$$

This equation contains the unknowns I_{1A}, I_{2A}, I_{3A} and Λ_A and the known quantities $I_{2\text{DEV}A}$ and $I_{3\text{DEV}A}$. Ultimately, we are looking for I_{1A}. From Equation (6.270), we know that $I_{3A} = J^2$. To eliminate I_{2A}, the following equation is used:

$$\text{TR}[(\text{DEV}\boldsymbol{A})^2] = \text{TR}(\boldsymbol{A}^2) - \tfrac{1}{3}[\text{TR}(\boldsymbol{A})]^2 \tag{6.277}$$

which can be obtained by simple expansion. Accordingly,

$$J_{2\text{DEV}A} = J_{2A} - \tfrac{1}{3}J_{1A}^2 \tag{6.278}$$

or, using Equations (6.266) to (6.268),

$$I_{2A} = \tfrac{1}{3}I_{1A}^2 + I_{2\text{DEV}A}. \tag{6.279}$$

Substitution of Equation (6.279) into Equation (6.276) yields

$$(\tfrac{1}{3}I_{1A})^3 + (\tfrac{1}{3}I_{1A})I_{2\text{DEV}A} + (I_{3\text{DEV}A} - J^2) = 0. \tag{6.280}$$

This is a cubic equation in $I_{1A} = \text{TR}\boldsymbol{C}^{\text{p}^{-1}}$, which can be solved explicitly (Abramowitz and Stegun 1972).

The condition in Equation (6.261) was actually also used to determine $\overline{\overline{\mu}}_{n+1}$ (see Equations (6.149)–(6.152)). Since at this point $\text{TR}\boldsymbol{C}^{\text{p}^{-1}}$ is not necessarily constant in time, the result of Equation (6.280) allows for an update of $\overline{\overline{\mu}}_{n+1}$ and an iterative procedure ensues.

6.9 Burst Calculation of a Compressor

Plasticity is an important phenomenon in the deformation of metallic materials. Because of plasticity, high linear elastic stresses at notches and other geometric discontinuities are relaxed and redistributed. In the present application, the rotational speed of a radial compressor is increased till burst. The compressor is made of an aluminum alloy with Young's modulus $E = 75\,000$ MPa and a Poisson coefficient $\nu = 0.3$. The isotropic hardening curve is bilinear and described in Table 6.1. The geometry of the compressor can be downloaded from the CalculiX® Homepage (CalculiX GraphiX examples, (CalculiX 2003)).

Figure 6.6 shows the equivalent plastic strain at a location at the bore (inner radius) and the rim (outer radius) of the disk. It is well known from the theory of elasticity (Timoshenko and Goodier 1970) that the stresses are highest in the bore region and that is where plastic flow starts from. At a rotational speed of about 170 000 cycles/min, the disk collapses. This

Table 6.1 Isotropic hardening curve.

von Mises stress (MPa)	Equivalent plastic strain (%)
290	0
347	6
347	100

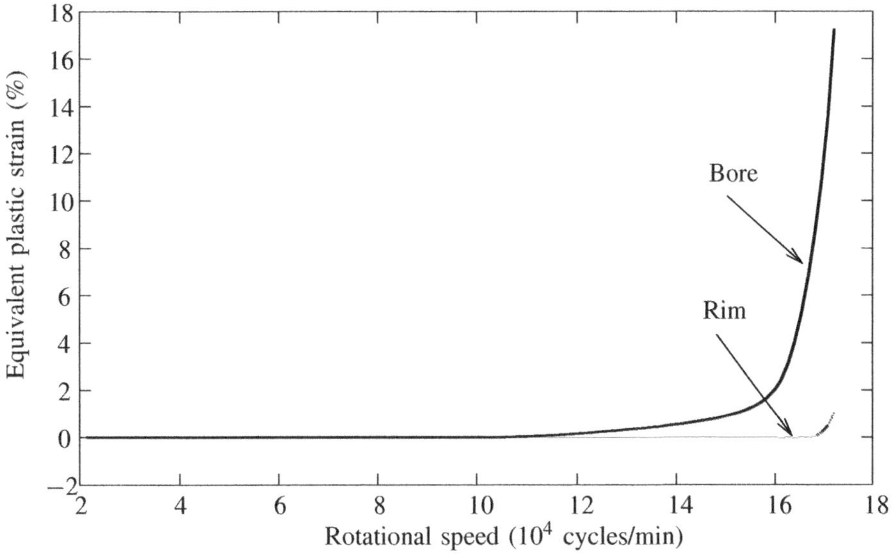

Figure 6.6 Equivalent plastic strain in the disk

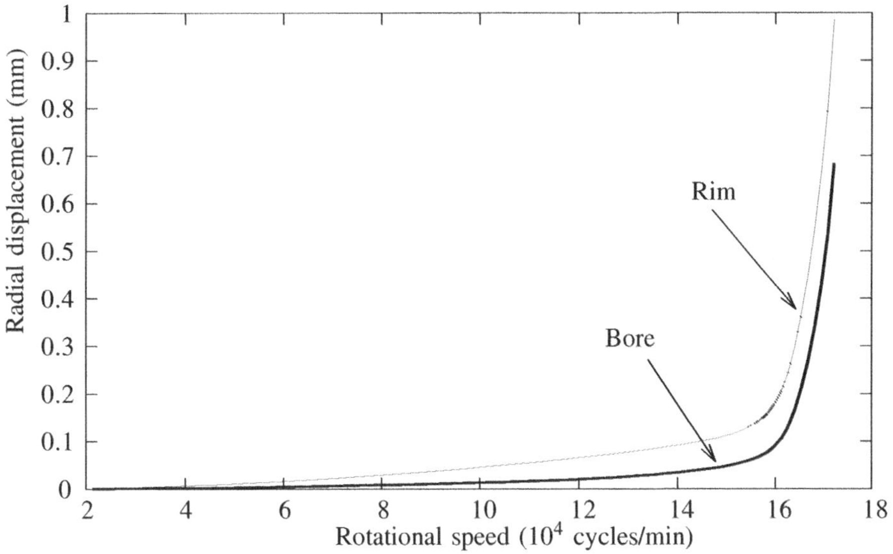

Figure 6.7 Radial displacements in the disk

is clear from the asymptotic increase of the plastic flow in the bore region. At the same time, the inner and outer radii also increase significantly (Figure 6.7). The bore radius at rest is about 3.5 mm, the rim radius is 43.5 mm. The calculation allows us to determine the burst margin for a given operation point.

7

Heat Transfer

7.1 Introduction

So far, the temperature has been considered as known. This is generally not the case. Usually, one knows thermal boundary conditions such as the environmental temperature or the value of a heat source, but not the temperature field in the entire body. This is the subject of heat-transfer calculations. Often, heat-transfer calculations are performed independently of stress calculations: they yield the temperature field, which serves as an input to the stress calculations through the force term in Equation (2.23). Also, material properties such as Young's modulus and other stress–strain curve characteristics change with temperature. Nonuniform temperature fields, especially, often induce considerable stress. In a few cases, the converse also applies: deformations lead to a temperature rise, for example, in forging operations. Then, there is a true mutual interaction between stress/deformation and temperature, resulting in coupled calculations.

7.2 The Governing Equations

In the present derivation we will allow for plastic processes, but we assume small strains, that is, we start from a free energy potential of the form in Equation (5.11):

$$\Sigma = \Sigma(\boldsymbol{\epsilon} - \boldsymbol{\epsilon}^{\mathrm{p}}, \boldsymbol{\alpha}, \theta, \nabla\theta, \boldsymbol{X}). \tag{7.1}$$

Furthermore, rectangular coordinates are assumed throughout. Of course, the conservation laws still apply. In particular, the Clausius–Duhem inequality leads to (cf Chapter 5)

$$\boldsymbol{\sigma} = \frac{\rho}{\rho_0}\frac{\partial \Sigma}{\partial \boldsymbol{\epsilon}^{\mathrm{e}}} \tag{7.2}$$

$$\eta = -\frac{1}{\rho_0}\frac{\partial \Sigma}{\partial \theta} \tag{7.3}$$

The Finite Element Method for Three-dimensional Thermomechanical Applications Guido Dhondt
 ISBN: 0-470-85752-8

$$\boldsymbol{q}^i = -\frac{\partial \Sigma}{\partial \boldsymbol{\alpha}} \tag{7.4}$$

$$\frac{\partial \Sigma}{\partial \nabla \theta} = 0 \tag{7.5}$$

where $\boldsymbol{q}^i$ stands for the internal dynamic variables. The index 'i' was introduced to avoid confusion with the heat flux $\boldsymbol{q}$. Introducing a reference temperature θ_{ref}, we define the relative temperature

$$T := \theta - \theta_{\text{ref}}. \tag{7.6}$$

T is assumed to be small compared to θ_{ref}. We now expand Σ as a function of T as follows (Σ does not depend on $\nabla\theta$ because of Equation (7.5)):

$$\Sigma(\boldsymbol{\epsilon}^{\mathrm{e}}, \boldsymbol{\alpha}, \theta, \boldsymbol{X}) = \rho_0(\boldsymbol{X})\psi_0(\boldsymbol{\epsilon}^{\mathrm{e}}, \boldsymbol{\alpha}, \boldsymbol{X}) - \rho_0(\boldsymbol{X})\eta_0(\boldsymbol{\epsilon}^{\mathrm{e}}, \boldsymbol{X})T - \left[\frac{\rho_0(\boldsymbol{X})c(\boldsymbol{\epsilon}^{\mathrm{e}}, \theta, \boldsymbol{X})}{2\theta_{\text{ref}}}\right]T^2. \tag{7.7}$$

This is an equality, not an approximation: notice that c is a function of the temperature θ. It is assumed that the dependence on $\boldsymbol{\alpha}$ does not depend on the temperature (only ψ_0 contains $\boldsymbol{\alpha}$). Applying Equations (7.2) to (7.4) and keeping the linear terms only (T is assumed to be small) leads to

$$\boldsymbol{\sigma} = \rho\frac{\partial \psi_0}{\partial \boldsymbol{\epsilon}^{\mathrm{e}}} - \rho\frac{\partial \eta_0}{\partial \boldsymbol{\epsilon}^{\mathrm{e}}}T + O(T^2) \tag{7.8}$$

$$\eta = \eta_0 + \frac{cT}{\theta_{\text{ref}}} \tag{7.9}$$

$$\boldsymbol{q}^i = -\rho_0\frac{\partial \psi_0}{\partial \boldsymbol{\alpha}} \tag{7.10}$$

where $\rho_0(\boldsymbol{X})$, $\psi_0(\boldsymbol{\epsilon}^{\mathrm{e}}, \boldsymbol{\alpha}, \boldsymbol{X})$, $\eta_0(\boldsymbol{\epsilon}^{\mathrm{e}}, \boldsymbol{X})$ and $c(\boldsymbol{\epsilon}^{\mathrm{e}}, \theta, \boldsymbol{X})$, as in Equation (7.7). Equation (7.8) splits the stresses into a mechanical part and a thermal part.

The internal energy satisfies (cf Equation (1.387):

$$\varepsilon = \frac{\Sigma}{\rho_0} + \theta\eta. \tag{7.11}$$

Substitution of Equations (7.7) and (7.9) into Equation (7.11) yields

$$\varepsilon = \psi_0 - \eta_0 T + \theta\eta_0 + \frac{cT}{\theta_{\text{ref}}}\theta + O(T^2) \tag{7.12}$$

which can be further simplified to

$$\varepsilon = \psi_0(\boldsymbol{\epsilon}^{\mathrm{e}}, \boldsymbol{\alpha}, \boldsymbol{X}) + \eta_0(\boldsymbol{\epsilon}^{\mathrm{e}}, \boldsymbol{X})\theta_{\text{ref}} + c(\boldsymbol{\epsilon}^{\mathrm{e}}, \theta, \boldsymbol{X})T + O(T^2). \tag{7.13}$$

The conservation of energy requires (Equation (1.355), spatial form)

$$\rho\dot{\varepsilon} = \dot{\boldsymbol{\epsilon}} : \boldsymbol{\sigma} - q^k{}_{,k} + \rho h \tag{7.14}$$

which yields after the use of Equation (7.13)

$$\rho\frac{\partial\psi_0}{\partial\boldsymbol{\epsilon}^{\mathrm{e}}}:\dot{\boldsymbol{\epsilon}}^{\mathrm{e}}+\rho\frac{\partial\psi_0}{\partial\boldsymbol{\alpha}}:\dot{\boldsymbol{\alpha}}+\rho\frac{\partial\eta_0}{\partial\boldsymbol{\epsilon}^{\mathrm{e}}}:\dot{\boldsymbol{\epsilon}}^{\mathrm{e}}\theta_{\mathrm{ref}}+\rho\frac{\partial c}{\partial\boldsymbol{\epsilon}^{\mathrm{e}}}:\dot{\boldsymbol{\epsilon}}^{\mathrm{e}}T$$
$$+\rho\frac{\partial c}{\partial T}\dot{T}T+\rho c\dot{T}-(\dot{\boldsymbol{\epsilon}}^{\mathrm{e}}+\dot{\boldsymbol{\epsilon}}^{\mathrm{p}}):\boldsymbol{\sigma}+q^{k}{}_{,k}-\rho h=0. \quad (7.15)$$

The first term in Equation (7.15) is a linear approximation to $\boldsymbol{\sigma}:\boldsymbol{\epsilon}^{\mathrm{e}}$ (cf Equation (7.8)), the second term corresponds to $-\boldsymbol{q}^{i}:\dot{\boldsymbol{\alpha}}$ (cf Equation (7.10)) and the fourth and fifth terms are quadratic (T and $\dot{\boldsymbol{\epsilon}}^{\mathrm{e}}$ are both small). Accordingly, Equation (7.15) reduces to

$$\rho c\dot{T}=-q^{k}{}_{,k}+\rho h+\boldsymbol{\sigma}:\dot{\boldsymbol{\epsilon}}^{\mathrm{p}}+\boldsymbol{q}^{i}:\dot{\boldsymbol{\alpha}}-\boldsymbol{\beta}:\dot{\boldsymbol{\epsilon}}^{\mathrm{e}}\theta_{\mathrm{ref}} \quad (7.16)$$

where

$$\boldsymbol{\beta}:=\rho\frac{\partial\eta_0}{\partial\boldsymbol{\epsilon}^{\mathrm{e}}} \quad (7.17)$$

is the stress reduction per temperature increase (cf Equations (7.8) and (1.413)). Equation (7.16) expresses that a temperature increase can result from heat flux, heat sources, plastic dissipation, internal-variable dissipation and the work rate of the thermal stresses or any combination. The last three terms depend on the deformation and embody the influence of the deformation (mechanical action) on the temperature. The conservation of energy in the form of Equation (7.16) is the governing equation in heat-transfer calculations.

7.3 Weak Form of the Energy Equation

To obtain the weak form of Equation (7.16), we proceed as explained in Section 1.12. Multiplying by an infinitesimal perturbation of the temperature δT and integrating over V yields

$$\int_V\rho c\dot{T}\delta T\,\mathrm{d}v=-\int_V q^{k}{}_{,k}\delta T\,\mathrm{d}v+\int_V\left(\rho h+\boldsymbol{\sigma}:\dot{\boldsymbol{\epsilon}}^{\mathrm{p}}+\boldsymbol{q}^{i}:\dot{\boldsymbol{\alpha}}-\boldsymbol{\beta}:\dot{\boldsymbol{\epsilon}}^{\mathrm{e}}\theta_{\mathrm{ref}}\right)\delta T\,\mathrm{d}v. \quad (7.18)$$

Integrating the first term on the right-hand side by parts, one obtains

$$-\int_V q^{k}{}_{,k}\delta T\,\mathrm{d}v=-\int_A q^{k}\delta T\,\mathrm{d}a_k+\int_V q^{k}\delta T_{,k}\,\mathrm{d}v \quad (7.19)$$

leading to

$$\int_V\rho c\dot{T}\delta T\,\mathrm{d}v-\int_V q^{k}\delta T_{,k}\,\mathrm{d}v$$
$$=-\int_A q^{k}\delta T\,\mathrm{d}a_k+\int_V\left(\rho h+\boldsymbol{\sigma}:\dot{\boldsymbol{\epsilon}}^{\mathrm{p}}+\boldsymbol{q}^{i}:\dot{\boldsymbol{\alpha}}-\boldsymbol{\beta}:\dot{\boldsymbol{\epsilon}}^{\mathrm{e}}\theta_{\mathrm{ref}}\right)\delta T\,\mathrm{d}v. \quad (7.20)$$

The entropy inequality, Equation (5.17), requires that

$$\boldsymbol{q}\cdot\nabla T\geq 0 \quad (7.21)$$

which implies that $\boldsymbol{q}$ must be at least a linear function of ∇T:

$$q^k = -\kappa^{kl}(T)T,l \tag{7.22}$$

that is, for a zero-temperature gradient, there is no heat flux. In Equation (7.22), the coefficients κ^{kl} are generally a function of the temperature. It is a nonlinear equation of the temperature.

The flux in the first term on the right-hand side of Equation (7.20) is the heat flux entering the body through its surface. It consists of three parts:

1. a convective part, which is more or less linear in T:

$$q^k_{\text{conv}} = h(T)(T - T_{\text{e}})n^k \tag{7.23}$$

where T_{e} is the environmental temperature, $h(T)$ is the convective coefficient and $\boldsymbol{n}$ is the normal to the surface.

2. a radiation part, which is highly nonlinear (Incropera and DeWitt 2002)

$$q^k_{\text{rad}} = A(T)(\theta^4 - \theta^4_{\text{e}})n^k \tag{7.24}$$

where θ_{e} is the absolute environmental temperature (in Kelvin) and $A(T)$ is the product of the Stefan–Boltzmann constant σ with the emissivity $\epsilon(T)$:

$$A(T) = \sigma\epsilon(T). \tag{7.25}$$

The emissivity is a property of the surface and takes values between zero and one. It is a measure of how well the surface emits radiation. For a perfect black body, $\epsilon = 1$.

3. any other known flux

$$\overline{q}^k = \overline{q}n^k. \tag{7.26}$$

Summarizing, the heat equation for small strains and small temperature deviations from a reference temperature yields

$$\begin{aligned}\int_V \rho c(T)\dot{T}\delta T\,\mathrm{d}v &+ \int_V \kappa^{kl}T_{,l}\delta T_{,k}\,\mathrm{d}v \\ &= -\int_A h(T)(T - T_{\text{e}})\delta T n^k\,\mathrm{d}a_k - \int_A A(T)(\theta^4 - \theta^4_{\text{e}})\delta T n^k\,\mathrm{d}a_k \\ &\quad - \int_A \overline{q}\delta T n^k\,\mathrm{d}a_k + \int_V \left(\rho h + \boldsymbol{\sigma} : \dot{\boldsymbol{\epsilon}}^{\text{p}} + \boldsymbol{q}^i : \dot{\boldsymbol{\alpha}} - \boldsymbol{\beta} : \dot{\boldsymbol{\epsilon}}^{\text{e}}\theta\right)\delta T\,\mathrm{d}v.\end{aligned} \tag{7.27}$$

In the last term, θ_{ref} was replaced by θ, which also corresponds to a second-order correction. Indeed,

$$\boldsymbol{\beta} : \dot{\boldsymbol{\epsilon}}^{\text{e}}\theta = \boldsymbol{\beta} : \dot{\boldsymbol{\epsilon}}^{\text{e}}\theta_{\text{ref}} + \boldsymbol{\beta} : \dot{\boldsymbol{\epsilon}}^{\text{e}}T \tag{7.28}$$

where $\boldsymbol{\beta} : \dot{\boldsymbol{\epsilon}}^{\text{e}}T = O(\|\boldsymbol{\epsilon}\|T)$. Equation (7.27) is highly nonlinear because of the radiation term. Furthermore, the temperature dependence of the materials constants in Equation (7.27) cannot be neglected and must be taken into account through an iterative procedure.

7.4 Finite Element Procedure

Similar to the discretization procedure for the displacements in Section 2.1, the temperatures are interpolated within an element between the nodal values by shape functions

$$T(\xi, \eta, \zeta, t) = \sum_{i=1}^{N} \varphi_i(\xi, \eta, \zeta) T_i(t). \tag{7.29}$$

The time derivative yields

$$\dot{T}(\xi, \eta, \zeta, t) = \sum_{i=1}^{N} \varphi_i(\xi, \eta, \zeta) \dot{T}_i(t) \tag{7.30}$$

and similarly

$$\delta T(\xi, \eta, \zeta, t) = \sum_{i=1}^{N} \varphi_i(\xi, \eta, \zeta) \delta T_i(t). \tag{7.31}$$

Substituting these expressions into Equation (7.27) and breaking down the volume integration on the element level yields

$$\begin{aligned}
&\sum_{\mathrm{e}} \sum_{i=1}^{N} \sum_{j=1}^{N} \left[\int_{V_\mathrm{e}} \rho c(T) \varphi_j \varphi_i \, \mathrm{d}v_\mathrm{e} \right] \dot{T}_j \delta T_i + \sum_{\mathrm{e}} \sum_{i=1}^{N} \sum_{j=1}^{N} \left[\int_{V_\mathrm{e}} \kappa^{kl}(T) \varphi_{j,l} \varphi_{i,k} \, \mathrm{d}v_\mathrm{e} \right] T_j \delta T_i \\
&= - \sum_{\mathrm{e}} \sum_{i=1}^{N} \left[\int_{A_\mathrm{e}} h(T)(T - T_\mathrm{e}) \varphi_i \, \mathrm{d}a_\mathrm{e} \right] \delta T_i - \sum_{\mathrm{e}} \sum_{i=1}^{N} \left[\int_{A_\mathrm{e}} A(T)(\theta^4 - \theta_\mathrm{e}^4) \varphi_i \, \mathrm{d}a_\mathrm{e} \right] \delta T_i \\
&- \sum_{\mathrm{e}} \sum_{i=1}^{N} \left[\int_{A_\mathrm{e}} \bar{q} \varphi_i \, \mathrm{d}a_\mathrm{e} \right] \delta T_i + \sum_{\mathrm{e}} \sum_{i=1}^{N} \left[\int_{V_\mathrm{e}} \rho h + \boldsymbol{\sigma} : \dot{\boldsymbol{\epsilon}}^\mathrm{p} + \boldsymbol{q}^i : \dot{\boldsymbol{\alpha}} - \boldsymbol{\beta} : \dot{\boldsymbol{\epsilon}}^\mathrm{e} \theta \Big) \varphi_i \, \mathrm{d}v_\mathrm{e} \right] \delta T_i.
\end{aligned} \tag{7.32}$$

Defining for each element a vector containing the nodal temperatures

$$\{T\}_\mathrm{e} := \begin{Bmatrix} T_1 \\ T_2 \\ \vdots \\ T_N \end{Bmatrix} \tag{7.33}$$

Equation (7.32) can be written as

$$\sum_{\mathrm{e}} \delta \{T\}_\mathrm{e}^\mathrm{T} [C]_\mathrm{e} \frac{D}{Dt} \{T\}_\mathrm{e} + \sum_{\mathrm{e}} \delta \{T\}_\mathrm{e}^\mathrm{T} [K]_\mathrm{e} \{T\}_\mathrm{e} = \sum_{\mathrm{e}} \delta \{T\}_\mathrm{e}^\mathrm{T} \{Q\}_\mathrm{e} \tag{7.34}$$

where

$$[C]_{\mathrm{e}ij} = \int_{V_\mathrm{e}} \rho c(T) \varphi_i \varphi_j \, \mathrm{d}v_\mathrm{e} \tag{7.35}$$

$$[K]_{\mathrm{e}ij} = \int_{V_\mathrm{e}} \kappa^{kl}(T) \varphi_{i,k} \varphi_{j,l} \, \mathrm{d}v_\mathrm{e} \tag{7.36}$$

$$\{Q\}_{ei} = -\int_{A_e} h(T)(T - T_e)\varphi_i \, da_e - \int_{A_e} A(T)(\theta^4 - \theta_e^4)\varphi_i \, da_e - \int_{A_e} \bar{q}\varphi_i \, da_e$$

$$+ \int_{V_e} \rho h + \boldsymbol{\sigma} : \dot{\boldsymbol{\epsilon}}^{\mathrm{p}} + \boldsymbol{q}^i : \dot{\boldsymbol{\alpha}} - \boldsymbol{\beta} : \dot{\boldsymbol{\epsilon}}^{\mathrm{e}}\theta \Big) \varphi_i \mathrm{d}v_e. \quad (7.37)$$

$[C]_e$ is the element capacity matrix and $[K]_e$ is the element conduction matrix. Both are symmetric matrices (κ^{kl} is a symmetric tensor). Defining the localization matrix $[L]_e$ that localizes element "e" within the structure by

$$\{T\}_e = [L]_e \{T\} \quad (7.38)$$

where $\{T\}$ contains the temperatures of all nodes, Equation (7.34) now reads

$$\delta \{T\}^{\mathrm{T}} [C] \frac{D}{Dt} \{T\} + \delta \{T\}^{\mathrm{T}} [K] \{T\} = \delta \{T\}^{\mathrm{T}} \{Q\} \quad (7.39)$$

where

$$[C] = \sum_{e} [L]_e^{\mathrm{T}} [C]_e [L]_e \quad (7.40)$$

$$[K] = \sum_{e} [L]_e^{\mathrm{T}} [K]_e [L]_e \quad (7.41)$$

$$\{Q\} = \sum_{e} [L]_e^{\mathrm{T}} \{Q\}_e . \quad (7.42)$$

Since Equation (7.39) must apply for any $\delta \{T\}^{\mathrm{T}}$, one finally arrives at the following governing set of finite element equations:

$$[C] \frac{D}{Dt} \{T\} + [K] \{T\} = \{Q\} . \quad (7.43)$$

Although Equation (7.43) looks linear in the temperature, it is not linear at all. Indeed, both $[C]$ and $[K]$ are a function of the temperature, since the capacity and conduction coefficients are temperature-dependent. Furthermore, the driving flux $\{Q\}$ (units of power) is highly nonlinear because of the radiation terms.

7.5 Time Discretization and Linearization of the Governing Equation

Equation (7.43) is an ordinary differential equation in t. For the time discretization, a backward Euler scheme is taken. Accordingly,

$$\frac{D}{Dt} \{T\}_{n+1} \approx \frac{1}{\Delta t} \left[\{T\}_{n+1} - \{T\}_n \right] . \quad (7.44)$$

Evaluating Equation (7.43) at $t = t_{n+1}$ leads to

$$\frac{1}{\Delta t} [C]_{n+1} \ \{T\}_{n+1} - \{T\}_n \Big) + [K]_{n+1} \{T\}_{n+1} = \{Q\}_{n+1} . \quad (7.45)$$

This nonlinear equation will be solved in an iterative way. Assume $\{T\}_n$ is known and we want to determine $\{T\}_{n+1}$. In the iteration $k+1$, we have an approximation $\{T\}_{n+1}^{(k)}$ for $\{T\}_{n+1}$, and we seek a better approximation $\{T\}_{n+1}^{(k+1)}$ that satisfies

$$\{T\}_{n+1}^{(k+1)} = \{T\}_{n+1}^{(k)} + \{\Delta T\}_{n+1}^{(k)} . \tag{7.46}$$

Substitution of the approximation $\{T\}_{n+1}^{(k)}$ into $[C]$ will be denoted $[C]_{n+1}^{(k)}$. Linearization of the improved value $[C]_{n+1}^{(k+1)}$ leads to

$$[C]_{n+1}^{(k+1)} = [C]_{n+1}^{(k)} + \left[\frac{\partial [C]}{\partial \{T\}}\right]_{n+1}^{(k)} \{\Delta T\}_{n+1}^{(k)} . \tag{7.47}$$

Similar expressions apply to $[K]_{n+1}^{(k+1)}$ and $\{F\}_{n+1}^{(k+1)}$. Evaluation of Equation (5.42) in the iteration $k+1$ yields

$$\begin{aligned} \frac{1}{\Delta t} \left([C]_{n+1}^{(k)} + \left[\frac{\partial [C]}{\partial \{T\}}\right]_{n+1}^{(k)} \{\Delta T\}_{n+1}^{(k)} \right) \left(\{T\}_{n+1}^{(k)} + \{\Delta T\}_{n+1}^{(k)} - \{T\}_n \right) \\ + \left([K]_{n+1}^{(k)} + \left[\frac{\partial [K]}{\partial \{T\}}\right]_{n+1}^{(k)} \{\Delta T\}_{n+1}^{(k)} \right) \left(\{T\}_{n+1}^{(k)} + \{\Delta T\}_{n+1}^{(k)} \right) \\ = \{Q\}_{n+1}^{(k)} + \left[\frac{\partial \{Q\}}{\partial \{T\}}\right]_{n+1}^{(k)} \{\Delta T\}_{n+1}^{(k)} . \end{aligned} \tag{7.48}$$

Collecting terms and neglecting quadratic contributions yields

$$\begin{aligned} \left\{ \frac{1}{\Delta t} \left[[C]_{n+1}^{(k)} + \left[\frac{\partial [C]}{\partial \{T\}}\right]_{n+1}^{(k)} \left(\{T\}_{n+1}^{(k)} - \{T\}_n \right) \right] \right. \\ \left. + \left[[K]_{n+1}^{(k)} + \left[\frac{\partial [K]}{\partial \{T\}}\right]_{n+1}^{(k)} \{T\}_{n+1}^{(k)} \right] - \left[\frac{\partial \{Q\}}{\partial \{T\}}\right]_{n+1}^{(k)} \right\} \{\Delta T\}_{n+1}^{(k)} \\ = -\frac{1}{\Delta t} [C]_{n+1}^{(k)} \left(\{T\}_{n+1}^{(k)} - \{T\}_n \right) - [K]_{n+1}^{(k)} \{T\}_{n+1}^{(k)} + \{Q\}_{n+1}^{(k)} . \end{aligned} \tag{7.49}$$

The right-hand side is the residual $\{R\}_{n+1}^{(k)}$ of Equation (7.45) in iteration (k). The dependence of the capacity and conduction terms on the temperature is usually benign, and the corresponding temperature-derivative terms in Equation (7.49) are often neglected. In this way, Equation (7.49) reduces to

$$\left(\frac{1}{\Delta t} [C]_{n+1}^{(k)} + [K]_{n+1}^{(k)} - \left[\frac{\partial \{Q\}}{\partial \{T\}}\right]_{n+1}^{(k)} \right) \{\Delta T\}_{n+1}^{(k)} = \{R\}_{n+1}^{(k)} . \tag{7.50}$$

The only term that needs further analysis is the derivative of the driving flux with respect to the temperature. Equations (7.38) and (7.42) yield

$$\left[\frac{\partial\{Q\}}{\partial\{T\}}\right]_{n+1}^{(k)} = \sum_{\mathrm{e}}[L]_{\mathrm{e}}^{\mathrm{T}}\left[\frac{\partial\{Q\}_{\mathrm{e}}}{\partial\{T\}}\right]_{n+1}^{(k)} = \sum_{\mathrm{e}}[L]_{\mathrm{e}}^{\mathrm{T}}\left[\frac{\partial\{Q\}_{\mathrm{e}}}{\partial\{T\}_{\mathrm{e}}}\right]_{n+1}^{(k)}[L]_{\mathrm{e}}. \tag{7.51}$$

The derivative of $\{Q\}_{\mathrm{e}}$ with respect to the temperature reduces to the derivative of any of its entries in Equation (7.37). Concentrating on the first term on the right-hand side of Equation (7.37),

$$\begin{aligned}\left[\frac{\partial\{Q\}_{\mathrm{e}i}^{1}}{\partial\{T\}_{\mathrm{e}j}}\right]_{n+1}^{(k)} &= -\left[\frac{\partial}{\partial T_j}\int_{A_{\mathrm{e}}} h\left(\sum_{k=1}^{N}\varphi_k T_k\right)\left(\sum_{l=1}^{N}\varphi_l T_l - T_{\mathrm{e}}\right)\varphi_i \,\mathrm{d}a_{\mathrm{e}}\right]_{n+1}^{(k)} \\ &= -\int_{A_{\mathrm{e}}}\left[\frac{\partial h}{\partial T}\right]_{n+1}^{(k)}(T_{n+1}^{(k)} - T_{\mathrm{e}})\varphi_i\varphi_j \,\mathrm{d}a_{\mathrm{e}} - \int_{A_{\mathrm{e}}} h(T_{n+1}^{(k)})\varphi_i\varphi_j \,\mathrm{d}a_{\mathrm{e}}. \end{aligned}\tag{7.52}$$

In a similar way, one finds for the second term on the right-hand side of Equation (7.37),

$$\begin{aligned}\left[\frac{\partial\{Q\}_{\mathrm{e}i}^{2}}{\partial\{T_{\mathrm{e}j}\}}\right]_{n+1}^{(k)} &= -\left[\frac{\partial}{\partial T_j}\int_{A_{\mathrm{e}}} A\left(\sum_{k=1}^{N}\varphi_k T_k\right)\left[\left(\theta_{\mathrm{ref}} + \sum_{l=1}^{N}\varphi_l T_l\right)^4 - \theta_{\mathrm{e}}^4\right]\varphi_i \,\mathrm{d}a_{\mathrm{e}}\right]_{n+1}^{(k)} \\ &= -\int_{A_{\mathrm{e}}}\left[\frac{\partial A}{\partial T}\right]_{n+1}^{(k)}\left[(\theta_{\mathrm{ref}} + T_{n+1}^{(k)})^4 - \theta_{\mathrm{e}}^4\right]\varphi_i\varphi_j \,\mathrm{d}a_{\mathrm{e}} \\ &\quad - \int_{A_{\mathrm{e}}} A(T_{n+1}^{(k)})4(\theta_{\mathrm{ref}} + T_{n+1}^{(k)})^3\varphi_i\varphi_j \,\mathrm{d}a_{\mathrm{e}}. \end{aligned}\tag{7.53}$$

The dependence of h and A on T is usually benign, such that the first terms in Equations (7.52) and (7.53) are frequently dropped. The dependence on T of the third and fourth terms in Equation (7.37) is usually also small. If not, their derivative must also be included. Summarizing, one obtains

$$\left[\frac{\partial\{Q\}_{\mathrm{e}i}}{\partial\{T\}_{\mathrm{e}j}}\right]_{n+1}^{(k)} = -\int_{A_{\mathrm{e}}} h(T_{n+1}^{(k)})\varphi_i\varphi_j \,\mathrm{d}a_{\mathrm{e}} - \int_{A_{\mathrm{e}}} A(T_{n+1}^{(k)})4(\theta_{\mathrm{ref}} + T_{n+1}^{(k)})^3\varphi_i\varphi_j \,\mathrm{d}a_{\mathrm{e}}. \tag{7.54}$$

This yields a contribution of the convection and radiation fluxes to the "stiffness" matrix in Equation (7.50), comparable to the stiffness contribution of the centrifugal forces and traction forces in Section 3.3. Notice that the resulting equation, Equation (7.50), does not contain any explicit reference to θ_{ref}. Consequently, we can freely choose θ_{ref}, for example, as absolute zero.

7.6 Forced Fluid Convection

In most cases, the flux boundary conditions are made up of the terms in Equations (7.23) and (7.24). Equation (7.23) can also be written as

$$q_{\mathrm{conv}}^{k} = h(\theta)(\theta - \theta_{\mathrm{e}})n^{k} \tag{7.55}$$

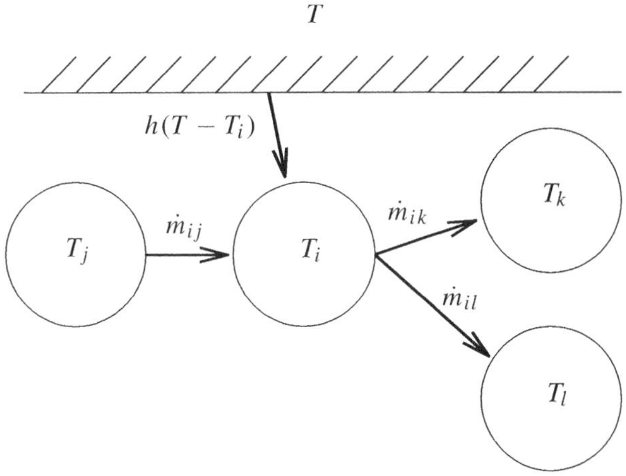

Figure 7.1 Heat fluxes from and toward location i

where θ_e is the absolute temperature of the surrounding fluid. In some applications, such as in tubes with internal flow, this temperature is itself also an unknown, depending on the fluid temperature at the entry of the tube. In such cases, the fluid temperature can be calculated using a simple network. For applications in which the fluid is meshed with finite elements, see (Reddy and Gartling 2001).

Consider the part of the tube wall shown in Figure 7.1. The relative material temperature T interacts through convection with the gas temperature T_i at location i. At that location, mass flow arrives from location j, whose temperature is T_j, and the mass flow leaves to locations k and l, which are at temperature T_k and T_l respectively. The energy equation for gases is (Equation (1.554)),

$$\rho\theta\frac{\partial^2\psi}{\partial\theta^2}\dot{\theta} + \theta\frac{\partial^2\psi}{\partial\rho^{-1}\partial\theta}\boldsymbol{d} : \boldsymbol{I} - \nabla\cdot\boldsymbol{q} + \rho h = 0 \tag{7.56}$$

where $\psi(\rho^{-1}, \theta)$. Similar to Equation (7.7), we expand ψ as a function of the temperature T:

$$\psi(\rho^{-1}, \theta) = \psi_0(\rho^{-1}) - \eta_0(\rho^{-1})T - \frac{c_v(\rho^{-1}, \theta)}{2\theta_{\text{ref}}}T^2 \tag{7.57}$$

where

$$c_v := \frac{\partial\varepsilon}{\partial\theta} \tag{7.58}$$

is the specific heat at constant volume for an ideal gas (Anderson 1989). Hence,

$$\frac{\partial\psi}{\partial\theta} = -\eta_0 - c_v\frac{T}{\theta_{\text{ref}}} + O(T^2), \quad T \to 0 \tag{7.59}$$

$$\frac{\partial^2\psi}{\partial\theta\partial\rho^{-1}} = -\frac{\partial\eta_0}{\partial\rho^{-1}} + O(T), \quad T \to 0 \tag{7.60}$$

$$\frac{\partial^2\psi}{\partial\theta^2} = -\frac{c_v}{\theta_{\text{ref}}} + O(T), \quad T \to 0. \tag{7.61}$$

Substituting into Equation (7.56) leads to (keeping only first-order terms)

$$\rho c_{\mathrm{v}} \dot{T} + \theta_{\mathrm{ref}} \frac{\partial \eta_0}{\partial \rho^{-1}} \boldsymbol{d} : \boldsymbol{I} = -\nabla \cdot \boldsymbol{q} + \rho h. \tag{7.62}$$

The function η_0 can be further specified if we take the gas equation of state into account. For an ideal gas, we have as the equation of state

$$p = R\rho(\theta_{\mathrm{ref}} + T) \tag{7.63}$$

where R is the specific gas constant, and

$$p = -\frac{\partial \psi}{\partial \rho^{-1}} = -\frac{\partial \psi_0}{\partial \rho^{-1}} + \frac{\partial \eta_0}{\partial \rho^{-1}} T. \tag{7.64}$$

Accordingly,

$$\frac{\partial \psi_0}{\partial \rho^{-1}} = R\rho\theta_{\mathrm{ref}} \quad \Rightarrow \quad \psi_0 = R\theta_{\mathrm{ref}} \ln \rho + C_1 \tag{7.65}$$

$$\frac{\partial \eta_0}{\partial \rho^{-1}} = R\rho \quad \Rightarrow \quad \eta_0 = -R \ln \rho + C_2. \tag{7.66}$$

The heat equation, Equation (7.60), now yields

$$\rho c_{\mathrm{v}} \dot{T} + R\rho\theta_{\mathrm{ref}} \boldsymbol{d} : \boldsymbol{I} = -\nabla \cdot \boldsymbol{q} + \rho h. \tag{7.67}$$

Using Equation (1.517), this can also be written as

$$\rho c_{\mathrm{v}} \dot{T} = -\nabla \cdot \boldsymbol{q} + \rho h + R\theta_{\mathrm{ref}} \dot{\rho} \tag{7.68}$$

or, since $\rho = 1/v$ and $\theta_{\mathrm{ref}} \approx \theta$,

$$\rho c_{\mathrm{v}} \dot{T} = -\nabla \cdot \boldsymbol{q} + \rho h - \rho p \dot{v}. \tag{7.69}$$

Accordingly, a temperature increase can be obtained through heat influx or through mechanical work (Anderson 1991). If we assume that the pressure p is constant, we have

$$R\rho\theta = \text{constant} \tag{7.70}$$

or

$$R\dot{\rho}\theta + R\rho\dot{\theta} = 0. \tag{7.71}$$

This can be transformed into

$$R\theta_{\mathrm{ref}} \dot{\rho} \approx R\theta \dot{\rho} = -R\rho\dot{\theta} = -R\rho\dot{T} \tag{7.72}$$

and

$$\rho c_{\mathrm{v}} \dot{T} - R\theta_{\mathrm{ref}} \dot{\rho} \approx \rho \dot{T}(c_{\mathrm{v}} + R) = \rho \dot{T} c_{\mathrm{p}} \tag{7.73}$$

since the specific heat at constant pressure, c_{p}, satisfies

$$R = c_{\mathrm{p}} - c_{\mathrm{v}}. \tag{7.74}$$

Consequently, the energy equation for a gas reduces to

$$\rho c_{\mathrm{p}} \dot{T} = -q^{k}{}_{,k} + \rho h. \tag{7.75}$$

The derivative of the temperature on the left-hand side is the total derivative consisting of the local variation and the change due to convection:

$$\dot{T} = \frac{DT}{Dt} = \frac{\partial T}{\partial t} + T_{,k} v^{k}. \tag{7.76}$$

So far, we dealt with solids, for which the convective term can be neglected. This is not so for fluids and gases. Accordingly,

$$\rho c_{\mathrm{p}} \left(\frac{\partial T}{\partial t} + T_{,k} v^{k} \right) = -q^{k}{}_{,k} + \rho h. \tag{7.77}$$

The conservation of mass requires (Equation (1.223))

$$\frac{\partial \rho}{\partial t} + (\rho v^{k})_{,k} = 0. \tag{7.78}$$

Combining Equations (7.77) and (7.78) yields

$$c_{\mathrm{p}} \frac{\partial \rho T}{\partial t} + c_{\mathrm{p}} (T \rho v^{k})_{,k} = -q^{k}{}_{,k} + \rho h. \tag{7.79}$$

The gas nodes $i, j, k, l, \cdots$ stand for a given control volume that is fixed in space and assigned to them. Integrating Equation (7.79) for node i yields

$$\int_{V_i} c_{\mathrm{p}} \frac{\partial \rho T}{\partial t} \mathrm{d}v + \int_{V_i} c_{\mathrm{p}} (T \rho v^{k})_{,k} \, \mathrm{d}v = -\int_{V_i} q^{k}{}_{,k} \, \mathrm{d}v + \int_{V_i} \rho h \, \mathrm{d}v. \tag{7.80}$$

Transforming the volume integrals for the divergence terms to surface integrals (assuming c_{p} to be constant over the volume),

$$\int_{V_i} c_{\mathrm{p}} \frac{\partial \rho T}{\partial t} \, \mathrm{d}v + \int_{A_i} c_{\mathrm{p}} T \rho v^{k} \, \mathrm{d}a_k = -\int_{A_i} q^{k} \, \mathrm{d}a_k + \int_{V_i} \rho h \, \mathrm{d}v. \tag{7.81}$$

We assume that the integrands of the volume integrals are constant across the volume:

$$\int_{V_i} c_{\mathrm{p}} \frac{\partial \rho T}{\partial t} \mathrm{d}v = c_{\mathrm{p}}(T_i) \frac{\partial \rho(T_i) T_i}{\partial t} V_i \tag{7.82}$$

$$\int_{V_i} \rho h \mathrm{d}v = \rho(T_i) h_i V_i. \tag{7.83}$$

The area of the convective term is split into areas with inflow and areas with outflow. For both types, c_{p} and T are assumed to be constant across the area. For inflow, T is the temperature of the neighboring node providing the flow; for outflow it is the temperature of node i. Hence,

$$\int_{A_i} c_p T \rho v^k \, da_k = \sum_{j \in \text{in}} c_p(T_j) T_j \int_{A_{ij}} \rho v^k \, da_k + \sum_{j \in \text{out}} c_p(T_i) T_i \int_{A_{ij}} \rho v^k \, da_k. \tag{7.84}$$

The mass flow between node i and node j is defined by

$$\dot{m}_{ij} = \pm \int_{A_{ij}} \rho v^k \, da_k. \tag{7.85}$$

The plus sign applies to the outflow and the minus sign to the inflow. Accordingly,

$$\int_{A_i} c_p T \rho v^k \, da_k = -\sum_{j \in \text{in}} c_p(T_j) T_j \dot{m}_{ij} + \sum_{j \in \text{out}} c_p(T_i) T_i \dot{m}_{ij}. \tag{7.86}$$

The first term on the right-hand side of Equation (7.81) relates to the convection from the wall (surface A_{iw}) and the conduction in the fluid (surface A_{if}, $A_i = A_{if} \cup A_{iw}$).

$$\int_{A_i} q^k \, da_k = \int_{A_{iw}} q^k_{\text{conv}} \, da_k + \int_{A_{if}} q^k \, da_k. \tag{7.87}$$

The conduction in the fluid is neglected. Hence,

$$\int_{A_i} q^k \, da_k = -[h(T_i, T)(T - T_i)] A_{iw}. \tag{7.88}$$

Summarizing, one obtains the following equation:

$$c_p(T_i) \frac{\partial [\rho(T_i) T_i]}{\partial t} V_i = \sum_{j \in \text{in}} c_p(T_j) T_j \dot{m}_{ij} - c_p(T_i) T_i \sum_{j \in \text{out}} \dot{m}_{ij} + \overline{h}(T_i, T)(T - T_i) + m_i h_i \tag{7.89}$$

where

$$m_i = \rho(T_i) V_i \tag{7.90}$$

is the mass in the control volume and

$$\overline{h}(T_i, T) = h(T_i, T) A_{iw} \tag{7.91}$$

$$\overline{A}(T_i, T) = A(T_i, T) A_{iw}. \tag{7.92}$$

Equation (7.89) expresses that the change of heat energy at node i is caused by influx from the other nodes, plus convection from the wall, minus outflux to the other nodes. In reality, the inertia of the gas is small compared to the inertia of the wall. Consequently, the term on the left-hand side of Equation (7.89) is usually neglected leading to

$$0 = \sum_{j \in \text{in}} c_p(T_j) T_j \dot{m}_{ij} - c_p(T_i) T_i \sum_{j \in \text{out}} \dot{m}_{ij} + \overline{h}(T_i, T)(T - T_i) + m_i h_i. \tag{7.93}$$

This is a weakly nonlinear equation in the temperature:

$$\left[\overline{h}(T_i, T) + c_{\text{p}}(T_i) \sum_{j \in \text{out}} \dot{m}_{ij}\right] T_i - \sum_{j \in \text{in}} \left[c_{\text{p}}(T_j)\dot{m}_{ij}\right] T_j - \overline{h}(T_i, T)T = m_i h_i. \tag{7.94}$$

Equation (7.94) can be considered as a nonlinear multiple-point constraint in the temperature, analogous to the nonlinear displacement multiple-point constraints in Chapter 3. It allows for the calculation of the gas temperatures as soon as the structural temperatures are known.

7.7 Cavity Radiation

In the present section, we examine what happens if radiation is exchanged among several surfaces. Radiation is a rather complicated subject meriting a much more extensive treatise. For more details, the reader is referred to (Incropera and DeWitt 2002).

7.7.1 Governing equations

Consider a differential surface $\text{d}A_1$ emitting radiation toward a differential surface $\text{d}A_2$ at a distance r (Figure 7.2). The surface $\text{d}A_2$ is perpendicular to the line connecting $\text{d}A_1$ with $\text{d}A_2$ and covers a spatial angle $\text{d}\omega$ satisfying

$$\text{d}\omega = \frac{\text{d}A_2}{r^2}. \tag{7.95}$$

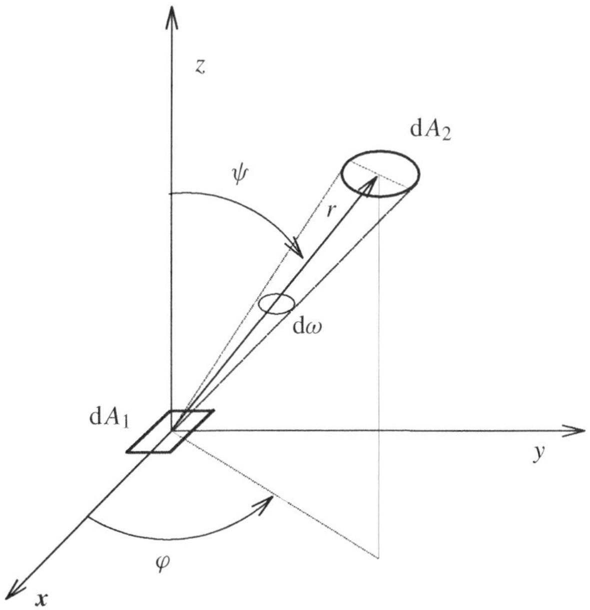

Figure 7.2 Radiation of surface $\text{d}A_1$ onto $\text{d}A_2$

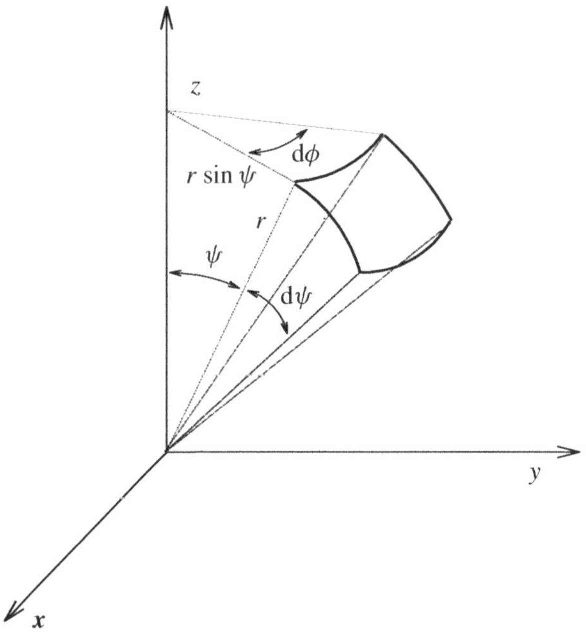

Figure 7.3 Infinitesimal surface element

The spectral intensity I_E in a certain direction (φ, ψ) is defined as the radiation power per unit solid angle $d\omega$ about this direction, per unit wavelength $d\lambda$, per unit emitting area perpendicular to this direction:

$$I_E = \frac{dP_E}{d\omega\, d\lambda\, dA_1 \cos\psi}. \tag{7.96}$$

Accordingly, for the radiation power between two infinitesimal areas dA_1 and dA_2, the area dA_1 enters in I_E in the form of the projected area $dA_1 \cos\psi$, whereas dA_2 enters in the form of the spatial angle $d\omega$. Furthermore, I_E depends on the wavelength of emission. The spectral, hemispherical emissive power E_λ, is defined as the radiation power in all directions of a hemisphere per unit wavelength $d\lambda$ per unit emitting area (not projected!). Hence,

$$E_\lambda = \int_{\text{hemisphere}} I_E \cos\psi\, d\omega. \tag{7.97}$$

An infinitesimal solid angle can be written as (Figure 7.3)

$$d\omega = \sin\psi\, d\varphi\, d\psi \tag{7.98}$$

leading to

$$E_\lambda = \int_0^{2\pi} \int_0^{\pi/2} I_E \cos\psi \sin\psi\, d\psi\, d\varphi. \tag{7.99}$$

If the emission does not depend on the direction (φ, ψ), it is called *diffuse emission*. Here and in the section that follows, we assume that we deal with diffuse emitters. In that case, I_E is no function of φ and ψ, and Equation (7.99) reads

$$E_\lambda = I_E \int_0^{2\pi} \int_0^{\pi/2} \cos\psi \sin\psi \, d\psi \, d\varphi = \pi I_E. \tag{7.100}$$

A special kind of diffuse emitter is a blackbody. Its properties are as follows:

1. It emits diffuse, that is, the spectral intensity only depends on the wavelength and temperature, not on the emission angle.
2. No body can emit more energy than a blackbody for a given wavelength and temperature.
3. All incident radiation is completely absorbed, no reflection takes place.

A blackbody is classically symbolized by a cavity at a uniform temperature with a small aperture. The spectral intensity of blackbody radiation was first determined by Planck, and satisfies

$$I_{E,b} = \frac{2hc_0^2}{\lambda^5[\exp(hc_0/\lambda k\theta) - 1]} \tag{7.101}$$

where $h = 6.6256 \times 10^{-34}$ Js is the Planck constant, $k = 1.3805 \times 10^{-23}$ J/K is the Boltzmann constant, $c_0 = 2.998 \times 10^8$ m/s is the speed of light in vacuum and θ is the temperature of the blackbody in Kelvin. Since a blackbody is a diffuse emitter, one obtains for the spectral emissive power

$$E_{\lambda,b} = \pi I_{E,b}. \tag{7.102}$$

The total emissive power is the power emitted per unit of emitting area and satisfies

$$E_b = \int_0^\infty E_{\lambda,b} \, d\lambda. \tag{7.103}$$

Substituting Equations (7.101) and (7.102) into Equation (7.103) and performing the integration, one obtains the Stefan–Boltzmann law

$$E_b = \sigma\theta^4 \tag{7.104}$$

where $\sigma = 5.67 \times 10^{-8}$ W/m^{2}K^4 is the Stefan–Boltzmann constant.

The blackbody is an ideal emitter. Real bodies will emit less. The spectral, directional emissivity is defined as the ratio of the real spectral, directional radiation intensity to the spectral blackbody intensity at the same temperature:

$$\epsilon_{\lambda,\omega} := \frac{I_E}{I_{E,b}}. \tag{7.105}$$

Here, we will assume to deal with diffuse emitters and work with averages over all wavelengths. Therefore, we define the total hemispherical emissivity as the ratio of the total emissive power to the emissive power of a blackbody at the same temperature:

$$\epsilon(\theta) := \frac{E}{E_b}. \tag{7.106}$$

The total emissive power is a function of the radiating surface and the temperature. Using Equations (7.96), (7.97), (7.103) and (7.106), one can write the radiation power as

$$\mathrm{d}P = \epsilon E_{\mathrm{b}} \mathrm{d}A_1 \tag{7.107}$$

and since radiation power and flux are related by

$$\mathrm{d}P = q\, \mathrm{d}A_1 \tag{7.108}$$

the flux satisfies

$$q = \epsilon(\theta) E_{\mathrm{b}} = \epsilon(\theta)\sigma\theta^4. \tag{7.109}$$

Comparing Equation (7.109) with Equation (7.24) for $\theta_{\mathrm{e}} = 0$ (no irradiation) reveals that

$$A(\theta) = \epsilon(\theta)\sigma. \tag{7.110}$$

In reality, we not only have radiation leaving the body but also irradiation entering the body. The spectral, directional irradiation intensity I_I in a certain direction (φ, ψ) is defined as the irradiation power per unit solid angle $\mathrm{d}\omega$ about this direction, per unit wavelength $\mathrm{d}\lambda$, per unit receiving area perpendicular to this direction:

$$I_I = \frac{\mathrm{d}P_I}{\mathrm{d}\omega\, \mathrm{d}\lambda\, \mathrm{d}A_1 \cos\psi}. \tag{7.111}$$

Likewise, the total hemispherical irradiation power G is defined as the irradiation power per unit receiving area:

$$G = \int_0^\infty \int_{\mathrm{hemisphere}} I_I \cos\psi\, \mathrm{d}\omega\, \mathrm{d}\lambda. \tag{7.112}$$

A part of the irradiation power is absorbed (αG), a part of it is reflected (ρG) and a part of it is transmitted (τG) (Figure 7.4).

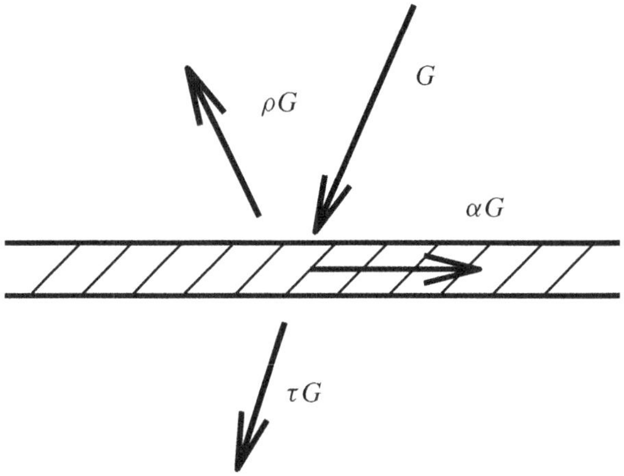

Figure 7.4 Absorption, reflection and transmission of irradiation

Energy conservation requires that $\alpha + \rho + \tau = 1$. We assume that we are dealing with opaque materials, that is, materials for which there is no transmission. Accordingly, $\tau = 0$ and $\tau = 0$. Accordingly,

$$\alpha + \rho = 1. \tag{7.113}$$

α is the total hemispherical absorptivity and ρ is the total hemispherical reflectivity. In reality, α and ρ are dependent on the irradiation angle and its spectrum. Therefore, α and ρ are averaged values in the same sense as ϵ is an averaged value of $\epsilon_{\lambda,\omega}$.

Here and in the section that follows, we assume that we deal with

1. diffuse surfaces, that is, ϵ and α are independent of the radiation and irradiation direction;

2. gray surfaces, that is, ϵ and α are independent of the wavelength for the actual range of interest.

Under these conditions, the important relationship

$$\alpha = \epsilon \tag{7.114}$$

applies (Incropera and DeWitt 2002), that is, the absorptivity equals the emissivity. Looking at Figure 7.4, the total radiation leaving the surface is the sum of the total emissive power E and the reflected total irradiation power. This is called the *total radiosity* J:

$$J = E + \rho G. \tag{7.115}$$

Now we arrive at the view-factor concept. The view factor F_{ij} is defined as the fraction of the radiation power leaving surface i that is intercepted by surface j. It is assumed that the surface A_i is characterized by a uniform radiosity J_i. The total radiation leaving the surface A_i amounts to

$$R = J_i A_i. \tag{7.116}$$

Since the radiosity is assumed to be uniform, the directional radiosity $J_{\omega,i}$ satisfies

$$J_{\omega,i} = \frac{J_i}{\pi} \tag{7.117}$$

and the radiosity leaving $\mathrm{d}A_i$ and reaching surface $\mathrm{d}A_j$ yields (Figure 7.5)

$$\mathrm{d}R_{ij} = J_{\omega,i}\, \mathrm{d}A_i \cos\psi_i \omega_{ij} \tag{7.118}$$

where ω_{ij} is the view angle covered by $\mathrm{d}A_j$ seen by $\mathrm{d}A_i$:

$$\omega_{ij} = \frac{\mathrm{d}A_j \cos\psi_j}{R^2}. \tag{7.119}$$

Accordingly,

$$\begin{aligned} F_{ij} &= \frac{1}{A_i J_i} \int_{A_i} \int_{A_j} \left(\frac{J_{\omega,i} \cos\psi_i \cos\psi_j}{R^2} \right) \mathrm{d}A_i\, \mathrm{d}A_j \\ &= \frac{1}{A_i} \int_{A_i} \int_{A_j} \left(\frac{\cos\psi_i \cos\psi_j}{\pi R^2} \right) \mathrm{d}A_i\, \mathrm{d}A_j. \end{aligned} \tag{7.120}$$

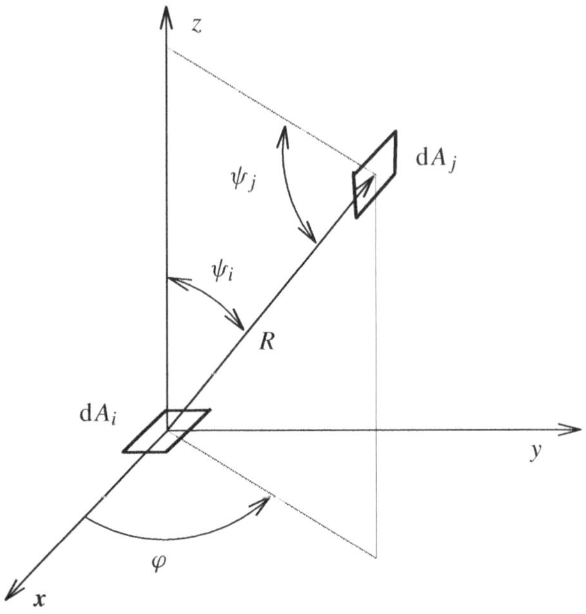

Figure 7.5 Geometry for the view-factor calculation

Important relations are the reciprocity relation

$$A_i F_{ij} = A_j F_{ji} \tag{7.121}$$

and the summation rule for enclosures

$$\sum_{j=1}^{N} F_{ij} = 1. \tag{7.122}$$

Now consider N surfaces A_i interacting with each other. From Figure 7.4, we obtain the relationships

$$q_i = E_i - \alpha_i G_i = E_i - \epsilon_i G_i \tag{7.123}$$

$$J_i = E_i + \rho_i G_i = E_i + (1 - \epsilon_i) G_i. \tag{7.124}$$

Hence, eliminating G_i from Equations (7.123) and (7.124),

$$q_i = \frac{E_i - \epsilon_i J_i}{1 - \epsilon_i} = \frac{\epsilon_i (E_{bi} - J_i)}{1 - \epsilon_i} \tag{7.125}$$

where E_{bi} stands for the blackbody radiation of surface i. Eliminating E_i from Equations (7.123) and (7.124) leads to

$$q_i = J_i - G_i. \tag{7.126}$$

G_i is the irradiation from all other bodies. Conservation of energy requires

$$A_i G_i = \sum_{\substack{j=1 \\ j \neq i}}^{N} F_{ji} A_j J_j. \tag{7.127}$$

Using the reciprocity rule, Equation (7.127) can be rewritten as

$$G_i = \sum_{\substack{j=1 \\ j \neq i}}^{N} F_{ij} J_j. \tag{7.128}$$

Accordingly, Equation (7.126) now reads

$$q_i = J_i - \sum_{\substack{j=1 \\ j \neq i}}^{N} F_{ij} J_j. \tag{7.129}$$

Equating Equations (7.125) and (7.129) yields

$$J_i - (1 - \epsilon_i) \sum_{\substack{j=1 \\ j \neq i}}^{N} F_{ij} J_j = \epsilon_i E_{\mathrm{b}i}. \tag{7.130}$$

If the temperatures of all the participating surfaces are known, Equation (7.130) constitutes a set of N linear equations in the N unknowns J_i. This set is not necessarily symmetric. After solving for J_i, the fluxes q_i can be obtained through Equations (7.125) or (7.129). From q_i, an equivalent environmental temperature can be derived for each surface A_i using Equation (7.24):

$$\theta_{\mathrm{e}i} = \left[\theta_i^4 - \frac{q_i}{A_i(\theta_i)} \right]^{1/4} \tag{7.131}$$

where θ_i is the mean temperature of surface i. Sometimes a cavity is not completely closed and part of the radiation escapes to the environment. Considering this environment to behave as a blackbody and attributing it to the surface k, one obtains

$$\epsilon_k = 1 \Rightarrow J_k = E_{\mathrm{b}k} = E_{\mathrm{b,environment}} \tag{7.132}$$

and Equation (7.130) now yields

$$J_i - (1 - \epsilon_i) \sum_{\substack{j=1 \\ j \neq i,k}}^{N} F_{ij} J_j - (1 - \epsilon_i) F_{ik} E_{\mathrm{b}k} = \epsilon_i E_{\mathrm{b}i} \tag{7.133}$$

or, since

$$F_{ik} = 1 - \sum_{\substack{j=1 \\ j \neq i,k}}^{N} F_{ij} \tag{7.134}$$

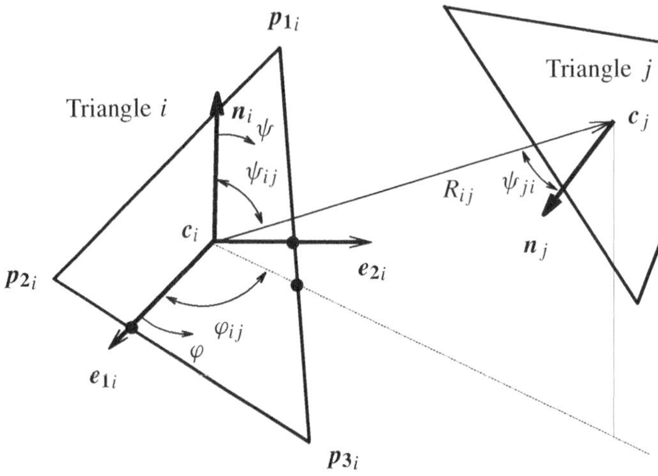

Figure 7.6 Local coordinate system in triangle i

one obtains

$$J_i - (1 - \epsilon_i) \sum_{\substack{j=1 \\ j \neq i,k}}^{N} F_{ij} J_j = \epsilon_i E_{bi} + (1 - \epsilon_i)(1 - \sum_{\substack{j=1 \\ j \neq i,k}}^{N} F_{ij}) E_{\text{b,environment}}. \tag{7.135}$$

7.7.2 Numerical aspects

The time-consuming part in generating Equation (7.130) is the calculation of the view factors. The method proposed here consists of the following steps:

1. Triangulate the free surface of the structure by defining linear triangles within the element faces without generating any new nodes. For instance, a face of a 20-node brick element is divided in six triangles, a face of an 8-node brick element in two triangles and a face of a 10-node tetrahedral element in four triangles. Number the nodes within each triangle in counterclockwise direction when viewed from outside the body.

2. For each triangle i, determine the following:

   (a) The center of gravity $\boldsymbol{c}_i$.

   (b) The normal $\boldsymbol{n}_i$, pointing away from the body (Figure 7.6):

$$\boldsymbol{n}_i = \frac{(\boldsymbol{p}_{2i} - \boldsymbol{p}_{1i}) \times (\boldsymbol{p}_{3i} - \boldsymbol{p}_{2i})}{\|(\boldsymbol{p}_{2i} - \boldsymbol{p}_{1i}) \times (\boldsymbol{p}_{3i} - \boldsymbol{p}_{2i})\|}. \tag{7.136}$$

   (c) The area A_i.

(d) The unit vector $\boldsymbol{e}_{1_i}$ satisfying

$$\boldsymbol{e}_{1_i} = \frac{(\boldsymbol{p}_{2_i} - \boldsymbol{p}_{1_i})}{\|\boldsymbol{p}_{2_i} - \boldsymbol{p}_{1_i}\|} \tag{7.137}$$

(e) The unit vector $\boldsymbol{e}_{2_i} = \boldsymbol{n}_i \times \boldsymbol{e}_{1_i}$. The basis $(\boldsymbol{e}_{1_i}, \boldsymbol{e}_{2_i}, \boldsymbol{n})$ defines a right-handed rectangular coordinate system.

(f) The scalar

$$d_i = -\boldsymbol{p}_{1_i} \cdot \boldsymbol{n}_i. \tag{7.138}$$

A point $\boldsymbol{p}$ lies in the plane of triangle i if

$$\boldsymbol{p} \cdot \boldsymbol{n}_i + d_i = 0. \tag{7.139}$$

It is visible from triangle i if and only if (assuming no other triangles block the view)

$$\boldsymbol{p} \cdot \boldsymbol{n}_i + d_i \geq 0. \tag{7.140}$$

3. For each triangle i:

(a) Perform a loop over all triangles $j \neq i$ with the following actions:

(i) Check whether $\boldsymbol{c}_j$ is visible from triangle i. If it is not, that is, if

$$\boldsymbol{c}_j \cdot \boldsymbol{n}_i + d_i < 0 \tag{7.141}$$

cycle

(ii) Check whether $\boldsymbol{c}_i$ is visible from triangle j. If it is not, that is, if

$$\boldsymbol{c}_i \cdot \boldsymbol{n}_j + d_j < 0 \tag{7.142}$$

cycle. Only those triangles j remain from which triangle i can be seen and which are visible from triangle i (assuming no other triangles block the view). In the remainder of the text, they will be called *visible triangles*.

(iii) Calculate the distance

$$R_{ij} = \|\boldsymbol{c}_j - \boldsymbol{c}_i\| \tag{7.143}$$

and the unit vector

$$\boldsymbol{\xi}_{ij} = (\boldsymbol{c}_j - \boldsymbol{c}_i)/R_{ij}. \tag{7.144}$$

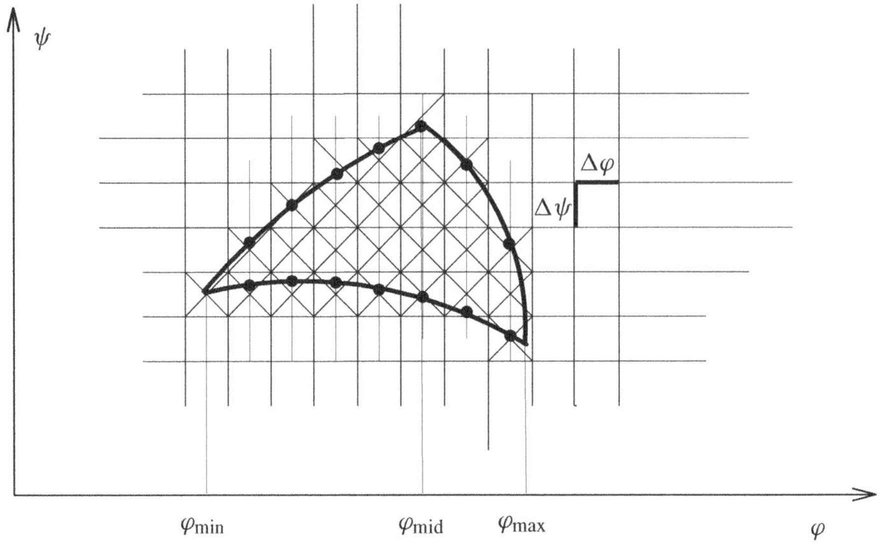

Figure 7.7 $\phi - \psi$ grid

(b) Generate a rectangular grid with φ on the x-axis and ψ on the y-axis. A (φ, ψ) pair uniquely defines a direction in the local $(\boldsymbol{e}_{1i}, \boldsymbol{e}_{2i}, \boldsymbol{n})$ system, where $0 < \varphi < 2\pi, 0 < \psi < \pi/2$ (cf Figure 7.6). The (φ, ψ)-range is meshed with an $N \times M$ rectangular grid (Figure 7.7). Let k and l be functions such that $k(\varphi)$ and $l(\psi)$ denote the discrete grid element to which (φ, ψ) belongs. If $\Delta\varphi = 2\pi/N$ and $\Delta\psi = \pi/(2M)$, then the functions satisfy

$$k(\varphi) = \text{int}(\varphi/\Delta\varphi) + 1 \tag{7.145}$$

$$l(\psi) = \text{int}(\psi/\Delta\psi) + 1 \tag{7.146}$$

where int(x) is the largest integer smaller than or equal to x. Initialize by assuming that all grid elements are uncovered.

(c) For all visible triangles, order R_{ij} in ascending order.

(d) Perform a loop over all visible triangles $j \neq i$ in ascending R_{ij}-order with the following actions:

(i) Calculate the coordinates of $\boldsymbol{\xi}_{ij}$ in the local $(\boldsymbol{e}_{1i}, \boldsymbol{e}_{2i}, \boldsymbol{n})$ system and determine the angles φ_{ij} and ψ_{ij} by inverting the relations

$$\boldsymbol{\xi}_{ij} \cdot \boldsymbol{e}_{1i} = \sin\psi_{ij} \cos\varphi_{ij} \tag{7.147}$$

$$\boldsymbol{\xi}_{ij} \cdot \boldsymbol{e}_{2i} = \sin\psi_{ij} \sin\varphi_{ij} \tag{7.148}$$

$$\boldsymbol{\xi}_{ij} \cdot \boldsymbol{n}_i = \cos\psi_{ij}. \tag{7.149}$$

(ii) Determine the grid element $k(\varphi_{ij}), l(\psi_{ij})$ and check whether it was already covered. If so, cycle.

(iii) Calculate the view factor

$$F_{ij} = \frac{\cos\psi_{ij}\cos\psi_{ji}A_j}{\pi R_{ij}^2} \tag{7.150}$$

where

$$\cos\psi_{ji} = -\boldsymbol{\xi}_{ij}\cdot\boldsymbol{n}_j. \tag{7.151}$$

(iv) Determine which grid elements are covered by triangle j. To that end, calculate the unit vectors $\boldsymbol{q}_{k_{ij}}$ connecting $\boldsymbol{c}_i$ with $\boldsymbol{p}_{k_j}$, $k = 1, 2, 3$:

$$\boldsymbol{q}_{k_{ij}} = \frac{(\boldsymbol{p}_{k_j} - \boldsymbol{c}_i)}{\|\boldsymbol{p}_{k_j} - \boldsymbol{c}_i\|}. \tag{7.152}$$

If

$$\boldsymbol{n}_{31_{ij}} = \boldsymbol{q}_{3_{ij}} \times \boldsymbol{q}_{1_{ij}} \tag{7.153}$$

$$\boldsymbol{n}_{12_{ij}} = \boldsymbol{q}_{1_{ij}} \times \boldsymbol{q}_{2_{ij}} \tag{7.154}$$

$$\boldsymbol{n}_{23_{ij}} = \boldsymbol{q}_{2_{ij}} \times \boldsymbol{q}_{3_{ij}} \tag{7.155}$$

then the equations of the planes connecting the edges of triangle j with $\boldsymbol{c}_i$ satisfy

$$\boldsymbol{p}\cdot\boldsymbol{n}_{31_{ij}} = 0 \tag{7.156}$$

$$\boldsymbol{p}\cdot\boldsymbol{n}_{12_{ij}} = 0 \tag{7.157}$$

$$\boldsymbol{p}\cdot\boldsymbol{n}_{23_{ij}} = 0, \tag{7.158}$$

(see Figure 7.8). A unit vector $\boldsymbol{p}$ with coordinates ($\sin\psi\cos\varphi$, $\sin\psi\sin\varphi$, $\cos\psi$) lies in the plane defined by $\boldsymbol{p}_{3_j}$, $\boldsymbol{p}_{1_j}$ and $\boldsymbol{c}_i$ if

$$\psi = -\tan^{-1}\left[\frac{(\boldsymbol{n}_{31_{ij}}\cdot\boldsymbol{n}_i)}{(\boldsymbol{n}_{31_{ij}}\cdot\boldsymbol{e}_{1_i})\cos\varphi + (\boldsymbol{n}_{31_{ij}}\cdot\boldsymbol{e}_{2_i})\sin\varphi}\right] =: f_{3/1}(\varphi) \tag{7.159}$$

and likewise for the other planes. The spatial angle covered by triangle j corresponds to a triangle with curved sides in the φ, ψ plane (Figure 7.7). Its vertices are made up of $\left(\varphi(\boldsymbol{p}_{1_j}), \psi_i(\boldsymbol{p}_{1_j})\right)$, $\left(\varphi(\boldsymbol{p}_{2_j}), \psi_i(\boldsymbol{p}_{2_j})\right)$ and $\left(\varphi(\boldsymbol{p}_{3_j}), \psi_i(\boldsymbol{p}_{3_j})\right)$. Now determine $\varphi_{i\min}$ and $\varphi_{i\max}$ defined by

$$\varphi_{\min} = \min\{\varphi(\boldsymbol{p}_{1_j}), \varphi(\boldsymbol{p}_{2_j}), \varphi(\boldsymbol{p}_{3_j})\} \tag{7.160}$$

$$\varphi_{\max} = \max\{\varphi(\boldsymbol{p}_{1_j}), \varphi(\boldsymbol{p}_{2_j}), \varphi(\boldsymbol{p}_{3_j})\}. \tag{7.161}$$

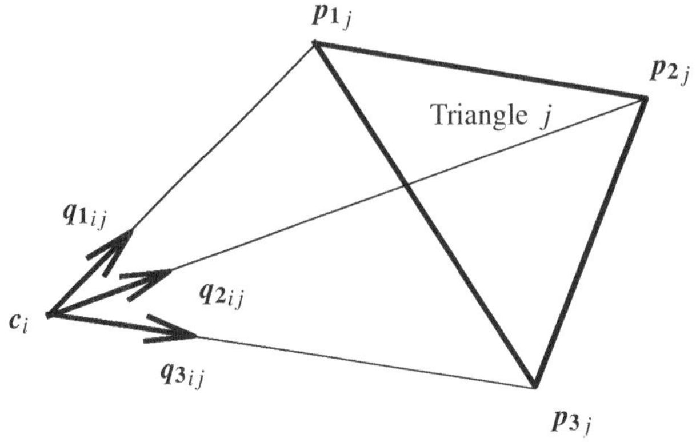

Figure 7.8 Spatial angle covered by triangle j

The one that is left is called φ_{mid}. Let ψ_{min} be the ψ-value corresponding to φ_{min} and ψ_{max} the ψ-value corresponding to φ_{max}. First, the grid elements $(k(\varphi_{\text{min}}), l(\psi_{\text{min}}))$ and $(k(\varphi_{\text{max}}), l(\psi_{\text{max}}))$ are marked as covered. Then, for $k(\varphi_{\text{min}}) \leq m \leq k(\varphi_{\text{mid}})$ and $\varphi_{\text{min}} \leq (m - \frac{1}{2})\Delta\varphi \leq \varphi_{\text{max}}$, those grid elements are marked as covered for which

$$\min\left\{l\{f_{\text{min/max}}[(m-\tfrac{1}{2})\Delta\varphi]\}, l\{f_{\text{min/mid}}[(m-\tfrac{1}{2})\Delta\varphi]\}\right\} \leq n \leq \max\left\{l\{f_{\text{min/max}}[(m-\tfrac{1}{2})\Delta\varphi]\}, l\{f_{\text{min/mid}}[(m-\tfrac{1}{2})\Delta\varphi]\}\right\} \quad (7.162)$$

and for $k(\varphi_{\text{mid}}) \leq m \leq k(\varphi_{\text{max}})$ and $\varphi_{\text{min}} \leq (m - \frac{1}{2})\Delta\varphi \leq \varphi_{\text{max}}$ the elements for which

$$\min\left\{l\{f_{\text{min/max}}[(m-\tfrac{1}{2})\Delta\varphi]\}, l\{f_{\text{mid/max}}[(m-\tfrac{1}{2})\Delta\varphi]\}\right\} \leq n \leq \max\left\{l\{f_{\text{min/max}}[(m-\tfrac{1}{2})\Delta\varphi]\}, l\{f_{\text{mid/max}}[(m-\tfrac{1}{2})\Delta\varphi]\}\right\}. \quad (7.163)$$

This corresponds to the crossed elements in Figure 7.8.

(v) If less than ϵ% of the grid elements are marked as uncovered, exit the loop.

4. Solve Equations (7.135) to obtain J_i, or, equivalently, $\boldsymbol{q}_i$. The corresponding temperatures θ_{ei} (Equation (7.131)), can be used as the thermal boundary condition in the next iteration. Since the structure deforms, the view factor has to be recalculated in each iteration. However, the changes are usually small and the algorithm can be accelerated by making diligent use of the visible triangles from the last iteration.

References

ABAQUS 1997 *Theory Manual*, Hibbitt, Karlsson & Sorensen, Inc, Pawtucket, Rhode Island.

Abramowitz M and Stegun IA 1972 *Handbook of Mathematical Functions*, AMS 55, National Bureau of Standards, US Department of Commerce.

Anderson JD 1989 *Introduction to Flight*, McGraw-Hill.

Anderson JD 1991 *Fundamentals of Aerodynamics*, McGraw-Hill.

Antman 1983 Regular and singular problems for large elastic deformations of tubes, wedges, and cylinders. *Arch. Rational Mech. Anal.* **83**, 1–52.

Arruda EM and Boyce MC 1993 A three-dimensional constitutive model for the large stretch behavior of rubber elastic materials. *J. Mech. Phys. Solids* **41**, 389–412.

Ashcraft C, Grimes RG, Pierce DJ and Wah DK 1999 *The User Manual for SPOOLES, Release 2.2: An Object Oriented Software Library for Solving Sparse Linear Systems of Equations*, Boeing Shared Services Group, P.O Box 24346, Mail Stop 7L-22, Seattle, Washington 98124 USA, see also http://netlib.bell-labs.com/netlib/linalg/spooles/spooles.2.2.html.

Ball JM 1977 Convexity conditions and existence theorems in nonlinear elasticity. *Arch. Rational Mech. Anal.* **63**, 337–403.

Barlow J 1976 Optimal stress locations in finite element models. *Int. J. Numer. Methods Eng.* **10**, 243–251.

Bathe K-J 1995 *Finite Element Procedures*, Prentice Hall.

Beatty MF 1987 Topics in finite elasticity: hyperelasticity of rubber, elastomers, and biological tissues –with examples. *Appl. Mech. Rev.* **40**, 1699–1734.

Belytschko T, Liu WK and Moran B 2000 *Nonlinear Finite Elements for Continua and Structures*, Wiley.

Belytschko T and Bindeman LP 1993 Assumed strain stabilization of the eight node hexahedral element. *Comput. Methods Appl. Mech. Eng.* **105**, 225–260.

Belytschko T and Mish K 2001 Computability in non-linear solid mechanics. *Int. J. Numer. Methods Eng.* **52**, 3–21.

Bischoff M and Ramm E 1999 Solid-like shell or shell-like solid formulation? A personal view. In *ECCM '99 Proc. European Conference on Computational Mechanics*, Munich, Germany, 31 Aug–3 Sept 1999.

Bonet J and Wood RD 1997 *Nonlinear Continuum Mechanics for Finite Element Analysis*, Cambridge University Press.

CalculiX 2003 *CalculiX: A Three-Dimensional Structural Finite Element Program*, Dhondt G and Klaus W (programmers) "www.calculix.de".

Chung J, Cho EH and Choi K 2003 A priori error estimator of the generalized–α method for structural dynamics. *Int. J. Numer. Methods Eng.* **57**, 537–554.

The Finite Element Method for Three-dimensional Thermomechanical Applications Guido Dhondt
 ISBN: 0-470-85752-8

Ciarlet PG 1993 *Mathematical Elasticity, Volume 1: Three Dimensional Elasticity*, North Holland.

Crisfield MA 1983 An arc-length method including line searches and accelerations. *Int. J. Numer. Methods Eng.* **19**, 1269–1289.

Dacorogna B 1989 *Direct Methods in Calculus of Variations*, Applied Mathematical Sciences 78, Springer-Verlag, Berlin.

Dhondt G 1993 General behaviour of collapsed 8-node 2-D and 20-node 3-D isoparametric elements. *Int. J. Numer. Methods Eng.* **36**, 1223–1243.

Dhondt G 2002 Mixed-mode K-calculations in anisotropic materials. *Eng. Fract. Mech.* **69**, 909–922.

Doll S and Schweizerhof K 2000 On the development of volumetric strain energy functions. *J. Appl. Mech.* **67**, 17–21.

Düster A, Bröker H and Rank E 2001 The p-version of the finite element method for three-dimensional curved thin walled structures. *Int. J. Numer. Methods Eng.* **52**, 673–703.

Düster A, Hartmann S and Rank E 2003 p-FEM applied to finite isotropic hyperelastic bodies. *Comput. Methods Appl. Mech. Eng.* **192**, 5147–5166.

Eringen AC (ed.) 1975 *Continuum Physics, Volume II: Continuum Mechanics of Single Substance Bodies*, Academic Press, New York.

Eringen AC (ed.) 1976 *Continuum Physics, Volume IV: Polar and Nonlocal Field Theories*, Academic Press, New York.

Eringen AC 1980 *Mechanics of Continua*, Robert E. Krieger Publishing Company, Huntington, New York.

Fedelich B 2002 A microstructural model for the monotonic and the cyclic mechanical behavior of single crystals of superalloys at high temperatures. *Int. J. Plast.* **18**, 1–49.

Flores GF and Oñate E 2001 A basic thin shell triangle with only translational DOFs for large strain plasticity. *Int. J. Numer. Methods Eng.* **51**, 57–83.

Freitag LA and Knupp PM 2002 Tetrahedral mesh improvement via optimization of the element condition number. *Int. J. Numer. Methods Eng.* **53**, 1377–1391.

Gabaldón F and Goicolea JM 2002 Linear and non-linear finite element error estimation based on assumed strain fields. *Int. J. Numer. Methods Eng.* **55**, 413–429.

Geers MGD, Ubachs RLJM and Engelen RAB 2003 Strongly non-local gradient-enhanced finite strain elastoplasticity. *Int. J. Numer. Methods Eng.* **56**, 2039–2068.

George P-L and Borouchaki H 1998 *Delauney Triangulation and Meshing: Application to Finite Elements*, Kogan Page.

Gradshteyn IS and Ryzhik IM 1980 *Table of Integrals, Series and Products*, Academic Press.

Graff KK 1975 *Wave Motion in Elastic Solids*, Ohio State University Press.

Greenberg MD 1978 *Foundations of Applied Mathematics*, Prentice Hall, Englewood Cliffs, New Jersey.

Halphen B and Nguyen Quoc Son 1975 Sur les matériaux standards généralisés. *Journal de Mécanique* **14**, 39–63.

Hartmann S 2001a Numerical studies on the identification of the material parameters of Rivlin's hyperelasticity using tension-torsion tests. *Acta Mech.* **148**, 129–155.

Hartmann S 2001b Parameter estimation of hyperelasticity relations of generalized polynomial-type with constraint conditions. *Int. J. Solids Struct.* **38**, 7999–8018.

Hartmann S and Neff P 2003 Polyconvexity of generalized polynomial-type hyperelastic strain energy functions for near-incompressibility. *Int. J. Solids Struct.* **40**, 2767–2791.

Hartmann S 2003 *Finite-Elemente Berechnung inelastischer Kontinua*, Habilitationsschrift, Berichte des Instituts für Mechanik (Bericht 1/2003), Universität Kassel, Germany.

Havner KS 1992 *Finite Plastic Deformation of Crystalline Solids*, Cambridge University Press.

Hilber HM 1976 *Analysis and Design of Numerical Integration Methods in Structural Dynamics*, PhD Thesis, University of California, Berkeley.

Hilber HM and Hughes TJR 1978 Collocation, dissipation and 'overshoot' for time integration schemes in structural dynamics *Earthquake Eng. Struct. Dyn.* **6**, 99–117.

Hilber HM, Hughes TJR and Taylor RL 1977 Improved numerical dissipation for time integration algorithms in structural dynamics. *Earthquake Eng. Struct. Dyn.* **5**, 283–292.

Holzapfel GA 2000 *Nonlinear Solid Mechanics*, Wiley.

Holzapfel GA, Gasser TC and Ogden RW A 2000 new constitutive framework for arterial wall mechanics and a comparative study of material models *J. Elasticity* **61**, 1–48.

Hughes TJR 2000 *The Finite Element method*, Dover, New York.

Hulbert GM and Hughes TJR 1987 An error analysis of truncated starting conditions in step-by-step time integration: consequences for structural dynamics *Earthquake Eng. Struct. Dyn.* **15**, 901–910.

Incropera FP and DeWitt DP 2002 *Fundamentals of Heat and Mass Transfer*, Wiley.

Itskov M 2001 A generalized orthotropic hyperelastic material model with application to incompressible shells. *Int. J. Numer. Methods Eng.* **50**, 1777–1799.

Kachanov LM 1971 *Foundations of the Theory of Plasticity*, North Holland Publishing Company.

Kaliske M and Rothert H On the finite element implementation of rubber-like materials at finite strains. *Eng. Comp.* **14**, 216–232.

Kim JB, Yoon JW and Yang DY 2003 Investigation into the wrinkling behaviour of thin sheets in the cylindrical cup deep drawing process using bifurcation theory. *Int. J. Numer. Methods Eng.* **56**, 1673–1705.

Koiter WT 1960 General theorems for elastic-plastic solids. In *Progress in Solid Mechanics 6*, Sneddon IN and Hill R (eds), North Holland, Amsterdam, 167–221.

Lee EH and Liu DT 1967 Finite-strain elastic-plastic theory of application to plane wave analysis. *J. Appl. Phys.* **38**, 19–27.

Lee EH 1969 Elastic plastic deformation at finite strains. *J. Appl. Mech.* **36**, 1–6.

Lehoucq RB, Sorensen DC and Yang C 1998 *ARPACK Users' Guide*, SIAM.

Lemaitre J and Chaboche JL 1990 *Mechanics of Solid Materials*, Cambridge University Press, Cambridge, UK.

Liew KM and Rajendran S 2002 New superconvergent points of the 8-node serendipity plane element for patch recovery. *Int. J. Numer. Methods Eng.* **54**, 1103–1130.

Lubliner J 1985 A model of rubber viscoelasticity. *Mech. Res. Commun.* **12**, 93–99.

Luenberger DG 1989 *Linear and Nonlinear Programming*, Addison-Wesley Publishing Company, Reading, Massachusetts.

MLührs G, Hartmann S and Haupt P 1997 On the numerical treatment of finite deformations in elastoviscoplasticity. *Comput. Methods Appl. Mech. Eng.* **144**, 1–21.

Mackinnon RJ and Carey GF 1989 Superconvergent derivatives: a Taylor series analysis. *Int. J. Numer. Methods Eng.* **28**, 489–509.

Mang HA, Hellmich C, Lackner R and Pichler B 2001 Computational structural mechanics. *Int. J. Numer. Methods Eng.* **52**, 569–587.

Marsden JE and Hughes TJR 1983 *Mathematical Foundations of Elasticity*, Dover, New York.

Matthies H and Strang G 1979 The solution of nonlinear finite element equations. *Int. J. Numer. Methods Eng.* **14**, 1613–1626.

Meirovitch L 1967 *Analytical Methods in Vibrations*, The MacMillan Company, London.

Meissonnier FT, Busso EP and O'Dowd NP 2001 Finite element implementation of a generalised non-local rate-dependent crystallographic formulation for finite strains. *Int. J. Plast.* **17**, 601–640.

Méric L and Cailletaud G 1991 Single crystal modeling for structural calculations: Part 2: Finite element implementation. *J. Eng. Mater. Tech.* **113**, 171–182.

Méric L, Poubanne P and Cailletaud G 1991 Single crystal modeling for structural calculations: Part 1 –Model presentation. *J. Eng. Mater. Tech.* **113**, 162–170.

Miehe C 1996a Exponential map algorithm for stress updates in anisotropic multiplicative elastoplasticity for single crystals. *Int. J. Numer. Methods Eng.* **39**, 3367–3390.

Miehe C 1996b Multisurface thermoplasticity for single crystals at large strains in terms of Eulerian vector updates. *Int. J. Solids Struct.* **33**, 3103–3130.

Miranda I, Ferencz RM and Hughes TJR 1989 An improved implicit-explicit time integration method for structural dynamics. *Earthquake Eng. Struct. Dyn.* **18**, 643–653.

Muğan A and Hulbert GM 2001a Frequency-domain analysis of time-integration methods for semidiscrete finite element equations –Part I: Parabolic problems. *Int. J. Numer. Methods Eng.* **51**, 333–350.

Muğan A and Hulbert GM 2001b Frequency-domain analysis of time-integration methods for semidiscrete finite element equations –Part II: Hyperbolic and parabolic-hyperbolic problems. *Int. J. Numer. Methods Eng.* **51**, 351–376.

Ogden RW 1984 *Non Linear Elastic Deformations*, Dover, New York.

Popov EP 1968 *Introduction to Mechanics of Solids*, Prentice Hall, Englewood Cliffs, New Jersey.

Press WH, Flannery BP, Teukolsky SA and Vetterling WT 1990 *Numerical Recipes in Pascal*, Cambridge University Press.

Prudhomme S, Oden T, Westermann T, Bass J and Botkin ME 2003 Practical methods for a posteriori error estimation in engineering applications. *Int. J. Numer. Methods Eng.* **56**, 1193–1224.

Puso MA 2000 A highly efficient enhanced assumed strain physically stabilized hexahedral element. *Int. J. Numer. Methods Eng.* **49**, 1029–1064.

Ramamurti V and Seshu P 1990 On the principle of cyclic symmetry in machine dynamics. *Commun. Appl. Numer. Methods* **6**, 259–268.

Reddy JN and Gartling DK 2001 *The Finite Element Method in heat Transfer and Fluid Dynamics*, CRC Press.

Reese S 1994 *Theorie und Numerik des Stabilitätsverhaltens hyperelastischer Festkörper*, Dissertation, Fachbereich Mechanik, Technische Hochschule Darmstadt, Germany.

Reese S, Raible T and Wriggers P 2001 Finite element modelling of orthotropic material behaviour in pneumatic membranes. *Int. J. Solids Struct.* **38**, 9525–9544.

Reese S 2001 *Thermomechanische Modellierung gummiartiger Polymerstrukturen*, Habilitationsschrift, Institut für Baumechanik und Numerische Mechanik (Bericht F01/4), Universität Hannover, Germany.

Reese S 2002 On the equivalence of mixed element formulations and the concept of reduced integration in large deformation problems. *Int. J. Nonlin. Sc. Num. Sim.* **3**, 1–33.

Reese S 2003a On a consistent hourglass stabilization technique to treat large inelastic deformations and thermo-mechanical coupling in plane strain problems. *Int. J. Numer. Methods Eng.* **57**, 1095–1127.

Reese S 2003b Meso-macro modelling of fibre-reinforced rubber-like composites exhibiting large elastoplastic deformation. *Int. J. Solids Struct.* **40**, 951–980.

Reese S and Svendsen B 2003 On the use of evolving structure tensors to model initial and induced elastic and inelastic anisotropy at finite deformation. *J. Phys. IV France* **105**, 31–37.

Reese S and Wriggers P 2000 A stabilization technique to avoid hourglassing in finite elasticity. *Int. J. Numer. Methods Eng.* **48**, 79–109.

Riks E 1987 Progress in collapse analyses. *J. Press. Vessel Tech.* **109**, 33–41.

Rumpel T and Schweizerhof K 2003 Volume-dependent pressure loading and its influence on the stability of structures. *Int. J. Numer. Methods Eng.* **56**, 211–238.

Save MA and Massonnet CE 1972 *Plastic Analysis and Design of Plates, Shells and Disks*, North Holland Publishing Company.

Schröder J, Gruttmann F and Löblein J 2002 *A Simple Orthotropic Finite Elasto-Plasticity Model Based on generalized Stress-Strain measures*, Report No 2, Institut für Mechanik, Fachbereich 10, Universität Essen, 45117 Essen, Germany.

Schröder J and Neff P 2001 *Invariant Formulation of Hyperelastic Transverse Isotropy Based on Polyconvex Free Energy Functions*, Report No 1, Institut für Mechanik, Fachbereich 10, Universität Essen, 45117 Essen, Germany.

Simo JC 1988a A framework for finite strain elastoplasticity based on maximum plastic dissipation and the multiplicative decomposition: Part I: Continuum formulation. *Comput. Methods Appl. Mech. Eng.* **68**, 199–219.

Simo JC 1988b A framework for finite strain elastoplasticity based on maximum plastic dissipation and the multiplicative decomposition: Part II: Computational aspects. *Comput. Methods Appl. Mech. Eng.* **68**, 199–219.

Simo JC and Hughes TJR 1997 *Computational Inelasticity*, Springer-Verlag, New York.

Simo JC and Miehe C 1992 Associative coupled thermoplasticity at finite strains: formulation, numerical analysis and implementation. *Comput. Methods Appl. Mech. Eng.* **98**, 41–104.

Simo JC and Taylor RL 1991 Quasi-incompressible finite elasticity in principal stretches. Continuum basis and numerical algorithms. *Comput. Methods Appl. Mech. Eng.* **85**, 273–310.

Sloan SW 1989 A FORTRAN program for profile and wavefront reduction. *Int. J. Numer. Methods Eng.* **28**, 2651–2679.

Spencer AJM 1971 Theory of invariants. In *Continuum Physics, Volume I: Mathematics*, Eringen Cemal A (ed.), Academic Press, New York.

Storåkers B 1986 On material representation and constitutive branching in finite compressible elasticity. *J. Mech. Phys. Solids* **34**, 125–145.

Stroud AH 1971 *Approximate Calculation of Multiple Integrals*, Prentice Hall.

Sze KY, Chan WK and Pian THH 2002 An eight-node hybrid-stress solid-shell element for geometric non-linear analysis of elastic shells. *Int. J. Numer. Methods Eng.* **55**, 853–878.

Tautges TJ 2001 The generation of hexahedral meshed for assembly geometry: survey and progress. *Int. J. Numer. Methods Eng.* **50**, 2617–2642.

Timoshenko SP and Goodier JN *Theory of Elasticity*, International Student Edition, McGraw-Hill Kogakusha Ltd.

Treloar LRG 1975 *The Physics of Rubber Elasticity*, Third Edition, Oxford University Press.

van den Bogert PAJ and de Borst R 1994 On the behaviour of rubber-like materials in compression and shear. *Arch. Appl. Mech.* **64**, 136–146.

Verron E, Marckmann G and Peseux B 2001 Dynamic inflation of non-linear elastic and viscoelastic rubber-like membranes. *Int. J. Numer. Methods Eng.* **50**, 1233–1251.

Wriggers P, Eberlein R and Reese S 1996 A comparison of three-dimensional continuum and shell elements for finite plasticity. *Int. J. Solids Struct.* **33**, 3309–3326.

Zienkiewicz OC and Taylor RL 1989 *The Finite Element Method*, Fourth Edition, McGraw-Hill, London.

Zienkiewicz OC and Zhu JZ 1992a The superconvergent patch recovery and a posteriori error estimates. Part 1: The recovery technique. *Int. J. Numer. Methods Eng.* **33**, 1331–1364.

Zienkiewicz OC and Zhu JZ 1992b The superconvergent patch recovery and a posteriori error estimates. Part 2: Error estimates and adaptivity. *Int. J. Numer. Methods Eng.* **33**, 1365–1382.

Index

The Finite Element Method for Three-dimensional Thermomechanical Applications Guido Dhondt
 ISBN: 0-470-85752-8

Printed and bound in the UK by
CPI Antony Rowe, Eastbourne

Printed and bound by CPI Group (UK) Ltd, Croydon, CR0 4YY
16/04/2025
14658471-0001